Methods in Molecular Biology

Series Editor
John M. Walker
School of Life and Medical Sciences
University of Hertfordshire
Hatfield, Hertfordshire, AL10 9AB, UK

For further volumes:
http://www.springer.com/series/7651

Recombinant Protein Production in Yeast

Edited by

Brigitte Gasser

Department of Biotechnology, University of Natural Resources and Life Sciences (BOKU) and Austrian Centre of Industrial Biotechnology (acib), Vienna, Austria

Diethard Mattanovich

Department of Biotechnology, University of Natural Resources and Life Sciences (BOKU) and Austrian Centre of Industrial Biotechnology (acib), Vienna, Austria

Humana Press

Editors
Brigitte Gasser
Department of Biotechnology
University of Natural Resources
and Life Sciences (BOKU)
and Austrian Centre of Industrial
Biotechnology (acib)
Vienna, Austria

Diethard Mattanovich
Department of Biotechnology
University of Natural Resources
and Life Sciences (BOKU)
and Austrian Centre of Industrial
Biotechnology (acib)
Vienna, Austria

ISSN 1064-3745 ISSN 1940-6029 (electronic)
Methods in Molecular Biology
ISBN 978-1-4939-9023-8 ISBN 978-1-4939-9024-5 (eBook)
https://doi.org/10.1007/978-1-4939-9024-5

Library of Congress Control Number: 2018966883

This Humana Press imprint is published by the registered company Springer Science+Business Media, LLC, part of Springer Nature.
The registered company address is: 233 Spring Street, New York, NY 10013, U.S.A.

Preface

Since the 1980s with the first published record of the expression of heterologous genes in bakers' yeast *Saccharomyces cerevisiae*, and the first approved protein product made in yeast, this group of unicellular eukaryotic microorganisms has established an important role as a production platform for biopharmaceuticals and technical enzymes. Besides *S. cerevisiae*, a couple of non-conventional yeasts are used as production hosts due to their reported superior performance. For laboratories and companies who are not experienced in the work with alternative yeasts it is however a major hurdle to enter into their application.

This volume is therefore dedicated to provide state of the art background and methods for the work with the major protein producing yeast platforms, as well as case studies for special applications. It is aimed at scientists in various fields of biotechnology who want to initiate or improve production of recombinant proteins. Different to previous editions a special focus has been laid on the whole production process including protein purification as well as qualitative and quantitative analytics.

As a starting point, *Recombinant Protein Production in Yeasts* provides an overview of the main yeast production platforms used today as well as promising future yeast cell factories for recombinant protein production. For the most popular alternative yeast production platforms, methods how to generate production strains and how to select the best strains for production are described. Furthermore, *Recombinant Protein Production in Yeasts* covers various aspects of protein production in yeasts, including process design and purification as well as quantification of secreted recombinant proteins. Each chapter opens with a description of the background theory of the method being described, before giving a detailed step-by-step protocol as well as tips & tricks how to handle difficulties in the *Notes* section. Aside from describing basic concepts and methodology for yeast protein production, further chapters deal with successful approaches of genetic and metabolic engineering for improved production of specific protein classes including membrane proteins, co-factor containing proteins and virus-like particles. A special focus is also given to glycoproteins and their analysis, as glycosylation is among the most abundant post-translational modifications of proteins.

We are very grateful to all authors who were willing to share their protocols and insights with the scientific community. In total 64 authors from 13 countries contributed to this volume of *Methods in Molecular Biology*. Additionally, we want to thank all reviewers who served as the first readers and helped to ascertain that all necessary information to perform the described protocols was clearly outlined in the individual chapters. We also acknowledge support by Birgit Marckhgott from BOKU during editing of the chapters. Furthermore, we thank John M. Walker, Series Editor of *Methods in Molecular Biology*, and Springer for the opportunity to assemble a further edition of the successful series.

We hope that many researchers will find this outstanding collection of articles helpful when working on recombinant protein production in yeasts.

Vienna, Austria

Brigitte Gasser
Diethard Mattanovich

Contents

Contributors

FRIEDRICH ALTMANN • *Department of Chemistry, University of Natural Resources and Life Sciences (BOKU), Vienna, Austria*

ROSLYN M. BILL • *School of Life and Health Sciences, Aston University, Birmingham, UK*

FELIX BISCHOFF • *Leibniz Institute of Plant Genetics and Crop Plant Research (IPK), Gatersleben, Germany*

PINAR ÇALIK • *Biochemical Reaction Engineering Laboratory, Department of Chemical Engineering, Middle East Technical University, Ankara, Turkey; Industrial Biotechnology and Metabolic Engineering Laboratory, Department of Biotechnology, Graduate School of Natural and Applied Sciences, Middle East Technical University, Ankara, Turkey*

STEPHANIE P. CARTWRIGHT • *School of Life and Health Sciences, Aston University, Birmingham, UK*

EDA ÇELIK • *Department of Chemical Engineering, Hacettepe University, Ankara, Turkey; Bioengineering Division, Institute of Science, Hacettepe University, Ankara, Turkey*

EWELINA CELIŃSKA • *Department of Biotechnology and Food Microbiology, Poznan University of Life Sciences, Poznan, Poland*

ANITA EMMERSTORFER-AUGUSTIN • *Division of Biochemistry, Biophysics and Structural Biology, Department of Molecular and Cell Biology, University of California, Berkeley, CA, USA*

BURCU GÜNDÜZ ERGÜN • *Biochemical Reaction Engineering Laboratory, Department of Chemical Engineering, Middle East Technical University, Ankara, Turkey*

RENA FAR • *Department of Biochemistry and Molecular Biology, Cumming School of Medicine, University of Calgary, Calgary, AB, Canada*

PAU FERRER • *Luxembourg Institute of Science and Technology, Belvaux, Luxembourg; Department of Chemical, Biological and Environmental Engineering, Universitat Autònoma de Barcelona, Bellaterra (Cerdanyola del Vallès), Catalonia, Spain*

ALEXANDER D. FREY • *Department of Bioproducts and Biosystems, School of Chemical Engineering, Aalto University, Espoo, Finland*

BRIGITTE GASSER • *Department of Biotechnology, University of Natural Resources and Life Sciences (BOKU) and Austrian Centre of Industrial Biotechnology (acib), Vienna, Austria; Christian Doppler Laboratory for Growth-Decoupled Protein Production in Yeast, Department of Biotechnology, University of Natural Resources and Life Sciences, Vienna, Austria*

THOMAS GASSLER • *Department of Biotechnology, University of Natural Resources and Life Sciences (BOKU), Vienna, Austria; Austrian Centre of Industrial Biotechnology (acib), Vienna, Austria*

MARTIN GIERSBERG • *Leibniz Institute of Plant Genetics and Crop Plant Research (IPK), Gatersleben, Germany*

CLEMENS GRÜNWALD-GRUBER • *Austrian Centre of Industrial Biotechnology (acib), Vienna, Austria; Department of Chemistry, University of Natural Resources and Life Sciences (BOKU), Vienna, Austria*

LINA HEISTINGER • *Department of Biotechnology, University of Natural Resources and Life Sciences (BOKU), Vienna, Austria; Christian Doppler Laboratory for Innovative*

Immunotherapeutics at Department of Biotechnology, University of Natural Resources and Life Sciences (BOKU), Vienna, Austria

MELANIE HIRZ • *Institute of Molecular Biotechnology, Graz University of Technology, NAWI Graz, BioTechMed Graz, Graz, Austria*

DAMLA HÜCCETOĞULLARI • *Biochemical Reaction Engineering Laboratory, Department of Chemical Engineering, Middle East Technical University, Ankara, Turkey*

MEHMET INAN • *Food Safety and Agricultural Research Center, Akdeniz University, Antalya, Turkey; Department of Food Engineering, Akdeniz University, Antalya, Turkey*

OLENA P. ISHCHUK • *Department of Biology and Biological Engineering, Systems and Synthetic Biology, Chalmers University of Technology, Gothenburg, Sweden*

HYUN AH KANG • *Department of Life Science, College of Natural Science, Chung-Ang University, Seoul, Republic of Korea*

JAY KIM • *Department of Biochemistry and Molecular Biology, University of Calgary, Calgary, AB, Canada*

GOTTHARD KUNZE • *Leibniz Institute of Plant Genetics and Crop Plant Research (IPK), Gatersleben, Germany*

MACARENA LARROUDE • *Micalis Institute, INRA, AgroParisTech, Université Paris-Saclay, Jouy-en-Josas, France*

GABRIELE LHOTA • *Department of Biotechnology, University of Natural Resources and Life Sciences (BOKU), Vienna, Austria*

JUSTIN A. MACDONALD • *Department of Biochemistry and Molecular Biology, Cumming School of Medicine, University of Calgary, Calgary, AB, Canada*

JOSÉ L. MARTÍNEZ • *Department of Biology and Biological Engineering, Systems and Synthetic Biology, Chalmers University of Technology, Gothenburg, Sweden; Department of Biotechnology and Biomedicine, Section for Synthetic Biology, Technical University of Denmark, Kongens Lyngby, Denmark*

DIETHARD MATTANOVICH • *Department of Biotechnology, University of Natural Resources and Life Sciences (BOKU) and Austrian Centre of Industrial Biotechnology (acib), Vienna, Austria*

FALKO MATTHES • *Leibniz Institute of Plant Genetics and Crop Plant Research (IPK), Gatersleben, Germany*

MICHAEL M. MAURER • *FH Campus Wien—University of Applied Sciences, Vienna, Austria; Austrian Centre of Industrial Biotechnology (acib), Vienna, Austria*

HYE YUN MOON • *Department of Life Science, College of Natural Science, Chung-Ang University, Seoul, Republic of Korea*

JEAN-MARC NICAUD • *Micalis Institute, INRA, AgroParisTech, Université Paris-Saclay, Jouy-en-Josas, France*

SIBEL ÖZTÜRK • *Biochemical Reaction Engineering Laboratory, Department of Chemical Engineering, Middle East Technical University, Ankara, Turkey*

DINA PETRANOVIC • *Department of Biology and Biological Engineering, Systems and Synthetic Biology, Chalmers University of Technology, Gothenburg, Sweden; Novo Nordisk Foundation Center for Biosustainability, Chalmers University of Technology, Gothenburg, Sweden*

HARALD PICHLER • *Austrian Centre of Industrial Biotechnology (acib), Graz, Austria; Institute of Molecular Biotechnology, Graz University of Technology, NAWI Graz, BioTechMed Graz, Graz, Austria*

MARI A. PIIRAINEN • *Department of Bioproducts and Biosystems, School of Chemical Engineering, Aalto University, Espoo, Finland*

ROLAND PRIELHOFER • *Austrian Centre of Industrial Biotechnology (acib), Vienna, Austria; Department of Biotechnology, University of Natural Resources and Life Sciences (BOKU), Vienna, Austria*
URSULA RINAS • *Helmholtz Center for Infection Research, Braunschweig, Germany; Technical Chemistry—Life Science, Leibniz University of Hannover, Hannover, Germany*
TRISTAN ROSSIGNOL • *Micalis Institute, INRA, AgroParisTech, Université Paris-Saclay, Jouy-en-Josas, France*
FLORIAN RÜKER • *Christian Doppler Laboratory for Innovative Immunotherapeutics, Department of Biotechnology, University of Natural Resources and Life Sciences (BOKU), Vienna, Austria*
FLÁVIO SÁDIO • *Christian Doppler Laboratory for Innovative Immunotherapeutics, Department of Biotechnology, University of Natural Resources and Life Sciences (BOKU), Vienna, Austria*
HEIDI SALMINEN • *Department of Bioproducts and Biosystems, School of Chemical Engineering, Aalto University, Espoo, Finland*
HARALD SCHILLINGER • *3M Separation and Purification Sciences Division (SPSD), 3M Österreich GmbH, Vienna, Austria*
TOBIAS SCHWALENBERG • *Leibniz Institute of Plant Genetics and Crop Plant Research (IPK), Gatersleben, Germany*
PAUL SOUDIER • *Micalis Institute, INRA, AgroParisTech, Université Paris-Saclay, Jouy-en-Josas, France*
OLIVER SPADIUT • *Institute of Chemical, Environmental and Bioscience Engineering, Research Division Biochemical Engineering, Vienna University of Technology (TU Wien), Vienna, Austria*
KATHARINA STADLBAUER • *Christian Doppler Laboratory for Innovative Immunotherapeutics, Department of Biotechnology, University of Natural Resources and Life Sciences (BOKU), Vienna, Austria*
GERHARD STADLMAYR • *Christian Doppler Laboratory for Innovative Immunotherapeutics, Department of Biotechnology, University of Natural Resources and Life Sciences (BOKU), Vienna, Austria*
AYSUN TÜRKANOĞLU ÖZÇELIK • *Food Safety and Agricultural Research Center, Akdeniz University, Antalya, Turkey*
ANNEGRET ULKE-LEMÉE • *Department of Biochemistry and Molecular Biology, Cumming School of Medicine, University of Calgary, Calgary, AB, Canada*
KAROLA VORAUER-UHL • *Department of Biotechnology, University of Natural Resources and Life Sciences (BOKU), Vienna, Austria*
ROLAND WEIS • *VALIDOGEN (formerly VTU Technology) GmbH, Raaba-Grambach, Austria*
SEBASTIAN WORCH • *Leibniz Institute of Plant Genetics and Crop Plant Research (IPK), Gatersleben, Germany*
GORDANA WOZNIAK-KNOPP • *Christian Doppler Laboratory for Innovative Immunotherapeutics, Department of Biotechnology, University of Natural Resources and Life Sciences Vienna (BOKU), Vienna, Austria*
TAMARA WRIESSNEGGER • *Austrian Centre of Industrial Biotechnology (acib), Graz, Austria*
DAVID JOHANNES WURM • *Institute of Chemical, Environmental and Bioscience Engineering, Research Division Biochemical Engineering, Vienna University of Technology (TU Wien), Vienna, Austria*
SEMIRAMIS YILMAZ • *Department of Food Engineering, Akdeniz University, Antalya, Turkey*

SU JIN YOO • *Department of Life Science, College of Natural Science, Chung-Ang University, Seoul, Republic of Korea*

MARIA ZAHID • *Technical Chemistry—Life Science, Leibniz University of Hannover, Hannover, Germany; Department of Biology, Lahore University of Management Sciences, Lahore, Pakistan*

RICHARD J. ZAHRL • *Austrian Centre of Industrial Biotechnology (acib), Vienna, Austria; Department of Biotechnology, University of Natural Resources and Life Sciences (BOKU), Vienna, Austria*

GUENTHER ZELLNIG • *Institute of Plant Sciences, University of Graz, NAWI Graz, Graz, Austria*

Chapter 1

Established and Upcoming Yeast Expression Systems

Burcu Gündüz Ergün, Damla Hüccetoğulları, Sibel Öztürk, Eda Çelik, and Pınar Çalık

Abstract

Yeast was the first microorganism used by mankind for biotransformation of feedstock that laid the foundations of industrial biotechnology. Long historical use, vast amount of data, and experience paved the way for *Saccharomyces cerevisiae* as a first yeast cell factory, and still it is an important expression platform as being the production host for several large volume products. Continuing special needs of each targeted product and different requirements of bioprocess operations have led to identification of different yeast expression systems. Modern bioprocess engineering and advances in omics technology, i.e., genomics, transcriptomics, proteomics, secretomics, and interactomics, allow the design of novel genetic tools with fine-tuned characteristics to be used for research and industrial applications. This chapter focuses on established and upcoming yeast expression platforms that have exceptional characteristics, such as the ability to utilize a broad range of carbon sources or remarkable resistance to various stress conditions. Besides the conventional yeast *S. cerevisiae*, established yeast expression systems including the methylotrophic yeasts *Pichia pastoris* and *Hansenula polymorpha*, the dimorphic yeasts *Arxula adeninivorans* and *Yarrowia lipolytica*, the lactose-utilizing yeast *Kluyveromyces lactis*, the fission yeast *Schizosaccharomyces pombe*, and upcoming yeast platforms, namely, *Kluyveromyces marxianus*, *Candida utilis*, and *Zygosaccharomyces bailii*, are compiled with special emphasis on their genetic toolbox for recombinant protein production.

Key words Recombinant protein production, Yeast expression platform, Vector, Promoter, Secretion signal, Selection marker

1 Introduction

Yeasts are attractive hosts for the production of recombinant proteins (r-proteins) as they combine the advantages of microorganisms, e.g., ease of manipulation and rapid growth, with the ability to conduct the eukaryotic posttranslational modifications and to secrete the produced proteins efficiently. Moreover, their superior fermentation characteristics, such as high cell densities and protein yields, make the manufacturing processes more economical, while

Burcu Gündüz Ergün, Damla Hüccetoğulları, Sibel Öztürk contributed equally to this chapter.

Brigitte Gasser and Diethard Mattanovich (eds.), *Recombinant Protein Production in Yeast*, Methods in Molecular Biology, vol. 1923, https://doi.org/10.1007/978-1-4939-9024-5_1, © Springer Science+Business Media, LLC, part of Springer Nature 2019

the absence of pyrogens, pathogens, or viral inclusions guarantees the safety criteria of the synthesized biomolecules [1].

The first and best characterized yeast expression system *Saccharomyces cerevisiae* has been successfully employed for the production of several Food and Drug Administration (FDA)-approved molecules. However, even with this most established yeast species, a variety of limitations, i.e., plasmid instability, low protein yields, and non-native glycoforms such as hyperglycosylation or α-1,3-linked mannose-terminated *N*-linked glycosylation of proteins, may be encountered [2]. To solve the problems mentioned, research has focused on the development of alternative yeast expression systems including the two methylotrophic yeasts *Pichia pastoris* and *Hansenula polymorpha*; the two dimorphic yeasts *Arxula adeninivorans* and *Yarrowia lipolytica*; the lactose-assimilating yeast *Kluyveromyces lactis*; the fission yeast *Schizosaccharomyces pombe*; and some other budding yeasts such as *Kluyveromyces marxianus*, *Candida utilis*, and *Zygosaccharomyces bailii*.

In spite of distinct favorable characteristics of each yeast species, there is no single system which is optimal for the production of all r-proteins. In this context, a successful design initiates with the selection of the host strain and continues with the determination of the most appropriate vector, selection marker, promoter, and signal sequence. Therefore, to benefit from the current status of a wide range of available host-vector combinations and to design the most effective system, this chapter focuses on the accumulated knowledge and state-of-the-art developments in the established and upcoming yeast expression systems, providing a comparative analysis. Furthermore, future perspectives on the design of these expression systems for homologous and heterologous protein production are discussed, considering each genetic tool separately.

2 *Saccharomyces cerevisiae*

Saccharomyces cerevisiae, the first eukaryotic organism completely sequenced [3], is one of the best-understood and most powerful genetic model systems that turned into an exciting model for evolutionary genetics and genomics. It has a tiny 12 Mb genome, the sequence of which is better annotated than that of any other eukaryote. Multi-locus and whole genome sequencing have solved several persistent taxonomic problems, but many archaic and misleading synonyms are unfortunately still in use. The genus *Saccharomyces* is classified into seven groups, i.e., *S. cerevisiae*, *S. paradoxus*, *S. mikatae*, *S. kudriavzevii*, *S. arboricola*, *S. eubayanus*, and *S. uvarum* [4]. Among these, *S. cerevisiae* originated from the millennia-long evolution, and due to its use in baking and brewing since ancient times, it is also known as baker's yeast. Meanwhile since more than a decade, *S. cerevisiae* is an industrial organism with GRAS status and with intrinsic resistance to high osmolality, low pH, and various inhibitors, which enables cheap and easy fermentation processes with fast cell growth under aerobic and anaerobic conditions [5]. These advantages make it a primary choice for the production of many compounds through modification and/or redirection of metabolic pathway under the guidance of its well-known physiology and genetics, vector-based tools, and genomic integration techniques [6, 7]. It also opened the gates for production of important recombinant enzymes and therapeutic proteins [8]. Furthermore, the progress in "omics" sciences has resulted in generation of high-throughput data [9] of transcriptomics [10–12], proteomics [13, 14], metabolomics [15, 16], fluxomics [17, 18], interactomics [19–21], locasomics (subcellular localization) [22], and a genome-scale metabolic model (GSMM) [23, 24]. The application of metabolic engineering principles associated with genetic engineering and "omics" technologies in *S. cerevisiae* paved the way for the synthesis of high-value products such as steroid hydrocortisones, alkaloids, and terpenoids as well as biofuel production and biomass conversion [25]. Thus, *S. cerevisiae* has been a favorite organism for pioneering studies to enhance its capacity for the production of high value-added compounds [26] and to reveal the genetic and cellular processes in higher eukaryotes [27–29]. Also, it was on the forefront for developing novel technologies and for the construction of expression and cloning vectors.

The first FDA-approved r-protein produced by *S. cerevisiae*, hepatitis B surface antigen [30, 31], has led to various commercial biopharmaceuticals such as human serum albumin, hirudin, human transferrin, insulin precursor, glucagon, urate oxidase, and platelet-derived growth factor [32, 33]. To date, the marketed recombinant pharmaceuticals from microbial eukaryotic cells approved by FDA and EMEA are mainly produced by *S. cerevisiae* [2, 34, 35].

Despite many favorable characteristics, *S. cerevisiae* has secretion-specific bottlenecks due to endoplasmic reticulum (ER) misfolding or inefficient trafficking [8, 36] and also strain-specific bottlenecks resulting in low cell densities [37]. In addition to these, hyperglycosylation is among the problems of *S. cerevisiae*-based protein production, and efforts in humanization of the glycosylation pathway are ongoing [38, 39]. Nonetheless, *S. cerevisiae* is still an attractive eukaryotic model organism to study fundamental cellular processes as 30% of human proteins linked to genetic and degenerative disease have orthologs in yeast [40, 41].

2.1 Expression Vectors

For the production of homologous and/or heterologous proteins, yeast can be transformed either with a self-replicating vector or an integrative expression cassette [42]. These vectors have been developed and modified since the 1990s. As each strategy has its own advantages and disadvantages, thus, making the most appropriate choice according to the designed manufacturing process is essential. Autonomously replicating circular plasmids provide higher copy numbers, while they usually become unstable under nonselective conditions. On the other hand, when integration vectors are used, generally, only a single gene copy is stably integrated into the genome [43], but the difficulties of maintaining selective pressure in the culture medium, particularly in industrial-scale operations, are reduced. Therefore, the product levels obtained with these two types of expression systems were said to be close to each other.

Plasmid vectors designed as yeast expression systems are categorized into yeast episomal vector (YEp), yeast autonomously replicating plasmid (YRp), and yeast centromere plasmid (YCp). Although there are structural differences between them, the common feature of all is to be a shuttle vector that allows propagation in yeast and *E. coli* host cells, as well as the presence of a multiple cloning site (MCS) for the insertion of desired gene sequence [43]. On the other hand, the detailed classification [44] for YEp as the first designed shuttle vector has arisen from the full or partial 2 μm plasmid native to *S. cerevisiae* and other yeast species. The endogenous 2 μm circle episome of *S. cerevisiae* [45] is comprised of 6318 bp double-stranded DNA [46] and allows up to a hundred copies per cell in most *Saccharomyces* species [47]. The YEp of *S. cerevisiae* includes either its entire 2 μm circle episome [48, 49] or both 2 μm Ori and *REP3* gene [50–53]. The first episomal shuttle vector was a chimeric plasmid resulting from the successful combination of pBR322 as an *E. coli* cloning vector and 2 μm circle episome native to *S. cerevisiae* [54]. The yeast replicons originated from 2 μm plasmids also functioned well in non-*Saccharomyces* yeasts, i.e., *Debaryomyces cerevisiae*, *K. lactis*, *Pichia angusta*, *P. pastoris*, *Pichia stipitis*, and *Pachysolen tannophilus* [55]. Furthermore, the YEp confers the replicative function with high gene copy numbers via self-replicating elements of the native yeast episome [56]. The first YEp was relatively stable and enabled high-frequency transformation in the range of ten to hundred thousand transformants per μg DNA, whereas some other YEps were reported as being prone to instability after each generation [53]. The partial 2 μm plasmid vectors are more stable than the whole plasmid, however, transformation of partial plasmids resulted in lower copy numbers [57].

The autonomously replicating sequences (ARS), originated from either the yeast chromosome or mitochondria and used in YRp instead of the replication origin of the 2 μm episome in YEp, provided a significant increase in reproduction of the transformed plasmids but are intrinsically unstable due to their segregational loss without selection pressure [53]. In addition to the first intrinsic ARS of *S. cerevisiae* [58], also the autonomously replicating sequences of non-*Saccharomyces* yeasts, i.e., *Candida glabrata*,

Candida maltose, *D. occidentalis*, and *P. tannophilus*, are employed in YRp [55].

The YCp is an advanced YRp that ensures the plasmid stability due to the addition of yeast centromeric sequences (CEN) into YRp. Unlike the other vector plasmids, the YCp is not a convenient system for heterologous gene expression because of lower expression levels via drastically decreased copy number, i.e., 1–2 copies, per cell [59], but can be used to construct genomic DNA libraries due to its improved plasmid stability [53].

A variety of commercial yeast expression systems with their distinct features, pYES, and the original pRS vector series are supplied by Invitrogen and New England Biolabs, respectively. The pRS vector series comprising either of the 2 μm origin of replication or CEN sequences include *URA3*, *TRP1*, *HIS3*, and *LEU2* genes as auxotrophic selection markers [60, 61], along with the promoters of genes encoding glyceraldehyde-3-phosphate dehydrogenase *GPD/GAP/TDH3*, translation elongation factor *TEF1*, cytochrome C oxidase *CYC1*, alcohol dehydrogenase *ADH1* [62], or the copper-responsive *CUP1* gene [63]. The pYES vector systems have also the same origin of replication as the pRS vectors and various dominant selectable markers, e.g., *URA3*, *TRP1*, and ampicillin resistance gene. In addition, they contain the promoter of the galactose-inducible *GAL1* gene. The pESC vector series, a commercial product of Agilent Technologies, is derived from the 2 μm plasmid vector and includes a bidirectional *GAL1-GAL10* promoter cassette. In addition to the above-listed companies, the European *Saccharomyces cerevisiae* Archives for Functional Analysis (EUROSCARF) supplies diverse yeast expression vectors for many purposes.

2.2 Selection Markers

The existence of selection markers in all types of vectors is a prerequisite to assess whether or not the cells introduced the DNA sequences necessary to perform a specific task through deletion of endogenous genes and/or introduction of desired sequences into the yeast. Based on the vector type, the recombinant strain has to be maintained in either nonselective or selective conditions. The self-replicating gene expression requires marker gene-specific selective conditions to provide maintenance of the plasmid. Conversely, if the gene of interest (GOI) is integrated into host genome, the recombinant strain can be cultivated under nonselective conditions [64]; however, the selective pressure should be initially applied to select for positive transformants.

The first marker genes used for yeast transformation were endogenous prototrophic markers, which were later complemented by dominant markers [64]. The auxotrophic strains were usually derived from single gene perturbations in essential metabolic pathways, i.e., amino acid biosynthesis pathway and nucleotide base biosynthesis pathway, and complemented with the prototrophic markers to be able to grow on relevant amino acid- or nucleotide-deficient medium. In literature, these complementary markers are simply called auxotrophic markers.

The significant and common auxotrophic marker alleles in *S. cerevisiae* are localized in adenine, uracil, histidine, leucine, methionine, lysine, and tryptophan metabolism. The pioneer auxotrophic selection markers in *S. cerevisiae* took the advantages of the mutations in five genes, *HIS3* (imidazoleglycerol-phosphate dehydratase), *LEU2* (β-isopropylmalate dehydrogenase) [65], *URA3* (orotidine-5′-phosphate decarboxylase), *LYS2* (α-aminoadipate reductase), and *MET17* (O-acetyl homoserine-O-acetyl serine sulfhydrylase) [66]. The corresponding markers of *S. cerevisiae* were followed by the various markers exploiting the presence of mutant alleles in genes such as *ADE1* (N-succinyl-5-aminoimidazole-4-carboxamide ribotide synthetase), *ADE2* (phosphoribosylaminoimidazole carboxylase), *ADE8* (phosphoribosyl-glycinamide transformylase), *ECM31* (ketopantoate hydroxymethyltransferase), *HIS2* (histidinol-phosphatase), *LYS5* (phosphopantetheinyl transferase), and *TRP1* (phosphoribosylanthranilate isomerase).

Additionally, a variety of exogenous genes originating from different budding yeast species were used to construct auxotrophic markers, i.e., the *URA3* gene of *K. lactis*, *A. adeninivorans* [67], and *C. albicans*, the *LYS5* gene of *C. albicans*, and the *LEU2* of *K. lactis*. Furthermore, the auxotrophic markers derived from the fission yeast *Schizosaccharomyces pombe* were employed using *his5*$^+$ or *ura4*$^+$ genes, which are being equivalents of *HIS3* and of *URA3* genes in *S. cerevisiae*, respectively.

The construction of dominant markers is principally based on either carbon/nitrogen (C/N) source utilization or resistance to specific agents. C/N source-related markers take advantage of strain-specific ability to metabolize a certain carbon or nitrogen source. Apart from using endogenous genes such as nitrogen source-specific *FCY1* (cytosine deaminase) and *GAP1* (general amino acid permease) in *S. cerevisiae*, the C/N source-related dominant markers also employ nitrogen source-specific heterologous genes, i.e., *amdS* (*Aspergillus nidulans* acetamidase) and *FCA1* (*C. albicans* cytosine deaminase), and carbon source-specific heterologous genes, i.e., *LAC4/LAC12* (*K. lactis* β-galactosidase/lactose permease) and *LSD1* (*Lipomyces starkeyi* dextranase).

Lastly, in the case of selection mediated by the resistance markers, cell growth on media containing a resistance agent is promoted by the degradation of the agent through expression of corresponding endogenous or heterologous resistance genes. Endogenous genes only show resistance to specific agents when they are highly expressed via multiple copy integration or under the control of strong promoters [64]. Some of the genes used as endogenous resistance markers are *CUP1*, *ERG11*, *MPR1*, *SSU1*, *SFA1*, and *YPA1*. Beyond dominant endogenous markers, drug resistance markers have been constructed with heterologous genes such as *aroA* (*E. coli* 5-enolpyruvylshikimate-3-phosphate synthase), *ble* (Tn5 phleomycin-binding protein), *cat* (Tn9 acetyltransferase), *hph* (phosphotransferase), *kan* (Tn 903 phosphotransferase), *mdr3* (*Mus musculus* P-glycoprotein), *nat1* (*Streptomyces noursei* acetyltransferase), *pat* (*Streptomyces viridochromogenes* acetyltransferase), and *dhfr* (*E. coli* dihydrofolate reductase). The application of antibiotic markers in plasmid sets was firstly performed with *KanMX* (resistance to kanamycin) [68] and then was extended using other antibiotic resistance genes, i.e., *nat1* (for nourseothricin), *pat* (for phosphinothricin), and *hph* (for hygromycin) [69].

2.3 Promoters

The most straightforward approach for efficient expression of r-proteins is the choice of the promoter in terms of its strength and regulation as well as the potential toxicity, metabolic burden, and load of the produced r-protein on the host organism. The power of any promoter is related with its function and localization in the metabolic pathway. The glycolytic enzymes encoded by housekeeping genes are generally controlled by strong constitutive promoters, which allow the expression of relevant genes during the life cycle of the cell. Besides, it is important to note that the basal expression of the desired gene can be enhanced by the appropriate choice of various carbon sources. The ability to catabolize any other carbon source different from glucose highlights the existence of inducible promoter(s) responsible for the metabolism of the related carbon source. Accordingly, when considering the production of potentially toxic r-proteins, the choice of moderate or strong inducible promoters enables high yields in the production phase which is following the nonproductive growth phase. Nonetheless, constitutive promoter-based fermentations have advantages over inducible ones, due to the simplicity in the design of bioreactor operation conditions.

For the last three decades, the conventional constitutive and inducible promoters of *S. cerevisiae* have been used not only as endogenous promoters but also as exogenous promoters in the engineering of other yeast hosts for the production of r-proteins, by taking the advantage of well-established regulation and operation principles of these promoters (Table 1). The initially employed native constitutive promoters used for r-protein production in *S. cerevisiae* are strong promoters of the glycolytic pathway genes, i.e., P_{PGK1} [70–72], $P_{GAP491/THD3}$, or P_{GPD} [73]; of the fermentative pathway genes, i.e., P_{ADH1} [74] and P_{PDC1}; and of other genes such as P_{TRP1} [75] and $P_{MF\alpha 1}$ [76]. Out of these, the first yeast promoter implemented for the production of foreign proteins was P_{ADH1} [74, 77].

Among the native regulated promoters used in design of expression systems, P_{GAL10} is one of the strongest inducible promoters and

Table 1
Promoters commonly used in *S. cerevisiae*

Promoter	Protein	Regulation	Reference
Constitutive			
ACT1	Actin		[446]
ADH1	Alcohol dehydrogenase 1		[447–449]
CYC1	Cytochrome-c-oxidase iso-1		[450]
ENO	Enolase		[451]
GAP/TDH3	Glyceraldehyde-3-phosphate dehydrogenase		[452, 453]
HXT7	Hexose transporter		[454]
MFα1	Mating pheromone alpha-factor		[455]
PGK	3-Phosphoglycerate kinase		[456–459]
PYK1	Pyruvate kinase		[460]
TEF	Translation elongation factor		[461, 462]
TRPI	Phosphoribosylanthranilate isomerase		[75, 463]
Inducible			
ADH2	Alcohol dehydrogenase	Ethanol	[464]
CUP1	Metallothionein	Copper	[465–467]
GAL1	Galactokinase	Galactose	[468]
GAL7	α-D-galactose-1-phosphate uridyltransferase	Galactose	[468]
GAL10	Uridine diphosphoglucose epimerase	Galactose	[468]
ICL1	Isocitrate lyase	Ethanol, acetate	[469]
MET25	*O*-acetylhomoserine sulfhydrylase	Methionine	[470]
PHO5	Acid phosphatase	Phosphate (repressible)	[471]

has a well-characterized regulation mechanism. The regulation depends on the type of carbon source used, whereby the promoter activity is 1000-fold induced by galactose and repressed by glucose. The regulation mechanism of P_{GAL10} has also been observed in some other strong promoters, i.e., P_{GAL1} [78] and P_{GAL7}. The comparative analysis between the strength of P_{GAL1} and P_{MET25} in terms of the expression levels of various foreign proteins was reviewed by Rönicke et al. [79]. Recently, more carbon source-dependent inducible promoters of *S. cerevisiae* were reviewed by Weinhandl et al. [80]. Among them, P_{ICL1} and P_{CYC1} show 200-fold increase in expression when induced by ethanol and high dissolved oxygen concentrations, respectively, in the absence of glucose. P_{CYC1} is also triggered by lactate, however with a 20-fold lower expression level compared to that obtained with dissolved oxygen induction. Moreover, the promoter of acid phosphatase, P_{PHO5}, is active under phosphate starvation [81], whereas the other strongly inducible promoters such as the promoter of alcohol dehydrogenase, P_{ADH2}, and metallothionein, P_{CUP1}, are activated in the presence of ethanol and copper, respectively. Other native inducible promoters were less successful such as P_{PIS1} (phosphoinositol synthase), P_{GUT1} (glycerol kinase), P_{FBP1} (fructose-1,6-bisphosphatase), and P_{PCK1} (PEP carboxykinase). Hybrid promoters are also available that combine promoter elements of strong constitutive promoters with upstream activation sequence (UAS) from regulated promoters (reviewed by [43]).

New synthetic promoters were designed for *S. cerevisiae* through a computationally guided approach after investigation of the nucleosome architecture [82]. Stress-responsive promoters can be used to achieve dynamic control of pathway enzymes [83]. For example, by modifying the transcription factor binding sites in the UAS of the *YGP1* promoter, the low-pH performance was significantly increased [84].

2.4 Secretion Signals

For efficient secretion of r-proteins, the sequential processes comprising protein synthesis, folding, and secretory trafficking are required in a cell. A secretion signal, which should be compatible to both foreign sequence and yeast secretory pathway [85], is required for the first step into the secretory pathway mediating translocation of the proteins into the endoplasmic reticulum (ER).

Both endogenous and heterologous secretion signals have been used to secrete r-proteins from yeast (Table 2). The first employed signal sequences in *S. cerevisiae* were heterologous ones being native to the human interferons IFN-α1, IFN-α2, and IFN-γ [86]. Thereafter, the homologous signal sequence of *S. cerevisiae* proteins, i.e., α-mating factor (MFα1), invertase (Suc2), acid

Table 2
Secretion signals commonly used in *S. cerevisiae*.

Signal sequence	Source	Protein	Reference
Endogenous			
PHO5		Acid phosphatase	[472]
MFα1		α-Mating factor 1	[473]
SUC2		Invertase	[474]
Exogenous			
INU	*K. marxianus*	Inulosucrase	[85]
MEL	*S. carlsbergensis*	α-Galactosidase	[475]
Killer	*K. lactis*	Killer protein	[476]

phosphatase (Pho5), and killer toxin, has been exploited. Among them, the pre-pro-leader region of MFα1 has been the most widely used signal sequence due to its high secretion efficiency. Besides, there is a variety of employed heterologous signal sequences from other yeasts (Table 2).

It is indeed noteworthy to mention that as the secretion efficiency of a target protein highly depends on the secretion signal, the optimization of signal and leader peptides occasionally have been investigated to increase the yield of the secreted protein [87].

3 *Pichia pastoris*

P. pastoris is an ascosporous yeast that was first isolated from exudate of a chestnut tree in France in 1919 by Guilliermond. Around the mid-twentieth century, its methanol-utilizing ability as a sole carbon source was discovered [88]. Philips Petroleum Company employed *P. pastoris* to produce single-cell protein (SCP) as an animal feed additive; however, the dramatic increase in methanol prices because of the oil crisis in 1973 made SCP processes uneconomical. In the 1980s, *P. pastoris* was transformed into a heterologous host system to produce r-proteins under control of the strong and tightly regulated alcohol oxidase 1 promoter (P_{AOX1}) [89]. *P. pastoris* has attracted attention of the scientific community and currently has become the most prevalent yeast system in r-protein production research. More than 10 years ago, phylogenetic analyses led to reclassification of the species *P. pastoris* to the genus *Komagataella*, which was further split into *Komagataella pastoris* and *Komagataella phaffii* [90]. As both *K. pastoris* and *K. phaffii* strains have been used in research and industry since the beginning, for the sake of simplicity, the *Pichia* community continues to call the yeast as *P. pastoris*. The success of *P. pastoris* as an efficient heterologous host can be attributed to a number of advantages including its ability to grow rapidly to very high cell densities in defined minimal medium, rather straightforward genetic manipulations, and—like other yeast hosts—perform eukaryotic posttranslational modifications such as glycosylation, disulfide bond formation, proteolytic modification, and targeting to subcellular compartments [91]. Specifically, *P. pastoris* is able to reach high product yields and secretes r-proteins to the extracellular medium relatively pure of contaminations due to low-level secretion of its endogenous proteins. The N-glycans in *P. pastoris* are of the mannose-rich yeast type and usually contain 9–11 mannoses with terminal α-1,2-linkages. Unlike *S. cerevisiae*, *P. pastoris* does not hyperglycosylate therapeutic proteins and does not contain potentially immunogenic terminal α-1,3-linked mannoses. During the last 10 years, several humanized *P. pastoris* strains were developed and applied for r-protein production to allow for more homogeneous and proper glycosylation patterns [92–94].

In *P. pastoris*, high-level expression of exogenous genes has been predominantly controlled by the methanol-induced P_{AOX1}. Regarding methanol utilization, there are three possible phenotypes of *P. pastoris*, namely, Mut^+ (wild type methanol consumption; both of the alcohol oxidase enzymes, Aox1 and Aox2, are functional), Mut^S (slow methanol consumption; disrupted Aox1, functional Aox2), and Mut^- (no methanol consumption; both Aox1 and Aox2 are disrupted). The constitutive promoter of glyceraldehyde-3-phosphate dehydrogenase P_{GAP} provides comparable expression levels to P_{AOX1} and also facilitates the process management by eliminating the use of potentially hazardous inducers such as methanol [95].

Moreover, *P. pastoris* is a model organism used to study peroxisome biology as it possesses the peculiar methanol utilization metabolism, and for the secretory pathway as the structure of its ER exit sites and Golgi apparatus are similar to higher eukaryotes in contrast to *S. cerevisiae* [96]. A major breakthrough was the availability of genome sequence data of the commercial *P. pastoris* strain GS115 in 2009, which has provided researchers with a deeper insight into yeast physiology and genetics, and speeded up the developments of the genetic toolbox or engineered strains [97]. Later, sequence data of the type strain *P. pastoris DSMZ* 70382 [98, 99] and the original SCP production strain CBS7435 [100–102] was published. The Food and Drug Administration (FDA) gave GRAS (generally regarded as safe) status to phospholipase C [103] as a feed additive, and also in 2009, the first biopharmaceutical protein produced by a non-*Saccharomyces* yeast, Kalbitor®, a kallikrein inhibitor produced by *P. pastoris*, was approved by the FDA [104].

3.1 Expression Vectors

In general, yeast vectors can be either self-replicating or integrative. Self-replicating vectors should contain a sequence to direct DNA replication, and these sequences are usually derived from centromeres (CENs) or other autonomously replicating sequences (ARSs). Centromeric sequences of *P. pastoris* have been identified [102, 105] very recently but not used in vectors yet; however, a few ARS-based episomal vectors were developed for *P. pastoris* [89, 106–109]. Yeast ARS-based vectors have low replicative stability; thus without selective pressure, after ten generations plasmid loss ranges from 50% to 70% [108–110]. On the other hand, genetic stability can be achieved via integrative vectors that are flanked on one or both sides by a homologous sequence, and this is the most preferred method for r-protein expression studies in *P. pastoris*. However, unlike *S. cerevisiae* where homologous recombination is predominant, nonhomologous end joining (NHEJ) also occurs frequently in *P. pastoris*. Efficiency of homologous recombination can be increased via elongating the length of homologous regions that flank the expression cassette and/or by using a *ku70* deletion strain [111]. The most commonly used integration sites are P_{AOX1}, P_{GAP}, or the *AOX1* terminator regions.

Different commercial expression vectors for intracellular production, i.e., pPICZ, pGAPZ series, and pFLD1, and for secretory r-proteins production, i.e., series of pPICZα and pGAPZα, were developed for *P. pastoris* and commercialized by Life Technologies (Carlsbad, CA, USA). These vectors employ P_{AOX1} or P_{FLD1} for inducible expression and P_{GAP} for constitutive production and use the Zeocin™ resistance gene as a selective marker. Commercial vectors that contain different selective markers against blasticidin (pPIC6) and Geneticin® (pPIC9K and pPIC3.5 K) are also available by Life Technologies (Carlsbad, CA, USA) as well as a vector (pAO815) for cloning multiple copies of a target gene. In addition, the PichiaPink™ expression kit offers easy selection of multicopy clones based on color formation differences (Life Technologies, Carlsbad, CA, USA). Besides providing standard expression vectors and strains, BioGrammatics (Carlsbad, CA, USA) also commercializes GlycoSwith® strains and vectors for humanized glycosylation of heterologous proteins [112]. Additionally, ATUM (Newark, CA, USA) offers expression vectors with 11 different secretion signals for secretory production. Recently, several noncommercial *P. pastoris* vector systems such as the MoClo Pichia toolkit, the Golden*Pi*CS kit, and the CRISPi kit became available through Addgene [113–116].

3.2 Selection Markers

A variety of selection markers has become available for *P. pastoris* over the past few decades. Recessive markers, which are complementary markers of a mutation in the host genome and are usually obtained from amino acid or nucleotide synthesis pathways, namely, *HIS*, *ARG*, *ADE*, *URA*, and *MET*, were employed in *P. pastoris* expression system. The main drawback of auxotrophic markers is the risk of cross-feeding between marker-positive and marker-negative cells, especially in high-cell density liquid cultures. Alternatively, metabolic auxotrophies have been developed by focusing on utilization of a compound instead of synthesis of a substance, such as deficiency of formaldehyde utilization in the *FLD1* knockout strain [117] or glycerol utilization with *GUT1* knockout strain [111].

Dominant markers do not require any prior mutation in the host cell, and usually they are resistance genes against antibiotics or other toxic compounds. Dominant selection markers for *P. pastoris* have been developed against Zeocin™, Geneticin®, blasticidin, hygromycin, nourseothricin, and arsenic. Antibiotic selection markers also provide us with the advantage of selecting high copy transformants having several copies of the gene expression cassette integrated into the genome of the host cells. This situation is especially applicable for dose-dependent markers, namely, Zeocin™, Geneticin®, and hygromycin. Among recessive markers, *FLD1* may also be used to select high copy number transformants [117]. Multicopy strain selection was also achieved with an attenuated adenine auxotrophic marker that had a truncated promoter sequence, as multiple copies of the attenuated marker are needed to complement the auxotrophy [118]. Besides slow growth, *ade* auxotrophy leads to red pigment accumulation, so multicopy integrants can be easily identified with their white color and larger colony size [118].

3.3 Promoters

Transcription, the first step of protein synthesis, is a very crucial point in r-protein production; therefore, strong and controllable promoters are essential tools for efficient expression systems [119]. One of the most important features of methylotrophic yeasts is their strong and tightly regulated promoters obtained from the methanol utilization (MUT) pathway. Alcohol oxidase (AOX) catalyzes the first step of methanol utilization, conversion of methanol to formaldehyde. In *P. pastoris*, under methanol induction, AOX constitutes 5% of total mRNA and 33% of total cellular protein [120]. On the other hand, growth on repressing carbon sources, i.e., glucose, ethanol, and glycerol, leads to complete lack of AOX activity [121]. In bioprocess operations, tight regulation of P_{AOX1} provides the ability to divide cell growth and production phases and enables high cell densities without r-protein production hampering the cellular activities in the growth phase on glycerol or glucose. *P. pastoris* has two *AOX* genes, *AOX1* and *AOX2* [122]. As P_{AOX1} is significantly stronger and tightly regulated with carbon sources, it has become the most prevalently used promoter for r-protein production in *P. pastoris*. Up to 22 g/L intracellular [123] and 10.5 g/L secreted [124] heterologous protein production was achieved by P_{AOX1}. P_{AOX2} has a similar regulation profile; however, it provides around 10–20 times lower expression levels than P_{AOX1} [125]. Although P_{AOX1} is the pioneer promoter in r-protein production research, its inducer methanol is a toxic and flammable compound that can cause problems and safety concerns in industrial applications. Other important promoters of the MUT pathway are formaldehyde dehydrogenase (P_{FLD1}) and dihydroxyacetone synthase (P_{DAS1}) promoters. P_{DAS} shows similar regulatory pattern and production levels as P_{AOX1} [126], while P_{FLD1} has the opportunity to be induced by either methanol or in a methanol-free system with methylamine [127]. Most of the commercial vectors for *P. pastoris* employ P_{AOX1}, but there are also P_{FLD1}-based vectors. Among alternative inducible promoters, P_{G1} is promising as it is induced by limiting the safe carbon source glucose and shows higher expression levels than P_{GAP} [128]. Other regulated promoters induced under phosphate limitation are the acid phosphatase promoter P_{PHO1} [129] and the promoter of the putative sodium-coupled phosphate symporter P_{PHO89} (also named as P_{NPS}) [130]. To develop methanol-free inducible systems, P_{PHO89} is also an alternative as it has similar strength as P_{GAP} [130]. Some years ago, novel promoters, P_{ICL1} (derepressed by glucose depletion and induced by ethanol), P_{ADH1} (induced by glycerol and ethanol), and P_{GUT1} (induced by glucose, glycerol, and ethanol), were patented [131]. Another methanol-free alternative, P_{THI11}, that is derived from a gene involved in thiamine biosynthesis, reached 70% of P_{GAP} activity in the absence of thiamine [132]. P_{AOD} (alternative oxidase) is a moderate promoter induced by glucose and repressed by methanol or upon glucose depletion but with limited applicability as it ceased its functioning at about half of the maximal activity of P_{GAP} [133].

For the production of toxic proteins, regulated expression is favorable, but it requires at least two production phases, the first one to obtain high cell yield under repressing conditions and the second one to perform expression of heterologous proteins, resulting in more handling efforts and longer process times. On the other hand, for nontoxic proteins, constitutive expression simplifies bioprocess operations and provides higher space-time yield. Since the glyceraldehyde-3-phosphate dehydrogenase promoter (P_{GAP}) is strong and constitutive (reviewed by Çalık et al. [95]), it can reach similar expression levels as P_{AOX1} [134], and it performs better under hypoxic conditions [135]. The highest expression levels are reached when cells are grown on glucose, but also comparable product yields are obtained with glycerol. The sorbitol dehydrogenase promoter P_{SDH} is an alternative to P_{GAP} as their strengths are reported to be similar [136]. Phosphoglycerate kinase promoter (P_{PGK1}), involved in glycolysis and gluconeogenesis, is a relatively weak promoter showing 10% activity of P_{GAP} [132]. Translation elongation factor 1 alpha is an essential module of the eukaryotic translation machinery; thus, its promoter P_{TEF1} exhibits a growth-associated regulatory pattern, with high activity during exponential phase and reduced performance in the stationary phase [137]. Under carbon-limited conditions, almost twofold higher expression levels were obtained with P_{TEF1} compared to P_{GAP} [137]; however, with different reporter genes and at different cultivation times, similar or less expression levels were observed with P_{TEF1} compared to P_{GAP} [132]. Other constitutive promoters obtained from the glycolytic pathway are P_{ENO1}, P_{GPM1}, and P_{TPI1}, and those obtained from stress response metabolism are P_{HSP82}, P_{KAR2}, and P_{SSA4} (Table 3). P_{PET9} exhibited different expression levels with different reporter genes, in comparison with P_{GAP} [132]. Promoter engineering studies and the design of synthetic core promoters have also widened the promoter toolbox for *P. pastoris* as they created novel promoters that cover a wide range of strength and have different regulation patterns [138–142].

Table 3
Promoters commonly used in *P. pastoris*

Promoter	Protein	Regulation	Reference
Constitutive			
GAP	Glyceraldehyde-3-phosphate dehydrogenase		[134]
GCW14	Potential glycosyl phosphatidyl inositol (GPI)-anchored protein		[477]
GPM1	Phosphoglycerate mutase		[132]
HSP82	Cytoplasmic chaperone (Hsp90 family)		[132]
ILV5	Acetohydroxy acid isomeroreductase		[111]
KAR2	ER resident hsp70 chaperone (binding protein BiP)		[132]
TEF1	Translation elongation factor 1 α		[137]
PET9	Major ADP/ATP carrier of the mitochondrial inner membrane		[132]
PGK1	3-Phosphoglycerate kinase		[478]
SDH	Sorbitol dehydrogenase		[136]
SSA4	Heat shock protein		[132]
TPI1	Triose phosphate isomerase		[132]
YPT1	GTPase involved in secretion		[116]
Inducible			
ADH1	Alcohol dehydrogenase 1	Ethanol, glycerol	[131]
AOD	Alternative oxidase	Glucose, antimycin A	[133]
AOX1	Alcohol oxidase 1	Methanol	[126]
AOX2	Alcohol oxidase 2	Methanol	[122, 479]
DAS	Dihydroxyacetone synthase	Methanol	[126, 480]
ENO1	Enolase	Glucose, glycerol, ethanol	[131]
FLD1	Formaldehyde dehydrogenase	Methanol, methylamine, choline	[127]
G1	High affinity glucose transporter	Glucose limitation	[128]
G6	Putative aldehyde dehydrogenase	Glucose limitation	[128]
GUT1	Glycerol kinase	Glycerol	[131]
ICL1	Isocitrate lyase	Ethanol, derepressed with glucose depletion	[481]
MET3	ATP sulfurylase	L-methionine (repressible)	[482]
PEX8	Peroxisomal matrix protein	Methanol, oleate	[143, 483]

(continued)

Table 3 (continued)

Promoter	Protein	Regulation	Reference
PIS1	Phosphatidylinositol synthase	Zinc (repressible)	[482]
PHO89 or *NSP*	Putative Na+/phosphate symporter	Phosphate limitation	[130]
SER1	3-Phosphoserine aminotransferase	L-serine (repressible)	[482]
THI11	thiamine biosynthesis	Thiamine (repressible)	[132]
THR1	Homoserine kinase	L-threonine, L-valine, L-leucine, and L-isoleucine (repressible)	[482]

For the co-expression of multiple proteins simultaneously, it is important to identify alternative promoters covering a wide range of strength and showing different regulatory patterns. Also, if correct folding or secretion steps are rate limiting in expression of active products, weaker expression levels can be favorable. P_{PEX8} (peroxin 8) [143] and P_{YPT1} (a GTPase function in secretion) [116] are alternative moderate inducible and constitutive promoters, when low expression levels are desirable [113, 144].

Few promoters from other microorganisms have been also employed in r-protein production in *P. pastoris* successfully. P_{ADH2} from *P. stipitis* [145, 146] and P_{PDC1} from *K. lactis* [147] were induced in *P. pastoris* via decreasing dissolved oxygen concentration. *S. cerevisiae* P_{CUP1} was also utilized in *P. pastoris* expression system, where expression was induced by copper [148].

3.4 Signal Sequences

One of the most prominent advantages of the *P. pastoris* expression system is its ability to secrete high amounts of active r-proteins and low amounts of endogenous proteins to the extracellular medium [98, 149]. For this purpose, a variety of endogenous and exogenous secretion signals have been successfully employed (Table 4). Among them, the most commonly used and successful one is the *S. cerevisiae* α-mating factor pre-pro-sequence. However, because of limited processing by Ste13 in *P. pastoris* [76], the residual two EA repeats at the N-terminus of secreted r-proteins are a common problem encountered with α-MF; on the other hand, to eliminate this issue, the leader without EA repeats could also be used for extracellular protein production.

Besides, few other signal sequences have been exploited with favorable outcomes. *P. pastoris* endogenous acid phosphatase signal sequence (*PHO1*) [129, 150] has been successfully used in recombinant expression of mouse 5-HT5A serotonin receptor [151], Hantaan virus G2 protein [152], levansucrase [153], porcine pepsinogen [154], and human midkine [155], despite of lower

Table 4
Secretion signals commonly used in *P. pastoris*

Signal sequence	Source	Protein	Reference
Endogenous			
DDDK		DDDK protein	[163]
DSE		Daughter cell-specific secreted protein with similarity to glucanases, endo-1,3-beta-glucanase	[161]
EPX1		Extracellular protein X	[164]
EXG		Major exo-1,3-beta-glucanase of the cell wall	[161]
FRP2		Hypothetical protein	[160]
PDI1		Protein disulfide isomerase	[160]
PHO1		Acid phosphatase	[129, 150]
PIR1		Proteins with internal repeats	[162]
SCW		Cell wall protein with similarity to glucanases	[161]
SSP120		Protein of unknown function	[160]
UTR2		Cell wall protein that functions in the transfer of chitin to beta(1–6)glucan	[160]
Exogenous			
α-Amy	*Bacillus licheniformis*	Alpha amylase	[157]
α-MF	*S. cerevisiae*	α-Mating factor	[76]
CLY and *CLY-L8*	Chicken	Lysozyme- and leucine-rich artificial peptide	[484]
CSN2	Bovine	Beta casein	[485]
gla	*A. niger*	Glucoamylase	[486]
HBFI and *HBFII*	*Trichoderma reesei*	Hydrophobins	[487]
HSA	Human	Serum albumin	[488]
IgG1	Murine	Immunoglobulin G1	[165]
K28 pptox	M28 virus	K28 pre-pro-toxin	[489]
pGKL	*K. lactis*	Killer toxin	[490]
PHA-E	*Phaseolus vulgaris*	Phytohemagglutinin	[491]
SED1	*S. cerevisiae*	Major glycosylphosphatidylinositol protein in the stationary phase	[492]
SUC2	*S. cerevisiae*	Invertase	[156]
	Pichia acaciae	Killer toxin	[493]
	Human	Midkine	[155]

secretion levels compared to α-MF. In addition, the signal sequence derived from *S. cerevisiae* invertase (*SUC2*) gene has been employed in secretion of invertase [156], thermostable α-amylase [157], and human antithrombin [158] in *P. pastoris.*

As performances of signal peptides are often product dependent and the choice is arbitrary, it is important to have a signal sequence library for heterologous expression hosts. Identification of endogenous signal peptides associated with highly secreted proteins has attracted research attention recently. Novel endogenous signal peptides were identified by in silico analysis using secretome data of *P. pastoris* [159], and among them, the leader sequence of protein disulfide isomerase, a multifunctional protein resident in the ER lumen, exhibited promising secretion level as being 70–80% of α-MF, while it is 73 aa shorter in length than α-MF [160]. Also, *P. pastoris* endogenous signal peptides, namely, Scw, Dse, and Exg, were tested for their secretion capacity, and all three showed better secretion for enhanced green fluorescent protein (EGFP) than α-MF; and even Exg outperformed α-MF for expression of *Candida antarctica* lipase B [161]. *P. pastoris* Pir1 (proteins with internal repeats) pre-pro-signal was employed for secretion of EGFP and human α1-antitrypsin [162]; however, comparative data on expression levels with known signal sequences were not provided. Further, the leader sequence of *P. pastoris* DDDK protein exhibited comparable secretion efficiency for porcine carboxypeptidase B and *Erythrina* trypsin inhibitor with α-MF [163]. *P. pastoris* experimental secretome data revealed extracellular protein X (Epx1) as being the most abundant secretory protein; Epx1 signal sequence variants exhibited higher secretion levels than commonly used signals in *P. pastoris* [164].

Additionally, many heterologous secretion leaders have been employed in *P. pastoris* (Table 4); however, for most cases, they do not reach similar high secretion level as α-MF. On the other hand, for particular cases, such as anti-HIV VRC01 antibody expression via the murine IgG1 leader sequence [165], better expression levels were obtained with exogenous secretion signals.

4 *Hansenula polymorpha*

H. polymorpha (*Ogataea polymorpha*, *Pichia angusta*) is one of the commonly used methylotrophic yeast species that is able to grow rapidly on simple media [166] and differs from other methylotrophs with its nitrate assimilation capability and thermotolerance up to 50 °C [167].

Owing to its strong and tunable promoters derived from the methanol utilization pathway [168], high secretion capacity [169], and lower hyperglycosylation activity compared to *S. cerevisiae* [2], *H. polymorpha* appears to be another promising alterative host for the production of r-proteins. Its relatively high optimum growth temperature makes this microorganism favorable particularly for the manufacturing of thermostable proteins to be used in crystallographic studies [1] and also for the synthesis of mammalian proteins [170]. Additionally, the excessive peroxisomal membrane of *H. polymorpha* is an excellent storage site for heterologous membrane proteins [171] produced with a suitable signal peptide [172].

Apart from r-protein production, *H. polymorpha* is a good candidate for bioethanol production, as it can utilize xylose and cellobiose as carbon and energy source [173], and further is used as a model organism in basic research, where methanol metabolism [174], peroxisomal function and biogenesis [174, 175], and nitrate transport and assimilation [176–178] are investigated.

To date, three basic strains of *H. polymorpha* with undefined relationships, different properties, and independent origins have been isolated and characterized. Strains CBS4732 and NCYY495 are able to mate, while strain DL-1 cannot be mated with the others [1]. Strain NCYC495 is generally used in classical genetic studies, whereas strains CBS4732 and DL-1 are frequently employed for r-protein production [179].

4.1 Expression Vectors

A variety of expression vectors designed for *H. polymorpha* can be used in transfection either as a circular plasmid or as a linearized expression cassette. Although the constructed circular plasmids generally contain the autonomously replicating sequence, *Hansenula ARS1* (*HARS1*), the genomic integration of episomal vectors takes place in a high frequency [180], when the transformants are kept at selective conditions for a prolonged period of time. In this case, heterologous gene copy number reaches up to 60 in a head-to-tail arrangement [181]. Contrary to previous findings [182–184], integration does not occur randomly; the circular plasmids recombine with the genome via genomic sequences located on the vector [1]. On the other hand, when targeting genome integration, the plasmids containing sequences homologous to a specific genome locus are linearized before transfection. The commonly used integration loci of *H. polymorpha* are the *MOX/TRP* locus [185–187], *HARS36* locus [188], ARS sequence [180, 188], *URA3* gene [177], *LEU2* gene [188], and the *GAP* promoter region [189]. Integration at rDNA clusters also seems to be a promising alternative due to the presence of conserved rDNA sequences in a wide range of yeasts [190–195]. Therefore, this strategy allows the use of only one vector, e.g., ComEd® [1], for the transformation of various species. When different parts of the *H. polymorpha* rDNA were investigated, transformation of all yeast species tested was achieved using the putative promoter and external transcribed spacer region of the 35S rDNA and the full-length 18S rDNA. The 25S rDNA region of this nonconventional yeast was also successfully implemented for the same purpose

[191, 193], and no difference was detected in the expression level of a firefly luciferase reporter gene, when *H. polymorpha* was transformed with either an 18S rDNA or 25S rDNA containing vector [193].

4.2 Selection Markers

Native *LEU2*, *URA3*, *TRP3*, and *ADE11* genes, *S. cerevisiae*-derived *LEU2* and *URA3* genes, and *Candida albicans*-derived *LEU2* gene have been commonly used as selective markers in *H. polymorpha* expression systems for the complementation of respective auxotrophic mutants [196–199]. In addition to these auxotrophic markers, a range of dominant selection markers has been utilized successfully, such as resistance to phleomycin [200], hygromycin B [201], Zeocin™ [187], and G418 [193], and in this way it becomes possible to identify multicopy strains by increasing the antibiotic concentration.

4.3 Promoters

There are various inducible and constitutive promoters that have been used to design *H. polymorpha* expression systems (Table 5). As this yeast species is able to metabolize methanol as a sole source of

Table 5
Promoters of *H. polymorpha*

Promoter	Protein	Regulation	Reference
Constitutive			
ACT	Actin		[201]
ALG2	Isocitrate lyase		[494]
GAP	Glyceraldehyde-3-phosphate dehydrogenase		[189]
PMA1	Plasma membrane H^+-ATPase		[190]
TPS1	Trehalose-6-phosphate synthase		[210]
Inducible			
MAL1	Maltase	Maltose or sucrose	[495]
MOX	Methanol oxidase	Methanol	[496]
FLD1	Formaldehyde dehydrogenase	Methanol, primary amines	[205]
FMD	Formate dehydrogenase	Methanol	[497]
DAS	Dihydroxyacetone synthase	Methanol	[498]
YNR1	Nitrate reductase	Nitrate	[176]
YNI1	Nitrite reductase	Nitrate	[207]
YNT1	Nitrate transporter	Nitrate	[208]
YNA1	$Zn(II)_2Cys_6$ transcriptional factor	Nitrate	[499]
PHO1	Acid phosphatase	Repressible	[206]

carbon and energy, the most commonly used promoters are derived from the genes of the methanol utilization pathway [169, 170], such as the genes of methanol oxidase (*MOX*), formate dehydrogenase (*FMD*), and dihydroxyacetone synthase (*DAS*). All the promoters mentioned above are fully repressed by glucose, derepressed by glycerol, and strongly induced by methanol [1, 166]. Therefore, in an efficiently designed operation strategy after sufficient cell density is obtained, the heterologous gene expression can be started by methanol induction. This is particularly important when the desired r-protein is toxic for the host cell [202]. Additionally, the possibility of achieving high promoter activity under glucose- or glycerol-limited conditions is unique for *H. polymorpha* among the methylotrophic yeasts [170, 203, 204]. P_{FLD1} is also methanol inducible, but it shows moderate activity when primary amines are used as a nitrogen source [205].

The repressible acid phosphatase promoter (P_{PHO1}) [205, 206] and the promoters derived from the genes of the nitrate assimilation pathway, e.g., P_{YNT1}, P_{YNR1}, and P_{YNL1} [176, 207, 208], are less frequently used inducible promoters of *H. polymorpha*. The promoters in the second group are induced by nitrate and repressed by ammonium [176]. *H. polymorpha* is the only methylotrophic yeast species that is capable of metabolizing maltose, and P_{MAL1} performs very well compared to widely used P_{MOX}, especially when it is induced by sucrose [80]. Moreover, this promoter is functional in another maltose-utilizing yeast species, *S. cerevisiae* [209].

Among *H. polymorpha*-derived constitutive promoters, such as P_{PMA1} [190], P_{TPS1} [210], P_{ACT} [201], and P_{GAP} [189], P_{PMA1} exhibits high transcriptional activity when it is compared with P_{AOX} [1]. In contrast to the other constitutive promoters, the activity of P_{TPS1} does not depend on the use of a particular carbon source, but rather it may increase at elevated temperatures [210], while P_{GAP} may be a good alternative to the large-scale P_{MOX}-based r-protein production systems due to the higher specific production rate provided in a technically less challenging fermentation process [189].

4.4 Secretion Signals

The available signal sequences of *H. polymorpha* (Table 6) are the targeting signals of the peroxisomal matrix proteins Pts1 and Pts2 [169] and the secretion signal of Pho1 [206]. In addition to these endogenous signal peptides, the *H. sapiens*-derived *HSA* [211], *S. occidentalis*-derived *GAM1* [169, 212], *Y. lipolytica*-derived *YlLip11* [213], and *S. cerevisiae*-derived *MFα1* signal sequences [170, 214–216] have been utilized to direct efficient extracellular protein production. Regarding the development of cell surface display systems in *H. polymorpha*, glycosylphosphatidylinositol (GPI)-anchoring motifs, which were derived from the GPI-anchored cell surface proteins, e.g., Sed1, Gas1, Tip1, Cwp1, Yps1, and Yps7 [217–219], have also been employed.

Table 6
Secretion signals of *H. polymorpha*

Signal sequence	Source	Protein	Reference
Endogenous			
PHO1		Acid phosphatase	[206]
PTS1		Peroxisomal targeting signal	[500]
PTS2		Peroxisomal targeting signal	[501]
Exogenous			
GAM1	*S. occidentalis*	Glucoamylase	[212]
HSA	*Human*	Serum albumin	[211]
YlLip11	*Y. lipolytica*	Lipase	[213]
MFα1	*S. cerevisiae*	α-Mating factor 1	[212]

Accordingly, by producing the recombinant glucose oxidase as a fusion protein with these anchoring motifs, most enzyme activity was detected at the cell surface [219].

5 *Arxula adeninivorans*

A. adeninivorans (*Blastobotrys adeninivorans*) is a nonconventional yeast species that has attracted significant attention recently with its completely sequenced genome [220] and unusual features such as thermotolerance, osmotolerance, xerotolerance, and nitrate assimilation capability [221]. The nonpathogenic yeast *A. adeninivorans* can grow at high rates on a wide variety of substrates [222], and due to having many degradative pathways, it is a prominent source of genes that encode industrially important enzymes, particularly used in food and feed manufacturing and bioremediation. Among ten *A. adeninivorans*-based products on the market [221], the first commercially available protein was a recombinant tannase [223].

The haploid yeast *A. adeninivorans* displays three morphological forms in a temperature-dependent manner: yeast cells reproduce by budding up to 42 °C, form pseudomycelia at 42 °C, and become mycelial above 42 °C [224, 225]. Although the exact mechanism behind this dimorphism has not been elucidated yet, the reversible morphological change can easily be controlled by altering the temperature. This is significant as many posttranslational modifications, protein expression, and secretion properties largely depend on the morphology, i.e., higher secretion capacity is observed in mycelial cultures [225] or more frequent *O*-linked glycosylation is performed by budding cells [226].

5.1 Expression Vectors

The expression systems initially developed for *A. adeninivorans* [227] integrated randomly into the yeast chromosomes or remained episomally in the cytoplasm. As transformation with circular plasmids did not bring success, linearized cassettes carrying the heterologous DNA were stably integrated into rDNA by homologous recombination; and the first successful *A. adeninivorans* transformation was reported with this targeting approach [228]. Since then, the integration into the 25S rDNA of *A. adeninivorans* by gap repair has been extensively studied. In this design strategy, the copy number of the heterologous gene is very low, i.e., 1–3 copies [229, 230], due the large size of the plasmid making the specific insertion difficult [230]. To solve this problem, cells may be co-transformed with two rDNA integration vectors harboring two different selection markers [166].

The commercial transfection/expression platforms, Xplor1® [231], and its derivative, Xplor2® [232], are hybrid vectors that contain fragments derived from *A. adeninivorans*, *E. coli*, and *S. cerevisiae*. Owing to this feature, both plasmids have been used not only for *A. adeninivorans* but also for some other industrially important yeast species, such as *S. cerevisiae*, *P. pastoris*, and *H. polymorpha*, as recently reviewed by Malak et al. [221].

5.2 Selection Markers

The dominant selection marker used in *A. adeninivorans* expression systems is the *E. coli*-derived *hph* gene which provides resistance against hygromycin B [228, 231]. Auxotrophic selection markers that eliminate the problems related with the use of toxic compounds or antibiotics in industrial biotechnology are also available for this nonconventional yeast species. Accordingly, in the first transformation system designed for *A. adeninivorans*, the *LYS2* gene from *A. adeninivorans* or *S. cerevisiae* was employed for this purpose [227]. Thereafter, the *AILV1* gene was identified as a selection marker for the transformation of *A. adeninivorans ailv1* mutant strains [233]. However, since an atypical anamorphic phenotype, causing inefficient centrifugal separation of the transformants from the media was characterized for *ailv1* mutant strains, a variety of auxotrophic *A. adeninivorans* mutants with a typical growth phenotype have been described, such as the *aleu2* mutants to be used in the *A. adeninivorans*-derived *ALEU2*-based complementation studies [191].

5.3 Promoters

Heterologous genes have been mostly expressed in *A. adeninivorans* under the control of constitutive promoters (Table 7). Accordingly, r-protein production by this nonconventional yeast species started with the use of P_{AILV1} upstream of the *Pseudomonas putida XylE* gene encoding the catechol 2,3-dioxygenase [227]. Due to relatively low expression levels of heterologous genes by the promoter mentioned above, the second r-protein, β-galactosidase, encoded by the *lacZ* gene from *E. coli*,

Table 7
Promoters of *A. adeninivorans*.

Promoter	Protein	Regulation	Reference
Constitutive			
AHSB4	Histone H4		[237]
AILV1	Threonine deaminase		[233]
ALEU2	β-isopropylmalate dehydrogenase		[230]
TEF1	Translation elongation factor EF-1α		[502]
Inducible			
AINV	Invertase	Sucrose	[245]
ALIP1	Lipase	Tween 20	[246]
ATAN1	Tannase	Tannic acid, gallic acid	[248]
AXDH	Xylitol dehydrogenase	Xylitol, D-sorbitol, D-xylose	[503]
GAA	Glucoamylase	Maltose	[504]

was synthesized under the control of the inducible P_{GAA} [234]. Thereafter, using the strong constitutive P_{TEF1} [233, 235], high expression levels were obtained for two model proteins, GFP and HSA, even when the gene copy number was low [230]. As this promising promoter has been found to be highly active as well in some other yeast species, e.g., *S. cerevisiae*, *P. pastoris*, *H. polymorpha*, *Debaryomyces hansenii*, and *Debaryomyces polymorphus*, it was utilized to develop a commercial transformation/ expression platform Xplor1® [231]. Moreover, in the first *A. adeninivorans* cell-based estrogen biosensor designed for the specific, sensitive, and reproducible estrogen detection in wastewater, recombinant strains containing the human estrogen receptor α (hERα) gene under the control of P_{TEF1} and the *Klebsiella* sp. phytase reporter gene under the control of P_{GAA} were employed [236]. In addition to P_{TEF1}, another strong constitutive promoter, P_{AHSB4}, was proved to be suitable for heterologous gene expression with similar expression levels as with P_{TEF1}. Therefore, P_{AHSB4m} was designed by modifying P_{AHSB4} sequence to facilitate the integration into the expression vectors [237].

5.4 Secretion Signals

For extracellular r-protein production by *A. adeninivorans*, the native signal peptides of the heterologous genes, e.g., the *H. sapiens*-derived *HAS* [229, 230, 237, 238], *Y. lipolytica*-derived *YlLip11* [213], *B. amyloliquefaciens*-derived *amyA* [239–241], *Klebsiella*-derived *phyK* [232, 236, 242], and *S. occidentalis*-derived *SFfase* [243], have been mostly used (Table 8). As a highly potent alternative, the signal sequence of mating factor can also be

Table 8
Secretion signals of *A. adeninivorans*

Signal sequence	Source	Protein	Reference
Endogenous			
ACUT1, ACUT2, ACUT3		Cutinase	[249]
AINV		Invertase	[245]
ALIP		Lipase	[246]
APHO1		Acid phosphatase	[247]
ATAN1		Tannase	[248]
Exogenous			
amyA	*B. amyloliquefaciens*	α-Amylase	[239]
HSA	*Human*	Serum albumin	[229]
YlLip11	*Y. lipolytica*	Lipase	[213]
MFα1	*S. cerevisiae*	α-Mating factor 1	[244]
phyK	*Klebsiella*	Phytase	[236]
SFfase	*S. occidentalis*	Fructofuranosidase	[243]

employed for the secretion of proteins as in various other yeast species. Accordingly, extracellular production of human interleukin-6 (IL-6) [244] and human interferon α (IFNα2a) [232, 243] in *A. adeninivorans* was achieved by the MFα1 secretion signal of *S. cerevisiae*. Additionally, a variety of endogenous signal sequences have been identified for *A. adeninivorans*, since it is able to metabolize a wide range of substrates via efficiently secreted degrading enzymes. In this context, the putative secretion signals of the invertase-encoding *AINV* gene [245], lipase-encoding *ALIP1* gene [246], acid phosphatase-encoding *APHO1* gene [247], tannase-encoding *ATAN1* gene [248], and cutinase-encoding *ACUT1*, *ACUT2*, and *ACUT3* genes [249] were described. In addition to their high potential for extracellular r-protein production, these signal peptides have been used for the secretion of overexpressed homologous proteins such as tannase, which was expressed by the *ATAN1* gene and secreted under the regulation of its own signal sequence [223]. On the other hand, despite the secretion signal of the *APHO1* gene used in front of the cloned gene, approximately 90% of total acid phosphatase remained cell bound [250]. Therefore, the widespread use of these putative signal sequences requires further investigation.

6 *Kluyveromyces lactis*

The genus *Kluyveromyces*, named in honor of the Dutch microbiologist Albert Jan Kluyver (1888–1956) in 1956, is one of the small ascomycetous yeast genera comprising 21 species [251, 252]. *Kluyveromyces marxianus* and *Kluyveromyces lactis* are the most commonly used species belonging to the genus *Kluyveromyces* when considering their applications in food industry. In contrast to other species, both of them have the ability to metabolize xylitol, cellobiose, and lactose in addition to pentose sugars containing xylose and arabinose. Among them, *K. lactis* has been presented as one of the first efficient hosts constructed for homologous and heterologous protein production since the 1980s, in addition to its use as single-cell protein (SCP) for nutritional supplement in food in the 1960s [253]. To date, approximately 100 heterologous proteins produced by *K. lactis* have been reported, while this expression host is best known for its use in large-scale production of the native β-galactosidase and recombinant bovine chymosine [252]. *K. lactis* has obvious advantages, i.e., the feasibility of easy genetic manipulation, the existence of an easy-to-use reagent kit, the availability of a fully sequenced genome [254], a low amount of secreted endogenous proteins, and the presence of commercial enzymes from *K. lactis* having GRAS status by the FDA. Additionally, *K. lactis* is a model organism to investigate gene regulation, mitochondrial biogenesis, and killer phenomena [255].

6.1 Expression Vectors

K. lactis is also an attractive host for r-protein production owing to the availability of both self-replicating and integrative vectors. Unlike many other yeasts, *K. lactis* has a pair of natural episomal vectors, pGKL1 and pGKL2, called as cytoplasmic killer plasmids consisting of linear dsDNA, that confer killer attributes [256]. Although the corresponding vectors allow the cytoplasmic maintenance of 100–200 copies per cell, they are not favorable for heterologous protein production because of their poor stability and low recognition capacity for nuclear promoters [252, 257]. Therefore, the episomes native to other yeasts, *S. cerevisiae* and *Kluyveromyces drosophilarum*, have been implemented as YEp. Among them, the most widely used episomal vector in *K. lactis* is the 1.6 μm pKD1 plasmid isolated from *K. drosophilarum* [258]; however, in *K. lactis*, an instability problem was observed related to *S. cerevisiae*-specific 2 μm circle episome [53]. The indigenous pKD1, analogous to the circular *S. cerevisiae* 2 μm plasmid with little similarity of the nucleotide sequence, is successfully transformed and also replicates stably up to 60–80 copies without selection pressure in *K. lactis* [259]. Nevertheless, in the case of heterologous gene expression, the copy number is decreased to approximately 20 copies due to *recombinase A* inactivation, because the GOI is integrated into the *cis*-acting stability locus (CSL) containing the *recombinase A* gene [260]. Furthermore, the use of the full pKD1 plasmid in *K. lactis* [261] was followed by the use of partial pKD1 [262] or its derivatives [263]. The corresponding plasmid has also high transformation frequency in other *Kluyveromyces* species, i.e., *Kluyveromyces vanudenii*, *Kluyveromyces dobzhanskii*, *Kluyveromyces phaseolosporus*, and some strains of *K. marxianus* [264].

Among self-replicating plasmids, the ARS-based vectors include diverse *K. lactis* ARS sequences (KARS), which were independently isolated from its chromosome [265–267]. The corresponding vector type has also been used for recombinant gene expression in *K. lactis*, e.g., β-galactosidase [268], α-amylase [269], and α-galactosidase [270], even though it is less stable and requires higher amount of DNA for transformation compared to episomal vectors [267].

The YCp, another type of self-replicating plasmid, comprises both *K. lactis*-specific CEN [271] and ARS sequence, since CEN sequences native to other yeast species are not compatible to *K. lactis* sequences. Furthermore, owing to enhanced plasmid maintenance of the YCp system, it has been employed for many genetic studies, e.g., construction of a fungal expression library [272] and investigation of gene function [273]. However, there is a bottleneck in copy number, with 1–2 copies per cell [267].

Even if the self-replicating plasmids are widely employed for many purposes, these vector systems are not feasible for overexpression of heterologous proteins in *K. lactis* because of potential instability problems. To investigate the efficiency of each vector system, a comparative study was performed for four systems, i.e., partial pKD1, whole pKD1, pYCp, and an integrative vector. In contrast to the cells that have the integrated gene, approximately 84% of the cells transformed by the episomal vectors lost their plasmids [274]. Accordingly, the integrative vectors using homologous recombination are an alternative to self-replicating plasmid-based expression systems. The yeast integrative expression system was preliminarily applied through a *LAC4* integration locus for the first r-protein production attempt in *K. lactis* [275] and then exploited through ribosomal DNA (rDNA) [270] and *TRP1* integration [276]. Despite the improved genetic stability of recombinant strains, a decreased copy number is the drawback of integration vector systems. To overcome this problem, a region of tandemly repeating DNA in the rDNA locus of *K. lactis* is targeted for multicopy integration since the repeats intrinsically reach up to 60 copies [270, 277]. These well-established expression systems have been used by several research groups successfully.

Among the various integrative vectors, the most currently and widely used ones are commercial pKLAC1 and pKLAC2 that are supplied by New England Biolabs. pKLAC2 differs from the other with universal multiple cloning site, while both vectors include a variant of the *LAC* promoter, pro-sequence of *K. lactis* α-mating factor, and *AMDS* gene as the selection marker.

6.2 Selection Markers

The expression systems of *K. lactis* employ two conventional selection strategies in addition to a more specific one. In the auxotrophic selection marker-based strategies, the auxotrophies, such as *ura3*, *leu2*, and *trp1*, have been applied to *K. lactis* strains. Similar to the detrimental effects of auxotrophic mutations observed in other yeasts, the reduced production yield of bovine enterokinase was obtained because of the availability of exogenous uracil and uradine [257].

In the dominant selectable marker-based strategies, the genes conferring resistance to the following antibiotics, i.e., geneticin, hygromycin B, G418, and Zeocin™, have been employed in vectors. Bleocin has recently been used as an inexpensive alternative to Zeocin™ [252].

The *K. lactis*-specific selection strategy is mainly applied in the integrative expression vectors, pKLAC1 and pKLAC2. The dominant nitrogen source selection-based strategy aims at the breakdown of acetamide to ammonia by the expression of *Aspergillus nidulans* acetamidase gene (*amdS*) under the control of *ADH2* promoter. Only the transformed cells can grow on the medium containing acetamide as a sole nitrogen source. This method has successfully been used to create *K. lactis* GG799 strains that produce bovine enterokinase, maltose-binding protein, ovalbumin, cellulose, and mouse transthyretin, using the integration vector pKLAC1, which is supplied with the commercial kit [278].

6.3 Promoters

Heterologous gene expression in *K. lactis* has been mainly performed using inducible and constitutive promoters originated from both *K. lactis* and *S. cerevisiae* [257] (Table 9). Among them, the first employed native promoter for heterologous expression of prochymosin [275] is the tightly controlled *LAC4*

Table 9
Promoters of *K. lactis*

Promoter	Protein	Regulation	Reference
Constitutive			
PDC1	Pyruvate decarboxylase		[505]
Inducible			
ADH4	Mitochondrial alcohol dehydrogenase	Ethanol	[506]
LAC4	β-galactosidase	Lactose/galactose	[280]
PHO5	Acid phosphatase	Phosphate (repressible)	[507]

promoter (P_{LAC4}) [257], which is induced by lactose or galactose and not completely repressed by glucose [279]. This promoter regulates the expression of native lactase (β-galactosidase) in *K. lactis* [280]. Since the cross-recognition resulting from the similarity between the sequences of P_{LAC4} and Pribnow box transcriptional element of bacterial promoters, the expression of the desired gene initiates constitutively in *E. coli*, before the transformation of plasmid DNA to yeast cells [278]. For this reason, the semisynthetic $P_{LAC-PBI}$ as a variant of P_{LAC4} has been developed through mutagenesis of the native lactase promoter. A variant displays full activity in *K. lactis*, whereas the gene expression is greatly inhibited in bacterial cells [257, 278, 281]. To date, the commercially available P_{LAC4}- and $P_{LAC4-PBI}$-driven expression systems have been frequently used due to their strength and regulatory capacity for the production of various recombinant biomolecules that are mainly comprised of enzymes and therapeutic proteins. Another endogenous, inducible promoter, P_{PHO5}, promotes the production of acid phosphatase through phosphate starvation in *K. lactis.* Thus, the expression of heterologous genes under the control of P_{PHO5} is strongly regulated by the low/high level of inorganic phosphate in the medium [261, 282]. Further, the inducible promoter of the *ADH4* gene encoding mitochondrial alcohol dehydrogenase is also one of first native promoters used for r-protein production, via ethanol induction. Similar to regulation of P_{LAC4}, P_{ADH4} allows gene expression in the simultaneous presence of ethanol and glucose because of its insensitivity to glucose repression. In contrast to the majority of native inducible promoters, the *PDC1* promoter is the only endogenous constitutive promoter used for r-protein production.

Concerning the use of exogenous promoters in *K. lactis*, the phosphoglycerate kinase (PGK) promoter, among the strong and constitutive glycolytic ones derived from *S. cerevisiae*, has been predominantly used for diverse r-protein production [261]. On the other hand, *PHO5* promoter of *S. cerevisiae* [282] and *INU1* promoter of *K. marxianus* [283] have rarely been applied for recombinant biomolecule production.

6.4 Secretion Signals

Several endogenous and exogenous yeast signal sequences as well as the ones targeting the killer toxins have been used to secrete r-proteins in *K. lactis* (Table 10). Among them, the pre-pro-sequence of the *S. cerevisiae* α-mating factor is still mostly used owing to its high secretion capacity on the recombinant cargo proteins. Also the *K. lactis* pre-pro-sequence of the α-mating factor has been widely employed, although it is reported not to be as strong as that of *S. cerevisiae*. The native signal sequences of the promoters are commonly used in expression vectors whether or not the promoter is derived from *K. lactis* [284, 285]. On the other hand, the pre-sequence of α-subunit signal peptide of killer toxin

Table 10
Secretion signals of *K. lactis*

Signal sequence	Source	Protein	Reference
Endogenous			
Killer		Killer protein	[508]
MFα		α-Mating factor	[278, 509]
PHO5		Acid phosphatase	[507]
Exogenous			
CBH1	*T. reesei*	Cellobiohydrolase I	[287]
MFα1	*S. cerevisiae*	α-Mating factor 1	[510]

and acid phosphatase [286] has been integrated into the expression vectors containing the promoters native to *K. lactis*.

Furthermore, owing to the secretion ability of filamentous fungi as good producers of extracellular enzymes, EGFP secretion mediated by several signal peptides from *Aspergillus niger*, *Aspergillus nidulans*, *Aspergillus terreus*, *Aspergillus awamori*, and *Trichoderma reesei* was achieved and resulted in its elevated secretion. Especially, the *CBH1* signal sequence of *T. reesei* gave higher yields compared to the native α-mating factor [287].

Additionally, a synthetic signal peptide-driven expression system was used to produce recombinant chymosin [275], while the native signal sequences of heterologous proteins have been applied in many studies [269, 288], irrespective of the host organism.

7 *Yarrowia lipolytica*

Y. lipolytica is an oleaginous dimorphic yeast that displays significant lipolytic and proteolytic activities. It naturally secretes enzymes, and under optimized conditions, several grams of alkaline extracellular protease per liter can be reached [289]. *Y. lipolytica* is a ubiquitous yeast, naturally found in a variety of habitats, namely, marine waters, soils, food systems, and mycorrhizae (as reviewed by Groenewald et al. [290]; Zinjarde [291]). *Y. lipolytica* is also a model organism to study dimorphic transition, lipid accumulation, peroxisome biogenesis, mitochondria, salt tolerance, and the secretory pathway (reviewed by Nicaud [292]).

Since the mid-nineteenth century, *Y. lipolytica* has been employed in industrial applications. Its industrial use was initiated to produce SCP as feed supply by British Petroleum (UK) and executed for this purpose until the 1970s oil crisis. Then, industrial applications of the yeast targeted production of commercially valuable metabolites such as citric acid (Pfizer Inc. and ADM, USA) and erythritol (Baolingbao Biology Co., China). As it is usually found in heavily hydrocarbon-contaminated soils, it attracted interest for bioremediation of oil-polluted soils and waters [293]. A starter composed of wild-type *Y. lipolytica* and lipase was produced for wastewater bioremediation (Artechno, Belgium). Since the 2000s, genetically modified *Y. lipolytica* strains were developed to produce eicosapentaenoic acid (EPA)-rich biomass [294] for ω-3 supplement in fish feed or as dietary supplement (New Harvest™). *Y. lipolytica* has also been engineered to produce carotenoids, for which safety evaluations have been recently disclosed [295]. *Y. lipolytica* strains are also used for heterologous enzyme production for enzyme replacement therapies (ERTs), namely, human lysosomal enzymes [296] for lysosomal storage disease and lipase [297] that can be used for exocrine pancreatic insufficiency. Large-scale industrial utilization of *Y. lipolytica,* with several processes classified as GRAS by FDA, led to the acquisition of extensive knowledge on its fermentation characteristics.

The use of *Y. lipolytica* as a microbial cell factory for r-protein production was started with the foundation of transformation methods by INRA and Pfizer Inc. (USA) [298, 299]. Up to now, more than 130 r-proteins have been produced by *Y. lipolytica* species [300]. As a host system, *Y. lipolytica* has many advantages including high secretion efficiency of high molecular weight proteins, low glycosylation modifications, secretion by the co-translational translocation pathway [301], and the possibility of high cell density fermentation [302]. It can utilize glucose (but not sucrose), acetate, alcohols, and hydrophobic compounds such as fatty acids, alkanes, and oils as carbon sources [303]. The genome sequence of three *Y. lipolytica* strains, CLIB122, WSH-Z06, and Po1f, are available online (http://www.ncbi.nlm.nih.gov/genome/genomes/194) [304].

7.1 Expression Vectors

Y. lipolytica strains naturally do not bear an episome; but for replicative vector design, chromosomal autonomously replicating sequence/centromere replication origins have been used [305, 306]. Application of episomal vectors is limited due to their low copy number (1–3 plasmids/cell under selective pressure) and high loss frequency [307]. The addition of the centromeric sequence upstream region resulted in 80% increase in copy number [308], demonstrating the improvement potential of these vectors. Still, integrative vectors are the main tools used in r-protein production in *Y. lipolytica*. Through integration of the recombinant vector DNA, *Y. lipolytica* mainly employs nonhomologous end joining (NHEJ) rather than homologous recombination [309]. However, homologous recombination via single crossover can occur at rates of over 80%, unless 0.5–1 kb homologous flanking regions are present [300]. *Y. lipolytica ku70* deleted strain raised the rate of homologous recombination to 43% by double crossing-over with short homologous regions as 50 bp [310]. Multicopy integration methods in *Y. lipolytica* have mainly targeted either ribosomal DNA loci or Ylt1 retrotransposon flanked by long terminal repeats named as zeta sites, in the genome.

Commercial expression vectors pYLEX1 and pYLSC1, employing a hybrid promoter that contain four tandem copies of P_{XPR2} UAS1B element with minimal P_{LEU2}, were provided by Yeastern Biotech (Taiwan). Another vector series, containing a hybrid promoter UAS1B$_{8/16}$-TEF library in which 8 or 16 tandem UAS1B elements are fused to P_{TEF} regions of various sizes [311], is available at Addgene (USA).

7.2 Selection Markers

Wild-type *Y. lipolytica* strains cannot utilize sucrose as a carbon source. *S. cerevisiae SUC2* gene was used as a selective dominant marker [312], but selection was impaired with the residual growth of the yeast because of the impurities found in selective sucrose plates [289]. Even though *SUC2* gene could not be used as a selection marker, this approach has paved the way for invertase-producing strains such as Po1d and E129 derivatives that can grow on sucrose and molasses. *Y. lipolytica* is naturally resistant to most antibiotics but sensitive to hygromycin B and the bleomycin/phleomycin group which are commonly used as dominant selection markers for *Y. lipolytica* [313, 314].

Auxotrophic markers are the mainly used selection tools for *Y. lipolytica*. Among them, the most commonly used are *URA3* and *LEU2* as the auxotrophic recipient strains harboring non-leaky and non-reverting *ura3* and *leu2* mutations [289, 312]. For the selection of multiple integration events, defective promoter variants of the *URA3* marker gene, i.e., *ura3d4*, have been designed [315]. Single copy integration of *ura3d4* cannot complement the auxotrophy, while about ten copies provide normal growth on uracil-free media [315].

7.3 Promoters

A variety of promoters have been used for heterologous protein production in *Y. lipolytica*. Among them, the most commonly used promoters are inducible P_{XPR2}, P_{POX2}, and P_{ICL1}, constitutive P_{TEF}, bidirectional P_{MTP}, and synthetic P_{hp4d} (Table 11).

Under proper conditions, *Y. lipolytica* secretes very high amounts of alkaline extracellular protease (AEP) (1–2 g/L) encoded by the *XPR2* gene [316, 317]. In the development of *Y. lipolytica* heterologous systems, P_{XPR2} has great historical importance. Although it was extensively studied [235, 318–321], its complex regulation prevented its industrial use. The strong and inducible P_{XPR2} is active at $pH > 6$, on media lacking certain carbon and nitrogen sources. In addition, high concentrations of peptone are required for full induction [317]. To eliminate these difficulties, a synthetic hybrid promoter, P_{hp4d}, was developed by

Table 11
Promoters of *Y. lipolytica*

Promoter	Protein	Regulation	Reference
Constitutive			
FBA1	Fructose 1,6-bisphosphate aldolase		[511]
$FBA1_{IN}$	Intron included P_{FBA1}		[511]
GPM1	Phosphoglycerate mutase		[511]
RPS7	Ribosomal protein S7		[235]
TEF1	Translation elongation factor-1α		[235]
UAS1B8/16-TEF	Hybrid P_{TEF1} with UAS1B		[311]
$TEF1_{IN}$	Intron included P_{TEF1}		[512]
Inducible			
hp4d and *hp4i*	Four direct/inverted copies of UAS1B with minimal P_{LEU}	Growth phase dependent	[322]
G3P	Glycerol-3-phosphate dehydrogenase	Glycerol	[325]
ICL1	Isocitrate lyase	Fatty acids, alkanes, ethanol, and acetate	[325]
LEU2	β-Isopropylmalate dehydrogenase	Leucine precursor	[513]
LIP2	Lipase	Fatty acids	[514]
MTP (bidirectional)	Metallothionein 1 and 2	Metal salts	[324]
POT1	3-oxo-acyl-CoA thiolase	Fatty acids and alkanes	[325]
POX1 and *POX5*	Acyl-CoA oxidases	Alkanes	[325]
POX2	Acyl-CoA oxidase	Fatty acids and alkanes	[325]
UAS1Bn-Leum	Hybrid UAS1B with P_{LEU}	Growth phase dependent	[311]
XPR2	Alkaline extracellular protease	Peptone	[317, 513]

fusing four tandem copies of P_{XPR2} UAS1B region to minimal P_{LEU2} that leads to the same high expression efficiency as fully induced P_{XPR2}. Additionally, P_{hp4d} is not repressed by carbon and nitrogen sources and does not require peptone for induction [322]. P_{hp4d} exhibits growth-phase-dependent activity, as it is usually active at the beginning of the stationary phase [323], that allows partial disassociation of growth and expression phases. The hybrid promoter, P_{hp4d}, has been employed in more than half of the studies in the literature for r-protein production [300]. Also, a number of hybrid and strong promoter variants have been developed via addition of 1–32 tandem copies of UAS1B elements to minimal core regions of P_{LEU2} or P_{TEF} regions of various sizes, the strongest variant exhibited eightfold higher expression efficiency than any known *Y. lipolytica* endogenous promoter [311].

Also another inducible bidirectional promoter, P_{MTP} (of metallothioneins 1 and 2), was identified, but the need of metal salts for induction hindered its industrial use [324]. The capability of *Y. lipolytica* to use hydrophobic substrates led to identification of many promoters from this pathway, i.e., P_{ICL} (isocitrate lyase), P_{POT1} (3-oxo-acyl-CoA thiolase), P_{POX1}, P_{POX2}, and P_{POX5} (acyl-CoA oxidases) [325]. When compared to native *Y. lipolytica* promoters P_{XPR2} and P_{G3P} (glycerol-3-phosphate dehydrogenase) and synthetic hybrid promoter P_{hp4d}, the strongest regulated endogenous promoters of *Y. lipolytica* were found to be P_{ICL}, P_{POT1}, and P_{POX2} [325], which are induced by alkanes and fatty acids and repressed with glucose and glycerol. P_{ICL} is also induced by ethanol and acetate but not completely repressed by glycerol and glucose. However, these promoters still have limitations as they need hydrophobic inducers. In addition to these, P_{ICL} has leaky expression with certain carbon sources.

7.4 Secretion Signals

The *Y. lipolytica* secretion system mainly performs via the co-translational pathway [301]. Most commonly used endogenous signal peptides are derived from *Y. lipolytica XPR2* and *LIP2* (extracellular lipase) genes, but also few recombinant extracellular proteins from plants and fungi have been produced efficiently via their own signal sequences (Table 12).

Table 12
Secretion signals of *Y. lipolytica*

Signal sequence	Source	Protein	Reference
Endogenous			
XPR2 pre-pro		Alkaline extracellular protease (AEP)	[513]
XPR2 pre		AEP	[513]
XPR2 pre + dipeptides		AEP	[515]
LIP2 pre-pro		Lipase	[297]
LIP2 pre + dipeptides		Lipase	[516]
Exogenous			
αAmy	*Oryza sativa*	α-Amylase	[517]
LAC1	*Pycnoporus cinnabarinus*	Laccase I	[518]
LACIIIb	*T. versicolor*	Laccase IIIb	[519]
EG1	*T. reesei*	Endoglucanase I	[520]
SUC2	*S. cerevisiae*	Invertase	[511]
MAN1	*A. aculeatus*	Endo-β-1,4-mannanase	[521]
ROL	*R. oryzae*	Lipase	[522]
ROL-XPR2	Hybrid	Lipase	[522]

Alkaline extracellular protease is synthesized as a pre-pro-protein with a 13 aa length pre-sequence, followed by a dipeptide region and 120 aa length pro-region [312]. The *XPR2* pro-region harbors a glycosylation site and functions as an internal chaperone [326, 327]. The *LIP2* signal sequence is similar to that of *XPR2* with a 13 aa length pro-sequence followed by 4 dipeptides and 10 aa length pro-region [297].

8 *Schizosaccharomyces pombe*

S. pombe is a Crabtree positive fission yeast. Industrial application of *S. pombe* in heterologous protein production is still limited; however, its similarity to higher eukaryotes in terms of cellular processes such as posttranslational modifications, mRNA splicing, RNA interference, cell cycle control, and transcription initiation makes it a promising platform for the production of mammalian proteins [35]. It was the sixth eukaryote that had its genome fully sequenced and the third eukaryotic organism, whose complete proteome data is available in UniProtKB/Swiss-Prot [328, 329]. *S. pombe* has one of the smallest genomes among single-celled eukaryotes; furthermore, genome-scale engineering via a large-scale gene deletion approach showed an important potential for the development of host strains that have enhanced r-protein production ability [330]. Additionally, its Golgi apparatus is morphologically well characterized and can introduce galactose to both *O*- and *N*-linked glycans [331].

8.1 Expression Vectors

A variety of different plasmids including episomes, expression vectors, epitope-tagging plasmids, and integration vectors have been developed for molecular studies of *S. pombe* [332]. The early *S. pombe* plasmids, derived from *S. cerevisiae* vectors, showed low efficiency in fission yeast. Most self-replicating vectors of fission yeast contain an *ars1* replication origin that allows efficient transformation and maintenance of the plasmid. Earliest plasmids that have 2 μm *ori* are less stable and have higher rearrangement frequencies than plasmids with the native *ars1* [333]. On the other hand, *S. pombe* can be transformed with vectors lacking origin of replication, but their transformation efficiency is low, and also they are maintained as heterologous integrants or unstable multimers [334]. Currently, there are a number of vectors that have been specifically designed for *S. pombe*, and these are mainly preferred for expression studies. As centromeres of *S. pombe* are too large, it is impossible to construct stable and low copy number CEN plasmids, and even under selective pressure, their frequency of loss is fairly high [335]. Most current vectors of *S. pombe* are derived from a few common plasmid backbones containing common multiple cloning sites in a *lacZ*-derived polylinker region, which provide blue-white screening in bacteria. The pAL1, pAL19, pBG1, and pUR18/19 plasmids are all derivatives of pUC19 cloning vectors [336–338].

To achieve stable and defined copy number strains, integration vectors have been usually employed. The best approach for integration is to target plasmid DNA to a homologous region in the genome. Efficiency of homologous recombination can differ at different loci, and the chance of homologous recombination increases with increased targeting homology [332]. For instance, the pJK148 and pJK210 vectors were used for integration of GOI into the *leu1* or *ura4* loci, whereby more than 1.5 kb homologous regions resulted in extremely efficient homologous recombination [339]. The fission yeast can also perform random integration of the plasmid DNA at reasonable efficiency in the absence of homologous sequences.

8.2 Selection Markers

The *ura4* gene encoding orotidine monophosphate decarboxylase is the most frequently used marker in *S. pombe* [340]. The *ura4*$^{-}$ auxotrophy is weakly complemented by the *S. cerevisiae URA3* gene, so its use in vectors is discouraged. Another commonly used marker in fission yeast is *leu1* encoding β-isopropylmalate dehydrogenase [341]. As its auxotrophy is complemented by the *S. cerevisiae* orthologue *LEU2*, this marker gene is commonly used in vectors. However, complementation by a single copy of *S. cerevisiae LEU2* is weaker. Therefore, *S. pombe leu1* is the choice for integration or disruption vectors. As the *ade6* gene is a large DNA fragment and also contains many restriction enzyme recognition sites, it has very limited use. However, this marker can be very convenient if a visual selection is necessary to observe plasmid maintenance, as cells form pink or red colonies due to accumulation of P-ribosylaminoimidazole in *ade6* mutants that is then oxidized to a red pigment [333]. Currently available auxotrophy-complementing markers include *ade7*, *his1*, *his2*, *his3*, *his5*, *arg3*, *arg12*, *lys1*, *lys2*, and *tyr1*; however, *ura4*, *LEU2*, and *his3* are still the most commonly used markers for selection of multicopy plasmids [342]. Additionally, for fission yeast, genes conferring resistance to kanamycin/G418 [343], hygromycin B [344], phleomycin/bleomycin, and nourseothricin/ClonNat [345] are effective dominant markers.

8.3 Promoters

The most common constitutive promoters of *S. pombe* are the strong P_{adh1} [346] and a P_{CaMV} variant that was engineered to be induced by tetracycline. P_{CaMV} is a derivative of plant viral cauliflower mosaic virus 35S promoter and functions constitutively in fission yeast [347]. Tet repressor binding sites were introduced near the TATA box of P_{CaMV}. When Tet^{R} is expressed in *S. pombe*, transcription from the engineered promoter is blocked; otherwise, upon addition of tetracycline, $P_{CaMV\text{-}tet}$ is strongly induced up to 400-fold [348]. Inducible promoter options are quite limited for

Table 13
Promoters of *S. pombe*

Promoter	Protein	Regulation	Reference
Constitutive			
adh1^{+}	Alcohol dehydrogenase		[346]
CaMV	Cauliflower mosaic virus 35S promoter		[347]
SV40	Simian virus 40 small-T antigen		[523]
Inducible			
CaMV-tet	Derivative of cauliflower mosaic virus 35S promoter	Tetracycline	[348]
gld1^{+}	Glycerol dehydrogenase	Ethanol and l-propanol	[524]
fbp1	Fructose-1,6-bisphosphatase	Glucose (repressible)	[352]
nmt1	No message in thiamine	Thiamine (repressible)	[350]
REP41 and *REP81*	Attenuated nmt1 with mutations in the TATA box	Thiamine (repressible)	[351]
ctr4^{+}	Copper transporter	Copper (repressible)	[355]
urg1	Uracil regulatable gene	Uracil	[354]
in v1^{+}	Invertase	Glucose (repressible)	[353]
pho1	Acid phosphatase	Adenine, phosphate (repressible)	[525]

S. pombe; the most popular one is P_{nmt1} (no message in thiamine), repressed in the presence of thiamine [349, 350]. The TATA box of P_{nmt1} has been modified to tune its overall strength; and intermediate strength P_{nmt41} and weak strength P_{nmt81} were generated [351]. Another regulated promoter of *S. pombe* is P_{fbp1}, the promoter of the fructose-1,6-bisphosphatase gene, which is repressed by glucose [352]. Other inducible promoters of fission yeast are P_{inv1} (invertase gene promoter) [353], P_{urg1} (uracil-regulatable gene promoter) [354], and P_{ctr4} (copper transporter gene promoter) [355] (Table 13).

8.4 Secretion Signals

Two secretion pathways were discovered in fission yeast related to its two peptide pheromones: (a) the classic secretion pathway or pheromone P pathway that employs endoplasmic reticulum and Golgi apparatus in protein transport and (b) pheromone M pathway that uses an ABC transporter system. Few *S. pombe* endogenous signal peptides, namely, P3 signal sequence (P pheromone precursor, product of *map 2* gene) [356], pho1 signal sequence (acid phosphatase) [357], and cpy1 signal sequence (carboxypeptidase Y, *cpy1*) [358], have been used for the secretion

Table 14
Secretion signals of *S. pombe*

Signal sequence	Source	Protein	Reference
Endogenous			
P3		Map2, P-factor precursor	[356]
Cpy1		Carboxypeptidase Y	[358]
Dpp1		Putative dipeptidyl protease	[358]
pho1$^+$		Acid phosphatase	[357]
mfm1		M pheromone precursor	[359]
Exogenous			
CBH2	*T. reesei*	Cellobiohydrolase II	[526]
EGIII	*T. reesei*	Endo-β-1,4-glucanase	[527]
K28 pptox	K28 killer virus	K28 pre-pro-toxin	[489]
pGKL	*K. lactis*	Killer toxin	[528]
SUC2	*S. cerevisiae*	Invertase	[529]
OXDC	*Collybia velutipes*	Oxalate decarboxylase	[530]
ATIII	Human	Antithrombin III	[531]
hGL	Human	Gastric lipase	[532]
PLAP	Human	Placental alkaline phosphatase	[533]

of heterologous proteins via the classic secretion pathway. The ABC transporter system exists both in prokaryotes and eukaryotes. This system is located in the plasma membrane and translocates proteins at the expense of ATP. As being the natural secretion system of M pheromone, the secretion signal of the pheromone was used to excrete green fluorescent protein [359]. Additionally, many heterologous genes such as *T. reesei* cellulose, *S. cerevisiae* invertase, *Collybia velutipes* oxalate decarboxylase, and human antithrombin III were successfully secreted with their own leader sequences in the *S. pombe* secretion system (Table 14).

9 *Candida utilis*

Candida is one of the largest and extremely heterogeneous yeast genera, containing about 200 different ascomycetous species [360–362]. *C. utilis* has a nomenclatural confusion ever since Wilhelm Henneberg first identified it as *Torula utilis* in several German yeast factories in 1926. The following names in publications probably refer to the same organism: *Saccharomyces jadinii*

(1932), *Torulopsis utilis* (1934), *Hansenula jadinii* (1951), *Candida utilis* (1952), and *Hansenula jadinii* and *Pichia jadinii* as its teleomorphs (1984) [363]. *Cyberlindnera jadinii* or *Lindnera jadinii* is also its actual name in accordance with the current taxonomic classification (UniProt).

The edible yeast *C. utilis* has been approved as GRAS by the FDA [364]. However, it was lesser known but worthwhile for the food and feed industry after World War I resulted in the shortage of protein sources [365–367]. Accordingly, for more than six decades, it has been commercially marketed as a food additive to human diet and as a nutritional supplement in animal feed through the large-scale production of SCP owing to its ability to utilize various cost-effective and biomass-derived wastes such as molasses, brewing by-products, spent sulfite liquor from paper industry, and the pentose xylose [368]. Taking the diversity of ingredients involved in complex media into consideration, the food and fodder yeast *C. utilis* has the ability to assimilate a variety of carbon sources, i.e., pentose sugars, organic acids, and alcohols, and of nitrogen sources such as urea, nitrate, ammonium salts, pyrimidine, and various amino acids [369]. Besides, in contrast to *S. cerevisiae*, *C. utilis* as Crabtree-negative yeast allows for high cell density culture under both strict aerobic and anaerobic conditions, by virtue of the lack of ethanol production [364, 370]. The industrially important yeast *C. utilis* is not only significant for SCP production but also outstanding for the isolation of endogenous compounds such as invertase [371, 372], glutathione [373, 374], ribonucleic acids [364], glucomannan [375, 376], phospholipase B [377], or biotin [378] and for the production of heterologous proteins. However, compared to *S. cerevisiae*, *C. utilis* is a relatively less developed recombinant host, due to its polyploid genome and the lack of a sexual phase, besides the limited availability of selectable genetic markers and genetic transformation system [367, 379].

9.1 Expression Vectors

For *C. utilis*, initially two integrative vectors were designed simultaneously, targeting either the L41 locus or rDNA to obtain high-level r-protein expression through multiple copy integration [380]. Ever since, common intracellular r-protein production processes in *C. utilis* have been carried out by integrative vectors based on various chromosomal loci, i.e., rDNA[381], 18S rDNA [382], and *URA3* [381, 383]. Additionally, the intracellular production of biomolecules in recombinant *C. utilis* has been carried out by vectors integrated into corresponding loci, i.e., 18S rDNA [378], rDNA [378, 384, 385], L41 [384], *URA3* [386, 387], *HIS3* [378], and *PDC1* [388]. On the other hand, the integration stability and copy number depend on the chromosomal locus for both intracellular [380] and secreted r-proteins [389, 390] produced by *C. utilis*. The loci of *URA3* [381] and *HIS3* [389] serve as stable regions in contrast to the rDNA locus [381]. Additionally, the integration at the *URA3* locus yields about 90 copies [391] compared to the locus of L41 leading to two-copy integration [380] and loci of *THD3* and *HIS3* allowing single-copy integration [389, 390]. The stability of the *THD3* locus varies from study to study [389, 390]. On the other hand, the ARS-based vectors are not currently desirable for instability problem as well as low copy number, although two ARS sequences have been defined in *C. utilis* [392].

9.2 Selection Markers

Among selection markers, the initially employed type was a dominant selection marker, L41, that is an endogenous mutant gene conferring cycloheximide resistance in *C. utilis* [380]. Ever since, a variety of heterologous genes, i.e., *aph*, *hph*, *nat*, and *ble*, have been widely used to provide resistance to other antibiotic types, i.e., G418, hygromycin B [385], nourseothricin [389], and Zeocin™ [393], respectively. In contrast to commonly used dominant markers, an auxotrophic marker is rarely implemented to complement *ura3*Δ auxotrophy created by the Cre-loxP system [379].

9.3 Promoters

A few constitutive and inducible promoters have been identified (Table 15) for *C. utilis* because of the bottlenecks mentioned previously. Among strong promoters encoding housekeeping genes in glycolytic pathways, the most common one is the *TDH3* promoter (P_{TDH3}), which controls constitutive expression of glyceraldehyde-3-phosphate dehydrogenase [367]. The first expression system, which encoded the gene of single-chain monellin under the P_{TDH3} with high copy integration, yielded high production levels of

Table 15
Promoters of *C. utilis*

Promoter	Protein	Regulation	Reference
Constitutive			
GAP1/TDH3	Glyceraldehyde-3-phosphate dehydrogenase		[381, 534]
PGK1	3-phosphoglyceraldehyde kinase		[534]
PDC1	Pyruvate decarboxylase		[388]
ATPasePma 1	Plasma membrane ATPase		[534]
Rpl29/Rpl31	Ribosomal large subunit protein		[394]
L41	Ribosomal protein		[380]
Inducible			
GXS1	Glucose/xylose symporter	Xylose	[389]
XDH1	NAD-xylitol dehydrogenase	Xylose	[386]

intracellular r-protein [381]. Thereafter, high concentrations of bacterial intracellular α-amylase was attained under the control of same promoter [383]. In addition to intracellular production, almost all secreted heterologous proteins in *C. utilis* have been produced using P_{TDH3} [367, 389, 390]. As an alternative to P_{TDH3}, strong constitutive promoters P_{PGK1} [381] and P_{PDC1} [392] and the promoters of the following genes, ATPase Pma1 [384], Rpl29/Rpl31 [394], and Rpl41 [380], have been employed for various purposes.

Unlike the majority of constitutive promoters identified in *C. utilis*, there are two known inducible promoters, namely, P_{XDH1} and P_{GXS1}, which regulate the expression of NAD-xylitol dehydrogenase and a glucose/xylose symporter, respectively. Since both of these promoters are induced by xylose and repressed by glucose, high expression level of heterologous protein was achieved by the induction with xylose following the glucose depletion [389].

9.4 Secretion Signals

Since the discovery that the secretome is protease free and small [369], *C. utilis* has been recently revealed as a promising secretion host through the first endogenous and exogenous secretion signal sequences employed in expression vectors (Table 16). The homologous signal peptide of invertase as one of the most abundant protein in secretome analysis of *C. utilis* yielded higher secretion levels compared to that of the secretion signal native to model r-protein, lipase B, which is originated from *Candida antarctica* [389]. Later, the secretion of heterologous protein was accomplished in *C. utilis* and resulted in the highest secretion level by using heterologous pre-pro-region of α-mating factor, derived from *S. cerevisiae* [390].

Table 16
Secretion signals of *C. utilis*

Signal sequence	Source	Protein	Reference
Endogenous			
Inv1		Invertase	[535]
Exogenous			
CalB	*C. antarctica*	Lipase B	[389]
MFα1	*S. cerevisiae*	α-Mating factor 1	[390]

10 *Kluyveromyces marxianus*

K. marxianus is a thermotolerant yeast species that is able to grow up to 52 °C [395]. While the genome of *K. marxianus* CBS 6556 was completely sequenced [396], a great diversity was reported for numerous other strains that mate and generate fertile hybrids being not only intraspecies but also interspecies [397, 398]. This non-conventional yeast has been used to produce commercially important enzymes, e.g., inulinase [399] and β-galactosidase [400] and other r-proteins [283, 401–404], as it is able to grow rapidly even in a wide variety of industrial byproducts, e.g., corn silage juice, molasses, and whey powder [167]. *K. marxianus* is also a promising bioethanol producer that is capable of utilizing pentose sugars and fermenting glucose up to 45 °C [405]. Moreover, this microorganism seems to be a good candidate to investigate the mechanism of autophagy [406].

10.1 Expression Vectors

Since *K. marxianus* does not have any natural plasmids, the development of the new expression vectors derived from other yeast species is required despite the metabolic burden that may occur due to the presence of chimeric plasmids in the host cell [407]. The 2 μm circular plasmid of *S. cerevisiae* is host limited and unstable in *Kluyveromyces* [408]. Therefore, the derivatives of the resident 1.6 μm circular plasmid pKD1 of *K. drosophilarum* were used for the transfection of several *Kluyveromyces* strains. Unfortunately, most of the designed expression systems brought success just in *K. lactis*, while they did not demonstrate sufficient stability in *K. marxianus* [409]. After the first stable autonomously replicating system was successfully constructed for *K. marxianus* using pKD1 [408], some other episomal vectors were also tested [402, 410–412].

In spite of high copy numbers reported for the transfection with the replicative plasmids mentioned above, e.g., 50 copies per cell [413], integrating the expression cassette into the chromosome may be necessary to reduce the variation in expression levels [402]. In this context, random gene integration has widely been employed for *K. marxianus* transfection [414–416]. However, when multiple genes were randomly integrated into the *K. marxianus* chromosome in a single step using the technique developed by Heo et al. [417], a variation in the hexanoic acid production level was detected due to the differences in the positions where the genes inserted during the design of engineered biosynthetic pathway [418].

Nonhomologous recombination is dominant in *K. marxianus*, i.e., 35-fold higher nonhomologous end joining activity was reported for *K. marxianus* than for *S. cerevisiae* [419]. To minimize the variation of heterologous gene expression, efforts have been made to achieve integration to the selected locus [404]. Homologous recombination is simpler for controlling the gene copy number as well as for targeting particular loci for insertion or disruption [420]. Consequently, designed integration cassettes have regions of homology with the genomic sequences, such as the *LAC4* locus [403, 420], *HO* locus [421], *LEU2* locus [404, 422], and *URA3* locus [422]. As a universal target for gene integration, rDNA, particularly 18 rDNA, has also been utilized in the design of *K. marxianus* integration systems [422–424]. Besides its highly conserved sequences allowing the design of common expression systems which can be used in different yeast species, its high copy number within the genome makes multicopy integration possible [231, 239, 425].

10.2 Selection Markers

The auxotrophic selection markers extensively used in *K. marxianus* expression systems for the complementation of respective defects are *URA3*, *LEU2*, and *TRP1* from *K. marxianus*, *URA3*, *LEU2*, *TRP1*, and *HIS3* from *S. cerevisiae* and *URA4* from *S. pombe* [426]. As dominant antibiotic resistance markers offer particular advantages over auxotrophic markers, aureobasidin A, nurseothricin, G418 [426], and hygromycin B [423] resistance markers have also been employed to select *K. marxianus* transformants. In this case, auxotrophic strain development is no longer necessary, and this is particularly important when various strains are to be investigated or an interspecies screening is to be conducted.

10.3 Promoters

Heterologous protein production by *K. marxianus* has been generally carried out using constitutive promoters (Table 17). Thus, making comparison between the production capacities of different yeast species becomes possible, as the GOI can be expressed under the same conditions without the specific induction requirements that might affect the growth patterns of some species more than others [423]. Among the most commonly used constitutive promoters, i.e., P_{GAP}, P_{PGK}, and P_{ADH1}, P_{GAP} was fused to the *eng1* gene, encoding a thermostable endo-β-1,4-glucanase, to develop a recombinant *K. marxianus* strain being able to grow on cellobiose or carboxymethyl cellulose [422]. This promoter was also employed for the heterologous expression of xylitol dehydrogenase, xylulokinase, and xylose reductase genes from different origins to enhance the xylose fermentation capability of *K. marxianus* by improving the intracellular redox balance [427–429]. For the same purpose, glutamine synthetase (*GLN1*) or glutamate synthetase (*GLT1*) genes were overexpressed using P_{PGK} [429]. For lactic acid production in yeasts, the genes encoding L-lactate dehydrogenase (*LDH*) were isolated from different microorganisms and expressed under the control of P_{GAP} in *K. marxianus* as well [416].

Table 17
Promoters of *K. marxianus*

Promoter	Protein	Regulation	Reference
Constitutive			
ADH1	Cytoplasmic alcohol dehydrogenase		[536]
CTR1	Transporter for copper ions		[537]
GAP	Glyceraldehyde-3-phosphate dehydrogenase		[538]
GLK1	Glucokinase		[539]
HXK1	Hexokinase		[539]
PGK	Phosphoglycerate kinase		[404]
PIR1	Protein with internal repeats 1		[537]
Inducible			
INU1	Inulinase	Fructose, inulin, sucrose	[540]

A widely used inducible *K. marxianus* promoter is derived from the *INU1* gene that encodes inulinase, the enzyme responsible for fructose hydrolysis [80]. Due to the close relation between *INU1* and *SUC2* genes, the expression under the control of P_{INU1} is activated by the addition of inulin or sucrose [409]. To date, P_{INU1} has been employed for the recombinant production of *Aspergillus niger* glucose oxidase [283], *Thermus thermophilus* esterase [401], and *Thermoascus aurantiacus* β-glucosidase [430] as well as for the overexpression of native inulinase [421, 424, 431].

Most if not all of exogenous promoters used for r-protein production in *K. marxianus* have been originated from *S. cerevisiae*, such as P_{ScPGK}, P_{ScGPD}, P_{ScGAL1}, and P_{ScGAP} [284, 401, 402, 404, 418, 432]. When the thermophilic esterase gene from *T. thermophilus* was expressed in *K. marxianus* under the control of P_{ScPGK} instead of the native P_{INU1}, almost equal enzyme activity was achieved [401]. The promoters that have been widely used to develop metabolically engineered *K. marxianus* strains are not well characterized to date, especially at elevated temperatures with different carbon sources. Therefore, the constitutive promoters P_{ADH1}, P_{GAP}, and P_{PGK} of *K. marxianus* and their counterparts from *S. cerevisiae* were evaluated at various temperatures, carbon sources, and oxygen transfer conditions by comparing the expression levels of the Zeocin-resistant gene, *SH BLE*; the β-glucuronidase gene, *GUSA*; and the β-galactosidase gene, *KlLAC4* [404]. Accordingly, *K. marxianus*-derived promoters were reported to be stronger compared to their corresponding promoters from *S. cerevisiae* at different temperatures probably due to higher suitability of *K. marxianus* transcription machinery for its own promoters [404]. Additionally, although the transcriptional strength of all promoters decreased with the increase in temperature, P_{PGK}, particularly P_{KmPGK}, further maintained its activity with elevated temperatures. While both of the *PGK* promoters showed relatively stronger activity with all carbon sources tested, i.e., glucose, xylose, and glycerol, the promoting capabilities of both the *ADH1* promoters were increased using glycerol as the carbon source [404]. Interestingly, the expression strength of the evaluated promoters was not affected by the applied oxygen transfer condition which was either aerobic or oxygen limited [404].

10.4 Secretion Signals

The native signal peptides of *K. marxianus* have been rarely utilized for the secretion of expressed r-proteins (Table 18). For the extracellular production of thermotolerant β-glucosidase from *Thermoascus aurantiacus* (*TaBGLI*), the promoter and signal sequence of the *K. marxianus INU1* gene was employed [430]. Related to the use of exogenous secretion signals in *K. marxianus* expression systems, the signal sequence of the *Rhizopus oryzae* glucoamylase gene was fused with the *Aspergillus aculeatus* β-glucosidase gene (*BGL1*) in the designed cell surface display system [414]. The secretion of a thermophilic esterase from *Thermus thermophilus* HB27 was driven not only by the signal sequence of the *K. marxianus INU1* gene but also by the signal sequence of the *K. lactis* killer toxin Kl [401]. The latter coding region was fused to the *K. lactis* Cu/Zn SOD gene (*SOD1*) for extracellular superoxide dismutase production in *K. marxianus* as well [411]. The signal peptide of the *K. lactis* alpha-mating factor was also used to secrete the heterologous cellulases [420, 433] and dengue virus type 1 nonstructural protein 1 (NS1) [403]. The own signal sequences of recombinant genes have also been tested for extracellular production of heterologous proteins, such as the buffalo rumen fungus *Neocallimastix patriciarum*-derived β-glucosidase gene [434] and the *HSA* gene [411, 412].

Table 18
Secretion signals of *K. marxianus*

Signal sequence	Source	Protein	Reference
Endogenous			
INU1		Inulinase	[540]
Exogenous			
amyA	*R. oryzae*	Glucoamylase	[414]
k1	*K. lactis*	Killer toxin	[411]
α-MF	*K. lactis*	α-Mating factor	[420]
npabgs	*N. patriciarum*	β-glucosidase	[434]
HSA	*Human*	Serum albumin	[411]

11 *Zygosaccharomyces bailii*

Z. bailii is a diploid budding yeast that exhibits exceptional resistance to environmental stresses like high sugar concentrations (osmotolerant), acidic environments, and relatively high temperatures [435] and can survive in the presence of chemical preservatives [436, 437]. As these unusual characteristics allow growth under constrained conditions, it can facilitate industrial fermentations via simplifying the process and also prevent bacterial contamination, thus increasing economic value. Moreover, the high specific growth rate and biomass yield of *Z. bailii* make it an attractive candidate for biomass, protein, or metabolite production processes. Additionally, the secretion capacity of *Z. bailii* was found to be superior to that of *S. cerevisiae* [438]. However, *Z. bailii* is still poorly characterized in terms of its genetics, physiology, and molecular biology. Indeed, it is widely known as a food spoilage microorganism [439, 440].

11.1 Expression Vectors and Selection Markers

Two different endogenous plasmids, namely, pSB1 and pSB2, structurally and functionally similar to *S. cerevisiae* 2 μm plasmid, have been isolated from *Z. bailii* strains IFO 0488 and IFO1047, respectively [441]. By exploiting the endogenous pSB2, multicopy vectors were obtained that resulted in fivefold increase in production level compared to centromeric vectors [438]. For expression of heterologous proteins in *Z. bailii*, few episomal vectors were developed using *S. cerevisiae* pYX022 (R&D Systems, Wiesbaden, Germany) as a backbone vector. As a selection marker, kanamycin/geneticin resistance and hygromycin resistance genes were introduced into designed pZ_3 and pZ_4 plasmids, respectively [442]. Leucine auxotrophy was also used as a recessive selection marker in r-protein production with *Z. bailii* [438]. Derivatives of pZ_3 were employed in recombinant production of β-galactosidase, GFP, human interleukin 1-β, glucoamylase, lactate dehydrogenase, *Candida rugose* lipase CRL1, and *Y. lipolytica* protease XPR2 [442–445].

11.2 Promoters

Only one endogenous promoter, P_{TPI} (triose phosphate isomerase) [442], and few heterologous promoter genes have been used for r-protein production in *Z. bailii* (Table 19). In the study of Branduardi et al. [442], homologous P_{TPI} and a variety of heterologous promoters, i.e., *S. cerevisiae* P_{TPI}, *S. cerevisiae* P_{ADH}, *S. cerevisiae* P_{GAL}, and *K. lactis* P_{PDC}, were tested for their functionality and potential in *Z. bailii*. The highest expression level was obtained with homologous P_{TPI}. When compared to P_{ScTPI} four to five times, higher activity was obtained with P_{ZbTPI}; even in *S. cerevisiae*, P_{ZbTPI} exhibited a higher expression level [442]. Heterologous P_{ScTPI} has been employed in recombinant human

Table 19
Promoters of *Z. bailii*

Promoter	Protein	Source	Regulation	Reference
Constitutive				
ADH	Alcohol dehydrogenase	*S. cerevisiae*		[442]
LEU	β-isopropylmalate dehydrogenase	*S. cerevisiae*		[438]
PDC	Pyruvate decarboxylase	*K. lactis*		[147, 442]
TPI	Triose phosphate isomerase	*Z. bailii*		[442]
TPI	Triose phosphate isomerase	*S. cerevisiae*		[442–444]
Inducible				
GAL	Galactokinase	*S. cerevisiae*	Galactose (inducible)	[442]

Table 20
Secretion signals of *Z. bailii*

Signal sequence	Source	Protein	Reference
Endogenous			
ZKL	*Z. bailii*	Zygocin killer toxin	[442]
Exogenous			
pGKL	*K. lactis*	K1 Killer toxin	[442, 443, 445]
MFα-1	*S. cerevisiae*	α-Mating factor	[442]
Gamy	*A. adeninivorans*	Glucoamylase	[442]
Amy	*S. diastaticus*	Amylase	[442]

interleukin 1-β production in *Z. bailii* [443, 444]. P_{KlPDC}, functioning under hypoxic conditions, was also used in *Z. bailii* [147]. Additionally, the *S. cerevisiae* leucine promoter was found to be functional in *Z. bailii*, as the *S. cerevisiae LEU* expression cassette containing P_{LEU} upstream of the *LEU* gene complemented the auxotrophy in *Z. bailii* [438].

11.3 Secretion Signals

Up to date, the functionality of only a few secretion signals was tested in *Z. bailii* (Table 20). The endogenous leader sequence of zygocin killer toxin and the leader sequences of *K. lactis* killer toxin and *S. cerevisiae* α-MF were employed in heterologous expression of interleukin in *Z. bailii* [442]. All signal sequences were functional with the *Z. bailii* expression system. Among them, the *K. lactis* killer toxin leader sequence led to the highest secretion level [442]. Furthermore, the leader sequences of *A. adeninivorans* glucoamylase and *Streptomyces diastaticus* amylase genes are functional in *Z. bailii*, although very low secretion levels were obtained. Cell wall engineering of *Z. bailii* also led to higher secretion levels: *Z. bailii* ΔGas1 (cell wall-defective mutant) exhibited significantly higher secretion levels of *Candida rugosa* lipase CRL1 and *Y. lipolytica* protease XPR2 via K1 killer toxin secretion signal, compared to wild-type yeast [445].

12 Conclusion and Future Perspectives

To date, important genetic and biochemical insights into promoter architectures, transcriptional engineering, and eukaryotic ribosome assembly have been obtained using primarily the model organism *S. cerevisiae* as well as *P. pastoris*. These two yeast platforms have also been recruited for producing complex natural products through the reconstruction of biosynthetic pathways taken from plants or human, by transferring genes that encode enzymes of the metabolic pathway from the organism that naturally produces the molecule of interest into yeast. Besides, a number of yeast expression systems have been successfully employed for r-protein production over the last few decades. Their simple growth requirements, easily performed genetic modifications, ability to perform eukaryotic post-translational modifications, and efficient secretion systems are key factors for that. The availability of large culture collections, a wide range of vectors, recessive/dominant markers, promoter and secretion signal sequences, efficient transfection protocols as well as recently revealed data on genome, transcriptome, proteome, and secretome have supported widespread use and engineering of yeast expression systems.

A reliable comparison among yeast expression systems is challenging, since there is not one ideal platform for expression of all proteins. It is crucial to analyze both the inherent advantages and limitations of each yeast expression platform and the requirements of a target protein during the decision-making. In this chapter, besides the conventional yeast *S. cerevisiae*, some established yeast expression systems, namely, *P. pastoris*, *H. polymorpha*, *A. adeninivorans*, *K. lactis*, *Y. lipolytica*, and *S. pombe*, and upcoming yeast expression systems, namely, *K. marxianus*, *C. utilis*, and *Z. bailii*, are briefly described in terms of their genetic tools for heterologous protein production. All yeasts evaluated have exceptional characteristics, such as the ability to utilize a broad range of carbon sources or remarkable resistance to stress conditions. Hence, they harbor special metabolic pathways, where genes serve as potential sources of promoter and signal sequence elements with outstanding features.

Recent systems biology studies have revealed the regulation of genes on transcriptional and translational levels as a result of an input such as methanol, glucose, or glycerol. Transcriptome and proteome data have the potential to find the most prominent promoter elements that have strong and regulated functioning at a certain condition. Alternatively, new promoters have been developed via promoter engineering approaches, such as random or saturation mutagenesis, upstream activation sequence element addition, and modification of transcription factor binding sites. These promoter engineering approaches enable to tune promoter strength and regulation and create large promoter libraries to be used for r-protein production, co-expression studies, and introduction of a new metabolic pathway. In addition to engineered promoters, transcriptional engineering has paved the way for rational cell design to improve the expression strength of recombinant yeast systems. Additionally, secretome data provide a vast amount of knowledge on identification of novel signal sequences and construction of signal peptide libraries to test them individually for different proteins. Moreover, synthetic signal sequences have been designed using in silico tools. For yeast expression systems, an increase in synthetic biology toolboxes is anticipated via more frequent use of genome, transcriptome, proteome, and secretome data, which will further boost the applications of yeast expression platforms for r-protein production. Without doubt, with the ever-growing biopharmaceutical and biorefinery markets, the yeast expression systems will continue to thrive.

References

1. Gellissen G, Kunze G, Gaillardin C et al (2005) New yeast expression platforms based on methylotrophic *Hansenula polymorpha* and *Pichia pastoris* and on dimorphic *Arxula adeninivorans* and *Yarrowia lipolytica*—a comparison. FEMS Yeast Res 5:1079–1096. https://doi.org/10.1016/j.femsyr.2005.06.004
2. Çelik E, Çalık P (2012) Production of recombinant proteins by yeast cells. Biotechnol Adv 30:1108–1118. https://doi.org/10.1016/j.biotechadv.2011.09.011
3. Goffeau A, Barrell BG, Bussey H et al (1996) Life with 6000 genes. Science 274:546–567
4. Hittinger CT (2013) Saccharomyces diversity and evolution: a budding model genus. Trends Genet 29:309–317. https://doi.org/10.1016/j.tig.2013.01.002
5. Hasunuma T, Ishii J, Kondo A (2015) Rational design and evolutional fine tuning of *Saccharomyces cerevisiae* for biomass breakdown. Curr Opin Chem Biol 29:1–9. https://doi.org/10.1016/j.cbpa.2015.06.004
6. Shen MWY, Fang F, Sandmeyer S, Da Silva NA (2012) Development and characterization of a vector set with regulated promoters for systematic metabolic engineering in *Saccharomyces cerevisiae*. Yeast 29:495–503. https://doi.org/10.1002/yea.2930
7. Siddiqui MS, Thodey K, Trenchard I, Smolke CD (2012) Advancing secondary metabolite biosynthesis in yeast with synthetic biology tools. FEMS Yeast Res 12:144–170. https://doi.org/10.1111/j.1567-1364.2011.00774.x
8. Shin MK, Yoo HS (2013) Animal vaccines based on orally presented yeast recombinants. Vaccine 31:4287–4292. https://doi.org/10.1016/j.vaccine.2013.07.029
9. Herrgård MJ, Swainston N, Dobson P et al (2008) A consensus yeast metabolic network reconstruction obtained from a community approach to systems biology. Nature Biotechnol 26(10):1155–1160. https://doi.org/10.1038/nbt1492
10. Nookaew I, Papini M, Pornputtapong N et al (2012) A comprehensive comparison of RNA-Seq-based transcriptome analysis from reads to differential gene expression and cross-comparison with microarrays: a case study in *Saccharomyces cerevisiae*. Nucleic Acids Res 40:10084–10097. https://doi.org/10.1093/nar/gks804

11. Ibanez C, Perez-Torrado R, Morard M et al (2017) RNAseq-based transcriptome comparison of *Saccharomyces cerevisiae* strains isolated from diverse fermentative environments. Int J Food Microbiol 257:262–270. https://doi.org/10.1016/j.ijfoodmicro.2017.07.001
12. Nagalakshmi U, Wang Z, Waern K et al (2008) The transcriptional landscape of the yeast genome defined by RNA sequencing. Science 320:1344–1349. https://doi.org/10.1126/science.1158441
13. Usaite R, Wohlschlegel J, Venable JD et al (2008) Characterization of global yeast quantitative proteome data generated from the wild-type and glucose repression *Saccharomyces cerevisiae* strains: the comparison of two quantitative methods. J Proteome Res 7:266–275. https://doi.org/10.1021/pr700580m
14. Paulo JA, O'Connell JD, Gaun A, Gygi SP (2015) Proteome-wide quantitative multiplexed profiling of protein expression: carbon-source dependency in *Saccharomyces cerevisiae*. Mol Biol Cell 26:4063–4074. https://doi.org/10.1091/mbc.E15-07-0499
15. Jewett MC, Hofmann G, Nielsen J (2006) Fungal metabolite analysis in genomics and phenomics. Curr Opin Biotechnol 17:191–197. https://doi.org/10.1016/j.copbio.2006.02.001
16. Villas-Boas SG, Moxley JF, Akesson M et al (2005) High-throughput metabolic state analysis: the missing link in integrated functional genomics of yeasts. Biochem J 388:669–677. https://doi.org/10.1042/BJ20041162
17. Sauer U (2006) Metabolic networks in motion: 13 C-based flux analysis. Mol Syst Biol 2:1–10. https://doi.org/10.1038/msb4100109
18. Wasylenko TM, Stephanopoulos G (2015) Metabolomic and (13)C-metabolic flux analysis of a xylose-consuming *Saccharomyces cerevisiae* strain expressing xylose isomerase. Biotechnol Bioeng 112:470–483. https://doi.org/10.1002/bit.25447
19. Uetz P, Giot L, Cagney G et al (2000) A comprehensive analysis of protein–protein interactions in *Saccharomyces cerevisiae*. Nature 403:623–627. https://doi.org/10.1038/35001009
20. Lee TI, Rinaldi NJ, Robert F et al (2002) Transcriptional regulatory networks in *Saccharomyces cerevisiae*. Science 298:799–805. https://doi.org/10.1126/science.1075090
21. Harbison CT, Gordon DB, Lee TI et al (2004) Transcriptional regulatory code of a eukaryotic genome. Nature 431:99–104. https://doi.org/10.1038/nature02800
22. Huh W, Falvo JV, Gerke LC et al (2003) Global analysis of protein localization in budding yeast. Nature 425:686–691. https://doi.org/10.1038/nature02026
23. Förster J, Famili I, Fu P et al (2003) Genome-scale reconstruction of the *Saccharomyces cerevisiae* metabolic network. Genome Res 13:244–253. https://doi.org/10.1101/gr.234503
24. Österlund T, Nookaew I, Bordel S, Nielsen J (2013) Mapping condition-dependent regulation of metabolism in yeast through genome-scale modeling. BMC Syst Biol 7:36. https://doi.org/10.1186/1752-0509-7-36
25. Sun J, Shao Z, Zhao H et al (2012) Cloning and characterization of a panel of constitutive promoters for applications in pathway engineering in *Saccharomyces cerevisiae*. Biotechnol Bioeng 109:2082–2092. https://doi.org/10.1002/bit.24481
26. Hammer SK, Avalos JL (2017) Harnessing yeast organelles for metabolic engineering. Nat Chem Biol 13:823–832. https://doi.org/10.1038/nchembio.2429
27. Botstein D, Fink GR (2011) Yeast: an experimental organism for 21st century biology. Genetics 189:695–704. https://doi.org/10.1534/genetics.111.130765
28. Natter K, Kohlwein SD (2013) Yeast and cancer cells—common principles in lipid metabolism. Biochim Biophys Acta 1831:314–326. https://doi.org/10.1016/j.bbalip.2012.09.003
29. Sun S, Yang F, Tan G et al (2016) An extended set of yeast-based functional assays accurately identifies human disease mutations. Genome Res 26:670–680. https://doi.org/10.1101/gr.192526.115
30. Valenzuela P, Medina A, Rutter WJ et al (1982) Synthesis and assembly of hepatitis B virus surface antigen particles in yeast. Nature 298:347–350. https://doi.org/10.1038/298347a0
31. Sanchez-Garcia L, Martín L, Mangues R et al (2016) Recombinant pharmaceuticals from microbial cells: a 2015 update. Microb Cell Factories 15:33. https://doi.org/10.1186/s12934-016-0437-3
32. Demain AL, Vaishnav P (2009) Production of recombinant proteins by microbes and higher organisms. Biotechnol Adv 27:297–306

33. Walsh G (2014) Biopharmaceutical benchmarks 2014. Nat Biotechnol 32:992–1000. https://doi.org/10.1038/nbt.3040
34. Huang C, Lowe AJ, Batt CA (2010) Recombinant immunotherapeutics: current state and perspectives regarding the feasibility and market. Appl Microbiol Biotechnol 87:401–410. https://doi.org/10.1007/s00253-010-2590-7
35. Kim H, Yoo SJ, Kang HA (2015) Yeast synthetic biology for the production of recombinant therapeutic proteins. FEMS Yeast Res 15:1–16. https://doi.org/10.1111/1567-1364.12195
36. Liu Z, Hou J, Martínez JL et al (2013) Correlation of cell growth and heterologous protein production by *Saccharomyces cerevisiae*. Appl Microbiol Biotechnol 97:8955–8962. https://doi.org/10.1007/s00253-013-4715-2
37. Berlec A, Štrukelj B (2013) Current state and recent advances in biopharmaceutical production in *Escherichia coli*, yeasts and mammalian cells. J Ind Microbiol Biotechnol 40:257–274. https://doi.org/10.1007/s10295-013-1235-0
38. Parsaie Nasab F, Aebi M, Bernhard G, Frey AD (2013) A combined system for engineering glycosylation efficiency and glycan structure in *Saccharomyces cerevisiae*. Appl Environ Microbiol 79:997–1007. https://doi.org/10.1128/AEM.02817-12
39. Xu S, Zhang G-Y, Zhang H et al (2016) Effects of Rho1, a small GTPase on the production of recombinant glycoproteins in *Saccharomyces cerevisiae*. Microb Cell Factories 15:179. https://doi.org/10.1186/s12934-016-0575-7
40. Galao RP, Scheller N, Alves-Rodrigues I et al (2007) *Saccharomyces cerevisiae*: a versatile eukaryotic system in virology. Microb Cell Factories 6:32. https://doi.org/10.1186/1475-2859-6-32
41. Petranovic D, Nielsen J (2008) Can yeast systems biology contribute to the understanding of human disease ? Trends Biotechnol 26:584–590. https://doi.org/10.1016/j.tibtech.2008.07.008
42. Billingsley JM, DeNicola AB, Tang Y (2016) Technology development for natural product biosynthesis in *Saccharomyces cerevisiae*. Curr Opin Biotechnol 42:74–83. https://doi.org/10.1016/j.copbio.2016.02.033
43. Da Silva NA, Srikrishnan S (2012) Introduction and expression of genes for metabolic engineering applications in *Saccharomyces cerevisiae*. FEMS Yeast Res 12:197–214. https://doi.org/10.1111/j.1567-1364.2011.00769.x
44. Parent SA, Fenimore CM, Bostian KA (1985) Vector systems for the expression, analysis and cloning of DNA sequences in *S cerevisiae*. Yeast 1:83–138. https://doi.org/10.1002/yea.320010202
45. Clark-Walker GD, Miklos GLG (1974) Localization and quantification of circular DNA in yeast. Eur J Biochem 41:359–365. https://doi.org/10.1111/j.1432-1033.1974.tb03278.x
46. Hartley JL, Donelson JE (1980) Nucleotide sequence of the yeast plasmid. Nature 286:860–864. https://doi.org/10.1038/286860a0
47. Broach JR (1983) Construction of high copy yeast vectors using 2-um circle sequences. In: Wu R, Grossman L, Moldave K (eds) Methods in enzymology. Academic Press, New York, pp 307–325
48. Futcher AB, Cox BS (1983) Maintenance of the 2 microns circle plasmid in populations of *Saccharomyces cerevisiae*. J Bacteriol 154:612–622
49. Futcher AB, Cox BS (1984) Copy number and the stability of 2-um circle-based artificial plasmids of *Saccharomyces cerevisiae*. J Bacteriol 157:283–290
50. Jayaram M, Li Y-Y, Broach JR (1983) The yeast plasmid 2 circle encodes components required for its high copy propagation. Cell 34:95–104. https://doi.org/10.1016/0092-8674(83)90139-3
51. Kikuchi Y (1983) Yeast plasmid requires a cis-acting locus and two plasmid proteins for its stable maintenance. Cell 35:487–493. https://doi.org/10.1016/0092-8674(83)90182-4
52. Sturley SL, Young TW (1986) Genetic manipulation of commercial yeast strains. Biotechnol Genet Eng Rev 4:1–38. https://doi.org/10.1080/02648725.1986.10647821
53. Mishra S, Baranwal R (2009) Yeast genetics and biotechnological applications. In: Satyanarayana T, Kunze G (eds) Yeast biotechnology: diversity and applications. Springer Netherlands, Dordrecht, pp 323–355
54. Beggs JD (1978) Transformation of yeast by a replicating hybrid plasmid. Nature 275:104–109. https://doi.org/10.1038/275104a0
55. Wang T-T, Choi Y-J, Lee BH (2001) Transformation systems of non-*Saccharomyces* yeasts. Crit Rev Biotechnol 21:177–218

56. Hill JE, Myers AM, Koerner TJ, Tzagoloff A (1986) Yeast/*E. coli* shuttle vectors with multiple unique restriction sites. Yeast 2:163–167. https://doi.org/10.1002/yea.320020304
57. Romanos MA, Scorer CA, Clare JJ (1992) Foreign gene expression in yeast: a review. Yeast 8:423–488. https://doi.org/10.1002/yea.320080602
58. Kuo C, Campbell JL (1983) Cloning of *Saccharomyces cerevisiae* DNA replication genes: isolation of the CDC8 gene and two genes that compensate for the cdc8-1 mutation. Am Soc Microbiol 3:1730–1737
59. Waterham HR, Wanders RJ (2007) *Saccharomyces cerevisiae* as a tool for human gene function discovery. In: Stansfield I, Stark MJ (eds) Methods in microbiology. Academic Press, New York, pp 577–595
60. Sikorski RS, Hieter P (1989) A system of shuttle vectors and yeast host strains designed for efficient manipulation of DNA in Saccharomyces cerevisiae. Genetics 122:19–27
61. Christianson TW, Sikorski RS, Dante M et al (1992) Multifunctional yeast high-copy-number shuttle vectors. Gene 110:119–122. https://doi.org/10.1016/0378-1119(92)90454-W
62. Mumberg D, Müller R, Funk M (1995) Yeast vectors for the controlled expression of heterologous proteins in different genetic backgrounds. Gene 156:119–122. https://doi.org/10.1016/0378-1119(95)00037-7
63. Labbe S, Thiele DJ (1999) Copper ion inducible and repressible Systems in Yeast. In: Glorioso JC, Schmidt MC (eds) Methods in enzymology. Academic Press, New York, pp 145–153
64. Siewers V (2014) An overview on selection marker genes for transformation of *Saccharomyces cerevisiae*. In: Mapelli V (ed) Yeast metabolic engineering. Springer New York, New York, NY
65. Hinnen A, Hicks JB, Fink GR (1978) Transformation of yeast. Proc Natl Acad Sci U S A 75:1929–1933. https://doi.org/10.1073/pnas.75.4.1929
66. Mülleder M, Campbell K, Matsarskaia O et al (2016) *Saccharomyces cerevisiae* single-copy plasmids for auxotrophy compensation, multiple marker selection, and for designing metabolically cooperating communities. F1000Res 5:2351. https://doi.org/10.12688/f1000research.9606.1
67. Giersberg M, Degelmann A, Bode R et al (2012) Production of a thermostable alcohol dehydrogenase from *Rhodococcus ruber* in three different yeast species using the Xplor? 2 transformation/expression platform. J Ind Microbiol Biotechnol 39:1385–1396. https://doi.org/10.1007/s10295-012-1134-9
68. Wach A, Brachat A, Pöhlmann R, Philippsen P (1994) New heterologous modules for classical or PCR-based gene disruptions in *Saccharomyces cerevisiae*. Yeast 10:1793–1808. https://doi.org/10.1002/yea.320101310
69. Goldstein AL, McCusker JH (1999) Three new dominant drug resistance cassettes for gene disruption in *Saccharomyces cerevisiae*. Yeast 15:1541–1553. https://doi.org/10.1002/(SICI)1097-0061(199910)15:14<1541::AID-YEA476>3.0.CO;2-K
70. Tuite MF, Dobson MJ, Roberts NA, King RM (1982) Regulated high efficiency expression of human interferon-alpha in *Saccharomyces cerevisiae*. EMBO J 1:603–608
71. Derynck R, Singh A, Goeddel DV (1983) Expression of the human interferon-y cDNA in yeast. Nucleic Acids Res 1:1819–1837
72. Mellor J, Dobson MJ, Roberts NA et al (1983) Efficient synthesis of enzymatically active calf chymosin in *Saccharomyces cerevisiae*. Gene 24:1–14. https://doi.org/10.1016/0378-1119(90)90159-O
73. Bitter GA, Egan KM (1984) Expression of heterologous genes in *Saccharomyces cerevisiae* from vectors utilizing the glyceraldehyde3-phosphate dehydrogenase gene promoter. Gene 32:263–274. https://doi.org/10.1016/0378-1119(84)90002-7
74. Hitzeman RA, Hagie FE, Levine HL et al (1981) Expression of a human gene for interferon in yeast. Nature 293:717–722. https://doi.org/10.1038/293717a0
75. Dobson MJ, Tuite MF, Mellor J et al (1983) Expression in *Saccharomyces cerevisiae* of human interferon-alpha directed by the TRP1 5′ region. Nucleic Acids Res 11:2287–2302. https://doi.org/10.1093/nar/11.8.2287
76. Brake AJ, Merryweather JP, Coit DG et al (1984) Alpha-factor-directed synthesis and secretion of mature foreign proteins in *Saccharomyces cerevisiae*. Proc Natl Acad Sci U S A 81:4642–4646
77. Ammerer G (1983) Expression of genes in yeast using the ADC1 promoter. In: Wu R, Grossman L, Moldave K (eds) Methods in enzymology. Academic Press, New York, pp 192–201
78. Goff CG, Moir DT, Kohno T et al (1984) Expression of calf prochymosin in

Saccharomyces cerevisiae. Gene 27:35–46. https://doi.org/10.1016/0378-1119(84)90236-1

79. Rönicke V, Graulich W, Mumberg D et al (1997) Use of conditional promoters for expression of heterologous proteins in *Saccharomyces cerevisiae*. Methods Enzymol 283:313–322
80. Weinhandl K, Winkler M, Glieder A, Camattari A (2014) Carbon source dependent promoters in yeasts. Microb Cell Factories 13:5. https://doi.org/10.1186/1475-2859-13-5
81. Miyanohara A, Toh-E A, Nozaki C et al (1983) Expression of hepatitis B surface antigen gene in yeast. Proc Natl Acad Sci U S A 80:1–5
82. Curran KA, Crook NC, Karim AS et al (2014) Design of synthetic yeast promoters via tuning of nucleosome architecture. Nat Commun 5:1–20. https://doi.org/10.1038/ncomms5002
83. Dahl RH, Zhang F, Alonso-gutierrez J et al (2013) Engineering dynamic pathway regulation using stress-response promoters. Nat Biotechnol 31:1039–1046. https://doi.org/10.1038/nbt.2689
84. Rajkumar AS, Liu G, Bergenholm D et al (2016) Engineering of synthetic, stress-responsive yeast promoters. Nucleic Acids Res 44:e136. https://doi.org/10.1093/nar/gkw553
85. Chung BH, Nam SW, Kim BM, Park YH (1996) Highly efficient secretion of heterologous proteins from *Saccharomyces cerevisiae* using inulinase signal peptides. Biotechnol Bioeng 49:473–479. https://doi.org/10.1002/(SICI)1097-0290(19960220)49:4<473::AID-BIT15>3.0.CO;2-B
86. Hitzeman R, Leung D, Perry L et al (1983) Secretion of human interferons by yeast. Science 219:620–625. https://doi.org/10.1126/science.6186023
87. Mori A, Hara S, Sugahara T et al (2015) Signal peptide optimization tool for the secretion of recombinant protein from *Saccharomyces cerevisiae*. J Biosci Bioeng 120:518–525. https://doi.org/10.1016/j.jbiosc.2015.03.003
88. Ogata K, Nishikawa H, Ohsugi M (1969) A yeast capable of utilizing methanol. Agric Biol Chem 33:1519–1520. https://doi.org/10.1080/00021369.1969.10859497
89. Cregg JM, Barringer KJ, Hessler AY, Madden KR (1985) *Pichia pastoris* as a host system for transformations. Mol Cell Biol 5:3376–3385
90. Kurtzman CP (2009) Biotechnological strains of *Komagataella* (*Pichia*) *pastoris* are *Komagataella phaffii* as determined from multigene sequence analysis. J Ind Microbiol Biotechnol 36:1435–1438. https://doi.org/10.1007/s10295-009-0638-4
91. Ergun BG, Calik P (2016) Lignocellulose degrading extremozymes produced by *Pichia pastoris*: current status and future prospects. Bioprocess Biosyst Eng 39:1–36. https://doi.org/10.1007/s00449-015-1476-6
92. Choi B-K, Bobrowicz P, Davidson RC et al (2003) Use of combinatorial genetic libraries to humanize N-linked glycosylation in the yeast *Pichia pastoris*. Proc Natl Acad Sci U S A 100:5022–5027
93. Hamilton SR, Bobrowicz P, Bobrowicz B et al (2003) Production of complex human glycoproteins in yeast. Science 301:1244–1246. https://doi.org/10.1126/science.1088166
94. Vervecken W, Kaigorodov V, Callewaert N et al (2004) In vivo synthesis of mammalian-like, hybrid-type N-glycans in *Pichia pastoris*. Appl Environ Microbiol 70:2639–2646
95. Çalık P, Ata Ö, Güneş H et al (2015) Recombinant protein production in *Pichia pastoris* under glyceraldehyde-3-phosphate dehydrogenase promoter: from carbon source metabolism to bioreactor operation parameters. Biochem Eng J 95:20–36. https://doi.org/10.1016/j.bej.2014.12.003
96. Gasser B, Prielhofer R, Marx H et al (2013) *Pichia pastoris*: protein production host and model organism for biomedical research. Future Microbiol 8:191–208. https://doi.org/10.2217/fmb.12.133
97. De Schutter K, Lin Y-C, Tiels P et al (2009) Genome sequence of the recombinant protein production host *Pichia pastoris*. Nat Biotechnol 27:561–566. https://doi.org/10.1038/nbt.1544
98. Mattanovich D, Graf A, Stadlmann J et al (2009) Genome, secretome and glucose transport highlight unique features of the protein production host *Pichia pastoris*. Microb Cell Factories 8:29. https://doi.org/10.1186/1475-2859-8-29
99. Love KR, Shah KA, Whittaker CA et al (2016) Comparative genomics and transcriptomics of *Pichia pastoris*. BMC Genomics 17:550. https://doi.org/10.1186/s12864-016-2876-y
100. Kuberl A, Schneider J, Thallinger GG et al (2011) High-quality genome sequence of *Pichia pastoris* CBS7435. J Biotechnol 154:312–320. https://doi.org/10.1016/j.jbiotec.2011.04.014
101. Valli M, Tatto NE, Peymann A et al (2016) Curation of the genome annotation of *Pichia*

pastoris (*Komagataella phaffii*) CBS7435 from gene level to protein function. FEMS Yeast Res 16. https://doi.org/10.1093/femsyr/fow051

102. Sturmberger L, Chappell T, Geier M et al (2016) Refined *Pichia pastoris* reference genome sequence. J Biotechnol 235:121–131. https://doi.org/10.1016/j.jbiotec.2016.04.023
103. Ciofalo V, Barton N, Kreps J et al (2006) Safety evaluation of a lipase enzyme preparation, expressed in *Pichia pastoris*, intended for use in the degumming of edible vegetable oil. Regul Toxicol Pharmacol 45:1–8. https://doi.org/10.1016/j.yrtph.2006.02.001
104. Thompson CA (2010) FDA approves kallikrein inhibitor to treat hereditary angioedema. Am J Health Syst Pharm 67:93
105. Coughlan AY, Hanson SJ, Byrne KP, Wolfe KH (2016) Centromeres of the yeast *Komagataella phaffii* (*Pichia pastoris*) have a simple inverted-repeat structure. Genome Biol Evol 8:2482–2492. https://doi.org/10.1093/gbe/evw178
106. Lee CC, Williams TG, Wong DWS, Robertson GH (2005) An episomal expression vector for screening mutant gene libraries in *Pichia pastoris*. Plasmid 54:80–85. https://doi.org/10.1016/j.plasmid.2004.12.001
107. Uchima CA, Arioka M (2012) Expression and one-step purification of recombinant proteins using an alternative episomal vector for the expression of N-tagged heterologous proteins in *Pichia pastoris*. Biosci Biotechnol Biochem 76:368–371. https://doi.org/10.1271/bbb.110628
108. Liachko I, Dunham MJ (2014) An autonomously replicating sequence for use in a wide range of budding yeasts. FEMS Yeast Res 14:364–367. https://doi.org/10.1111/1567-1364.12123
109. Camattari A, Goh A, Yip LY et al (2016) Characterization of a panARS-based episomal vector in the methylotrophic yeast *Pichia pastoris* for recombinant protein production and synthetic biology applications. Microb Cell Factories 15:139. https://doi.org/10.1186/s12934-016-0540-5
110. Mattanovich D, Sauer M, Gasser B (2017) Industrial microorganisms: *Pichia pastoris*. In: Industrial biotechnology. Wiley-VCH Verlag GmbH & Co. KGaA, pp 687–714
111. Naatsaari L, Mistlberger B, Ruth C et al (2012) Deletion of the *Pichia pastoris KU70* homologue facilitates platform strain generation for gene expression and synthetic biology. PLoS One 7:e39720. https://doi.org/10.1371/journal.pone.0039720
112. Jacobs PP, Geysens S, Vervecken W et al (2009) Engineering complex-type N-glycosylation in *Pichia pastoris* using GlycoSwitch technology. Nat Protoc 4:58–70. https://doi.org/10.1038/nprot.2008.213
113. Prielhofer R, Barrero JJ, Steuer S et al (2017) Golden*Pi*CS: a Golden Gate-derived modular cloning system for applied synthetic biology in the yeast *Pichia pastoris*. BMC Syst Biol 11:123. https://doi.org/10.1186/s12918-017-0492-3
114. Obst U, Lu TK, Sieber V (2017) A modular toolkit for generating *Pichia pastoris* secretion libraries. ACS Synth Biol 6:1016–1025. https://doi.org/10.1021/acssynbio.6b00337
115. Soderholm J, Bevis BJ, Glick BS (2001) Vector for pop-in/pop-out gene replacement in *Pichia pastoris*. Biotechniques 31:306–310
116. Sears IB, O'Connor J, Rossanese OW, Glick BS (1998) A versatile set of vectors for constitutive and regulated gene expression in *Pichia pastoris*. Yeast 14:783–790. https://doi.org/10.1002/(SICI)1097-0061(19980615)14:8<783::AID-YEA272>3.0.CO;2-Y
117. Sunga AJ, Cregg JM (2004) The *Pichia pastoris* formaldehyde dehydrogenase gene (*FLD1*) as a marker for selection of multicopy expression strains of *P. pastoris*. Gene 330:39–47. https://doi.org/10.1016/j.gene.2003.12.015
118. Du M, Battles MB, Nett JH (2012) A color-based stable multi-copy integrant selection system for *Pichia pastoris* using the attenuated *ADE1* and *ADE2* genes as auxotrophic markers. Bioeng Bugs 3:32–37. https://doi.org/10.4161/bbug.3.1.17936
119. Porro D, Sauer M, Branduardi P, Mattanovich D (2005) Recombinant protein production in yeasts. Mol Biotechnol 31:245–259. https://doi.org/10.1385/MB:31:3:245
120. Couderc R, Baratti J (1980) Oxidation of methanol by the yeast, *Pichia pastoris*. Purification and properties of the alcohol oxidase. Agric Biol Chem 44:2279–2289. https://doi.org/10.1080/00021369.1980.10864320
121. Inan M, Meagher MM (2001) Non-repressing carbon sources for alcohol oxidase (AOX1) promoter of *Pichia pastoris*. J Biosci Bioeng 92:585–589
122. Cregg JM, Madden KR, Barringer KJ et al (1989) Functional characterization of the two alcohol oxidase genes from the yeast *Pichia pastoris*. Mol Cell Biol 9:1316–1323
123. Hasslacher M, Schall M, Hayn M et al (1997) High-level intracellular expression of hydroxynitrile lyase from the tropical rubber tree *Hevea brasiliensis* in microbial hosts. Protein

Expr Purif 11:61–71. https://doi.org/10.1006/prep.1997.0765

124. Schotte P, Dewerte I, De Groeve M et al (2016) *Pichia pastoris* Mut(S) strains are prone to misincorporation of O-methyl-l-homoserine at methionine residues when methanol is used as the sole carbon source. Microb Cell Factories 15:98. https://doi.org/10.1186/s12934-016-0499-2
125. Macauley-Patrick S, Fazenda ML, McNeil B, Harvey LM (2005) Heterologous protein production using the *Pichia pastoris* expression system. Yeast 22:249–270. https://doi.org/10.1002/yea.1208
126. Tschopp JF, Brust PF, Cregg JM et al (1987) Expression of the lacZ gene from two methanol-regulated promoters in *Pichia pastoris*. Nucleic Acids Res 15:3859–3876. https://doi.org/10.1093/nar/15.9.3859
127. Shen S, Sulter G, Jeffries TW, Cregg JM (1998) A strong nitrogen source-regulated promoter for controlled expression of foreign genes in the yeast *Pichia pastoris*. Gene 216:93–102
128. Prielhofer R, Maurer M, Klein J et al (2013) Induction without methanol: novel regulated promoters enable high-level expression in *Pichia pastoris*. Microb Cell Factories 12:5. https://doi.org/10.1186/1475-2859-12-5
129. Payne WE, Gannon PM, Kaiser CA (1995) An inducible acid phosphatase from the yeast *Pichia pastoris*: characterization of the gene and its product. Gene 163:19–26
130. Ahn J, Hong J, Park M et al (2009) Phosphate-responsive promoter of a *Pichia pastoris* sodium phosphate symporter. Appl Environ Microbiol 75:3528–3534. https://doi.org/10.1128/AEM.02913-08
131. Cregg JM, Tolstorukov II (2012) *P. pastoris* ADH promoter and use thereof to direct expression of proteins. US Patent 8222386 B2. 2
132. Stadlmayr G, Mecklenbrauker A, Rothmuller M et al (2010) Identification and characterisation of novel *Pichia pastoris* promoters for heterologous protein production. J Biotechnol 150:519–529. https://doi.org/10.1016/j.jbiotec.2010.09.957
133. Kern A, Hartner FS, Freigassner M et al (2007) *Pichia pastoris* "just in time" alternative respiration. Microbiology 153:1250–1260. https://doi.org/10.1099/mic.0.2006/001404-0
134. Waterham HR, Digan ME, Koutz PJ et al (1997) Isolation of the *Pichia pastoris* glyceraldehyde-3-phosphate dehydrogenase gene and regulation and use of its promoter. Gene 186:37–44
135. Baumann K, Maurer M, Dragosits M et al (2008) Hypoxic fed-batch cultivation of *Pichia pastoris* increases specific and volumetric productivity of recombinant proteins. Biotechnol Bioeng 100:177–183. https://doi.org/10.1002/bit.21763
136. Periyasamy S, Govindappa N, Sreenivas S, Sastry K (2013) Isolation, characterization and evaluation of the *Pichia pastoris* sorbitol dehydrogenase promoter for expression of heterologous proteins. Protein Expr Purif 92:128–133. https://doi.org/10.1016/j.pep.2013.09.008
137. Ahn J, Hong J, Lee H et al (2007) Translation elongation factor 1-alpha gene from *Pichia pastoris*: molecular cloning, sequence, and use of its promoter. Appl Microbiol Biotechnol 74:601–608. https://doi.org/10.1007/s00253-006-0698-6
138. Ata Ö, Prielhofer R, Gasser B et al (2017) Transcriptional engineering of the glyceraldehyde-3-phosphate dehydrogenase promoter for improved heterologous protein production in *Pichia pastoris*. Biotechnol Bioeng 114(10):2319–2327. https://doi.org/10.1002/bit.26363
139. Portela RMC, Vogl T, Kniely C et al (2017) Synthetic core promoters as universal parts for fine-tuning expression in different yeast species. ACS Synth Biol 6:471–484. https://doi.org/10.1021/acssynbio.6b00178
140. Vogl T, Ruth C, Pitzer J et al (2014) Synthetic core promoters for *Pichia pastoris*. ACS Synth Biol 3:188–191. https://doi.org/10.1021/sb400091p
141. Hartner FS, Ruth C, Langenegger D et al (2008) Promoter library designed for fine-tuned gene expression in *Pichia pastoris*. Nucleic Acids Res 36:e76. https://doi.org/10.1093/nar/gkn369
142. Öztürk S, Gündüz-Ergün B, Çalık P (2017) Double promoter expression systems for recombinant protein production by industrial microorganisms. Appl Microbiol Biotechnol 101(20):7459–7475. https://doi.org/10.1007/s00253-017-8487-y
143. Liu H, Tan X, Russell KA et al (1995) PER3, a gene required for peroxisome biogenesis in *Pichia pastoris*, encodes a peroxisomal membrane protein involved in protein import. J Biol Chem 270:10940–10951
144. Vogl T, Sturmberger L, Kickenweiz T et al (2016) A toolbox of diverse promoters related to methanol utilization: functionally verified parts for heterologous pathway expression in *Pichia pastoris*. ACS Synth Biol 5:172–186. https://doi.org/10.1021/acssynbio.5b00199

145. Passoth V, Hahn-Hägerdal B (2000) Production of a heterologous endo-1,4-β-xylanase in the yeast *Pichia stipitis* with an O_2-regulated promoter. Enzym Microb Technol 26:781–784. https://doi.org/10.1016/S0141-0229(00)00171-X
146. Chien L-J, Lee C-K (2005) Expression of bacterial hemoglobin in the yeast, *Pichia pastoris*, with a low O_2-induced promoter. Biotechnol Lett 27:1491–1497. https://doi.org/10.1007/s10529-005-1324-x
147. Camattari A, Bianchi MM, Branduardi P et al (2007) Induction by hypoxia of heterologous-protein production with the KlPDC1 promoter in yeasts. Appl Environ Microbiol 73:922–929. https://doi.org/10.1128/AEM.01764-06
148. Koller A, Valesco J, Subramani S (2000) The *CUP1* promoter of *Saccharomyces cerevisiae* is inducible by copper in *Pichia pastoris*. Yeast 16:651–656. https://doi.org/10.1002/(SICI)1097-0061(200005)16:7<651::AID-YEA580>3.0.CO;2-F
149. Huang C-J, Damasceno LM, Anderson KA et al (2011) A proteomic analysis of the *Pichia pastoris* secretome in methanol-induced cultures. Appl Microbiol Biotechnol 90:235–247. https://doi.org/10.1007/s00253-011-3118-5
150. Laroche Y, Storme V, De Meutter J et al (1994) High-level secretion and very efficient isotopic Labeling of tick anticoagulant peptide (TAP) expressed in the methylotrophic yeast, *Pichia pastoris*. Nat Biotechnol 12:1119–1124
151. Weiss HM, Haase W, Michel H, Reilander H (1995) Expression of functional mouse 5-HT5A serotonin receptor in the methylotrophic yeast *Pichia pastoris*: pharmacological characterization and localization. FEBS Lett 377:451–456
152. Ha SH, Park JJ, Kim JW et al (2001) Molecular cloning and high-level expression of G2 protein of hantaan (HTN) virus 76-118 strain in the yeast *Pichia pastoris* KM71. Virus Genes 22:167–173
153. Trujillo LE, Arrieta JG, Dafhnis F et al (2001) Fructo-oligosaccharides production by the Gluconacetobacter diazotrophicus levansucrase expressed in the methylotrophic yeast *Pichia pastoris*. Enzym Microb Technol 28:139–144
154. Yoshimasu MA, Ahn J-K, Tanaka T, Yada RY (2002) Soluble expression and purification of porcine pepsinogen from *Pichia pastoris*. Protein Expr Purif 25:229–236
155. Murasugi A, Tohma-Aiba Y (2001) Comparison of three signals for secretory expression of recombinant human midkine in *Pichia pastoris*. Biosci Biotechnol Biochem 65:2291–2293. https://doi.org/10.1271/bbb.65.2291
156. Tschopp JF, Sverlow G, Kosson R et al (1987) High-level secretion of glycosylated invertase in the methylotrophic yeast, *Pichia pastoris*. Nat Biotechnol 5:1305–1308
157. Paifer E, Margolles E, Cremata J et al (1994) Efficient expression and secretion of recombinant alpha amylase in *Pichia pastoris* using two different signal sequences. Yeast 10:1415–1419. https://doi.org/10.1002/yea.320101104
158. Kuwae S, Ohyama M, Ohya T et al (2005) Production of recombinant human antithrombin by *Pichia pastoris*. J Biosci Bioeng 99:264–271. https://doi.org/10.1263/jbb.99.264
159. Massahi A, Calik P (2015) *In-silico* determination of *Pichia pastoris* signal peptides for extracellular recombinant protein production. J Theor Biol 364:179–188. https://doi.org/10.1016/j.jtbi.2014.08.048
160. Massahi A, Calik P (2016) Endogenous signal peptides in recombinant protein production by *Pichia pastoris*: from *in-silico* analysis to fermentation. J Theor Biol 408:22–33. https://doi.org/10.1016/j.jtbi.2016.07.039
161. Liang S, Li C, Ye Y, Lin Y (2013) Endogenous signal peptides efficiently mediate the secretion of recombinant proteins in *Pichia pastoris*. Biotechnol Lett 35:97–105. https://doi.org/10.1007/s10529-012-1055-8
162. Khasa YP, Conrad S, Sengul M et al (2011) Isolation of *Pichia pastoris PIR* genes and their utilization for cell surface display and recombinant protein secretion. Yeast 28:213–226. https://doi.org/10.1002/yea.1832
163. Govindappa N, Hanumanthappa M, Venkatarangaiah K et al (2014) A new signal sequence for recombinant protein secretion in *Pichia pastoris*. J Microbiol Biotechnol 24:337–345
164. Heiss S, Puxbaum V, Gruber C et al (2015) Multistep processing of the secretion leader of the extracellular protein Epx1 in *Pichia pastoris* and implications for protein localization. Microbiology 161:1356–1368. https://doi.org/10.1099/mic.0.000105
165. Aw R, McKay PF, Shattock RJ, Polizzi KM (2017) Expressing anti-HIV VRC01

antibody using the murine IgG1 secretion signal in *Pichia pastoris*. AMB Express 7:70. https://doi.org/10.1186/s13568-017-0372-7

166. Nel S, Labuschagne M, Albertyn J (2009) Advances in gene expression in non-conventional yeasts BT—yeast biotechnology: diversity and applications. In: Satyanarayana T, Kunze G (eds) . Springer Netherlands, Dordrecht, pp 369–403
167. Radecka D, Mukherjee V, Mateo RQ et al (2015) Looking beyond *Saccharomyces*: the potential of non-conventional yeast species for desirable traits in bioethanol fermentation. FEMS Yeast Res 15:1–13. https://doi.org/10.1093/femsyr/fov053
168. Hartner FS, Glieder A (2006) Regulation of methanol utilisation pathway genes in yeasts. Microb Cell Factories 5:39. https://doi.org/10.1186/1475-2859-5-39
169. van Dijk R, Faber KN, Kiel JAKW et al (2000) The methylotrophic yeast *Hansenula polymorpha*: a versatile cell factory. Enzym Microb Technol 26:793–800. https://doi.org/10.1016/S0141-0229(00)00173-3
170. Gellissen G (2000) Heterologous protein production in methylotrophic yeasts. Appl Microbiol Biotechnol 54:741–750. https://doi.org/10.1007/s002530000464
171. Veenhuis M, Kram AM, Kunau WH, Harder W (1990) Excessive membrane development following exposure of the methylotrophic yeast *Hansenula polymorpha* to oleic acid-containing media. Yeast 6:511–519. https://doi.org/10.1002/yea.320060608
172. Baerends RJS, Faber KN, Kram AM et al (2000) A stretch of positively charged amino acids at the N terminus of *Hansenula polymorpha* Pex3p is involved in incorporation of the protein into the peroxisomal membrane. J Biol Chem 275:9986–9995. https://doi.org/10.1074/jbc.275.14.9986
173. Ryabova OB, Chmil OM, Sibirny AA (2003) Xylose and cellobiose fermentation to ethanol by the thermotolerant methylotrophic yeast *Hansenula polymorpha*. FEMS Yeast Res 4:157–164. https://doi.org/10.1016/S1567-1356(03)00146-6
174. van Zutphen T, Baerends RJ, Susanna KA et al (2010) Adaptation of *Hansenula polymorpha* to methanol: a transcriptome analysis. BMC Genomics 11:1. https://doi.org/10.1186/1471-2164-11-1
175. Veenhuis M, van der Klei IJ, Titorenko V, Harder W (1992) *Hansenula polymorpha*: an attractive model organism for molecular studies of peroxisome biogenesis and function. FEMS Microbiol Lett 100:393–403. https://doi.org/10.1111/j.1574-6968.1992.tb14068.x
176. Ávila J, Pérez MD, Brito N et al (1995) Cloning and disruption of the YNR1 gene encoding the nitrate reductase apoenzyme of the yeast *Hansenula polymorpha*. FEBS Lett 366:137–142. https://doi.org/10.1016/0014-5793(95)00511-7
177. Brito N, Pérez MD, Perdomo G et al (1999) A set of *Hansenula polymorpha* integrative vectors to construct lacZ fusions. Appl Microbiol Biotechnol 53:23–29. https://doi.org/10.1007/s002530051609
178. Siverio JM (2002) Assimilation of nitrate by yeasts. FEMS Microbiol Rev 26:277–284. https://doi.org/10.1016/S0168-6445(02)00100-6
179. Kunze G, Kang HA, Gellissen G (2009) Hansenula polymorpha (Pichia angusta): biology and applications. In: Satyanarayana T, Kunze G (eds) Yeast biotechnology: diversity and applications. Springer Netherlands, Dordrecht, pp 47–64
180. Sohn JH, Choi ES, Kim CH et al (1996) A novel autonomously replicating sequence (ARS) for multiple integration in the yeast *Hansenula polymorpha* DL-1. J Bacteriol 178:4420–4428. https://doi.org/10.1128/jb.178.15.4420-4428.1996
181. Gellissen G, Hollenberg CP (1997) Application of yeasts in gene expression studies: a comparison of *Saccharomyces cerevisiae*, *Hansenula polymorpha* and *Kluyveromyces lactis*—a review. Gene 190:87–97. https://doi.org/10.1016/S0378-1119(97)00020-6
182. Janowicz ZA, Melber K, Merckelbach A et al (1991) Simultaneous expression of the S and L surface antigens of hepatitis B, and formation of mixed particles in the methylotrophic yeast, *Hansenula polymorpha*. Yeast 7:431–443. https://doi.org/10.1002/yea.320070502
183. Gatzke R, Weydemann U, Janowicz ZA, Hollenberg CP (1995) Stable multicopy integration of vector sequences in *Hansenula polymorpha*. Appl Microbiol Biotechnol 43:844–849. https://doi.org/10.1007/BF02431917
184. Gellissen G, Hollenberg CP, Janowicz ZA (1995) Gene expression in methylotrophic yeasts. Bioprocess Technol 22:195–239
185. Agaphonov MO, Beburov MY, Ter-Avanesyan MD, Smirnov VN (1995) A disruption-replacement approach for the targeted integration of foreign genes in

Hansenula polymorpha. Yeast 11:1241–1247. https://doi.org/10.1002/yea.320111304

186. Machín F, Perdomo G, Pérez MD et al (2001) Evidence for multiple nitrate uptake systems in the yeast *Hansenula polymorpha*. FEMS Microbiol Lett 194:171–174. https://doi.org/10.1016/S0378-1097(00)00524-3
187. Song H, Li Y, Fang W et al (2003) Development of a set of expression vectors in *Hansenula polymorpha*. Biotechnol Lett 25:1999–2006. https://doi.org/10.1023/B:BILE.0000004392.87179.29
188. Agaphonov MO, Trushkina PM, Sohn J-H et al (1999) Vectors for rapid selection of integrants with different plasmid copy numbers in the yeast *Hansenula polymorpha* DL1. Yeast 15:541–551. https://doi.org/10.1002/(SICI)1097-0061(199905)15:7<541::AID-YEA392>3.0.CO;2-G
189. Heo JH, Hong WK, Cho EY et al (2003) Properties of the *Hansenula polymorpha*-derived constitutive GAP promoter, assessed using an HSA reporter gene. FEMS Yeast Res 4:175–184. https://doi.org/10.1016/S1567-1356(03)00150-8
190. Cox H, Mead D, Sudbery P et al (2000) Constitutive expression of recombinant proteins in the methylotrophic yeast *Hansenula polymorpha* using the *PMA1* promoter. Yeast 16:1191–1203. https://doi.org/10.1002/1097-0061(20000930)16:13<1191::AID-YEA589>3.0.CO;2-2
191. Klabunde J, Diesel A, Waschk D et al (2002) Single-step co-integration of multiple expressible heterologous genes into the ribosomal DNA of the methylotrophic yeast *Hansenula polymorpha*. Appl Microbiol Biotechnol 58:797–805. https://doi.org/10.1007/s00253-002-0957-0
192. Klabunde J, Kunze G, Gellissen G, Hollenberg CP (2003) Integration of heterologous genes in several yeast species using vectors containing a *Hansenula polymorpha*-derived rDNA-targeting element. FEMS Yeast Res 4:185–193. https://doi.org/10.1016/S1567-1356(03)00148-X
193. Liu Y, Li Y, Liu L et al (2005) Design of vectors for efficient integration and transformation in *Hansenula polymorpha*. Biotechnol Lett 27:1529–1534. https://doi.org/10.1007/s10529-005-1469-7
194. Klabunde J, Kunze G, Gellissen G, Hollenberg CP (2005) Wide-range integrative expression vectors for fungi, based on ribosomal DNA elements. In: Gellissen G (ed) Production of recombinant proteins. Wiley-VCH Verlag GmbH & Co. KGaA, Weinheim, FRG, pp 273–286
195. Ilgen C, Lin-Cereghino J, Cregg JM (2005) Pichia pastoris. In: Gellissen G (ed) Production of recombinant proteins. Wiley-VCH Verlag GmbH & Co. KGaA, Weinheim, FRG, pp 143–162
196. Roggenkamp R, Hansen H, Eckart M et al (1986) Transformation of the methylotrophic yeast *Hansenula polymorpha* by autonomous replication and integration vectors. Mol Gen Genet MGG 202:302–308. https://doi.org/10.1007/BF00331655
197. Merckelbach A, Gödecke S, Janowicz ZA, Hollenberg CP (1993) Cloning and sequencing of the ura3 locus of the methylotrophic yeast *Hansenula polymorpha* and its use for the generation of a deletion by gene replacement. Appl Microbiol Biotechnol 40:361–364. https://doi.org/10.1007/BF00170393
198. Agaphonov MO, Poznyakovski AI, Bogdanova AI, Ter-Avanesyan MD (1994) I. Yeast sequencing reports. Isolation and characterization of the LEU2 gene of Hansenula polymorpha. Yeast 10:509–513. https://doi.org/10.1002/yea.320100410
199. Bogdanova AI, Agaphonov MO, Ter-Avanesyan MD (1995) Plasmid reorganization during integrative transformation in *Hansenula polymorpha*. Yeast 11:343–353. https://doi.org/10.1002/yea.320110407
200. Zurek C, Kubis E, Keup P et al (1996) Production of two aprotinin variants in *Hansenula polymorpha*. Process Biochem 31:679–689. https://doi.org/10.1016/S0032-9592(96)00018-0
201. Kang HA, Hong WK, Sohn JH et al (2001) Molecular characterization of the actin-encoding gene and the use of its promoter for a dominant selection system in the methylotrophic yeast *Hansenula polymorpha*. Appl Microbiol Biotechnol 55:734–741. https://doi.org/10.1007/s002530100605
202. Rezaee A (2003) Construction of a suitable vector for Lacz gene expression in *Hansunela polymorpha*. Pakistan J Biol Sci 6:1361–1364. https://doi.org/10.3923/pjbs.2003.1361.1364
203. Kang HA, Gellissen G (2005) Hansenula polymorpha. Wiley-VCH Verlag GmbH & Co. KGaA, Weinheim, FRG
204. Suckow M, Gellissen G (2005) The expression platform based on *H. polymorpha* strain RB11 and its derivatives—history, status and perspectives. In: Gellissen G (ed) Hansenula

polymorpha. Wiley-VCH Verlag GmbH & Co. KGaA, Weinheim, FRG, pp 105–123

205. Baerends RJS, Sulter GJ, Jeffries TW et al (2002) Molecular characterization of the *Hansenula polymorpha* FLD1 gene encoding formaldehyde dehydrogenase. Yeast 19:37–42. https://doi.org/10.1002/yea.805
206. Phongdara A, Merckelbach A, Keup P et al (1998) Cloning and characterization of the gene encoding a repressible acid phosphatase (*PHO1*) from the methylotrophic yeast *Hansenula polymorpha*. Appl Microbiol Biotechnol 50:77–84. https://doi.org/10.1007/s002530051259
207. Brito N, Avila J, Perez MD et al (1996) The genes *YNI1* and *YNR1*, encoding nitrite reductase and nitrate reductase respectively in the yeast *Hansenula polymorpha*, are clustered and coordinately regulated. Biochem J 317:89–95. https://doi.org/10.1042/bj3170089
208. Pérez MD, González C, Ávila J et al (1997) The *YNT1* gene encoding the nitrate transporter in the yeast *Hansenula polymorpha* is clustered with genes *YNI1* and *YNR1* encoding nitrite reductase and nitrate reductase, and its disruption causes inability to grow in nitrate. Biochem J 321:397–403. https://doi.org/10.1042/bj3210397
209. Alamäe T, Pärn P, Viigand K, Karp H (2003) Regulation of the *Hansenula polymorpha* maltase gene promoter in *H. polymorpha* and *Saccharomyces cerevisiae*. FEMS Yeast Res 4:165–173. https://doi.org/10.1016/S1567-1356(03)00142-9
210. Amuel C, Gellissen G, Hollenberg CP, Suckow M (2000) Analysis of heat shock promoters in *Hansenula polymorpha*: the *TPS1* promoter, a novel element for heterologous gene expression. Biotechnol Bioprocess Eng 5:247–252. https://doi.org/10.1007/BF02942181
211. Kang HA, Kang W, Hong W-K et al (2001) Development of expression systems for the production of recombinant human serum albumin using the MOX promoter in *Hansenula polymorpha* DL-1. Biotechnol Bioeng 76:175–185. https://doi.org/10.1002/bit.1157
212. Weydemann U, Keup P, Piontek M et al (1995) High-level secretion of hirudin by *Hansenula polymorpha* — authentic processing of three different preprohirudins. Appl Microbiol Biotechnol 44:377–385. https://doi.org/10.1007/BF00169932
213. Kumari A, Baronian K, Kunze G, Gupta R (2015) Extracellular expression of YlLip11 with a native signal peptide from *Yarrowia lipolytica* MSR80 in three different yeast hosts. Protein Expr Purif 110:138–144. https://doi.org/10.1016/j.pep.2015.02.016
214. Eilert E, Rolf T, Heumaier A et al (2013) Improved processing of secretory proteins in *Hansenula polymorpha* by sequence variation near the processing site of the alpha mating factor prepro sequence. J Biotechnol 167:94–100. https://doi.org/10.1016/j.jbiotec.2012.08.024
215. Qian W, Aguilar F, Wang T, Qiu B (2013) Secretion of truncated recombinant rabies virus glycoprotein with preserved antigenic properties using a co-expression system in *Hansenula polymorpha*. J Microbiol 51:234–240. https://doi.org/10.1007/s12275-013-2337-0
216. Talebkhan Y, Samadi T, Samie A et al (2016) Expression of granulocyte colony stimulating factor (GCSF) in *Hansenula polymorpha*. Iran J Microbiol 8:21–28
217. Sohn MJ, Oh D-B, Kim EJ et al (2012) Hp*YPS1* and Hp*YPS7* encode functional aspartyl proteases localized at the cell surface in the thermotolerant methylotrophic yeast *Hansenula polymorpha*. Yeast 29:1–16. https://doi.org/10.1002/yea.1912
218. Cheon SA, Jung J, Choo JH et al (2014) Characterization of putative glycosylphosphatidylinositol-anchoring motifs for surface display in the methylotrophic yeast *Hansenula polymorpha*. Biotechnol Lett 36:2085–2094. https://doi.org/10.1007/s10529-014-1582-6
219. Kim S-Y, Sohn J-H, Pyun Y-R, Choi E-S (2002) A cell surface display system using novel GPI-anchored proteins in *Hansenula polymorpha*. Yeast 19:1153–1163. https://doi.org/10.1002/yea.911
220. Kunze G, Gaillardin C, Czernicka M et al (2014) The complete genome of *Blastobotrys* (*Arxula*) *adeninivorans* LS3 - a yeast of biotechnological interest. Biotechnol Biofuels 7:66. https://doi.org/10.1186/1754-6834-7-66
221. Malak A, Baronian K, Kunze G (2016) *Blastobotrys* (*Arxula*) *adeninivorans*: a promising alternative yeast for biotechnology and basic research. Yeast 33:535–547. https://doi.org/10.1002/yea.3180
222. Middelhoven WJ, Niet MCH-T, Rij NJWK-V (1984) *Trichosporon adeninovorans* sp. nov., a yeast species utilizing adenine, xanthine, uric acid, putrescine and primary n-alkylamines as the sole source of carbon, nitrogen and energy. Antonie Van Leeuwenhoek

50:369–378. https://doi.org/10.1007/BF00394651

223. Böer E, Breuer FS, Weniger M et al (2011) Large-scale production of tannase using the yeast *Arxula adeninivorans*. Appl Microbiol Biotechnol 92:105–114. https://doi.org/10.1007/s00253-011-3320-5
224. Wartmann T, Krüger A, Adler K et al (1995) Temperature-dependent dimorphism of the yeast *Arxula adeninivorans* Ls3. Antonie Van Leeuwenhoek 68:215–223. https://doi.org/10.1007/BF00871818
225. Wartmann T, Erdmann J, Kunze I, Kunze G (2000) Morphology-related effects on gene expression and protein accumulation of the yeast *Arxula adeninivorans* LS3. Arch Microbiol 173:253–261. https://doi.org/10.1007/s002030000137
226. Wartmann T, Stephan UW, Bube I et al (2002) Post-translational modifications of the A*FET3* gene product-a component of the iron transport system in budding cells and mycelia of the yeast *Arxula adeninivorans*. Yeast 19:849–862. https://doi.org/10.1002/yea.880
227. Kunze G, Kunze I (1996) *Arxula adeninivorans*. In: Wolf K (ed) Nonconventional yeasts in biotechnology. Springer Berlin Heidelberg, Berlin, Heidelberg, pp 389–409
228. Rösel H, Kunze G (1998) Integrative transformation of the dimorphic yeast *Arxula adeninivorans* LS3 based on hygromycin B resistance. Curr Genet 33:157–163. https://doi.org/10.1007/s002940050322
229. Wartmann T, Böer E, Pico AH et al (2002) High-level production and secretion of recombinant proteins by the dimorphic yeast *Arxula adeninivorans*. FEMS Yeast Res 2:363–369. https://doi.org/10.1016/S1567-1356(02)00086-7
230. Wartmann T, Stoltenburg R, Böer E et al (2003) The A*LEU2* gene—a new component for an *Arxula adeninivorans*-based expression platform. FEMS Yeast Res 3:223–232. https://doi.org/10.1016/S1567-1356(02)00190-3
231. Terentiev Y, Pico AH, Böer E et al (2004) A wide-range integrative yeast expression vector system based on *Arxula adeninivorans*-derived elements. J Ind Microbiol Biotechnol 31:223–228. https://doi.org/10.1007/s10295-004-0142-9
232. Böer E, Piontek M, Kunze G (2009) Xplor® 2—an optimized transformation/expression system for recombinant protein production in the yeast *Arxula adeninivorans*. Appl Microbiol Biotechnol 84:583–594. https://doi.org/10.1007/s00253-009-2167-5
233. Wartmann T, Rösel H, Kunze I et al (1998) A*ILV1* gene from the yeast *Arxula adeninivorans* LS3—a new selective transformation marker. Yeast 14:1017–1025. https://doi.org/10.1002/(SICI)1097-0061(199808)14:11<1017::AID-YEA314>3.0.CO;2-0
234. Wartmann T, Kunze G (2000) Genetic transformation and biotechnological application of the yeast *Arxula adeninivorans*. Appl Microbiol Biotechnol 54:619–624. https://doi.org/10.1007/s002530000444
235. Müller S, Sandal T, Kamp-Hansen P et al (1998) Comparison of expression systems in the yeasts *Saccharomyces cerevisiae*, *Hansenula polymorpha*, *Kluyveromyces lactis, Schizosaccharomyces pombe* and *Yarrowia lipolytica*. Cloning of two novel promoters from *Yarrowia lipolytica*. Yeast 14:1267–1283. https://doi.org/10.1002/(SICI)1097-0061(1998100)14:14<1267::AID-YEA327>3.0.CO;2-2
236. Hahn T, Tag K, Riedel K et al (2006) A novel estrogen sensor based on recombinant *Arxula adeninivorans* cells. Biosens Bioelectron 21:2078–2085. https://doi.org/10.1016/j.bios.2005.10.019
237. Wartmann T, Bellebna C, Böer E et al (2003) The constitutive A*HSB4* promoter—a novel component of the *Arxula adeninivorans*-based expression platform. Appl Microbiol Biotechnol 62:528–535. https://doi.org/10.1007/s00253-003-1323-6
238. El Fiki A, El Metabteb G, Bellebna C et al (2007) The *Arxula adeninivorans* ATAL gene encoding transaldolase-gene characterization and biotechnological exploitation. Appl Microbiol Biotechnol 74:1292–1299. https://doi.org/10.1007/s00253-006-0785-8
239. Steinborn G, Gellissen G, Kunze G (2005) Assessment of *Hansenula polymorpha* and *Arxula adeninivorans*-derived rDNA-targeting elements for the design of *Arxula adeninivorans* expression vectors. FEMS Yeast Res 5:1047–1054. https://doi.org/10.1016/j.femsyr.2005.07.005
240. Steinborn G, Wartmann T, Gellissen G, Kunze G (2007) Construction of an *Arxula adeninivorans* host-vector system based on trp1 complementation. J Biotechnol 127:392–401. https://doi.org/10.1016/j.jbiotec.2006.07.026
241. Steinborn G, Gellissen G, Kunze G (2007) A novel vector element providing multicopy vector integration in *Arxula adeninivorans*. FEMS Yeast Res 7:1197–1205. https://doi.org/10.1111/j.1567-1364.2007.00280.x
242. Steinborn G, Böer E, Scholz A et al (2006) Application of a wide-range yeast vector

(CoMed™) system to recombinant protein production in dimorphic *Arxula adeninivorans*, methylotrophic *Hansenula polymorpha* and other yeasts. Microb Cell Factories 5:33. https://doi.org/10.1186/1475-2859-5-33

243. Álvaro-Benito M, Fernández-Lobato M, Baronian K, Kunze G (2013) Assessment of *Schwanniomyces occidentalis* as a host for protein production using the wide-range Xplor®2 expression platform. Appl Microbiol Biotechnol 97:4443–4456. https://doi.org/10.1007/s00253-012-4527-9
244. Böer E, Steinborn G, Matros A et al (2007) Production of interleukin-6 in *Arxula adeninivorans*, *Hansenula polymorpha* and *Saccharomyces cerevisiae* by applying the wide-range yeast vector (CoMed™) system to simultaneous comparative assessment. FEMS Yeast Res 7:1181–1187. https://doi.org/10.1111/j.1567-1364.2007.00254.x
245. Böer E, Wartmann T, Luther B et al (2004) Characterization of the *AINV* gene and the encoded invertase from the dimorphic yeast *Arxula adeninivorans*. Antonie Van Leeuwenhoek 86:121–134. https://doi.org/10.1023/B:ANTO.0000036135.69810.df
246. Böer E, Mock HP, Bode R et al (2005) An extracellular lipase from the dimorphic yeast *Arxula adeninivorans*: molecular cloning of the ALIP1 gene and characterization of the purified recombinant enzyme. Yeast 22:523–535. https://doi.org/10.1002/yea.1230
247. Kaur P, Lingner A, Singh B et al (2007) APHO1 from the yeast *Arxula adeninivorans* encodes an acid phosphatase of broad substrate specificity. Antonie Van Leeuwenhoek 91:45–55. https://doi.org/10.1007/s10482-006-9094-6
248. Böer E, Bode R, Mock H-P et al (2009) Atan1p-an extracellular tannase from the dimorphic yeast *Arxula adeninivorans*: molecular cloning of the ATAN1 gene and characterization of the recombinant enzyme. Yeast 26:323–337. https://doi.org/10.1002/yea.1669
249. Bischoff F, Litwińska K, Cordes A et al (2015) Three new cutinases from the yeast *Arxula adeninivorans* that are suitable for biotechnological applications. Appl Environ Microbiol 81:5497–5510. https://doi.org/10.1128/AEM.00894-15
250. Minocha N, Kaur P, Satyanarayana T, Kunze G (2007) Acid phosphatase production by recombinant *Arxula adeninivorans*. Appl Microbiol Biotechnol 76:387–393. https://doi.org/10.1007/s00253-007-1021-x
251. Rodicio R, Heinisch JJ (2013) Yeast on the milky way: genetics, physiology and biotechnology of *Kluyveromyces lactis*. Yeast 30:165–177. https://doi.org/10.1002/yea.2954
252. Spohner SC, Schaum V, Quitmann H, Czermak P (2016) *Kluyveromyces lactis*: an emerging tool in biotechnology. J Biotechnol 222:104–116. https://doi.org/10.1016/j.jbiotec.2016.02.023
253. Bonekamp FJ, Oosterom J (1994) On the safety of *Kluyveromyces lactis*-a review. Appl Microbiol Biotechnol 41:1–3. https://doi.org/10.1007/BF00166072
254. Dujon B, Sherman D, Fischer G et al (2004) Genome evolution in yeasts. Nature 430:35–44. https://doi.org/10.1038/nature02579
255. Chen XJ (1996) Low- and high-copy-number yeast *Kluyveromyces lactis* shuttle vectors for replication in the budding. Gene 172:131–136
256. Gunge N, Tamaru A, Ozawa F, Sakaguchi K (1981) Isolation and characterization of linear deoxyribonucleic acid plasmids from *Kluyveromyces lactis* and the plasmid-associated killer character. J Bacteriol 145:382–390
257. van Ooyen AJJ, Dekker P, Huang M et al (2006) Heterologous protein production in the yeast *Kluyveromyces lactis*. FEMS Yeast Res 6:381–392. https://doi.org/10.1111/j.1567-1364.2006.00049.x
258. Chen XJ, Saliola M, Falcone C et al (1986) Sequence organization of the circular plasmid pKD1 from the yeast *Kluyveromyces drosophilarum*. Nucleic Acids Res 14:4471–4481
259. Bianchi MM, Falcone C, Re CX et al (1987) Transformation of the yeast *Kluyveromyces lactis* by new vectors derived from the 1.6 μm circular plasmid pKD1. Curr Genet 12:185–192. https://doi.org/10.1007/BF00436877
260. Bianchi MM (1992) Site-specific recombination of the circular 2 microns-like plasmid pKD1 requires integrity of the recombinase gene A and of the partitioning genes B and C. J Bacteriol 174:6703–6706
261. Fleer R, Chen XJ, Amellal N et al (1991) High-level secretion of correctly processed recombinant human interleukin-1 beta in *Kluyveromyces lactis*. Gene 107:285–295. https://doi.org/10.1016/0378-1119(91)90329-A
262. Hsieh H-P, Da Silva NA (1998) Partial-pKD1 plasmids provide enhanced structural stability for heterologous protein production in

Kluyveromyces lactis. Appl Microbiol Biotechnol 49:411–416. https://doi.org/10.1007/s002530051191

263. Morlino GB, Tizzani L, Fleer R et al (1999) Inducible amplification of gene copy number and heterologous protein production in the yeast *Kluyveromyces lactis*. Appl Environ Microbiol 65:4808–4813
264. Chen XJ, Bianchi MM, Suda K, Fukuhara H (1989) The host range of the pKD1-derived plasmids in yeast. Curr Genet 16:95–98. https://doi.org/10.1007/BF00393401
265. Das S, Hollenberg CP (1982) A high-frequency transformation system for the yeast *Kluyveromyces lactis*. Curr Genet 6:123–128. https://doi.org/10.1007/BF00435211
266. Fabiani L, Aragona M, Frontali L (1990) Isolation and sequence analysis of a *K. lactis* chromosomal DNA element able to autonomously replicate in *S. cerevisiae* and *K. lactis*. Yeast 6:69–76. https://doi.org/10.1002/yea.320060108
267. Sreekrishna K, Webster TD, Dickson RC (1984) Transformation of *Kluyveromyces lactis* with the kanamycin (G418) resistance gene of Tn905. Gene 28:73–81. https://doi.org/10.1016/0378-1119(84)90089-1
268. Das S, Breunig KD, Hollenberg CP (1985) A positive regulatory element is involved in the induction of the beta-galactosidase gene from *Kluyveromyces lactis*. EMBO J 4:793–798
269. Strasser AWM, Selk R, Dohmen RJ et al (1989) Analysis of the alpha-amylase gene of *Schwanniomyces occidentalis* and the secretion of its gene product in transformants of different yeast genera. Eur J Biochem 184:699–706. https://doi.org/10.1111/j.1432-1033.1989.tb15069.x
270. Bergkamp RJM, Kool IM, Geerse RH, Planta RJ (1992) Multiple-copy integration of the -galactosidase gene from *Cyamopsis tetragonoloba* into the ribosomal DNA of *Kluyveromyces lactis*. Curt Genet 21:365–370. https://doi.org/10.1007/BF00351696
271. Heus JJ, Zonneveld BJM, Steensma HY, Van den Berg JA (1990) Centromeric DNA of *Kluyveromyces lactis*. Curr Genet 18:517–522. https://doi.org/10.1007/BF00327022
272. van der Vlugt-Bergmans CJB, van Ooyen AJJ (1999) Expression cloning in *Kluyveromyces lactis*. Biotechnol Tech 13:87–92. https://doi.org/10.1023/A:1008864118362
273. Prior C, Mamessier P, Fukuhara H et al (1993) The hexokinase gene is required for transcriptional regulation of the glucose transporter gene *RAG1* in *Kluyveromyces lactis*. Mol Cell Biol 13:3882–3889. https://doi.org/10.1128/MCB.13.7.3882
274. Iwata T, Tanaka R, Suetsugu M et al (2004) Efficient secretion of human lysozyme from the yeast, *Kluyveromyces lactis*. Biotechnol Lett 26:1803–1808. https://doi.org/10.1007/s10529-004-4614-9
275. van den Berg JA, van der Laken KJ, van Ooyen AJJ et al (1990) *Kluyveromyces* as a host for heterologous gene expression: expression and secretion of prochymosin. Bio/Technology 8:135–139. https://doi.org/10.1038/nbt0290-135
276. Mustilli AC, Izzo E, Houghton M, Galeotti CL (1999) Comparison of secretion of a hepatitis C virus glycoprotein in *Saccharomyces cerevisiae* and *Kluyveromyces lactis*. Res Microbiol 150:179–187. https://doi.org/10.1016/S0923-2508(99)80034-5
277. Rossolini GM, Riccio ML, Gallo E, Galeotti CL (1992) *Kluyveromyces lactis* rDNA as a target for multiple integration by homologous recombination. Gene 119:75–81. https://doi.org/10.1016/0378-1119(92)90068-Z
278. Colussi PA, Taron CH (2005) *Kluyveromyces lactis LAC4* promoter variants that lack function in bacteria but retain full function in *K. lactis*. Appl Environ Microbiol 71:7092–7098. https://doi.org/10.1128/AEM.71.11.7092-7098.2005
279. Dickson RC, Markin JS (1980) Physiological studies of β-galactosidase induction in *Kluyveromyces lactis*. J Bacteriol 142:777–785
280. Dickson RC, Markin JS (1978) Molecular cloning and expression in *E. coli* of a yeast gene coding for beta-galactosidase. Cell 15:123–130
281. Wagner JM, Alper HS (2016) Synthetic biology and molecular genetics in non-conventional yeasts: current tools and future advances. Fungal Genet Biol 89:126–136. https://doi.org/10.1016/j.fgb.2015.12.001
282. Tokunaga M, Ishibashi M, Tatsuda D, Tokunaga H (1997) Secretion of mouse α-amylase from *Kluyveromyces lactis*. Yeast 13:699–706
283. Rocha SN, Abrahao-Neto J, Cerdan ME et al (2010) Heterologous expression of glucose oxidase in the yeast *Kluyveromyces marxianus*. Microb Cell Factories 9:4. https://doi.org/10.1186/1475-2859-9-4
284. Rocha SN, Abrahão-Neto J, Cerdán ME et al (2010) Heterologous expression of glucose oxidase in the yeast *Kluyveromyces marxianus*.

Microb Cell Factories 9:4. https://doi.org/10.1186/1475-2859-9-4

285. Madhavan A, Sukumaran RK (2015) Signal peptides from filamentous fungi efficiently mediate the secretion of recombinant proteins in *Kluyveromyces lactis*. Biochem Eng J 102:31–37. https://doi.org/10.1016/j.bej.2015.03.008
286. Fermiñán E, Domínguez A (1998) Heterologous protein secretion directed by a repressible acid phosphatase system of *Kluyveromyces lactis*: characterization of upstream region-activating sequences in the *KlPHO5* gene. Appl Environ Microbiol 64:2403–2408
287. Madhavan A, Sukumaran RK (2014) Promoter and signal sequence from filamentous fungus can drive recombinant protein production in the yeast *Kluyveromyces lactis*. Bioresour Technol 165:302–308. https://doi.org/10.1016/j.biortech.2014.03.002
288. Amore A, Amoresano A, Birolo L et al (2012) A family GH51 α-l-arabinofuranosidase from *Pleurotus ostreatus*: identification, recombinant expression and characterization. Appl Microbiol Biotechnol 94:995–1006. https://doi.org/10.1007/s00253-011-3678-4
289. Barth G, Gaillardin C (1996) *Yarrowia lipolytica* BT—nonconventional yeasts in biotechnology: a handbook. In: Wolf K (ed) . Springer Berlin Heidelberg, Berlin, Heidelberg, pp 313–388
290. Groenewald M, Boekhout T, Neuveglise C et al (2014) *Yarrowia lipolytica*: safety assessment of an oleaginous yeast with a great industrial potential. Crit Rev Microbiol 40:187–206. https://doi.org/10.3109/1040841X.2013.770386
291. Zinjarde SS (2014) Food-related applications of *Yarrowia lipolytica*. Food Chem 152:1–10. https://doi.org/10.1016/j.foodchem.2013.11.117
292. Nicaud J-M (2012) Yarrowia lipolytica. Yeast 29:409–418. https://doi.org/10.1002/yea.2921
293. Bankar AV, Kumar AR, Zinjarde SS (2009) Environmental and industrial applications of *Yarrowia lipolytica*. Appl Microbiol Biotechnol 84:847–865. https://doi.org/10.1007/s00253-009-2156-8
294. Xue Z, Sharpe PL, Hong S-P et al (2013) Production of omega-3 eicosapentaenoic acid by metabolic engineering of *Yarrowia lipolytica*. Nat Biotechnol 31:734–740
295. Grenfell-Lee D, Zeller S, Cardoso R, Pucaj K (2014) The safety of beta-carotene from *Yarrowia lipolytica*. Food Chem Toxicol 65:1–11. https://doi.org/10.1016/j.fct.2013.12.010
296. Tiels P, Baranova E, Piens K et al (2012) A bacterial glycosidase enables mannose-6-phosphate modification and improved cellular uptake of yeast-produced recombinant human lysosomal enzymes. Nat Biotechnol 30:1225–1231
297. Pignède G, Wang H-J, Fudalej F et al (2000) Autocloning and amplification of *LIP2* in *Yarrowia lipolytica*. Appl Environ Microbiol 66:3283–3289. https://doi.org/10.1128/AEM.66.8.3283-3289.2000
298. Davidow LSDJ (1984) Process for transformation of *Yarrowia lipolytica*. US Patent Application US4880741
299. Gaillardin C, Heslot HRA (1984) Vecteurs de transformation de la levure *Yarrowia lipolytica*, procédé de transformation et levure transformée. French Patent Application FR2566424
300. Madzak C (2015) *Yarrowia lipolytica*: recent achievements in heterologous protein expression and pathway engineering. Appl Microbiol Biotechnol 99:4559–4577. https://doi.org/10.1007/s00253-015-6624-z
301. Boisramé A, Kabani M, Beckerich J-M et al (1998) Interaction of Kar2p and Sls1p is required for efficient co-translational translocation of secreted proteins in the yeast *Yarrowia lipolytica*. J Biol Chem 273:30903–30908. https://doi.org/10.1074/jbc.273.47.30903
302. Kim JW, Park TJ, Ryu DD, Kim JY (2000) High cell density culture of *Yarrowia lipolytica* using a one-step feeding process. Biotechnol Prog 16:657–660. https://doi.org/10.1021/bp000037n
303. Barth G, Gaillardin C (1997) Physiology and genetics of the dimorphic fungus *Yarrowia lipolytica*. FEMS Microbiol Rev 19:219–237
304. Liu L, Alper HS (2014) Draft genome sequence of the oleaginous yeast *Yarrowia lipolytica* PO1f, a commonly used metabolic engineering host. Genome Announc 2: e00652–e00614. https://doi.org/10.1128/genomeA.00652-14
305. Fournier P, Abbas A, Chasles M et al (1993) Colocalization of centromeric and replicative functions on autonomously replicating sequences isolated from the yeast *Yarrowia lipolytica*. Proc Natl Acad Sci U S A 90:4912–4916
306. Matsuoka M, Matsubara M, Daidoh H et al (1993) Analysis of regions essential for the function of chromosomal replicator sequences from *Yarrowia lipolytica*. Mol Gen

Genet MGG 237:327–333. https://doi.org/10.1007/BF00279435

307. Vernis L, Abbas A, Chasles M et al (1997) An origin of replication and a centromere are both needed to establish a replicative plasmid in the yeast *Yarrowia lipolytica*. Mol Cell Biol 17:1995–2004
308. Liu L, Otoupal P, Pan A, Alper HS (2014) Increasing expression level and copy number of a *Yarrowia lipolytica* plasmid through regulated centromere function. FEMS Yeast Res 14:1124–1127. https://doi.org/10.1111/1567-1364.12201
309. Richard G-F, Kerrest A, Lafontaine I, Dujon B (2005) Comparative genomics of hemiascomycete yeasts: genes involved in DNA replication, repair, and recombination. Mol Biol Evol 22:1011–1023. https://doi.org/10.1093/molbev/msi083
310. Verbeke J, Beopoulos A, Nicaud J-M (2013) Efficient homologous recombination with short length flanking fragments in Ku70 deficient *Yarrowia lipolytica* strains. Biotechnol Lett 35:571–576. https://doi.org/10.1007/s10529-012-1107-0
311. Blazeck J, Liu L, Redden H, Alper H (2011) Tuning gene expression in *Yarrowia lipolytica* by a hybrid promoter approach. Appl Environ Microbiol 77:7905–7914. https://doi.org/10.1128/AEM.05763-11
312. Nicaud JM, Fabre E, Gaillardin C (1989) Expression of invertase activity in *Yarrowia lipolytica* and its use as a selective marker. Curr Genet 16:253–260
313. Gaillardin C, Ribet AM (1987) LEU2 directed expression of beta-galactosidase activity and phleomycin resistance in *Yarrowia lipolytica*. Curr Genet 11:369–375
314. Fickers P, Le Dall MT, Gaillardin C et al (2003) New disruption cassettes for rapid gene disruption and marker rescue in the yeast *Yarrowia lipolytica*. J Microbiol Methods 55:727–737
315. Le Dall MT, Nicaud JM, Gaillardin C (1994) Multiple-copy integration in the yeast *Yarrowia lipolytica*. Curr Genet 26:38–44
316. Tobe S, Takami T, Ikeda S, Mitsugi K (1976) Production and some enzymatic properties of alkaline proteinase of *Candida lipolytica*. Agric Biol Chem 40:1087–1092. https://doi.org/10.1080/00021369.1976.10862177
317. Ogrydziak DM, Scharf SJ (1982) Alkaline extracellular protease produced by *Saccharomycopsis lipolytica* CX161-1B. J Gen Microbiol 128:1225–1234. https://doi.org/10.1099/00221287-128-6-1225
318. Blanchin-Roland S, Cordero Otero RR, Gaillardin C (1994) Two upstream activation sequences control the expression of the *XPR2* gene in the yeast *Yarrowia lipolytica*. Mol Cell Biol 14:327–338
319. Davidow LS, DeZeeuw JR, Franke AE (1990) Expression and secretion of heterologous proteins by *Yarrowia lipolytica* transformants. US Patent 4937189A
320. Nicaud JM, Fournier P, La Bonnardiere C et al (1991) Use of ars18 based vectors to increase protein production in *Yarrowia lipolytica*. J Biotechnol 19:259–270
321. Hamsa PV, Chattoo BB (1994) Cloning and growth-regulated expression of the gene encoding the hepatitis B virus middle surface antigen in *Yarrowia lipolytica*. Gene 143:165–170
322. Madzak C, Treton B, Blanchin-Roland S (2000) Strong hybrid promoters and integrative expression/secretion vectors for quasi-constitutive expression of heterologous proteins in the yeast *Yarrowia lipolytica*. J Mol Microbiol Biotechnol 2:207–216
323. Nicaud J-M, Madzak C, van den Broek P et al (2002) Protein expression and secretion in the yeast *Yarrowia lipolytica*. FEMS Yeast Res 2:371–379
324. Dominguez A, Ferminan E, Sanchez M et al (1998) Non-conventional yeasts as hosts for heterologous protein production. Int Microbiol 1:131–142
325. Juretzek T, Wang H-J, Nicaud J-M et al (2000) Comparison of promoters suitable for regulated overexpression of β-galactosidase in the alkane-utilizing yeast *Yarrowia lipolytica*. Biotechnol Bioprocess Eng 5:320–326. https://doi.org/10.1007/BF02942206
326. Fabre E, Tharaud C, Gaillardin C (1992) Intracellular transit of a yeast protease is rescued by trans-complementation with its prodomain. J Biol Chem 267:15049–15055
327. Fabre E, Nicaud JM, Lopez MC, Gaillardin C (1991) Role of the proregion in the production and secretion of the *Yarrowia lipolytica* alkaline extracellular protease. J Biol Chem 266:3782–3790
328. Wood V, Gwilliam R, Rajandream M-A et al (2002) The genome sequence of *Schizosaccharomyces pombe*. Nature 415:871–880. https://doi.org/10.1038/nature724
329. Sasaki M, Idiris A, Tada A et al (2008) The gap-filling sequence on the left arm of chromosome 2 in fission yeast *Schizosaccharomyces pombe*. Yeast 25:673–679. https://doi.org/10.1002/yea.1613

330. Sasaki M, Kumagai H, Takegawa K, Tohda H (2013) Characterization of genome-reduced fission yeast strains. Nucleic Acids Res 41:5382–5399. https://doi.org/10.1093/nar/gkt233
331. Ballou CE, Ballou L, Ball G (1994) *Schizosaccharomyces pombe* glycosylation mutant with altered cell surface properties. Proc Natl Acad Sci U S A 91:9327–9331
332. Siam R, Dolan WP, Forsburg SL (2004) Choosing and using *Schizosaccharomyces pombe* plasmids. Methods 33:189–198. https://doi.org/10.1016/j.ymeth.2003.11.013
333. Moreno S, Klar A, Nurse P (1991) Molecular genetic analysis of fission yeast *Schizosaccharomyces pombe*. Methods Enzymol 194:795–823
334. Wright AP, Maundrell K, Shall S (1986) Transformation of *Schizosaccharomyces pombe* by non-homologous, unstable integration of plasmids in the genome. Curr Genet 10:503–508
335. Hayles J, Nurse P (1992) Genetics of the fission yeast *Schizosaccharomyces pombe*. Annu Rev Genet 26:373–402. https://doi.org/10.1146/annurev.ge.26.120192.002105
336. Burke JD, Gould KL (1994) Molecular cloning and characterization of the *Schizosaccharomyces pombe* his3 gene for use as a selectable marker. Mol Gen Genet MGG 242:169–176. https://doi.org/10.1007/BF00391010
337. Barbet N, Muriel WJ, Carr AM (1992) Versatile shuttle vectors and genomic libraries for use with *Schizosaccharomyces pombe*. Gene 114:59–66
338. Okazaki K, Okazaki N, Kume K et al (1990) High-frequency transformation method and library transducing vectors for cloning mammalian cDNAs by trans-complementation of *Schizosaccharomyces pombe*. Nucleic Acids Res 18:6485–6489
339. Keeney JB, Boeke JD (1994) Efficient targeted integration at leu1-32 and ura4-294 in *Schizosaccharomyces pombe*. Genetics 136:849–856
340. Bach ML (1987) Cloning and expression of the OMP decarboxylase gene *URA4* from *Schizosaccharomyces pombe*. Curr Genet 12:527–534
341. Kikuchi Y, Kitazawa Y, Shimatake H, Yamamoto M (1988) The primary structure of the leu1+ gene of *Schizosaccharomyces pombe*. Curr Genet 14:375–379
342. Fennessy D, Grallert A, Krapp A et al (2014) Extending the *Schizosaccharomyces pombe* molecular genetic toolbox. PLoS One 9:e97683
343. Bahler J, Wu JQ, Longtine MS et al (1998) Heterologous modules for efficient and versatile PCR-based gene targeting in *Schizosaccharomyces pombe*. Yeast 14:943–951. https://doi.org/10.1002/(SICI)1097-0061(199807)14:10<943::AID-YEA292>3.0.CO;2-Y
344. Burland TG, Pallotta D, Tardif MC et al (1991) Fission yeast promoter-probe vectors based on hygromycin resistance. Gene 100:241–245. https://doi.org/10.1016/0378-1119(91)90374-K
345. Hentges P, Van Driessche B, Tafforeau L et al (2005) Three novel antibiotic marker cassettes for gene disruption and marker switching in *Schizosaccharomyces pombe*. Yeast 22:1013–1019. https://doi.org/10.1002/yea.1291
346. Russell PR, Hall BD (1983) The primary structure of the alcohol dehydrogenase gene from the fission yeast *Schizosaccharomyces pombe*. J Biol Chem 258:143–149
347. Gmunder H, Kohli J (1989) Cauliflower mosaic virus promoters direct efficient expression of a bacterial G418 resistance gene in *Schizosaccharomyces pombe*. Mol Gen Genet 220:95–101
348. Faryar K, Gatz C (1992) Construction of a tetracycline-inducible promoter in *Schizosaccharomyces pombe*. Curr Genet 21:345–349
349. Maundrell K (1993) Thiamine-repressible expression vectors pREP and pRIP for fission yeast. Gene 123:127–130. https://doi.org/10.1016/0378-1119(93)90551-D
350. Maundrell K (1990) nmt1 of fission yeast. A highly transcribed gene completely repressed by thiamine. J Biol Chem 265:10857–10864
351. Basi G, Schmid E, Maundrell K (1993) TATA box mutations in the *Schizosaccharomyces pombe* nmt1 promoter affect transcription efficiency but not the transcription start point or thiamine repressibility. Gene 123:131–136. https://doi.org/10.1016/0378-1119(93)90552-E
352. Hoffman CS, Winston F (1989) A transcriptionally regulated expression vector for the fission yeast *Schizosaccharomyces pombe*. Gene 84:473–479. https://doi.org/10.1016/0378-1119(89)90523-4
353. Iacovoni JS, Russell P, Gaits F (1999) A new inducible protein expression system in fission yeast based on the glucose-repressed inv1 promoter. Gene 232:53–58. https://doi.org/10.1016/S0378-1119(99)00116-X

354. Watt S, Mata J, Lopez-Maury L et al (2008) urg1: a uracil-regulatable promoter system for fission yeast with short induction and repression times. PLoS One 3:e1428. https://doi.org/10.1371/journal.pone.0001428
355. Bellemare DR, Sanschagrin M, Beaudoin J, Labbe S (2001) A novel copper-regulated promoter system for expression of heterologous proteins in *Schizosaccharomyces pombe*. Gene 273:191–198
356. Giga-Hama Y (1997) Secretion of human interleukin-6 using the P-factor secretion ignal in *Schizosaccharomyces pombe*. In: Giga-Hama Y, Kumagai H (eds) Foreign gene expression in fission yeast: *Schizosaccharomyces pombe*. Springer Berlin Heidelberg, Berlin, Heidelberg, pp 159–178
357. Braspenning J, Meschede W, Marchini A et al (1998) Secretion of heterologous proteins from *Schizosaccharomyces pombe* using the homologous leader sequence of pho1+ acid phosphatase. Biochem Biophys Res Commun 245:166–171. https://doi.org/10.1006/bbrc.1998.8402
358. Kjaerulff S, Jensen MR (2005) Comparison of different signal peptides for secretion of heterologous proteins in fission yeast. Biochem Biophys Res Commun 336:974–982. https://doi.org/10.1016/j.bbrc.2005.08.195
359. Kjaerulff S, Muller S, Jensen MR (2005) Alternative protein secretion: the Mam1 ABC transporter supports secretion of M-factor linked GFP in fission yeast. Biochem Biophys Res Commun 338:1853–1859. https://doi.org/10.1016/j.bbrc.2005.10.156
360. Reed G, Nagodawithana TW (1991) Yeast technology. Van Nostrand Reinhold, New York
361. Johnson EA, Echavarri-erasun C (2011) Yeast biotechnology. In: The yeasts, a taxonomic study. Elsevier B.V, Amsterdam, pp 21–44
362. Spampinato C, Leonardi D (2013) Candida infections, causes, targets, and resistance mechanisms: traditional and alternative antifungal agents. 2013
363. Barnett JA (2004) A history of research on yeasts 8: taxonomy. Yeast 21:1141–1193. https://doi.org/10.1002/yea.1154
364. Boze H, Moutin G, Galzy P (1992) Production of food and fodder yeasts. Crit Rev Biotechnol 12:65–86
365. Inskeep GC, Wiley AJ, Holdberry JM, Hughes LP (1951) Food yeast from sulfite liquor. Ind Eng Chem 43:1702–1711. https://doi.org/10.1021/ie50500a013
366. Kurtzman CP, Johnson CJ, Smiley MJ (1979) Determination of conspecificity of *Candida utilis* and *Hansenula jadini* through DNA reassociation. Mycologia 71:844–847
367. Buerth C, Tielker D, Ernst JF (2016) *Candida utilis* and *Cyberlindnera* (*Pichia*) *jadinii*: yeast relatives with expanding applications. Appl Microbiol Biotechnol 100:6981–6990. https://doi.org/10.1007/s00253-016-7700-8
368. Bekatorou A, Psarianos C, Koutinas AA (2006) Production of food grade yeasts. Food Technol Biotechnol 44:407–415
369. Buerth C, Heilmann CJ, Klis FM et al (2011) Growth-dependent secretome of *Candida utilis*. Microbiology 157:2493–2503. https://doi.org/10.1099/mic.0.049320-0
370. Tamakawa H, Tomita Y, Yokoyama A et al (2013) Metabolomic and transcriptomic analysis for rate-limiting metabolic steps in xylose utilization by recombinant *Candida utilis*. Biosci Biotechnol Biochem 77:1441–1448. https://doi.org/10.1271/bbb.130093
371. Dworschack RG, Wickerham LJ (1961) Production of extracellular and total invertase by *Candida utilis*, *Saccharomyces cerevisiae*, and other yeasts. Appl Environ Microbiol 9:291–294
372. Belcarz A, Ginalska G, Lobarzewski J, Penel C (2002) The novel non-glycosylated invertase from *Candida utilis* (the properties and the conditions of production and purification). Biochim Biophys Acta 1594:40–53
373. Wei G, Li Y, Du G, Chen J (2003) Application of a two-stage temperature control strategy for enhanced glutathione production in the batch fermentation by *Candida utilis*. Biotechnol Lett 25:887–890
374. Liang G, Liao X, Du G, Chen J (2008) Elevated glutathione production by adding precursor amino acids coupled with ATP in high cell density cultivation of *Candida utilis*. J Appl Microbiol 105:1432–1440. https://doi.org/10.1111/j.1365-2672.2008.03892.x
375. Kogan G, Sandula J, SImkovicova V (1993) Glucomannan from *Candida utilis*—structural investigation. Folia Microbiol (Praha) 38:219–224
376. Ruszova E, Pavek S, Hajkova V et al (2008) Photoprotective effects of glucomannan isolated from *Candida utilis*. Carbohydr Res 343:501–511. https://doi.org/10.1016/j.carres.2007.11.010
377. Fujino S, Akiyama D, Akaboshi S et al (2006) Purification and characterization of phospholipase B from *Candida utilis*. Biosci

Biotechnol Biochem 70:377–386. https://doi.org/10.1271/bbb.70.377

378. Hong Y-R, Chen Y-L, Farh L et al (2006) Recombinant *Candida utilis* for the production of biotin. Appl Microbiol Biotechnol 71:211–221. https://doi.org/10.1007/s00253-005-0133-4
379. Ikushima S, Fujii T, Kobayashi O (2009) Efficient gene disruption in the high-ploidy yeast *Candida utilis* using the Cre-loxP system. Biosci Biotechnol Biochem 73:879–884. https://doi.org/10.1271/bbb.80799
380. Kondo K, Saito T, Kajiwara S et al (1995) A transformation system for the yeast *Candida utilis*: use of a modified endogenous ribosomal protein gene as a drug-resistant marker and ribosomal DNA as an integration target for vector DNA. J Bacteriol 177:7171–7177. https://doi.org/10.1128/jb.177.24.7171-7177.1995
381. Kondo K, Miura Y, Sone H et al (1997) High-level expression of a sweet protein, monellin, in the food yeast *Candida utilis*. Nat Biotechnol 15:453–457. https://doi.org/10.1038/nbt0597-453
382. Wei W, Hong-Lan Y, HuiFang B et al (2010) The effective expression of xylanase gene in *Candida utilis* by 18S rDNA targeted homologous recombination in pGLR9K. Mol Biol Rep 37:2615–2620. https://doi.org/10.1007/s11033-009-9786-x
383. Miura Y, Kettoku M, Kato M et al (1999) High level production of thermostable α-amylase from *Sulfolobus solfataricus* in high-cell density culture of the food yeast *Candida utilis*. J Mol Microbiol Biotechnol 1:129–134
384. Miura Y, Kondo K, Shimada H et al (1998) Production of lycopene by the food yeast, *Candida utilis* that does not naturally synthesize carotenoid. Biotechnol Bioeng 58:306–308. https://doi.org/10.1002/(SICI)1097-0290(19980420)58:2/3<306::AID-BIT29>3.0.CO;2-8
385. Shimada H, Kondo K, Fraser PD et al (1998) Increased carotenoid production by the food yeast *Candida utilis* through metabolic engineering of the isoprenoid pathway. Appl Environ Microbiol 64:2676–2680
386. Tamakawa H, Ikushima S, Yoshida S (2011) Ethanol production from xylose by a recombinant *Candida utilis* strain expressing protein- engineered xylose reductase and xylitol dehydrogenase ethanol production from xylose by a recombinant *Candida utilis*. Biosci Biotechnol Biochem 75:1994–2000. https://doi.org/10.1271/bbb.110426
387. Tamakawa H, Ikushima S, Yoshida S (2012) Efficient production of L-lactic acid from xylose by a recombinant *Candida utilis* strain. J Biosci Bioeng 113:73–75. https://doi.org/10.1016/j.jbiosc.2011.09.002
388. Ikushima S, Fujii T, Kobayashi O et al (2009) Genetic engineering of *Candida utilis* yeast for efficient production of L-lactic acid. Biosci Biotechnol Biochem 73:1818–1824. https://doi.org/10.1271/bbb.90186
389. Kunigo M, Buerth C, Tielker D, Ernst JF (2013) Heterologous protein secretion by *Candida utilis*. Appl Microbiol Biotechnol 97:7357–7368. https://doi.org/10.1007/s00253-013-4890-1
390. Kunigo M, Buerth C, Ernst JF (2015) Secreted xylanase XynA mediates utilization of xylan as sole carbon source in *Candida utilis*. Appl Microbiol Biotechnol 99:8055–8064. https://doi.org/10.1007/s00253-015-6703-1
391. Tamakawa H, Mita T, Yokoyama A et al (2013) Metabolic engineering of *Candida utilis* for isopropanol production. Appl Microbiol Biotechnol 97:6231–6239. https://doi.org/10.1007/s00253-013-4964-0
392. Ikushima S, Minato T, Kondo K (2009) Identification and application of novel autonomously replicating sequences (ARSs) for promoter-cloning and co-transformation in *Candida utilis*. Biosci Biotechnol Biochem 73:152–159. https://doi.org/10.1271/bbb.80568
393. Bonkova H, Osadska M, Krahulec J et al (2014) Upstream regulatory regions controlling the expression of the *Candida utilis* maltase gene. J Biotechnol 189:136–142. https://doi.org/10.1016/j.jbiotec.2014.09.006
394. Iwakiri R, Noda Y, Adachi H et al (2006) Isolation and characterization of promoters suitable for a multidrug-resistant marker *CuYAP1* in the yeast *Candida utilis*. Yeast 23:23–34. https://doi.org/10.1002/yea.1335
395. Banat IM, Nigam P, Marchant R (1992) Isolation of thermotolerant, fermentative yeasts growing at 52°C and producing ethanol at 45°C and 50°C. World J Microbiol Biotechnol 8:259–263. https://doi.org/10.1007/BF01201874
396. Jeong H, Lee DH, Kim SH et al (2012) Genome sequence of the thermotolerant yeast *Kluyveromyces marxianus* var. *marxianus* KCTC 17555. Eukaryot Cell 11:1584–1585. https://doi.org/10.1128/EC.00260-12

397. Belloch C, Barrio E, García MD, Querol A (1998) Inter- and intraspecific chromosome pattern variation in the yeast genus *Kluyveromyces*. Yeast 14:1341–1354. https://doi.org/10.1002/(SICI)1097-0061(199811)14:15<1341::AID-YEA328>3.0.CO;2-U

398. Lane MM, Burke N, Karreman R et al (2011) Physiological and metabolic diversity in the yeast *Kluyveromyces marxianus*. Antonie Van Leeuwenhoek 100:507–519. https://doi.org/10.1007/s10482-011-9606-x

399. Rouwenhorst RJ, Visser LE, van der Baan AA et al (1988) Production, distribution, and kinetic properties of inulinase in continuous cultures of *Kluyveromyces marxianus* CBS 6556. Appl Environ Microbiol 54:1131–1137

400. Martins DBG, de Souza CG Jr, Simões DA, de Morais MA Jr (2002) The β-galactosidase activity in *Kluyveromyces marxianus* CBS6556 decreases by high concentrations of galactose. Curr Microbiol 44:379–382. https://doi.org/10.1007/s00284-001-0052-2

401. Rocha SN, Abrahão-Neto J, Cerdán ME et al (2011) Heterologous expression of a thermophilic esterase in *Kluyveromyces* yeasts. Appl Microbiol Biotechnol 89:375–385. https://doi.org/10.1007/s00253-010-2869-8

402. Lee K-S, Kim J-S, Heo P et al (2013) Characterization of *Saccharomyces cerevisiae* promoters for heterologous gene expression in *Kluyveromyces marxianus*. Appl Microbiol Biotechnol 97:2029–2041. https://doi.org/10.1007/s00253-012-4306-7

403. Bragança CRS, Colombo LT, Roberti AS et al (2015) Construction of recombinant *Kluyveromyces marxianus* UFV-3 to express dengue virus type 1 nonstructural protein 1 (NS1). Appl Microbiol Biotechnol 99:1191–1203. https://doi.org/10.1007/s00253-014-5963-5

404. Yang C, Hu S, Zhu S et al (2015) Characterizing yeast promoters used in *Kluyveromyces marxianus*. World J Microbiol Biotechnol 31:1641–1646. https://doi.org/10.1007/s11274-015-1899-x

405. Nonklang S, Abdel-Banat BMA, Cha-aim K et al (2008) High-temperature ethanol fermentation and transformation with linear DNA in the thermotolerant yeast *Kluyveromyces marxianus* DMKU3-1042. Appl Environ Microbiol 74:7514–7521. https://doi.org/10.1128/AEM.01854-08

406. Yamamoto H, Shima T, Yamaguchi M et al (2015) The thermotolerant yeast *Kluyveromyces marxianus* is a useful organism for structural and biochemical studies of autophagy. J Biol Chem 290:29506–29518. https://doi.org/10.1074/jbc.M115.684233

407. de Souza CG Jr, Ledingham WM, de Morais MA Jr (2001) Utilisation of cheese whey as an alternative growth medium for recombinant strains of *Kluyveromyces marxianus*. Biotechnol Lett 23:1413–1416. https://doi.org/10.1023/A:1011617914709

408. Bartkevičiūtė D, Šiekštelė R, Sasnauskas K (2000) Heterologous expression of the *Kluyveromyces marxianus* endopolygalacturonase gene (*EPG1*) using versatile autonomously replicating vector for a wide range of host. Enzym Microb Technol 26:653–656. https://doi.org/10.1016/S0141-0229(00)00155-1

409. Bergkamp RJM, Bootsman TC, Toschka HY et al (1993) Expression of an α-galactosidase gene under control of the homologous inulinase promoter in *Kluyveromyces marxianus*. Appl Microbiol Biotechnol 40:309–317. https://doi.org/10.1007/BF00170386

410. Raimondi S, Uccelletti D, Matteuzzi D et al (2008) Characterization of the superoxide dismutase *SOD1* gene of *Kluyveromyces marxianus* L3 and improved production of SOD activity. Appl Microbiol Biotechnol 77:1269–1277. https://doi.org/10.1007/s00253-007-1270-8

411. Raimondi S, Uccelletti D, Amaretti A et al (2010) Secretion of *Kluyveromyces lactis* Cu/Zn SOD: strategies for enhanced production. Appl Microbiol Biotechnol 86:871–878. https://doi.org/10.1007/s00253-009-2353-5

412. Raimondi S, Zanni E, Amaretti A et al (2013) Thermal adaptability of *Kluyveromyces marxianus* in recombinant protein production. Microb Cell Factories 12:34. https://doi.org/10.1186/1475-2859-12-34

413. Iborra F (1993) High efficiency transformation of *Kluyveromyces marxianus* by a replicative plasmid. Curr Genet 24:181–183. https://doi.org/10.1007/BF00324685

414. Yanase S, Hasunuma T, Yamada R et al (2010) Direct ethanol production from cellulosic materials at high temperature using the thermotolerant yeast *Kluyveromyces marxianus* displaying cellulolytic enzymes. Appl Microbiol Biotechnol 88:381–388. https://doi.org/10.1007/s00253-010-2784-z

415. Goshima T, Negi K, Tsuji M et al (2013) Ethanol fermentation from xylose by metabolically engineered strains of *Kluyveromyces marxianus*. J Biosci Bioeng 116:551–554. https://doi.org/10.1016/j.jbiosc.2013.05.010

416. Lee JW, In JH, Park J-B et al (2017) Co-expression of two heterologous lactate dehydrogenases genes in *Kluyveromyces marxianus* for L-lactic acid production. J Biotechnol 241:81–86. https://doi.org/10.1016/j.jbiotec.2016.11.015
417. Heo P, Yang T-J, Chung S-C et al (2013) Simultaneous integration of multiple genes into the *Kluyveromyces marxianus* chromosome. J Biotechnol 167:323–325. https://doi.org/10.1016/j.jbiotec.2013.06.020
418. Cheon Y, Kim J-S, Park J-B et al (2014) A biosynthetic pathway for hexanoic acid production in *Kluyveromyces marxianus*. J Biotechnol 182–183:30–36. https://doi.org/10.1016/j.jbiotec.2014.04.010
419. Hoshida H, Murakami N, Suzuki A et al (2014) Non-homologous end joining-mediated functional marker selection for DNA cloning in the yeast *Kluyveromyces marxianus*. Yeast 31:29–46. https://doi.org/10.1002/yea.2993
420. Chang J-J, Ho C-Y, Ho F-J et al (2012) PGASO: a synthetic biology tool for engineering a cellulolytic yeast. Biotechnol Biofuels 5:53. https://doi.org/10.1186/1754-6834-5-53
421. Yuan W, Zhao X, Chen L, Bai F (2013) Improved ethanol production in Jerusalem artichoke tubers by overexpression of inulinase gene in *Kluyveromyces marxianus*. Biotechnol Bioprocess Eng 18:721–727. https://doi.org/10.1007/s12257-013-0026-9
422. Hong J, Wang Y, Kumagai H, Tamaki H (2007) Construction of thermotolerant yeast expressing thermostable cellulase genes. J Biotechnol 130:114–123. https://doi.org/10.1016/j.jbiotec.2007.03.008
423. Theron CW, Labuschagné M, Gudiminchi R et al (2014) A broad-range yeast expression system reveals *Arxula adeninivorans* expressing a fungal self-sufficient cytochrome P450 monooxygenase as an excellent whole-cell biocatalyst. FEMS Yeast Res 14:556–566. https://doi.org/10.1111/1567-1364.12142
424. Zhou H-X, Xin F-H, Chi Z et al (2014) Inulinase production by the yeast *Kluyveromyces marxianus* with the disrupted *MIG1* gene and the over-expressed inulinase gene. Process Biochem 49:1867–1874. https://doi.org/10.1016/j.procbio.2014.08.001
425. Juretzek T, Le Dall M-T, Mauersberger S et al (2001) Vectors for gene expression and amplification in the yeast *Yarrowia lipolytica*. Yeast 18:97–113. https://doi.org/10.1002/1097-0061(20010130)18:2<97::AID-YEA652>3.0.CO;2-U
426. Fonseca GG, Heinzle E, Wittmann C, Gombert AK (2008) The yeast *Kluyveromyces marxianus* and its biotechnological potential. Appl Microbiol Biotechnol 79:339–354. https://doi.org/10.1007/s00253-008-1458-6
427. Zhang B, Li L, Zhang J et al (2013) Improving ethanol and xylitol fermentation at elevated temperature through substitution of xylose reductase in *Kluyveromyces marxianus*. J Ind Microbiol Biotechnol 40:305–316. https://doi.org/10.1007/s10295-013-1230-5
428. Zhang J, Zhang B, Wang D et al (2014) Xylitol production at high temperature by engineered *Kluyveromyces marxianus*. Bioresour Technol 152:192–201. https://doi.org/10.1016/j.biortech.2013.10.109
429. Zhang J, Zhang B, Wang D et al (2015) Rapid ethanol production at elevated temperatures by engineered thermotolerant *Kluyveromyces marxianus* via the NADP(H)-preferring xylose reductase-xylitol dehydrogenase pathway. Metab Eng 31:140–152. https://doi.org/10.1016/j.ymben.2015.07.008
430. Matsuzaki C, Nakagawa A, Koyanagi T et al (2012) *Kluyveromyces marxianus*-based platform for direct ethanol fermentation and recovery from cellulosic materials under air-ventilated conditions. J Biosci Bioeng 113:604–607. https://doi.org/10.1016/j.jbiosc.2011.12.007
431. Kim HE, Qin R, Chae KS (2005) Increased production of exoinulinase in *Saccharomyces cerevisiae* by expressing the *Kluyveromyces marxianus INU1* gene under the control of the *INU1* promoter. J Microbiol Biotechnol 15:447–450
432. Almeida C, Queirós O, Wheals A et al (2003) Acquisition of flocculation phenotype by *Kluyveromyces marxianus* when overexpressing *GAP1* gene encoding an isoform of glyceraldehyde-3-phosphate dehydrogenase. J Microbiol Methods 55:433–440. https://doi.org/10.1016/S0167-7012(03)00189-1
433. Chang J-J, Ho F-J, Ho C-Y et al (2013) Assembling a cellulase cocktail and a cellodextrin transporter into a yeast host for CBP ethanol production. Biotechnol Biofuels 6:19. https://doi.org/10.1186/1754-6834-6-19
434. Chen H-L, Chen Y-C, Lu M-YJ et al (2012) A highly efficient β-glucosidase from the buffalo rumen fungus *Neocallimastix*

patriciarum W5. Biotechnol Biofuels 5:24. https://doi.org/10.1186/1754-6834-5-24

435. Thomas DS, Davenport RR (1985) *Zygosaccharomyces bailii*—a profile of characteristics and spoilage activities. Food Microbiol 2:157–169. https://doi.org/10.1016/S0740-0020(85)80008-3
436. Cole MB, Keenan MH (1986) Synergistic effects of weak-acid preservatives and pH on the growth of *Zygosaccharomyces bailii*. Yeast 2:93–100. https://doi.org/10.1002/yea.320020204
437. Praphailong W, Fleet GH (1997) The effect of pH, sodium chloride, sucrose, sorbate and benzoate on the growth of food spoilage yeasts. Food Microbiol 14:459–468. https://doi.org/10.1006/fmic.1997.0106
438. Dato L, Branduardi P, Passolunghi S et al (2010) Advances in molecular tools for the use of *Zygosaccharomyces bailii* as host for biotechnological productions and construction of the first auxotrophic mutant. FEMS Yeast Res 10:894–908. https://doi.org/10.1111/j.1567-1364.2010.00668.x
439. Zuehlke JM, Petrova B, Edwards CG (2013) Advances in the control of wine spoilage by *Zygosaccharomyces* and *Dekkera/Brettanomyces*. Annu Rev Food Sci Technol 4:57–78. https://doi.org/10.1146/annurev-food-030212-182533
440. Buchta V, Sláviková E, Vadkartiová R et al (1996) *Zygosaccharomyces bailii* as a potential spoiler of mustard. Food Microbiol 13:133–135. https://doi.org/10.1006/fmic.1996.0017
441. Toh-E A, Araki H, Utatsu I, Oshima Y (1984) Plasmids resembling 2-μm DNA in the osmotolerant yeasts *Saccharomyces bailii* and *Saccharomyces bisporus*. Microbiology 130:2527–2534
442. Branduardi P, Valli M, Brambilla L et al (2004) The yeast *Zygosaccharomyces bailii*: a new host for heterologous protein production, secretion and for metabolic engineering applications. FEMS Yeast Res 4:493–504
443. Vigentini I, Brambilla L, Branduardi P et al (2005) Heterologous protein production in *Zygosaccharomyces bailii*: physiological effects and fermentative strategies. FEMS Yeast Res 5:647–652. https://doi.org/10.1016/j.femsyr.2004.11.006
444. Paciello L, Landi C, De Alteriis E, Parascandola P (2012) Mathematical modeling as a tool to describe and optimize heterologous protein production by yeast cells in aerated fed-batch reactor. Chem Eng Trans 27:79–84. https://doi.org/10.3303/CET1227014
445. Passolunghi S, Riboldi L, Dato L et al (2010) Cloning of the *Zygosaccharomyces bailii GAS1* homologue and effect of cell wall engineering on protein secretory phenotype. Microb Cell Factories 9:7. https://doi.org/10.1186/1475-2859-9-7
446. Gallwitz D, Seidel R (1980) Molecular cloning of the actin gene from yeast *Saccharomyces cerevisiae*. Nucleic Acids Res 8:1043–1059
447. Williamson VM, Bennetzen J, Young ET et al (1980) Isolation of the structural gene for alcohol dehydrogenase by genetic complementation in yeast. Nature 283:214–216
448. Bennetzens JL, Hall BD (1982) The primary structure of the *Saccharomyces cerevisiae* gene for alcohol dehydrogenase I. J Biol Chem 257:3018–3025
449. Denis CL, Ferguson J, Young ET (1983) mRNA levels for the fermentative alcohol dehydrogenase of *Saccharomyces cerevisiae* decrease upon growth on a nonfermentable carbon source. J Biol Chem 258:1165–1171
450. Guarente L, Lalonde B, Gifford P et al (1984) Distinctly regulated tandem upstream activation sites mediate catabolite repression of the CYC1 gene of *S. cerevisiae*. Cell 36:503–511. https://doi.org/10.1016/0092-8674(84)90243-5
451. Holland MJ, Holland JP, Thillg GP, Jackson KA (1981) The primary structures of two yeast enolase genes. J Biol Chem 256:1385–1395
452. McAlisters L, Holland MJ (1985) Differential expression of the three yeast glyceraldehyde-3-phosphate dehydrogenase genes. J Biol Chem 260:15019–15027
453. McAlisters L, Holland MJ (1985) Isolation and characterization of yeast strains carrying mutations in the glyceraldehyde-3-phosphate dehydrogenase genes. J Biol Chem 260:15013–15018
454. Reifenberger E, Freidel K, Ciriacy M (1995) Identification of novel HXT genes in *Saccharomyces cerevisiae* reveals the impact of individual hexose glycolytic flux. Mol Microbiol 16:157–167. https://doi.org/10.1111/j.1365-2958.1995.tb02400.x
455. Singh A, Chen EY, Lugovoy JM et al (1983) *Saccharomyces cerevisiae* contains two discrete genes coding for the alpha-factor pheromone. Nucleic Acids Res 11:4049–4063
456. Holland MJ, Holland JP (1978) Isolation and identification of yeast messenger ribonucleic acids coding for enolase, glyceraldehyde-3-

phosphate dehydrogenase, and phosphoglycerate kinase. Biochemistry 17:4900–4907

457. Dobson MJ, Tuite MF, Roberts NA et al (1982) Conservation of high efficiency promoter sequences in *Saccharomyces cerevisiae*. Nucleic Acids Res 10:2625–2637
458. Hitzeman RA, Hagie FE, Hayflick JS et al (1982) The primary structure of the *Saccharomyces cerevisiae* gene for 3-phosphoglycerate kinase. Nucleic Acids Res 10:7791–7808
459. Ogden JE, Stanway C, Kim S et al (1986) Efficient expression of the *Saccharomyces cerevisiae* PGK gene depends on an upstream activation sequence but does not require TATA sequences. Mol Cell Biol 6:4335–4343
460. Nishizawa M, Araki R, Teranishi Y (1989) Identification of an upstream activating sequence and an upstream repressible sequence of the pyruvate kinase gene of the yeast *Saccharomyces cerevisiae*. Mol Cell Biol 9:442–451
461. Schirmaier F, Philippsen P (1984) Identification of two genes coding for the translation elongation factor EF-la of *S. cerevisiae*. EMBO J 3:3311–3315
462. Nagashima K, Kasai M, Nagata S, Kaziro Y (1986) Structure of the two genes for coding polypeptide chain elongation factor 1 (EF-1) from *Saccharomyces cerevisiae*. Gene 45:265–273. https://doi.org/10.1016/0378-1119(86)90024-7
463. Kim S, Mellor J, Kingsman AJ, Kingsman SM (1986) Multiple control elements in the *TRP1* promoter of *Saccharomyces cerevisiae*. Mol Cell Biol 6:4251–4258
464. Beier DR, Young ET (1982) Characterization of a regulatory region upstream of the *ADR2* locus of *S. cerevisiae*. Nature 300:724–728
465. Fogel S, Welch JW (1982) Tandem gene amplification mediates copper resistance in yeast. Proc Natl Acad Sci U S A 79:5342–5346
466. Butt TR, Sternberg EJ, Gormant JA et al (1984) Copper metallothionein of yeast, structure of the gene, and regulation of expression biochemistry. Proc Natl Acad Sci U S A 81:3332–3336
467. Karin M, Najariant R, Haslinger A et al (1984) Primary structure and transcription of an amplified genetic locus: the CUP1 locus of yeast. Proc Natl Acad Sci U S A 81:337–341
468. Johnston M, Davis RW (1984) Sequences that regulate the divergent GAL1-GAL10 promoter in *Saccharomyces cerevisiae*. Mol Cell Biol 4:1440–1448
469. Fang F, Salmon K, Shen MWY et al (2011) A vector set for systematic metabolic engineering in *Saccharomyces cerevisiae*. Yeast 28:123–136
470. Kerjan P, Cherest H, Surdin-Kerjan Y (1986) Nucleotide sequence of the *Saccharomyces cerevisiae* MET25 gene. Nucleic Acids Res 14:7861–7871
471. Nakao J, Miyanohara A, Toh-e A, Matsubara K (1986) *Saccharomyces cerevisiae* PH05 promoter region: location and function of the upstream activation site. Mol Cell Biol 6:2613–2623
472. Arima K, Oshima T, Kubota I et al (1983) The nucleotide sequence of the yeast PHO5-gene: a putative precursor of repressible acid phosphatase contains a signal peptide. Nucleic Acids Res 11:1657–1672. https://doi.org/10.1093/nar/11.6.1657
473. Kurjan J, Herskowitz I (1982) Structure of a yeast pheromone gene (MFα): a putative α-factor precursor contains four tandem copies of mature α-factor. Cell 30:933–943. https://doi.org/10.1016/0092-8674(82)90298-7
474. Taussig R, Carlson M (1983) Nucleotide sequence of the yeast *SUC2* gene for invertase. Nucleic Acids Res 11:1943–1954. https://doi.org/10.1093/nar/11.6.1943
475. Hofmann KJ, Schultz LD (1991) Mutations of the α-galactosidase signal peptide which greatly enhance secretion of heterologous proteins by yeast. Gene 101:105–111. https://doi.org/10.1016/0378-1119(91)90230-9
476. Baldari C, Murray JAH, Ghiara P et al (1987) A novel leader peptide which allows efficient secretion of a fragment of human interleukin 1β in *Saccharomyces cerevisiae*. EMBO J 6:229–234. https://doi.org/10.1002/j.1460-2075.1987.tb04743.x
477. Liang S, Zou C, Lin Y et al (2013) Identification and characterization of P *GCW14*: a novel, strong constitutive promoter of *Pichia pastoris*. Biotechnol Lett 35:1865–1871. https://doi.org/10.1007/s10529-013-1265-8
478. de Almeida JRM, de Moraes LMP, Torres FAG (2005) Molecular characterization of the 3-phosphoglycerate kinase gene (*PGK1*) from the methylotrophic yeast *Pichia pastoris*. Yeast 22:725–737. https://doi.org/10.1002/yea.1243
479. Koutz P, Davis GR, Stillman C et al (1989) Structural comparison of the *Pichia pastoris* alcohol oxidase genes. Yeast 5:167–177. https://doi.org/10.1002/yea.320050306

480. Ellis SB, Brust PF, Koutz PJ et al (1985) Isolation of alcohol oxidase and two other methanol regulatable genes from the yeast *Pichia pastoris*. Mol Cell Biol 5:1111–1121
481. Menendez J, Valdes I, Cabrera N (2003) The *ICL1* gene of *Pichia pastoris*, transcriptional regulation and use of its promoter. Yeast 20:1097–1108. https://doi.org/10.1002/yea.1028
482. Delic M, Mattanovich D, Gasser B (2013) Repressible promoters - a novel tool to generate conditional mutants in *Pichia pastoris*. Microb Cell Factories 12:6. https://doi.org/10.1186/1475-2859-12-6
483. Cereghino JL, Cregg JM (2000) Heterologous protein expression in the methylotrophic yeast *Pichia pastoris*. FEMS Microbiol Rev 24:45–66
484. Oka C, Tanaka M, Muraki M et al (1999) Human lysozyme secretion increased by alpha-factor pro-sequence in *Pichia pastoris*. Biosci Biotechnol Biochem 63:1977–1983. https://doi.org/10.1271/bbb.63.1977
485. He Z, Huang Y, Qin Y et al (2012) Comparison of alpha-factor preprosequence and a classical mammalian signal peptide for secretion of recombinant xylanase xynB from yeast *Pichia pastoris*. J Microbiol Biotechnol 22:479–483
486. Whittaker MM, Whittaker JW (2000) Expression of recombinant galactose oxidase by *Pichia pastoris*. Protein Expr Purif 20:105–111. https://doi.org/10.1006/prep.2000.1287
487. Kottmeier K, Ostermann K, Bley T, Rodel G (2011) Hydrophobin signal sequence mediates efficient secretion of recombinant proteins in *Pichia pastoris*. Appl Microbiol Biotechnol 91:133–141. https://doi.org/10.1007/s00253-011-3246-y
488. Xiong R, Chen J, Chen J (2008) Secreted expression of human lysozyme in the yeast *Pichia pastoris* under the direction of the signal peptide from human serum albumin. Biotechnol Appl Biochem 51:129–134. https://doi.org/10.1042/BA20070205
489. Eiden-Plach A, Zagorc T, Heintel T et al (2004) Viral preprotoxin signal sequence allows efficient secretion of green fluorescent protein by *Candida glabrata*, *Pichia pastoris*, *Saccharomyces cerevisiae*, and *Schizosaccharomyces pombe*. Appl Environ Microbiol 70:961–966. https://doi.org/10.1128/AEM.70.2.961-966.2004
490. Kato S, Ishibashi M, Tatsuda D et al (2001) Efficient expression, purification and characterization of mouse salivary alpha-amylase secreted from methylotrophic yeast, *Pichia pastoris*. Yeast 18:643–655. https://doi.org/10.1002/yea.714
491. Raemaekers RJ, de Muro L, Gatehouse JA, Fordham-Skelton AP (1999) Functional phytohemagglutinin (PHA) and *Galanthus nivalis* agglutinin (GNA) expressed in *Pichia pastoris* correct N-terminal processing and secretion of heterologous proteins expressed using the PHA-E signal peptide. Eur J Biochem 265:394–403
492. Inokuma K, Bamba T, Ishii J et al (2016) Enhanced cell-surface display and secretory production of cellulolytic enzymes with *Saccharomyces cerevisiae* Sed1 signal peptide. Biotechnol Bioeng 113:2358–2366. https://doi.org/10.1002/bit.26008
493. Crawford K, Zaror I, Bishop RJ, Innis MA (1997) Pichia secretory leader for protein expression. WO1997012044 A2
494. Berardi E, Gambini A, Bellu AR (2003) *ALG2*, the *Hansenula polymorpha* isocitrate lyase gene. Yeast 20:803–811. https://doi.org/10.1002/yea.1002
495. Vanoni M, Sollitti P, Goldenthal M, Marmur J (1989) Structure and regulation of the multigene family controlling maltose fermentation in budding yeast. Prog Nucleic Acid Res Mol Biol 37:281–322
496. Ledeboer AM, Edens L, Maat J et al (1985) Molecular cloning and characterization of a gene coding for methanol oxidase in *Hansenula polymorpha*. Nucleic Acids Res 13:3063–3082. https://doi.org/10.1093/nar/13.9.3063
497. Hollenberg CP, Janowiez ZA (1989) DNA-molecules coding for FMDH control regions and structured gene for a protein having FMDH-activity and their uses. EP 0299108 A1
498. Janowicz ZA, Eckart MR, Drewke C et al (1985) Cloning and characterization of the DAS gene encoding the major methanol assimilatory enzyme from the methylotrophic yeast *Hansenula polymorpha*. Nucleic Acids Res 13:3043–3062. https://doi.org/10.1093/nar/13.9.3043
499. Ávila J, González C, Brito N, Siverio JM (1998) Clustering of the *YNA1* gene encoding a Zn(II)2Cys6 transcriptional factor in the yeast *Hansenula polymorpha* with the nitrate assimilation genes *YNT1*, *YNI1* and *YNR1*, and its involvement in their transcriptional activation. Biochem J 335:647–652
500. Hansen H, Didion T, Thiemann A et al (1992) Targeting sequences of the two major peroxisomal proteins in the

methylotrophic yeast *Hansenula polymorpha*. Mol Gen Genomics 235:269–278

501. Faber KN, Haima P, Gietl C et al (1994) The methylotrophic yeast *Hansenula polymorpha* contains an inducible import pathway for peroxisomal matrix proteins with an N-terminal targeting signal (PTS2 proteins). Proc Natl Acad Sci U S A 91:12985–12989
502. Rösel H, Kunze G (1995) Cloning and characterization of a *TEF* gene for elongation factor 1α from the yeast *Arxula adeninivorans*. Curr Genet 28:360–366. https://doi.org/10.1007/BF00326434
503. Böer E, Wartmann T, Schmidt S et al (2005) Characterization of the *AXDH* gene and the encoded xylitol dehydrogenase from the dimorphic yeast *Arxula adeninivorans*. Antonie Van Leeuwenhoek 87:233–243. https://doi.org/10.1007/s10482-004-3832-4
504. Bui DM, Kunze I, Förster S et al (1996) Cloning and expression of an *Arxula adeninivorans* glucoamylase gene in *Saccharomyces cerevisiae*. Appl Microbiol Biotechnol 44:610–619. https://doi.org/10.1007/BF00172493
505. Bianchi MM, Tizzani L, Destruelle M et al (1996) The "petite-negative" yeast *Kluyveromyces lactis* has a single gene expressing pyruvate decarboxylase activity. Mol Microbiol 19:27–36. https://doi.org/10.1046/j.1365-2958.1996.346875.x
506. Walton JD, Paquin CE, Kaneko K, Williamson VM (1986) Resistance to antimycin A in yeast by amplification of *ADH4* on a linear, 42 kb palindromic plasmid. Cell 46:857–863. https://doi.org/10.1016/0092-8674(86)90067-X
507. Ferminan E, Dominguez A (1997) The *KlPHO5* gene encoding a repressible acid phosphatase in the yeast *Kluyveromyces lactis*: cloning, sequencing and transcriptional analysis of the gene, and purification and properties of the enzyme. Microbiology 143:2615–2625. https://doi.org/10.1099/00221287-143-8-2615
508. Stark MJR, Boyd A (1986) The killer toxin of *Kluyveromyces lactis*: characterization of the toxin subunits and identification of the genes which encode them. EMBO J 5:1995–2002
509. Ongay-Larios L, Navarro-Olmos R, Kawasaki L et al (2007) *Kluyveromyces lactis* sexual pheromones. Gene structures and cellular responses to α-factor. FEMS Yeast Res 7:740–747. https://doi.org/10.1111/j.1567-1364.2007.00249.x
510. Chen X, Gao B, Shi W, Li Y (1992) Expression and secretion of human interferon alpha A in yeast *Kluyveromyces lactis*. Yi Chuan Xue Bao 19:284–288
511. Hong S-P, Seip J, Walters-Pollak D et al (2012) Engineering *Yarrowia lipolytica* to express secretory invertase with strong $FBA1_{IN}$ promoter. Yeast 29:59–72. https://doi.org/10.1002/yea.1917
512. Tai M, Stephanopoulos G (2013) Engineering the push and pull of lipid biosynthesis in oleaginous yeast *Yarrowia lipolytica* for biofuel production. Metab Eng 15:1–9. https://doi.org/10.1016/j.ymben.2012.08.007
513. Franke A, Kaczmarek F, Eisenhard M et al (1988) Expression and secretion of bovine prochymosin in *Yarrowia lipolytica*. Dev Ind Microbiol 29:43–57
514. Sassi H, Delvigne F, Kar T et al (2016) Deciphering how *LIP2* and *POX2* promoters can optimally regulate recombinant protein production in the yeast *Yarrowia lipolytica*. Microb Cell Factories 15:159. https://doi.org/10.1186/s12934-016-0558-8
515. Tharaud C, Ribet AM, Costes C, Gaillardin C (1992) Secretion of human blood coagulation factor XIIIa by the yeast *Yarrowia lipolytica*. Gene 121:111–119
516. Gasmi N, Fudalej F, Kallel H, Nicaud J-M (2011) A molecular approach to optimize hIFN alpha2b expression and secretion in *Yarrowia lipolytica*. Appl Microbiol Biotechnol 89:109–119. https://doi.org/10.1007/s00253-010-2803-0
517. Park CS, Chang CC, Kim JY et al (1997) Expression, secretion, and processing of rice alpha-amylase in the yeast *Yarrowia lipolytica*. J Biol Chem 272:6876–6881
518. Madzak C, Otterbein L, Chamkha M et al (2005) Heterologous production of a laccase from the basidiomycete *Pycnoporus cinnabarinus* in the dimorphic yeast *Yarrowia lipolytica*. FEMS Yeast Res 5:635–646. https://doi.org/10.1016/j.femsyr.2004.10.009
519. Jolivalt C, Madzak C, Brault A et al (2005) Expression of laccase IIIb from the white-rot fungus *Trametes versicolor* in the yeast *Yarrowia lipolytica* for environmental applications. Appl Microbiol Biotechnol 66:450–456. https://doi.org/10.1007/s00253-004-1717-0
520. Park CS, Chang CC, Ryu DD (2000) Expression and high-level secretion of *Trichoderma reesei* endoglucanase I in *Yarrowia lipolytica*. Appl Biochem Biotechnol 87:1–15
521. Roth R, Moodley V, van Zyl P (2009) Heterologous expression and optimized production of an *Aspergillus aculeatus* endo-1,4-beta-mannanase in *Yarrowia lipolytica*. Mol

Biotechnol 43:112–120. https://doi.org/10.1007/s12033-009-9187-3

522. Yuzbashev TV, Yuzbasheva EY, Vibornaya TV et al (2012) Production of recombinant *Rhizopus oryzae* lipase by the yeast *Yarrowia lipolytica* results in increased enzymatic thermostability. Protein Expr Purif 82:83–89. https://doi.org/10.1016/j.pep.2011.11.014
523. Kaufer NF, Simanis V, Nurse P (1985) Fission yeast *Schizosaccharomyces pombe* correctly excises a mammalian RNA transcript intervening sequence. Nature 318:78–80
524. Matsuzawa T, Tohda H, Takegawa K (2013) Ethanol-inducible gene expression using gld1 (+) promoter in the fission yeast *Schizosaccharomyces pombe*. Appl Microbiol Biotechnol 97:6835–6843. https://doi.org/10.1007/s00253-013-4812-2
525. Schweingruber ME, Edenharter E, Zurlinden A, Stockmaier KM (1992) Regulation of pho1-encoded acid phosphatase of *Schizosaccharomyces pombe* by adenine and phosphate. Curr Genet 22:289–292
526. Okada H, Sekiya T, Yokoyama K et al (1998) Efficient secretion of *Trichoderma reesei* cellobiohydrolase II in *Schizosaccharomyces pombe* and characterization of its products. Appl Microbiol Biotechnol 49:301–308
527. Okada H, Tada K, Sekiya T et al (1998) Molecular characterization and heterologous expression of the gene encoding a low-molecular-mass endoglucanase from *Trichoderma reesei* QM9414. Appl Environ Microbiol 64:555–563
528. Tokunaga M, Kawamura A, Yonekyu S et al (1993) Secretion of mouse alpha-amylase from fission yeast *Schizosaccharomyces pombe*: presence of chymostatin-sensitive protease activity in the culture medium. Yeast 9:379–387. https://doi.org/10.1002/yea.320090408
529. Sánchez Y, Moreno S, Rodríguez L (1988) Synthesis of *Saccharomyces cerevisiae* invertase by *Schizosaccharomyces pombe*. FEBS Lett 234:95–99. https://doi.org/10.1016/0014-5793(88)81311-5
530. Azam M, Kesarwani M, Natarajan K, Datta A (2001) A secretion signal is present in the *Collybia velutipes* oxalate decarboxylase gene. Biochem Biophys Res Commun 289:807–812. https://doi.org/10.1006/bbrc.2001.6049
531. Bröker M, Ragg H, Karges HE (1987) Expression of human antithrombin III in *Saccharomyces cerevisiae* and *Schizosaccharomyces pombe*. Biochim Biophys Acta 908:203–213
532. Smerdon GR, Aves SJ, Walton EF (1995) Production of human gastric lipase in the fission yeast *Schizosaccharomyces pombe*. Gene 165:313–318
533. Sambamurti K (1997) Expression and secretion of mammalian proteins in Schizosaccharomyces pombe. In: Giga-Hama Y, Kumagai H (eds) Foreign gene expression in fission yeast: *Schizosaccharomyces pombe*. Springer Berlin Heidelberg, Berlin, Heidelberg, pp 149–158
534. Kondo K, Kajiwara S, Misawa N (1996) Transformant line of *Candida utilis* yeast and expression of heterogene therewith
535. Chávez FP, Pons T, Delgado JM, Rodríguez L (1998) Cloning and sequence analysis of the gene encoding invertase (*INV1*) from the yeast *Candida utilis*. Yeast 14:1223–1232. https://doi.org/10.1002/(SICI)1097-0061(19980930)14:13<1223::AID-YEA301>3.0.CO;2-3
536. Ladrière J-M, Delcour J, Vandenhaute J (1993) Sequence of a gene coding for a cytoplasmic alcohol dehydrogenase from *Kluyveromyces marxianus* ATCC 12424. Biochim Biophys Acta 1173:99–101. https://doi.org/10.1016/0167-4781(93)90252-9
537. Shisa N, Akada R, Hoshida H, et al (2015) Novel promoter and use thereof. US 20150031103 A1
538. Fernandes PA, Sena-Esteves M, Moradas-Ferreira P (1995) Characterization of the glyceraldehyde-3-phosphate dehydrogenase gene family from *Kluyveromyces marxianus*—polymerase chain reaction-single-strand conformation polymorphism as a tool for the study of multigenic families. Yeast 11:725–733. https://doi.org/10.1002/yea.320110804
539. Zhang G, Lu M, Wang J et al (2017) Identification of hexose kinase genes in *Kluyveromyces marxianus* and thermo-tolerant one step producing glucose-free fructose strain construction. Sci Rep 7:45104. https://doi.org/10.1038/srep45104
540. Laloux O, Cassart J-P, Delcour J et al (1991) Cloning and sequencing of the inulinase gene of *Kluyveromyces marxianus* var. *marxianus* ATCC 12424. FEBS Lett 289:64–68. https://doi.org/10.1016/0014-5793(91)80909-M

Chapter 2

Detection and Elimination of Cellular Bottlenecks in Protein-Producing Yeasts

Richard J. Zahrl, Brigitte Gasser, Diethard Mattanovich, and Pau Ferrer

Abstract

Yeasts are efficient cell factories and are commonly used for the production of recombinant proteins for biopharmaceutical and industrial purposes. For such products high levels of correctly folded proteins are needed, which sometimes requires improvement and engineering of the expression system. The article summarizes major breakthroughs that led to the efficient use of yeasts as production platforms and reviews bottlenecks occurring during protein production. Special focus is given to the metabolic impact of protein production. Furthermore, strategies that were shown to enhance secretion of recombinant proteins in different yeast species are presented.

Key words Yeasts, Protein production, Secretion, Chaperones, Protein degradation, Metabolism, Promoters

1 Introduction

Yeasts are efficient hosts for the production of recombinant proteins. Since the first product was approved in the 1980s, the repertoire of yeasts used as production hosts has expanded. Today, in addition to *Saccharomyces cerevisiae*, research and industry also apply other species including the methylotrophic yeasts *Pichia pastoris* (syn. *Komagataella* spp.) and *Hansenula polymorpha* (syn. *Ogataea polymorpha* and *O. parapolymorpha*), the oleaginous yeast *Yarrowia lipolytica*, *Kluyveromyces lactis*, and fission yeast *Schizosaccharomyces pombe*.

Protein production is regulated at several cellular levels, starting with transcription of the gene of interest and subsequent translation of the mRNA. After translation, posttranslational steps, most importantly protein folding and secretion, are taking place. Furthermore a sufficient supply of metabolic precursors and energy is required. Over the years several aspects of protein production have been improved, which are presented in the following.

Brigitte Gasser and Diethard Mattanovich (eds.), *Recombinant Protein Production in Yeast*, Methods in Molecular Biology, vol. 1923, https://doi.org/10.1007/978-1-4939-9024-5_2, © Springer Science+Business Media, LLC, part of Springer Nature 2019

2 Early Bottlenecks in Yeast Protein Production: Transcription and Translation

2.1 Transcription

With the development of recombinant protein production in general, and in bakers' yeast [1], it has been recognized that transcriptional efficiency, or in other words promoter strength, is a central issue for efficient production. Isolation of promoters from yeast genomes however was difficult at times when no genome sequence information was available. Therefore in the 1980s and 1990s, only few promoters were in use, mainly deriving from genes encoding metabolic enzymes, such as alcohol dehydrogenase [1], glyceraldehyde 3-phosphate dehydrogenase [2], or enolase [3]. These promoters are not strictly controllable in a bioprocess, so that regulated promoters were sought after. Two concepts of regulated promoters dominated in the early time of protein production with *S. cerevisiae*, being copper-regulated promoters of metallothionein genes (*CUP1-1*, *CUP1-2*, *CRS5*) [4], and promoters regulated by galactose (*GAL1*, *GAL10*) [5]. Both types of promoters are not ideal for efficient protein production. Firstly, they are not among the strongest yeast promoters, and secondly, their regulation and deregulation mechanisms are not ideal for large-scale production. *CUP* promoters require the addition of heavy metal ions to the fermentation broth which is not favored on a large scale due to obvious concerns of water pollution. *GAL* promoters, on the other hand, are repressed by glucose so that another, non-repressing carbon source (mostly raffinose) is used for the production phase, which increases media costs considerably.

The isolation of methanol inducible promoters of the alcohol oxidase and dihydroxyacetone synthase genes was a major breakthrough for the development of the *Pichia pastoris* protein production platform [6]. However, even with these strong promoters, the achievable recombinant protein levels lag behind the levels of the natively produced proteins. This fact, although party curable by increasing the gene copy number, indicates the consecutive appearance of several bottlenecks in the cellular protein production process, as discussed below.

With the availability of genome sequence data for the production hosts, it has become way easier to directly access promoter regions and test them for recombinant protein production. Transcriptome data, providing information on the expression levels and regulation of nearly all genes, are a great resource for the identification of new promoters with desired features. Based on DNA microarray data, Stadlmayr et al. [7] identified 24 promoters of *P. pastoris* with distinctly different expression levels. Prielhofer et al. [8] used transcriptomics to identify *P. pastoris* promoters of genes that were both highly expressed in glucose limited fed-batch cultures and strongly downregulated in carbon-rich batch cultures. Later, Prielhofer et al. [9] exploited transcriptional data from four different carbon sources [10] to generate a library of promoters and transcriptional terminators that can be used for protein production or for strain engineering purposes (available as Golden*Pi*CS Kit at Addgene #1000000133). Vogl et al. [11] used transcriptome data of *P. pastoris* cultures in glucose-rich conditions and upon glucose depletion, methanol addition, and glucose re-addition, respectively, to identify promoters with different regulatory properties (however with the disadvantage of sampling all consecutively from the same batch cultures). In *S. cerevisiae*, transcriptomics was applied to identify strong and upregulated promoters active on glycerol [12].

Engineering of promoters, that is, the modification and/or addition of transcription factor binding sites, has been proposed for bacteria and yeasts [13], potentially increasing or decreasing expression strength, as well as a potentially modifying or removing of regulation (reviewed by [14]). Hartner et al. [15] created a library of engineered *P. pastoris AOX1* promoter variants with partly stronger, mostly lower strength and a gradual loss of regulation. Transcriptional engineering was introduced by Ata et al. [16] as a concept to engineer promoter strength both at the level of transcription factor binding sites and tuning the expression of the respective transcription factors, as illustrated with the *P. pastoris TDH3* promoter (PGAP).

Synthetic promoters use many of the elements introduced above for engineered promoters. Leavitt et al. [17] described a hybrid *S. cerevisiae* promoter combined with an engineered transcription factor. The same group advanced this concept by creating synthetic promoters of *Yarrowia lipolytica* by combining disparate upstream activating sequences to a core promoter [18]. Similarly, more synthetic promoters inducible by alternative carbon sources such as erythritol and erythrulose were described for *Y. lipolytica* [19]. By using orthogonal transcription factor binding site/DNA binding domains of bacterial origin, Rantasalo et al. [20] developed a synthetic promoter concept that is nearly independent of the physiology of the yeast cell. Thus, many new yeast promoters have been described in the recent years. It should be noted, however, that most of them have not been tested yet for larger-scale production in a bioreactor.

It is well established that increased gene copy numbers can contribute to higher expression levels. In *S. cerevisiae* high copy numbers are usually achieved by using high copy episomal vectors (such as 2 μ-derived plasmids), while in many other yeasts genome integration is the method of choice due to the lack of stable episomal plasmids. Methods to increase gene copy number include post-transformational vector amplification (cultivation on increasing selection pressure, commonly used in *P. pastoris* and *H. polymorpha*, [21, 22], or integration in native multicopy-loci such as the *Y. lipolytica* Ylt1 retrotransposon or the "zeta" sequences [23, 24], or the ribosomal DNA NTS loci for several yeasts [25]. However there are reports that with the increase of gene copy numbers the relative expression levels per copy may decrease, which is probably due to a titration effect of transcription factors which are required to activate the respective promoter [26, 27].

2.2 Translation

Different to bacteria, where translation initiation signals are rather clearly defined, there are no such distinct signals in yeasts. One obvious translation signal is the Kozak sequence that has a (species-dependent) consensus sequence [9, 28] and is typically added 5′ to a coding sequence in gene constructs for recombinant expression. Depending on the expression vector used, either the native Kozak sequence of the applied promoter or a predetermined well-performing Kozak sequence is used.

The three-dimensional structure of the target gene's mRNA can potentially influence translation, especially when double-strand loops can form at the 5′ end of the mRNA. The chance of double-strand formation can be lowered during sequence design for synthesis of codon-optimized genes.

The impact of transcriptional terminators (TTs) on recombinant gene expression has been discussed [29]. In *S. cerevisiae* terminator activity was shown to vary on a genome scale by 70-fold [30]. Expression-enhancing terminators increased mRNA and protein levels more than tenfold in *S. cerevisiae* especially when combined with weaker promoters [31]. So far, such a marked difference has not been observed for different TTs in *P. pastoris* where all tested TTs displayed almost equal efficiency [9, 11]. Notably, the commonly used *CYC1*-TT is by far not the best TT in *S. cerevisiae*, *Y. lipolytica*, and *P. pastoris*. Apparently, transcription terminators have very conserved features and can be readily transferred from one yeast species to another as exemplified by the fact that synthetic TTs designed for *S. cerevisiae* can also be applied for *Y. lipolytica* [29].

3 Metabolic Limitations in Recombinant Protein Production in Yeast

It has long been recognized that high-level expression of heterologous proteins has a direct impact on host cell metabolism (also known as metabolic burden, [32]), often negatively affecting growth parameters such as growth rate, biomass yield, and specific substrate consumption rate [33–37], or accumulation of less- or nonproducing cell populations [38], thereby limiting the amount of foreign protein that can be produced from the organism. In principle, foreign gene expression leads to an increase in specific transcription and translation, which may become limiting at very high levels due to depletion of precursors and energy. Producing strains may not cope with the additional demand for ATP,

NADPH, and precursors for de novo biosynthesis of amino acids, thus leading to a suboptimal cell fitness and reduced production yields [39]. Amino acid supplementation of growth media has been reported to be beneficial for high-level heterologous protein production in yeast, supporting a priori the hypothesis of limited supply of precursors [40–42]. Nonetheless, the specific productivities—particularly for secreted proteins—often achieved in yeast systems are rather low (relative to the total cell protein), therefore suggesting that limitations in amino acid synthesis would not be the major bottleneck. Still, the energy demand can be significantly higher for secreted proteins, as folding, glycosylation, and secretion are energy-intensive pathways, particularly in terms of NADPH, which is required for disulfide bond formation and alleviating ER oxidative stress [43]. Moreover, cellular stress responses to unfolded proteins (unfolded protein response, UPR), which are often triggered upon overexpression of a secreted recombinant protein, further increase the metabolic demand of the folding and secretory processes. Indeed, a metabolic burden related to protein secretion has been observed in yeast, even at low to medium expression levels [44–46]. In this context, it is also important to note that large-scale production of heterologous proteins in yeast is often carried out in high-cell density cultivations operated in fed-batch mode. Such processes are typically performed at low growth rate when operated under substrate limiting conditions. The effects of a low growth rate have a big impact on the cell physiology and, consequently, on the specific productivity [47, 48], as growth rate regulates core processes such as protein synthesis and secretion, as well as stress response [49]. Cellular responses such as nutrient starvation may also be elicited, thereby contributing to the metabolic stress of the host cells. Notably, a substantial part of the substrate carbon is expended to meet maintenance-energy requirements under growth-limiting conditions such as those found in fed-batch processes. High maintenance requirements go at the expense of biomass and product formation and therefore are not desired in heterologous protein production [47].

Over the past 15 years, several physiological studies have brought new insights on the impact of recombinant protein production on the metabolic network operation of yeast. Global analysis of the host cell metabolism by means of omics analytical platforms such as transcriptomics, proteomics, metabolomics, and fluxomics can now be used to investigate the physiological effect of both environmental stresses and recombinant biosynthesis, guiding the identification of potential metabolic targets for cell engineering, selection of growth conditions, and cultivation strategies favoring recombinant protein production. In particular, ^{13}C-based metabolic flux analysis (^{13}C-MFA) and metabolomics studies have revealed a significant impact of synthesis and secretion of heterologous proteins on energy metabolism, resulting in altered metabolic flux distributions through the central carbon metabolism of yeast (reviewed in Ferrer and Albiol [50]; Klein et al. [51]).

The effect of different recombinant protein production levels was assessed with ^{13}C-MFA using a series of *P. pastoris* strains producing a model protein intracellularly (a bacterial β-aminopeptidase) in glucose-based fed-batch cultures, achieving up to 2 g/L of β-aminopeptidase [35, 40]. Using a strain expressing an intracellular recombinant protein enabled to discard a potential metabolic burden related to protein secretion. The recombinant strain showed an increased relative flux through the TCA cycle compared to the reference strain, resulting in significantly increased NADH and ATP regeneration rates. In addition, the recombinant strain showed a slight decrease in the biomass yield compared to the reference strain, which correlated with a lower pentose phosphate pathway (PPP) activity. These studies provided direct evidence for a direct response of *P. pastoris*' metabolic network to recombinant protein production, which could not be explained by the direct resources (in terms of building blocks) necessary to produce it. This became even more evident in the light of other ^{13}C-based metabolic flux analysis and metabolomics studies on the metabolic burden caused by secretion of a recombinant lipase from *Rhizopus oryzae* (Rol) in *P. pastoris* growing on glucose/methanol mixtures [44, 45]. Even though Rol is produced at low levels (mg/L range), MFA allowed the identification of a limited but significant metabolic flux redistribution. Specifically, the flux through the glycolysis, TCA cycle, and methanol dissimilatory pathway (generating NADH) was increased in the Rol-producing strains in relation to the reference strain. Remarkably, although the biomass yield of the Rol-producing strains was somewhat lower compared to the reference strain, the flux through the oxidative branch of the PPP appeared to be constant in all strains. This points at the hypothesis of increased NADPH supply through this pathway in the Rol-producing strains. Such effect could be the indirect consequence of methanol co-assimilation, acting as an auxiliary substrate [52].

Overall, these studies strongly suggest that yeasts compensate the additional resources required for recombinant protein production by redirection of intracellular fluxes resulting in increased energy supply (NADH, ATP) [35, 40, 44, 45]. Nonetheless, it also results in altered redox cofactor state and, specifically, a reduction in NADPH availability, reflected in reduced biomass yields [35, 40]. Interestingly, supplementing acetate to glucose or glycerol minimal media of recombinant *Schizosaccharomyces pombe* secreting a model protein (maltase) in aerobic chemostat cultures improved protein secretion. ^{13}C-based MFA revealed that acetate co-feeding allowed for an increased carbon flux through the TCA cycle as well as increased mitochondrial NADPH production [46], i.e., provoking an effect similar as methanol co-feeding in *P. pastoris* [52]. Moreover, a model-based study using a genome-scale metabolic model identified NADPH generating reactions as a major cell engineering target for improved protein production [53, 54]. Nocon et al. [53, 55] further validated such in silico predictions by overexpressing gene coding for enzymes of the oxidative branch of the PPP, obtaining higher productivities in heterologous protein secretion. More recently, Tomàs-Gamisans [56] has demonstrated that the ATP-mediated conversion of NADH to NADPH using a heterologous cytosolic NADH kinase in a recombinant *P. pastoris* strain leads to increased recombinant protein secretion. Notably, such an effect was boosted under hypoxic conditions, where the reduced oxygen availability for electron transport chain leads to cytosolic NADH excess (reflected in a higher NADH/NAD^+ ratio, [57]), increasing the flux to NADPH.

Besides NADPH requirements to cope with the ER-stress resulting from protein processing, NADPH is particularly required for biosynthesis of amino acids as building blocks of proteins. This suggests that increased NADPH levels may meet the extra demand for the synthesis of recombinant protein. Heyland et al. [40] showed that metabolically costly (in terms of NADPH requirements) amino acids constitute a bottleneck in the production of β-aminopeptidase in *P. pastoris*. Nonetheless, other metabolomics studies of *P. pastoris* secreting different model proteins do not provide clear supporting data for this hypothesis [45, 58, 59].

Interestingly, ^{13}C-based studies of *Aspergillus niger* revealed that production of a recombinant fructofuranosidase induced a significant redistribution of metabolic fluxes enabling an elevated supply of NADPH via activation of the cytosolic pentose phosphate pathway and the mitochondrial malic enzyme, whereas the flux through the TCA cycle was reduced [60]. This common finding (the changing contribution of the PPP and the TCA cycle) in yeast and filamentous fungi, although in opposite directions, points at a general feature of the underlying carbon metabolism, i.e., the metabolic flexibility of fungi to cope with different cellular burdens and environmental perturbations by modulating the fluxes through these pathways.

4 Enhancing Protein Folding and Secretion

Recombinant proteins can be either produced in the cytosol or secreted to the cell exterior. Secretory production has several benefits including easy purification of the product from the supernatant. For natively secreted products, which constitute a majority of pharmaceutical proteins and industrial enzymes, a correctly processed N-terminus, disulfide bond formation, and posttranslational modifications can be achieved in yeast production platforms. On their way to the cell exterior, secreted proteins have to traverse the secretory pathway, which poses further possible obstacles to the desired product. Thus, many studies aim at customizing the secretory machinery for high-level production. High-level overexpression of heterologous secretory proteins has repeatedly been shown to activate cellular stress response pathways, including the unfolded protein response (UPR) and ER-associated protein degradation (ERAD) (e.g., [61–64]).

Right after synthesis, the recombinant protein may be exposed to the cytosolic environment of the cell. The heat shock response (HSR) is a major cellular process regulating the expression of chaperones and other proteins assisting in protein folding or degradation of heat-denatured proteins in the cytosol. Targeted induction of the HSR by overexpression of its constitutively activated transcriptional regulator Hsf1 led to improved secretion of recombinant proteins in S. *cerevisiae* [65], with larger benefits for the larger of the two tested recombinant proteins (insulin precursor vs alpha-amylase). One possible explanation for the observed secretion-enhancing phenotype was that induction of the HSR reduced ER stress of the producing cells [65]. On the other hand, induction of the UPR by overexpression of the induced version of its transcriptional activator Hac1^{i} enhanced secretion of several recombinant proteins in *S. cerevisiae*, *P. pastoris*, *Y. lipolytica*, and other fungal hosts [66–70], which is somehow counterintuitive to the previous study. One possible explanation might be that UPR induction does not only impact the levels of ER-resident chaperones and foldases [71–73] but also leads to an enlarged ER size [74], thereby diminishing the possibility of protein aggregation in the ER. Indeed, expanding the ER size by deleting the lipid regulator Opi1 in *S. cerevisiae* led to fourfold higher secretion levels of a recombinant antibody [75]. Strikingly, this positive effect was only observed for complex glycosylated mammalian proteins, while the secretion of yeast endogenous proteins was rather decreased in the Δ*opi1* strain [76]. ER size and membrane composition were also reported to be a bottleneck for membrane protein production and could be overcome by redirecting the flux from storage lipids to (ER) membrane proliferation in *Y. lipolytica* [77] or humanization of the lipid composition in *P. pastoris* [78, 79].

For secretory proteins, the first step on the secretory pathway is translocation of the nascent protein into the ER. Recently, insufficient translocation has been described as a potential bottleneck during secretion of a recombinant fluorescent reporter protein in *S. cerevisiae* and *P. pastoris* [80]. A further study in *P. pastoris* revealed that heterologous Fab fragments accumulated prior to translocation and might be degraded by a specific form of ERAD [81]. This specific form of ERAD, termed pre-insertional ERAD, was shown to clears proteins clogging the translocon channel at the cytosolic side [82]. Attempts to engineer protein translocation in

S. cerevisiae showed that overexpressing folding and translocation aiding factors improved recombinant protein secretion [83], yet the effect depended on both, signal peptides and product proteins [84].

Once inside the ER, the nascent proteins need to be correctly folded and posttranslationally modified. Several ER resident chaperones and foldases involved in disulfide bond formation and peptidyl-prolyl isomerization are acting in close interplay with the glycosylation machinery to achieve these tasks. Especially disulfide bond formation has been reported as rate-limiting step; thus many attempts to improve secretory protein production relied on overexpression of protein disulfide isomerase Pdi1, either alone or together with its oxidase Ero1 (reviewed by [85–88]). By this approach higher titers could be obtained for several different recombinant proteins in several different yeast species including *S. cerevisiae*, *P. pastoris*, and *K. lactis*. Pdi1 overexpression was also combined with the overexpression of ER chaperones such as binding protein Kar2 (reviewed by [85–88]). Interestingly, there was no evidence that beneficial combinatorial or even synergistic effects can be discovered by simultaneous overexpression of two or more chaperone genes [75, 76].

More recently, overexpression peptidyl-prolyl isomerase Cpr5 (either from *S. cerevisiae* or from human origin) was reported to increase secretion of IgG, which are proline-rich recombinant proteins and require cis-trans isomerization of at least one of these prolines in order to attain the correct Ig-folds [75].

The expression of secretory recombinant proteins often triggers the UPR [89], and this, in turn, is supposed to be linked to the ERAD system [90]. The significance of intracellular degradation was shown repeatedly, and Pfeffer et al. [91] determined the amount of antibody fragment lost in this way to be 58%. Due to this and the major role of the UPR, the ERAD complex emerged as strain-engineering target. ERAD was disrupted in antibody producing *S. cerevisiae* by deleting the genes *HTM1*, *YOS9*, *HRD1*, *HRD3*, or *UBC7* [92]. The increase in antibody secretion was, if at all, only very minor compared to, e.g., folding helper overexpression. Similar effects were observed in *P. pastoris* when disrupting *HRD3*, *DER1*, *RPN4*, or *DOA1* [81]. Together these results indicate that removal of the recombinant protein from the ER does not seem to be a major limitation.

After successful folding and ER-quality control, the recombinant protein travels to the Golgi apparatus in COPII vesicles. Stimulating ER exit and COPII formation by overexpression of Sec16 (but not other factors) enhanced the secretion of three model proteins in *S. cerevisiae* but also led to a depletion of ER membrane [93]. The latter could be rescued by simultaneously overexpressing a component of the retrograde transport process, Glo3, which is involved in trafficking of COPI vesicles from the

Golgi to the ER [94]. In contrast to *S. cerevisiae*, overexpression of *SEC16* in *P. pastoris* does not increase the number of ER exit sites [95]. Another approach to increase the rate of anterograde transport is to overexpress components of the vesicle fusion machinery including SNAREs and Sec1/Munc18 (SM) proteins which are required for membrane-specific fusion events during protein trafficking [87, 96, 97]. Based on several studies in *S. cerevisiae*, it can be concluded that the effects of overexpressing ER-to-Golgi SNAREs and SM proteins seems to be rather specific for the recombinant protein to be produced. Besides ER-to-Golgi transport, strengthening Golgi-to-plasma membrane (PM) transport may have a more general positive effect on secretion of several model proteins. For example, overexpression of ER-to-Golgi SM Sly1 increased only the production of an α-amylase, but not insulin precursor, whereas the Golgi-to-PM SM Sec1 increased secretion of both recombinant proteins and also the secretion of the endogenous enzyme invertase [98]. Similarly, the overexpression of exocytic SNAREs such as Sso1/2 or Snc2 enhanced the secretion of α-amylase and several cellulolytic enzymes [99–102], whereas the single or concerted overexpression of ER-to-Golgi SNARE components had contrary effects on their secretion [102, 103]. Apart from the mostly beneficial impact of syntaxin (Sso1/2) overexpression (which enhances secretion on average by 10–50%), there is no clear picture which component of the vesicle fusion machinery is most rate-limiting based on these studies.

In the Golgi, further posttranslational modifications (such as elongation of glycans and proteolytic processing) take place, and the proteins are sorted toward their final destination. For recombinant proteins, especially the cleavage of the secretion leader by trans-Golgi endoprotease Kex2 was described as rate-limiting step that could be overcome by overexpression of Kex2 or a truncated variant thereof [104–106]. Another interesting aspect relates to the Golgi localized Ca^{2+}/Mn^{2+}-ATPase Pmr1, which is responsible for Ca^{2+} and Mn^{2+} import into the secretory pathway [107]. Disruption of Pmr1 alters calcium homeostasis in the secretory organelles and leads to incomplete outer-chain glycosylation. Nevertheless, *pmr1Δ* mutants have been described to display a "super-secretory" phenotype for several heterologous proteins in *S. cerevisiae* and *K. lactis* [100, 108–111]. The deletion likely causes a Ca^{2+} decrease in the ER, which in turn facilitates an ATP-dependent dissociation of BiP (Kar2) from the substrate. However, also several other cellular functions are affected by *pmr1Δ*. To overcome the reduced growth rate and low viability of the *pmr1Δ* strains, extracellular Ca^{2+} should be added to the growth media. Extracellular Ca^{2+} addition also (at least partially) rescues the glycosylation defect in *S. cerevisiae pmr1Δ*, but not in *K. lactis pmr1Δ* [107, 111], which makes the latter an interesting alternative when hyper-glycosylation is supposed to be prevented. In contrast, alternating results regarding protein secretion were obtained in *H. polymorpha*, *P. pastoris*, and *Y. lipolytica* disrupted for *PMR1*, which might be partly attributed to a reduced viability of the strains thus eventually masking the positive effects on secretion [112–115]. Furthermore, it was recently demonstrated that there are also Pmr1-independent routes of Ca^{2+} delivery to the secretory organelles operating in *H. polymorpha* [116].

Even though the default pathway for recombinant proteins should be toward the cell exterior, the next possible major branch-off on the secretory pathway is missorting from the Golgi to the vacuole and subsequent degradation. Yeast cells have a quality control system in which the vacuolar sorting receptor Vps10 targets misfolded proteins from the Golgi to the vacuole. By mutating Vps10, Fitzgerald and Glick [80] could prevent accumulation of msGFP in the vacuole of *S. cerevisiae*. *VPS10* deletion also increased the secretion of other recombinant proteins in several yeast species (*H. polymorpha* [117], *S. cerevisiae* [100], and *S. pombe* [118]). Kitagawa et al. found that the deletion of genes encoding subunits of vacuole protein-sorting complexes such as *VPS3*, *VPS16*, *YPT7*, and *VPS41* enhanced the secretion of recombinant endoglucanase in *S. cerevisiae*. Especially, the *VPS3* deletion increased secretion of all tested reporter proteins, highlighting its general importance [119]. High-throughput screening for enhanced IgG secretion revealed *VPS30* as potential bottleneck; its disruption enhanced IgG secretion or the secretion of acid phosphatase about twofold [120]. Also in the methylotrophic yeasts *H. polymorpha* and *P. pastoris*, disruption of *VPS* genes proved to be beneficial for secretory protein production [117, 121]. In fission yeast *S. pombe*, the knockout of several vacuolar protease genes was required to enhance productivities [118]. Yapsins are a family of proteases located in the late secretory pathway or at the cell surface of yeasts [122] and can thus harm even correctly folded recombinant proteins during the late steps of secretion and excretion. Disruption of yapsins has been successfully attempted to reduce proteolysis of degradation-prone products such as collagen polymers, human parathyroid hormone hPTH, human pre-elafin, or human serum albumin (HSA)-fusion proteins, while no effect was observed for other more stable recombinant proteins [123–128].

While there are many reports indicating that vacuolar protein sorting is posing a limitation in recombinant protein secretion, there is no clear teaching to which of the *VPS* genes should be deleted. While some explanations exist why deletion of the receptor Vps10 prevents missorting of specific recombinant proteins, the beneficial effects of most other vps mutants were found by high-throughput screening for enhanced secretion [117, 119, 120, 129]. Furthermore, it should be considered that some vps mutants also oversecrete vacuolar proteases, which can be harmful to the secreted product. Proteolysis caused by the mis-sorted vacuolar proteases may even mask the positive effect of preventing vacuolar transport of the recombinant protein, as reported by Marsalek et al. [121]. Only the combined disruption of Vps8 or Vps21 with vacuolar proteases such as Pep4 or Prb1 led to increased titers of secreted product.

The final step on the secretory pathway is the exocytic fusion of secretory vesicles with the plasma membrane, representing another possible bottleneck and therefore strain-engineering target. Early work indicated the beneficial effect of overexpression of Sso1/Sso2 on secretion of heterologous α-amylase and endogenous invertase in *S. cerevisiae* [101] and antibody Fab fragments in *P. pastoris* [130]. Also overexpression of Sec4, which is required for vesicle-mediated exocytic secretion, yielded enhanced secretion in both yeast species [131, 132]. The overexpression of components of the exocytic SNARE complex (Snc1/Snc2, Sso1/Sso2, and Sec9) in *S. cerevisiae* could improve the secretion of cellulolytic reporter proteins [99]. However, not all single-gene overexpressions could improve secretion, and the simultaneous overexpression of several components turned out to be product specific, yielding different best performing combinations. Along with SNARE component overexpression, Xu et al. obtained their best performing quadruple-modified strain (*vps10Δ/pmr1Δ/SSO1/PDI1/cel7AF*), by combing several targets in the secretory pathway [100]. Interestingly, while all their *pmr1Δ* strains showed enhanced secretion characteristics for the recombinant cellulase, no impact on invertase secretion was observed.

After being secreted from the plasma membrane, the recombinant protein will still have to diffuse through the cell wall, which may or may not be a barrier [133]. Especially, in the case of accumulation of the recombinant protein in this cell wall-bounded periplasmic space, endocytosis may significantly decrease productivity. Along this line of evidence, the concentration of α-amylase in the supernatant could be increased by conditional knock-down of *RVS161* and *END3*. However, no effect could be observed on the secretion of recombinant insulin [134].

Another approach to lessen the cell wall barrier is to disrupt cell-wall cross-linking proteins. Knockout of the major cell wall beta-1,3-glucanosyltransferase Gas1 has been shown to enhance the secretion of some recombinant proteins in several different yeasts [135–138]. Furthermore, Larsen et al. reported the positive impact of disruption of cell wall-related genes when screening for enhanced β-galactosidase secretion in *P. pastoris* [139].

Interestingly, some bottlenecks seem to be host-related rather than protein-specific. It was reported that human parathyroid hormone (hPTH) production required the inhibition or disruption of extracellular proteases in *S. cerevisiae* [140], with disruption of all five yapsins (quintuple disruptant *yps1Δ yps2Δ yps3Δ yps6Δ yps7Δ*) giving the best effects in fed-batch cultivations [124]. In contrast, reducing proteolytic activity was not required for hPTH production in *P. pastoris* [141], but overexpression of *PDI1* proved to be beneficial.

Acknowledgments

Work in the group of Pau Ferrer has been supported by the project CTQ2016-74959-R (AEI/FEDER, UE) of the Spanish Ministry of Economy, Industry and Competitiveness (MINECO), and the Catalan Government (Research Group 2017-SGR-1462 and Xarxa de Referència en Biotecnologia).

Research on yeast protein production at BOKU is supported by the Austrian Science Fund (FWF): Doctoral Program BioToP—Biomolecular Technology of Proteins (FWF W1224), the Austrian Federal Ministry for Digital and Economic Affairs (BMDW), the Federal Ministry of Traffic, Innovation and Technology (bmvit), the Styrian Business Promotion Agency SFG, the Standortagentur Tirol, the Government of Lower Austria and ZIT—Technology Agency of the City of Vienna through the COMET-Funding Program managed by the Austrian Research Promotion Agency FFG (RZ, BG, DM), and by the Christian Doppler Research Association, the Austrian Federal Ministry for Digital and Economic Affairs (BMDW) and the National Foundation for Research, Technology and Development (BG).

References

1. Hitzeman RA, Hagie FE, Levine HL, Goeddel DV, Ammerer G, Hall BD (1981) Expression of a human gene for interferon in yeast. Nature 293(5835):717–722
2. Weber JM, Ponti CG, Kappeli O, Reiser J (1992) Factors affecting homologous overexpression of the *Saccharomyces cerevisiae* lanosterol 14 alpha-demethylase gene. Yeast 8 (7):519–533. https://doi.org/10.1002/yea.320080704
3. Innis MA, Holland MJ, McCabe PC, Cole GE, Wittman VP, Tal R, Watt KW, Gelfand DH, Holland JP, Meade JH (1985) Expression, glycosylation, and secretion of an *Aspergillus* glucoamylase by *Saccharomyces cerevisiae*. Science 228(4695):21–26. https://doi.org/10.1126/science.228.4695.21
4. Butt TR, Sternberg EJ, Gorman JA, Clark P, Hamer D, Rosenberg M, Crooke ST (1984) Copper metallothionein of yeast, structure of the gene, and regulation of expression. Proc Natl Acad Sci U S A 81(11):3332–3336
5. Stepien PP, Brousseau R, Wu R, Narang S, Thomas DY (1983) Synthesis of a human insulin gene. VI. Expression of the synthetic proinsulin gene in yeast. Gene 24 (2–3):289–297
6. Tschopp JF, Brust PF, Cregg JM, Stillman CA, Gingeras TR (1987) Expression of the lacZ gene from two methanol-regulated promoters in *Pichia pastoris*. Nucleic Acids Res 15(9):3859–3876
7. Stadlmayr G, Mecklenbrauker A, Rothmuller M, Maurer M, Sauer M, Mattanovich D, Gasser B (2010) Identification and characterisation of novel *Pichia pastoris* promoters for heterologous protein production. J Biotechnol 150(4):519–529. https://doi.org/10.1016/j.jbiotec.2010.09.957
8. Prielhofer R, Maurer M, Klein J, Wenger J, Kiziak C, Gasser B, Mattanovich D (2013) Induction without methanol: novel regulated promoters enable high-level expression in

Pichia pastoris. Microb Cell Factories 12:5. https://doi.org/10.1186/1475-2859-12-5

9. Prielhofer R, Barrero JJ, Steuer S, Gassler T, Zahrl R, Baumann K, Sauer M, Mattanovich D, Gasser B, Marx H (2017) GoldenPiCS: a Golden Gate-derived modular cloning system for applied synthetic biology in the yeast *Pichia pastoris*. BMC Syst Biol 11 (1):123. https://doi.org/10.1186/s12918-017-0492-3
10. Prielhofer R, Cartwright SP, Graf AB, Valli M, Bill RM, Mattanovich D, Gasser B (2015) *Pichia pastoris* regulates its gene-specific response to different carbon sources at the transcriptional, rather than the translational, level. BMC Genomics 16:167. https://doi.org/10.1186/s12864-015-1393-8
11. Vogl T, Sturmberger L, Kickenweiz T, Wasmayer R, Schmid C, Hatzl AM, Gerstmann MA, Pitzer J, Wagner M, Thallinger GG, Geier M, Glieder A (2016) A toolbox of diverse promoters related to methanol utilization: functionally verified parts for heterologous pathway expression in *Pichia pastoris*. ACS Synth Biol 5(2):172–186. https://doi.org/10.1021/acssynbio.5b00199
12. Ho PW, Klein M, Futschik M, Nevoigt E (2018) Glycerol positive promoters for tailored metabolic engineering of the yeast *Saccharomyces cerevisiae*. FEMS Yeast Res 18(3). https://doi.org/10.1093/femsyr/foy019
13. Alper H, Fischer C, Nevoigt E, Stephanopoulos G (2005) Tuning genetic control through promoter engineering. Proc Natl Acad Sci U S A 102(36):12678–12683. https://doi.org/10.1073/pnas.0504604102
14. Blazeck J, Alper HS (2013) Promoter engineering: recent advances in controlling transcription at the most fundamental level. Biotechnol J 8(1):46–58. https://doi.org/10.1002/biot.201200120
15. Hartner FS, Ruth C, Langenegger D, Johnson SN, Hyka P, Lin-Cereghino GP, Lin-Cereghino J, Kovar K, Cregg JM, Glieder A (2008) Promoter library designed for fine-tuned gene expression in *Pichia pastoris*. Nucleic Acids Res 36(12):e76. https://doi.org/10.1093/nar/gkn369
16. Ata O, Prielhofer R, Gasser B, Mattanovich D, Calik P (2017) Transcriptional engineering of the glyceraldehyde-3-phosphate dehydrogenase promoter for improved heterologous protein production in *Pichia pastoris*. Biotechnol Bioeng 114 (10):2319–2327. https://doi.org/10.1002/bit.26363
17. Leavitt JM, Tong A, Tong J, Pattie J, Alper HS (2016) Coordinated transcription factor and promoter engineering to establish strong expression elements in *Saccharomyces cerevisiae*. Biotechnol J 11(7):866–876. https://doi.org/10.1002/biot.201600029
18. Blazeck J, Reed B, Garg R, Gerstner R, Pan A, Agarwala V, Alper HS (2013) Generalizing a hybrid synthetic promoter approach in *Yarrowia lipolytica*. Appl Microbiol Biotechnol 97 (7):3037–3052. https://doi.org/10.1007/s00253-012-4421-5
19. Trassaert M, Vandermies M, Carly F, Denies O, Thomas S, Fickers P, Nicaud JM (2017) New inducible promoter for gene expression and synthetic biology in *Yarrowia lipolytica*. Microb Cell Factories 16(1):141. https://doi.org/10.1186/s12934-017-0755-0
20. Rantasalo A, Czeizler E, Virtanen R, Rousu J, Lahdesmaki H, Penttila M, Jantti J, Mojzita D (2016) Synthetic transcription amplifier system for orthogonal control of gene expression in *Saccharomyces cerevisiae*. PLoS One 11(2): e0148320. https://doi.org/10.1371/journal.pone.0148320
21. Sunga AJ, Tolstorukov I, Cregg JM (2008) Posttransformational vector amplification in the yeast *Pichia pastoris*. FEMS Yeast Res 8 (6):870–876. https://doi.org/10.1111/j.1567-1364.2008.00410.x
22. Gatzke R, Weydemann U, Janowicz ZA, Hollenberg CP (1995) Stable multicopy integration of vector sequences in *Hansenula polymorpha*. Appl Microbiol Biotechnol 43 (5):844–849
23. Pignede G, Wang HJ, Fudalej F, Seman M, Gaillardin C, Nicaud JM (2000) Autocloning and amplification of *LIP2* in *Yarrowia lipolytica*. Appl Environ Microbiol 66 (8):3283–3289
24. Juretzek T, Le Dall M, Mauersberger S, Gaillardin C, Barth G, Nicaud J (2001) Vectors for gene expression and amplification in the yeast *Yarrowia lipolytica*. Yeast 18 (2):97–113. https://doi.org/10.1002/1097-0061(20010130)18:2<97::aid-yea652>3.0.co;2-u
25. Klabunde J, Kunze G, Gellissen G, Hollenberg CP (2003) Integration of heterologous genes in several yeast species using vectors containing a *Hansenula polymorpha*-derived rDNA-targeting element. FEMS Yeast Res 4 (2):185–193
26. Camara E, Landes N, Albiol J, Gasser B, Mattanovich D, Ferrer P (2017) Increased dosage of *AOX1* promoter-regulated expression cassettes leads to transcription attenuation of the methanol metabolism in *Pichia*

pastoris. Sci Rep 7:44302. https://doi.org/10.1038/srep44302

27. Aw R, Polizzi KM (2013) Can too many copies spoil the broth? Microb Cell Factories 12:128. https://doi.org/10.1186/1475-2859-12-128
28. Hamilton R, Watanabe CK, de Boer HA (2018) Compilation and comparison of the sequence context around the AUG startcodons in *Saccharomyces cerevisiae* mRNAs. Nucleic Acids Res 15(8):3581–3593. https://doi.org/10.1093/nar/15.8.3581
29. Curran KA, Morse NJ, Markham KA, Wagman AM, Gupta A, Alper HS (2015) Short synthetic terminators for improved heterologous gene expression in yeast. ACS Synth Biol 4(7):824–832. https://doi.org/10.1021/sb5003357
30. Yamanishi M, Ito Y, Kintaka R, Imamura C, Katahira S, Ikeuchi A, Moriya H, Matsuyama T (2013) A genome-wide activity assessment of terminator regions in *Saccharomyces cerevisiae* provides a "terminatome" toolbox. ACS Synth Biol 2(6):337–347. https://doi.org/10.1021/sb300116y
31. Curran KA, Karim AS, Gupta A, Alper HS (2013) Use of expression-enhancing terminators in *Saccharomyces cerevisiae* to increase mRNA half-life and improve gene expression control for metabolic engineering applications. Metab Eng 19:88–97. https://doi.org/10.1016/j.ymben.2013.07.001
32. Glick BR (1995) Metabolic load and heterologous gene expression. Biotechnol Adv 13 (2):247–261
33. Vigentini I, Brambilla L, Branduardi P, Merico A, Porro D, Compagno C (2005) Heterologous protein production in *Zygosaccharomyces bailii*: physiological effects and fermentative strategies. FEMS Yeast Res 5 (6–7):647–652. https://doi.org/10.1016/j.femsyr.2004.11.006
34. Cos O, Resina D, Ferrer P, Montesinos JL, Valero F (2005) Heterologous production of *Rhizopus oryzae* lipase in *Pichia pastoris* using the alcohol oxidase and formaldehyde dehydrogenase promoters in batch and fed-batch cultures. Biochem Eng J 26(2–3):86–94. https://doi.org/10.1016/j.bej.2005.04.005
35. Heyland J, Fu J, Blank LM, Schmid A (2010) Quantitative physiology of *Pichia pastoris* during glucose-limited high-cell density fed-batch cultivation for recombinant protein production. Biotechnol Bioeng 107 (2):357–368. https://doi.org/10.1002/bit.22836
36. Gorgens JF, van Zyl WH, Knoetze JH, Hahn-Hagerdal B (2001) The metabolic burden of the *PGK1* and *ADH2* promoter systems for heterologous xylanase production by *Saccharomyces cerevisiae* in defined medium. Biotechnol Bioeng 73(3):238–245
37. Krogh AM, Beck V, Christensen LH, Henriksen CM, Moller K, Olsson L (2008) Adaptation of *Saccharomyces cerevisiae* expressing a heterologous protein. J Biotechnol 137 (1–4):28–33. https://doi.org/10.1016/j.jbiotec.2008.07.1787
38. Kazemi Seresht A, Palmqvist EA, Schluckebier G, Pettersson I, Olsson L (2013) The challenge of improved secretory production of active pharmaceutical ingredients in *Saccharomyces cerevisiae*: a case study on human insulin analogs. Biotechnol Bioeng 110(10):2764–2774. https://doi.org/10.1002/bit.24928
39. Wu G, Yan Q, Jones JA, Tang YJ, Fong SS, Koffas MAG (2016) Metabolic burden: cornerstones in synthetic biology and metabolic engineering applications. Trends Biotechnol 34(8):652–664. https://doi.org/10.1016/j.tibtech.2016.02.010
40. Heyland J, Fu J, Blank LM, Schmid A (2011) Carbon metabolism limits recombinant protein production in *Pichia pastoris*. Biotechnol Bioeng 108(8):1942–1953. https://doi.org/10.1002/bit.23114
41. Gorgens JF, Passoth V, van Zyl WH, Knoetze JH, Hahn-Hagerdal B (2005) Amino acid supplementation, controlled oxygen limitation and sequential double induction improves heterologous xylanase production by *Pichia stipitis*. FEMS Yeast Res 5 (6–7):677–683. https://doi.org/10.1016/j.femsyr.2004.12.003
42. Gorgens JF, van Zyl WH, Knoetze JH, Hahn-Hagerdal B (2005) Amino acid supplementation improves heterologous protein production by *Saccharomyces cerevisiae* in defined medium. Appl Microbiol Biotechnol 67 (5):684–691. https://doi.org/10.1007/s00253-004-1803-3
43. Delic M, Rebnegger C, Wanka F, Puxbaum V, Haberhauer-Troyer C, Hann S, Kollensperger G, Mattanovich D, Gasser B (2012) Oxidative protein folding and unfolded protein response elicit differing redox regulation in endoplasmic reticulum and cytosol of yeast. Free Radic Biol Med 52 (9):2000–2012. https://doi.org/10.1016/j.freeradbiomed.2012.02.048
44. Jordà J, Jouhten P, Cámara E, Maaheimo H, Albiol J, Ferrer P (2012) Metabolic flux profiling of recombinant protein secreting

Pichia pastoris growing on glucose: methanol mixtures. Microb Cell Factories 11:57. https://doi.org/10.1186/1475-2859-11-57

45. Jordà J, Rojas HC, Carnicer M, Wahl A, Ferrer P, Albiol J (2014) Quantitative metabolomics and instationary ^{13}C-metabolic flux analysis reveals impact of recombinant protein production on trehalose and energy metabolism in *Pichia pastoris*. Metabolites 4 (2):281–299. https://doi.org/10.3390/metabo4020281
46. Klein T, Lange S, Wilhelm N, Bureik M, Yang TH, Heinzle E, Schneider K (2014) Overcoming the metabolic burden of protein secretion in *Schizosaccharomyces pombe*—a quantitative approach using ^{13}C-based metabolic flux analysis. Metab Eng 21:34–45. https://doi.org/10.1016/j.ymben.2013.11.001
47. Hensing MC, Rouwenhorst RJ, Heijnen JJ, van Dijken JP, Pronk JT (1995) Physiological and technological aspects of large-scale heterologous-protein production with yeasts. Antonie Van Leeuwenhoek 67(3):261–279
48. Maurer M, Kuhleitner M, Gasser B, Mattanovich D (2006) Versatile modeling and optimization of fed batch processes for the production of secreted heterologous proteins with *Pichia pastoris*. Microb Cell Factories 5:37
49. Rebnegger C, Graf AB, Valli M, Steiger MG, Gasser B, Maurer M, Mattanovich D (2014) In *Pichia pastoris*, growth rate regulates protein synthesis and secretion, mating and stress response. Biotechnol J 9(4):511–525. https://doi.org/10.1002/biot.201300334
50. Ferrer P, Albiol J (2014) ^{13}C-based metabolic flux analysis of recombinant *Pichia pastoris*. Methods Mol Biol 1191:291–313. https://doi.org/10.1007/978-1-4939-1170-7_17
51. Klein T, Niklas J, Heinzle E (2015) Engineering the supply chain for protein production/secretion in yeasts and mammalian cells. J Ind Microbiol Biotechnol 42(3):453–464. https://doi.org/10.1007/s10295-014-1569-2
52. Jordà J, Suarez C, Carnicer M, ten Pierick A, Heijnen JJ, van Gulik W, Ferrer P, Albiol J, Wahl A (2013) Glucose-methanol co-utilization in *Pichia pastoris* studied by metabolomics and instationary ^{13}C flux analysis. BMC Syst Biol 7:17. https://doi.org/10.1186/1752-0509-7-17
53. Nocon J, Steiger MG, Pfeffer M, Sohn SB, Kim TY, Maurer M, Russmayer H, Pflugl S, Ask M, Haberhauer-Troyer C, Ortmayr K, Hann S, Koellensperger G, Gasser B, Lee SY, Mattanovich D (2014) Model based engineering of *Pichia pastoris* central metabolism enhances recombinant protein production. Metab Eng 24:129–138. https://doi.org/10.1016/j.ymben.2014.05.011
54. Mattanovich D, Sauer M, Gasser B (2017) Industrial microorganisms: *Pichia pastoris*. Chapter 19. Wiley-VCH Verlag GmbH & Co. KGaA, Weinheim. https://doi.org/10.1002/9783527807796.ch19
55. Nocon J, Steiger M, Mairinger T, Hohlweg J, Russmayer H, Hann S, Gasser B, Mattanovich D (2016) Increasing pentose phosphate pathway flux enhances recombinant protein production in *Pichia pastoris*. Appl Microbiol Biotechnol 100(13):5955–5963. https://doi.org/10.1007/s00253-016-7363-5
56. Tomàs-Gamisans T (2017) Developing strategies for systems metabolic engineering of *Pichia pastoris*. PhD thesis, Universitat Autònoma de Barcelona
57. Carnicer M (2012) Systematic metabolic analysis of recombinant *Pichia pastoris* under different oxygen conditions. PhD thesis, Universitat Autònoma de Barcelona
58. Carnicer M, Ten Pierick A, van Dam J, Heijnen JJ, Albiol J, van Gulik W, Ferrer P (2012) Quantitative metabolomics analysis of amino acid metabolism in recombinant *Pichia pastoris* under different oxygen availability conditions. Microb Cell Factories 11:83. https://doi.org/10.1186/1475-2859-11-83
59. Russmayer H (2015) The impact of amino acid metabolism on recombinant protein production in *Pichia pastoris*. PhD thesis, BOKU University of Natural Resources and Life Sciences Vienna
60. Driouch H, Melzer G, Wittmann C (2012) Integration of *in vivo* and *in silico* metabolic fluxes for improvement of recombinant protein production. Metab Eng 14(1):47–58. https://doi.org/10.1016/j.ymben.2011.11.002
61. Kauffman KJ, Pridgen EM, Doyle FJ 3rd, Dhurjati PS, Robinson AS (2002) Decreased protein expression and intermittent recoveries in BiP levels result from cellular stress during heterologous protein expression in *Saccharomyces cerevisiae*. Biotechnol Prog 18 (5):942–950. https://doi.org/10.1021/bp025518g
62. Hohenblum H, Gasser B, Maurer M, Borth N, Mattanovich D (2004) Effects of gene dosage, promoters, and substrates on unfolded protein stress of recombinant *Pichia pastoris*. Biotechnol Bioeng 85(4):367–375. https://doi.org/10.1002/bit.10904

63. Whyteside G, Alcocer MJ, Kumita JR, Dobson CM, Lazarou M, Pleass RJ, Archer DB (2011) Native-state stability determines the extent of degradation relative to secretion of protein variants from *Pichia pastoris*. PLoS One 6(7):e22692. https://doi.org/10.1371/journal.pone.0022692
64. de Ruijter JC, Koskela EV, Nandania J, Frey AD, Velagapudi V (2018) Understanding the metabolic burden of recombinant antibody production in *Saccharomyces cerevisiae* using a quantitative metabolomics approach. Yeast 35(4):331–341. https://doi.org/10.1002/yea.3298
65. Hou J, Osterlund T, Liu Z, Petranovic D, Nielsen J (2012) Heat shock response improves heterologous protein secretion in *Saccharomyces cerevisiae*. Appl Microbiol Biotechnol. https://doi.org/10.1007/s00253-012-4596-9
66. Valkonen M, Penttila M, Saloheimo M (2003) Effects of inactivation and constitutive expression of the unfolded- protein response pathway on protein production in the yeast *Saccharomyces cerevisiae*. Appl Environ Microbiol 69(4):2065–2072
67. Valkonen M, Ward M, Wang H, Penttilä M, Saloheimo M (2003) Improvement of foreign-protein production in *Aspergillus niger* var. *awamori* by constitutive induction of the unfolded-protein response. Appl Environ Microbiol 69(12):6979–6986
68. Gasser B, Maurer M, Gach J, Kunert R, Mattanovich D (2006) Engineering of *Pichia pastoris* for improved production of antibody fragments. Biotechnol Bioeng 94(2):353–361. https://doi.org/10.1002/bit.20851
69. Guerfal M, Ryckaert S, Jacobs PP, Ameloot P, Van Craenenbroeck K, Derycke R, Callewaert N (2010) The *HAC1* gene from *Pichia pastoris*: characterization and effect of its overexpression on the production of secreted, surface displayed and membrane proteins. Microb Cell Factories 9:49. https://doi.org/10.1186/1475-2859-9-49
70. Vogl T, Thallinger GG, Zellnig G, Drew D, Cregg JM, Glieder A, Freigassner M (2014) Towards improved membrane protein production in *Pichia pastoris*: general and specific transcriptional response to membrane protein overexpression. New Biotechnol 31(6):538–552. https://doi.org/10.1016/j.nbt.2014.02.009
71. Travers KJ, Patil CK, Wodicka L, Lockhart DJ, Weissman JS, Walter P (2000) Functional and genomic analyses reveal an essential coordination between the unfolded protein response and ER-associated degradation. Cell 101(3):249–258
72. Graf A, Gasser B, Dragosits M, Sauer M, Leparc GG, Tuchler T, Kreil DP, Mattanovich D (2008) Novel insights into the unfolded protein response using *Pichia pastoris* specific DNA microarrays. BMC Genomics 9:390. https://doi.org/10.1186/1471-2164-9-390
73. Moon HY, Cheon SA, Kim H, Agaphonov MO, Kwon O, Oh DB, Kim JY, Kang HA (2015) *Hansenula polymorpha* Hac1p is critical to protein N-glycosylation activity modulation, as revealed by functional and transcriptomic analyses. Appl Environ Microbiol 81(20):6982–6993. https://doi.org/10.1128/aem.01440-15
74. Schuck S, Prinz WA, Thorn KS, Voss C, Walter P (2009) Membrane expansion alleviates endoplasmic reticulum stress independently of the unfolded protein response. J Cell Biol 187(4):525–536. https://doi.org/10.1083/jcb.200907074
75. de Ruijter JC, Koskela EV, Frey AD (2016) Enhancing antibody folding and secretion by tailoring the *Saccharomyces cerevisiae* endoplasmic reticulum. Microb Cell Factories 15:87. https://doi.org/10.1186/s12934-016-0488-5
76. Koskela EV, de Ruijter JC, Frey AD (2017) Following nature's roadmap: folding factors from plasma cells led to improvements in antibody secretion in *S. cerevisiae*. Biotechnol J 12(8). https://doi.org/10.1002/biot.201600631
77. Guerfal M, Claes K, Knittelfelder O, De Rycke R, Kohlwein SD, Callewaert N (2013) Enhanced membrane protein expression by engineering increased intracellular membrane production. Microb Cell Factories 12:122. https://doi.org/10.1186/1475-2859-12-122
78. Emmerstorfer A, Wriessnegger T, Hirz M, Pichler H (2014) Overexpression of membrane proteins from higher eukaryotes in yeasts. Appl Microbiol Biotechnol 98(18):7671–7698. https://doi.org/10.1007/s00253-014-5948-4
79. Hirz M, Richter G, Leitner E, Wriessnegger T, Pichler H (2013) A novel cholesterol-producing *Pichia pastoris* strain is an ideal host for functional expression of human Na,K-ATPase alpha3beta1 isoform. Appl Microbiol Biotechnol 97(21):9465–9478. https://doi.org/10.1007/s00253-013-5156-7
80. Fitzgerald I, Glick BS (2014) Secretion of a foreign protein from budding yeasts is enhanced by cotranslational translocation and by suppression of vacuolar targeting.

Microb Cell Factories 13(1):125. https://doi.org/10.1186/s12934-014-0125-0

81. Zahrl RJ, Mattanovich D, Gasser B (2018) The impact of ERAD on recombinant protein secretion in *Pichia pastoris* (syn *Komagataella* spp.). Microbiology 164(4):453–463. https://doi.org/10.1099/mic.0.000630
82. Ast T, Aviram N, Chuartzman SG, Schuldiner M (2014) A cytosolic degradation pathway, prERAD, monitors pre-inserted secretory pathway proteins. J Cell Sci 127 (14):3017–3023. https://doi.org/10.1242/jcs.144386
83. Toikkanen J, Sundqvist L, Keränen S (2004) *Kluyveromyces lactis SSO1* and *SEB1* genes are functional in *Saccharomyces cerevisiae* and enhance production of secreted proteins when overexpressed. Yeast 21 (12):1045–1055
84. Tang H, Bao X, Shen Y, Song M, Wang S, Wang C, Hou J (2015) Engineering protein folding and translocation improves heterologous protein secretion in *Saccharomyces cerevisiae*. Biotechnol Bioeng 112 (9):1872–1882. https://doi.org/10.1002/bit.25596
85. Puxbaum V, Mattanovich D, Gasser B (2015) Quo vadis? The challenges of recombinant protein folding and secretion in *Pichia pastoris*. Appl Microbiol Biotechnol 99 (7):2925–2938. https://doi.org/10.1007/s00253-015-6470-z
86. Delic M, Gongrich R, Mattanovich D, Gasser B (2014) Engineering of protein folding and secretion-strategies to overcome bottlenecks for efficient production of recombinant proteins. Antioxid Redox Signal 21(3):414–437. https://doi.org/10.1089/ars.2014.5844
87. Hou J, Tyo KE, Liu Z, Petranovic D, Nielsen J (2012) Metabolic engineering of recombinant protein secretion by *Saccharomyces cerevisiae*. FEMS Yeast Res 12(5):491–510. https://doi.org/10.1111/j.1567-1364.2012.00810.x
88. Idiris A, Tohda H, Kumagai H, Takegawa K (2010) Engineering of protein secretion in yeast: strategies and impact on protein production. Appl Microbiol Biotechnol 86 (2):403–417. https://doi.org/10.1007/s00253-010-2447-0
89. Mattanovich D, Gasser B, Hohenblum H, Sauer M (2004) Stress in recombinant protein producing yeasts. J Biotechnol 113 (1–3):121–135
90. Friedlander R, Jarosch E, Urban J, Volkwein C, Sommer T (2000) A regulatory link between ER-associated protein degradation and the unfolded-protein response. Nat Cell Biol 2(7):379–384. https://doi.org/10.1038/35017001
91. Pfeffer M, Maurer M, Kollensperger G, Hann S, Graf AB, Mattanovich D (2011) Modeling and measuring intracellular fluxes of secreted recombinant protein in *Pichia pastoris* with a novel ^{34}S labeling procedure. Microb Cell Factories 10:47. https://doi.org/10.1186/1475-2859-10-47
92. de Ruijter JC, Frey AD (2015) Analysis of antibody production in *Saccharomyces cerevisiae*: effects of ER protein quality control disruption. Appl Microbiol Biotechnol 99 (21):9061–9071. https://doi.org/10.1007/s00253-015-6807-7
93. Bao J, Huang M, Petranovic D, Nielsen J (2017) Moderate expression of *SEC16* increases protein secretion by *Saccharomyces cerevisiae*. Appl Environ Microbiol 83(14). ppi: e03400-16 https://doi.org/10.1128/aem.03400-16
94. Bao J, Huang M, Petranovic D, Nielsen J (2018) Balanced trafficking between the ER and the Golgi apparatus increases protein secretion in yeast. AMB Express 8(1):37. https://doi.org/10.1186/s13568-018-0571-x
95. Connerly P, Esaki M, Montegna E, Strongin D, Levi S, Soderholm J, Glick B (2005) Sec16 is a determinant of transitional ER organization. Curr Biol 15 (16):1439–1447
96. Delic M, Valli M, Graf AB, Pfeffer M, Mattanovich D, Gasser B (2013) The secretory pathway: exploring yeast diversity. FEMS Microbiol Rev 37(6):872–914. https://doi.org/10.1111/1574-6976.12020
97. Cai H, Reinisch K, Ferro-Novick S (2007) Coats, tethers, Rabs, and SNAREs work together to mediate the intracellular destination of a transport vesicle. Dev Cell 12 (5):671–682. https://doi.org/10.1016/j.devcel.2007.04.005
98. Hou J, Tyo K, Liu Z, Petranovic D, Nielsen J (2012) Engineering of vesicle trafficking improves heterologous protein secretion in *Saccharomyces cerevisiae*. Metab Eng 14 (2):120–127. https://doi.org/10.1016/j.ymben.2012.01.002
99. Van Zyl JH, Den Haan R, Van Zyl WH (2014) Over-expression of native *Saccharomyces cerevisiae* exocytic SNARE genes increased heterologous cellulase secretion. Appl Microbiol Biotechnol 98(12):5567–5578. https://doi.org/10.1007/s00253-014-5647-1

100. Xu L, Shen Y, Hou J, Peng B, Tang H, Bao X (2014) Secretory pathway engineering enhances secretion of cellobiohydrolase I from *Trichoderma reesei* in *Saccharomyces cerevisiae.* J Biosci Bioeng 117(1):45–52. https://doi.org/10.1016/j.jbiosc.2013.06.017
101. Ruohonen L, Toikkanen J, Tieaho V, Outola M, Soderlund H, Keranen S (1997) Enhancement of protein secretion in *Saccharomyces cerevisiae* by overproduction of Sso protein, a late-acting component of the secretory machinery. Yeast 13(4):337–351
102. Tang H, Song M, He Y, Wang J, Wang S, Shen Y, Hou J, Bao X (2017) Engineering vesicle trafficking improves the extracellular activity and surface display efficiency of cellulases in *Saccharomyces cerevisiae.* Biotechnol Biofuels 10:53. https://doi.org/10.1186/s13068-017-0738-8
103. Van Zyl JH, Den Haan R, Van Zyl WH (2016) Overexpression of native *Saccharomyces cerevisiae* ER-to-Golgi SNARE genes increased heterologous cellulase secretion. Appl Microbiol Biotechnol 100(1):505–518. https://doi.org/10.1007/s00253-015-7022-2
104. Sreenivas S, Krishnaiah SM, Govindappa N, Basavaraju Y, Kanojia K, Mallikarjun N, Natarajan J, Chatterjee A, Sastry KN (2015) Enhancement in production of recombinant two-chain Insulin Glargine by overexpression of Kex2 protease in *Pichia pastoris.* Appl Microbiol Biotechnol 99(1):327–336. https://doi.org/10.1007/s00253-014-6052-5
105. Gasser B, Prielhofer R, Marx H, Maurer M, Nocon J, Steiger M, Puxbaum V, Sauer M, Mattanovich D (2013) *Pichia pastoris*: protein production host and model organism for biomedical research. Future Microbiol 8:191–208. https://doi.org/10.2217/fmb.12.133
106. Lee J, Park JS, Moon JY, Kim KY, Moon HM (2003) The influence of glycosylation on secretion, stability, and immunogenicity of recombinant HBV pre-S antigen synthesized in *Saccharomyces cerevisiae.* Biochem Biophys Res Commun 303(2):427–432
107. Antebi A, Fink GR (1992) The yeast Ca^{2+}-ATPase homologue, PMR1, is required for normal Golgi function and localizes in a novel Golgi-like distribution. Mol Biol Cell 3(6):633–654. https://doi.org/10.1091/mbc.3.6.633
108. Harmsen M, Bruyne M, Raué H, Maat J (1996) Overexpression of binding protein and disruption of the *PMR1* gene synergistically stimulate secretion of bovine prochymosin but not plant thaumatin in yeast. Appl Microbiol Biotechnol 46(4):365–370
109. Harmsen MM, Langedijk AC, van Tuinen E, Geerse RH, Raue HA, Maat J (1993) Effect of a pmr1 disruption and different signal sequences on the intracellular processing and secretion of Cyamopsis tetragonoloba alpha-galactosidase by *Saccharomyces cerevisiae.* Gene 125(2):115–123
110. Feng Z, Ren J, Zhang H, Zhang L (2011) Disruption of *PMR1* in *Kluyveromyces lactis* improves secretion of calf prochymosin. J Sci Food Agric 91(1):100–103. https://doi.org/10.1002/jsfa.4156
111. Uccelletti D, Farina F, Mancini P, Palleschi C (2004) *KlPMR1* inactivation and calcium addition enhance secretion of non-hyperglycosylated heterologous proteins in *Kluyveromyces lactis.* J Biotechnol 109(1–2):93–101. https://doi.org/10.1016/j.jbiotec.2003.10.037
112. Agaphonov MO, Plotnikova TA, Fokina AV, Romanova NV, Packeiser AN, Kang HA, Ter-Avanesyan MD (2007) Inactivation of the *Hansenula polymorpha PMR1* gene affects cell viability and functioning of the secretory pathway. FEMS Yeast Res 7(7):1145–1152. https://doi.org/10.1111/j.1567-1364.2007.00247.x
113. Zhao HL, Xue C, Wang Y, Duan QF, Xiong XH, Yao XQ, Liu ZM (2008) Disruption of *Pichia pastoris PMR1* gene decreases its folding capacity on human serum albumin and interferon-alpha2b fusion protein. Yeast 25(4):279–286. https://doi.org/10.1002/yea.1589
114. Zhao HL, Xue C, Wang Y, Yao XQ, Liu ZM (2008) Increasing the cell viability and heterologous protein expression of *Pichia pastoris* mutant deficient in *PMR1* gene by culture condition optimization. Appl Microbiol Biotechnol 81(2):235–241. https://doi.org/10.1007/s00253-008-1666-0
115. Sohn YS, Park CS, Lee SB, Ryu DD (1998) Disruption of *PMR1*, encoding a Ca^{2+}-ATPase homolog in *Yarrowia lipolytica*, affects secretion and processing of homologous and heterologous proteins. J Bacteriol 180(24):6736–6742
116. Fokina AV, Chechenova MB, Karginov AV, Ter-Avanesyan MD, Agaphonov MO (2015) Genetic evidence for the role of the vacuole in supplying secretory organelles with Ca2+ in *Hansenula polymorpha.* PLoS One 10(12):e0145915. https://doi.org/10.1371/journal.pone.0145915

117. Agaphonov M, Romanova N, Sokolov S, Iline A, Kalebina T, Gellissen G, Ter-Avanesyan M (2005) Defect of vacuolar protein sorting stimulates proteolytic processing of human urokinase-type plasminogen activator in the yeast *Hansenula polymorpha*. FEMS Yeast Res 5(11):1029–1035. https://doi.org/10.1016/j.femsyr.2005.07.003
118. Idiris A, Tohda H, Sasaki M, Okada K, Kumagai H, Giga-Hama Y, Takegawa K (2010) Enhanced protein secretion from multiprotease-deficient fission yeast by modification of its vacuolar protein sorting pathway. Appl Microbiol Biotechnol 85(3):667–677. https://doi.org/10.1007/s00253-009-2151-0
119. Kitagawa T, Kohda K, Tokuhiro K, Hoshida H, Akada R, Takahashi H, Imaeda T (2011) Identification of genes that enhance cellulase protein production in yeast. J Biotechnol 151(2):194–203. https://doi.org/10.1016/j.jbiotec.2010.12.002
120. de Ruijter JC, Jurgens G, Frey AD (2017) Screening for novel genes of *Saccharomyces cerevisiae* involved in recombinant antibody production. FEMS Yeast Res 17(1). ppi: fow104 https://doi.org/10.1093/femsyr/fow104
121. Marsalek L, Gruber C, Altmann F, Aleschko M, Mattanovich D, Gasser B, Puxbaum V (2017) Disruption of genes involved in CORVET complex leads to enhanced secretion of heterologous carboxylesterase only in protease deficient *Pichia pastoris*. Biotechnol J 12(5). https://doi.org/10.1002/biot.201600584
122. Gagnon-Arsenault I, Tremblay J, Bourbonnais Y (2006) Fungal yapsins and cell wall: a unique family of aspartic peptidases for a distinctive cellular function. FEMS Yeast Res 6(7):966–978. https://doi.org/10.1111/j.1567-1364.2006.00129.x
123. Silva CI, Teles H, Moers AP, Eggink G, de Wolf FA, Werten MW (2011) Secreted production of collagen-inspired gel-forming polymers with high thermal stability in *Pichia pastoris*. Biotechnol Bioeng 108(11):2517–2525. https://doi.org/10.1002/bit.23228
124. Cho EY, Cheon SA, Kim H, Choo J, Lee DJ, Ryu HM, Rhee SK, Chung BH, Kim JY, Kang HA (2010) Multiple-yapsin-deficient mutant strains for high-level production of intact recombinant proteins in *Saccharomyces cerevisiae*. J Biotechnol 149(1–2):1–7. https://doi.org/10.1016/j.jbiotec.2010.06.014
125. Bourbonnais Y, Larouche C, Tremblay GM (2000) Production of full-length human pre-elafin, an elastase specific inhibitor, from yeast requires the absence of a functional yapsin 1 (Yps1p) endoprotease. Protein Expr Purif 20(3):485–491. https://doi.org/10.1006/prep.2000.1338
126. Yao XQ, Zhao HL, Xue C, Zhang W, Xiong XH, Wang ZW, Li XY, Liu ZM (2009) Degradation of HSA-AX15(R13K) when expressed in *Pichia pastoris* can be reduced via the disruption of *YPS1* gene in this yeast. J Biotechnol 139(2):131–136. https://doi.org/10.1016/j.jbiotec.2008.09.006
127. Wu M, Shen Q, Yang Y, Zhang S, Qu W, Chen J, Sun H, Chen S (2013) Disruption of *YPS1* and *PEP4* genes reduces proteolytic degradation of secreted HSA/PTH in *Pichia pastoris* GS115. J Ind Microbiol Biotechnol 40(6):589–599. https://doi.org/10.1007/s10295-013-1264-8
128. Sohn MJ, Oh DB, Kim EJ, Cheon SA, Kwon O, Kim JY, Lee SY, Kang HA (2012) *HpYPS1* and *HpYPS7* encode functional aspartyl proteases localized at the cell surface in the thermotolerant methylotrophic yeast *Hansenula polymorpha*. Yeast 29(1):1–16. https://doi.org/10.1002/yea.1912
129. Liu Z, Liu L, Osterlund T, Hou J, Huang M, Fagerberg L, Petranovic D, Uhlen M, Nielsen J (2014) Improved production of a heterologous amylase in *Saccharomyces cerevisiae* by inverse metabolic engineering. Appl Environ Microbiol 80(17):5542–5550. https://doi.org/10.1128/aem.00712-14
130. Gasser B, Sauer M, Maurer M, Stadlmayr G, Mattanovich D (2007) Transcriptomics-based identification of novel factors enhancing heterologous protein secretion in yeasts. Appl Environ Microbiol 73(20):6499–6507
131. Toikkanen JH, Miller KJ, Soderlund H, Jantti J, Keranen S (2003) The beta subunit of the Sec61p endoplasmic reticulum translocon interacts with the exocyst complex in *Saccharomyces cerevisiae*. J Biol Chem 278(23):20946–20953. https://doi.org/10.1074/jbc.M213111200
132. Liu SH, Chou WI, Lin SC, Sheu CC, Chang MD (2005) Molecular genetic manipulation of *Pichia pastoris SEC4* governs cell growth and glucoamylase secretion. Biochem Biophys Res Commun 336(4):1172–1180. https://doi.org/10.1016/j.bbrc.2005.08.234
133. De Nobel JG, Barnett JA (1991) Passage of molecules through yeast cell walls: a brief essay-review. Yeast 7(4):313–323. https://doi.org/10.1002/yea.320070402
134. Rodriguez-Limas WA, Tannenbaum V, Tyo KE (2015) Blocking endocytotic mechanisms

to improve heterologous protein titers in *Saccharomyces cerevisiae*. Biotechnol Bioeng 112 (2):376–385. https://doi.org/10.1002/bit.25360

135. Marx H, Sauer M, Resina D, Vai M, Porro D, Valero F, Ferrer P, Mattanovich D (2006) Cloning, disruption and protein secretory phenotype of the *GAS1* homologue of *Pichia pastoris*. FEMS Microbiol Lett 264(1):40–47. https://doi.org/10.1111/j.1574-6968.2006.00427.x
136. Vai M, Brambilla L, Orlandi I, Rota N, Ranzi BM, Alberghina L, Porro D (2000) Improved secretion of native human insulin-like growth factor 1 from gas1 mutant *Saccharomyces cerevisiae* cells. Appl Environ Microbiol 66 (12):5477–5479
137. Resina D, Maurer M, Cos O, Arnau C, Carnicer M, Marx H, Gasser B, Valero F, Mattanovich D, Ferrer P (2009) Engineering of bottlenecks in *Rhizopus oryzae* lipase production in *Pichia pastoris* using the nitrogen source-regulated FLD1 promoter. New Biotechnol 25(6):396–403. https://doi.org/10.1016/j.nbt.2009.01.008
138. Passolunghi S, Riboldi L, Dato L, Porro D, Branduardi P (2010) Cloning of the *Zygosaccharomyces bailii GAS1* homologue and effect of cell wall engineering on protein secretory phenotype. Microb Cell Factories 9:7. https://doi.org/10.1186/1475-2859-9-7
139. Larsen S, Weaver J, de Sa Campos K, Bulahan R, Nguyen J, Grove H, Huang A, Low L, Tran N, Gomez S, Yau J, Ilustrisimo T, Kawilarang J, Lau J, Tranphung M, Chen I, Tran C, Fox M, Lin-Cereghino J, Lin-Cereghino GP (2013) Mutant strains of *Pichia pastoris* with enhanced secretion of recombinant proteins. Biotechnol Lett 35(11):1925–1935. https://doi.org/10.1007/s10529-013-1290-7
140. Chung BH, Park KS (1998) Simple approach to reducing proteolysis during secretory production of human parathyroid hormone in *Saccharomyces cerevisiae*. Biotechnol Bioeng 57(2):245–249
141. Vad R, Nafstad E, Dahl L, Gabrielsen O (2005) Engineering of a *Pichia pastoris* expression system for secretion of high amounts of intact human parathyroid hormone. J Biotechnol 116(3):251–260

Chapter 3

Pichia pastoris Promoters

Aysun Türkanoğlu Özçelik, Semiramis Yılmaz, and Mehmet Inan

Abstract

The methylotrophic yeast *Pichia pastoris* (*Komagataella phaffii*) is used as an expression system for recombinant protein production for a variety of applications. It grows rapidly on inexpensive media containing methanol, glucose, glycerol, or ethanol as a sole carbon source. *P. pastoris* makes many posttranslational modifications and produces recombinant proteins either intracellularly or extracellularly. Because of these properties, *P. pastoris* has become a highly preferred host organism for biotechnology, pharmaceutical industry, and researchers.

Recombinant protein production is usually performed under the control of the promoter of the alcohol oxidase gene I (*AOX1*). The *AOX1* promoter is induced by methanol and repressed by glucose and ethanol. The regulation mechanisms of the *AOX1* promoter have been studied in recent years. Another promoter used in recombinant protein production is derived from glyceraldehyde 3-phosphate dehydrogenase (*GAP*). It is a constitutive promoter. Recent literature showed that newly identified promoters of *P. pastoris* are promising as well, in addition to pAOX1 and pGAP.

In this chapter, the regulation mechanisms of inducible pAOX1 and constitutive pGAP promoters are discussed. In addition, here we present an overview about the novel *ADH3* promoter and alternative promoters of *P. pastoris*.

Key words Methylotrophic yeast, *Pichia pastoris*, *AOX1* promoter, *ADH3* promoter, *GAP* promoter

1 Introduction

Proteins are arguably the most important molecules which are responsible for biological processes. They can be found in the human body mostly as enzymes, hormones, and cellular repair and structural proteins. To obtain pure protein, natural sources of proteins are limited and costly. In the middle of the 1970s, with the development of recombinant DNA technology, microorganisms were enabled to produce peptides and/or proteins that they could not produce in their natural state. The use of recombinant proteins produced industrially by various host organisms is increasing especially in the food, agriculture, and health sectors. Over the years, a wide variety of heterologous expression systems have been developed to meet the need for recombinant protein. For this aim,

Brigitte Gasser and Diethard Mattanovich (eds.), *Recombinant Protein Production in Yeast*, Methods in Molecular Biology, vol. 1923, https://doi.org/10.1007/978-1-4939-9024-5_3,

recombinant proteins are expressed in organisms such as bacteria, yeast, or animal cells. Prokaryotic expression systems are the first preference for recombinant protein production because of being simple and cheap. However, during production of eukaryotic proteins into prokaryotic systems, some problems may be encountered such as instability or lack of biological activity of the produced protein. The intracellular environment of yeasts is very suitable for the correct folding of eukaryotic proteins when compared to prokaryotic systems. The eukaryotic expression system yeasts can perform posttranslational modifications of eukaryotic proteins such as glycosylation and formation of disulfide bridges, when compared to prokaryotic organisms [1]. Genetic manipulations of yeasts are also as easy as bacteria. Because of all these properties, yeasts have great advantages for the production of recombinant proteins.

The production of recombinant proteins is divided into four stages:

1. Cloning of the gene of interest in a suitable vector under a right promoter.
2. Transformation of the expression cassette into host cells.
3. Expression of the gene of interest in a bioreactor under controlled conditions.
4. Downstream processes such as purification and recovery.

Saccharomyces cerevisiae has been the first manipulated yeast for recombinant protein production. There is a lot of knowledge about its physiology and molecular biology in the literature, but it could not always be used as a suitable expression host due to its disadvantages like plasmid loss in scale-up production, hyper-glycosylation, and reduced yield [2]. Methylotrophic yeasts *Pichia pastoris* (*Komagataella spp.*) [3–6], *Pichia methanolica* [7], *Candida boidinii* [8, 9], and *Hansenula polymorpha* [10, 11] have also been used as expression host systems for recombinant protein production. Especially the *P. pastoris* expression system is a perfect host for industrial recombinant protein production.

P. pastoris is a single cell methylotrophic yeast and grows rapidly on inexpensive media containing methanol, glucose, glycerol, or ethanol as a sole carbon source. It is capable of many posttranslational modifications, important for eukaryotic proteins, such as proteolytic processing, disulfide bridge formation, and glycosylation [12]. These properties make the *Pichia* expression system more advantageous than bacteria for recombinant production of eukaryotic proteins. Large amounts of correct folded and glycosylated recombinant protein production are achieved by the *P. pastoris* expression system. Compared to *S. cerevisiae*, it typically has better secretion efficiency [13, 14]. Unlike *S. cerevisiae*, *P. pastoris* is not a fermentative yeast; as a result of this, almost all glucose is converted to biomass. Glucose is not converted to ethanol and other organic acids, which allows *P. pastoris* to reach high cell densities under aerobic conditions. *P. pastoris* can use methanol as a carbon and energy source for its growth. The alcohol oxidase enzyme, responsible for degradation of methanol in the catabolic pathway, reaches a level that will compose 30% of the total cell protein during growth on methanol [15]. Alcohol oxidase has low affinity to oxygen, and in order to compete with this disadvantage, high amounts of alcohol oxidase are produced. Thus, high level expression by the *AOX1* promoter makes *P. pastoris* advantageous for producing of recombinant protein.

In recombinant protein production processes, choosing the right promoter is important besides choosing the right expression host. Inducible and constitutive promoters of yeasts are commonly used in recombinant protein production. Gene expression can be controlled by induction or repression during the recombinant protein production with inducible promoters. The advantage of inducible promoters in practical applications is that biomass accumulation and recombinant protein production phases are carried out in different stages, which also provides an advantage in the production of toxic proteins [14, 16, 17]. Constitutive expression of foreign protein can be toxic to yeast cells. For this reason constitutive promoters are not preferred for production of toxic proteins.

The inducible and constitutive promoters of *Pichia pastoris* discussed in this chapter are summarized in Table 1.

2 *AOX1* Promoter

The promoter of alcohol oxidase I (*AOX1*) gene is commonly used for recombinant protein expression in *P. pastoris. AOX1* encodes for the alcohol oxidase enzyme which catalyzes the conversion of methanol to formaldehyde in the methanol utilization pathway. *AOX1* is a strong and tightly regulated methanol-inducible promoter of *P. pastoris.*

P. pastoris needs the alcohol oxidase enzyme during the growth on media containing methanol. Two alcohol oxidase genes are found in the *P. pastoris* genome, namely, *AOX1* and *AOX2. AOX1* is responsible for 85% of alcohol oxidase enzyme activity in the cell. The expression of *AOX1* is regulated with repression, derepression, and induction mechanisms developed in response to different carbon sources. *P. pastoris* cells can grow on different carbon sources such as glucose, methanol, glycerol, ethanol, mannitol, sorbitol, trehalose, and alanine. However, the presence of glucose, ethanol, glycerol, and acetate represses the usage of methanol [19, 36]. Inan and Meagher showed that sorbitol, mannitol, trehalose, and alanine have no catabolite repression effect on *AOX1* promoter, but methanol is required for expression [37]. During

Table 1
***Pichia pastoris* promoters (adapted from Vogl and Glieder [18])**

Inducible promoters				
Gene name	**Gene product**	**Inducer**	**Expression level**	**References**
AOX1	Alcohol oxidase 1	Methanol	Strong	[13, 19, 20]
ADH3	Alcohol dehydrogenase	Ethanol	Strong	[21]
DAS	Dihydroxyacetone phosphate	Methanol	Strong	[19, 20, 22]
FLD1	Formaldehyde dehydrogenase	Methanol/ methylamine	Strong	[23]
PEX8	Peroxisomal matrix protein	Methanol/oleate	Weak	[13]
ICL1	Isocitrate lyase	Ethanol		[24]
LRA3	L-rhamnonate dehydratase	Rhamnose	~75% of pGAP	[25]
LRA4	L-KDR aldolase	Rhamnose	~22% of pGAP	[25]
THI11	Thiamine biosynthesis protein	Repressed by thiamine	~70% of pGAP	[26, 27]
GTH1	High affinity glucose transporter	Limited glucose		[29]
CUP1	Copper-binding metallothionein protein	Copper		[25]
PPLCC1	Laccase	Copper		[25]
Constitutive promoters				
Gene name	**Gene product**	**Expression level**		
GAP	Glyceraldehyde 3-phosphate dehydrogenase	Strong		[31]
YPT1	GTPase involved in secretion	Weak		[32]
TEF1	Translation elongation factor-1 alpha	Strong		[26, 33]
GCW14	Glycosylphosphatidylinositol	Strong		[34]
PGK1	Phosphoglycerate kinase	~10% of pGAP		[26, 35]

studies about catabolite repression and alcohol oxidase activity in *P. pastoris* and other methylotrophic yeasts, it has been observed that alcohol oxidase activity is suppressed by the presence of glucose and ethanol in the medium. Researchers reported that this suppressive effect is mediated by two regulation mechanisms, catabolite repression and catabolite inactivation [38–40]. Similarly, Cregg et al. reported that the methanol-inducible *AOX1* promoter of *P. pastoris* is repressed by carbon sources like glucose, ethanol, and glycerol [41].

The expression of *MOD1* of *P. methanolica* and *MOX* of *H. polymorpha* are not repressed by glycerol. Approximately 1.5–1.6 kb regions upstream of the 5′ ends of *MOD1* and *MOD2* genes have been isolated as promoter regions, using an acid phosphatase enzyme as a reporter enzyme. While *MOD1* is expressed in glycerol, *MOD1* and *MOD2* are repressed in the presence of glucose. Parallel results have been obtained with the previous MOD isozyme activity measurement. Low concentration of methanol induces *MOD1* expression; on the other hand, high concentration of methanol induces strongly both expressions of *MOD1* and *MOD2* genes. The regulation of *MOD2* is similar to the regulation of *P. pastoris AOX1*. *MOD2* is repressed by glucose, but methanol is needed for expression of the gene. Unlike *PpAOX1*, it is not expressed in media containing glycerol, but addition of methanol to the medium causes derepression and induces full expression [42].

Studies to elucidate the regulation mechanisms of AOX expression under the conditions of repression, derepression, and induction have been carried out by performing functional analysis of genes thought to be involved in the mechanism according to determined models of *S. cerevisiae*. So far, general information and a hypothesis have been obtained from chemostat studies in which all growth factors can be controlled. With the molecular studies carried out by deletion of promoter sequences, important regions (*cis*-acting regulatory sequence elements) in promoter regulatory mechanisms and proteins (*trans*-acting factors) bound to these regions have been defined, some of the proteins have been identified, and thus regulation mechanisms have been found out in more detail.

The determination of important regions that play a role in induction and repression of the *PpAOX1* promoter has been started with a study by Inan [43]. The 5′ region of the *AOX1* gene is divided into six fragments by using deletion analysis, and these fragments are named A, B, C, D, E, and F. Inan showed that each fragment has a significant effect on the promoter activity in deletion analysis experiments using the *E. coli lacZ* gene as a reporter. Electrophoretic mobility shift assays (EMSA) with cell extracts grown in methanol and methanol/ethanol showed that DNA-binding proteins bound to the A and C fragments. It has also been proven that these two fragments do not compete for the same protein. It has been suggested that the E fragment carries an upstream activating sequence (UAS) because the deletion of the E fragment reduced promoter activity by 80%. As a result of deletion of the D fragment, a 30% increase in promoter activity has been observed suggesting that it carries an upstream repressive sequences (URS). While only proteins from cell extract grown on methanol alone bound to the A fragment, the C fragment made a complex with the cell extract grown on both methanol and methanol/

ethanol containing medium. A 10% increase in reporter enzyme activity has been observed due to deletion of the C fragment, and the amount of protein bound of extracts grown in methanol medium was higher than that of extracts of methanol/ethanol grown cells. According to this knowledge, it was concluded that protein bound to the C fragment plays a role in ethanol repression [43].

Hartner et al. have performed in silico analysis of the *AOX1* promoter to identify possible transcription factor-binding sites and identified regions that resulted in a 50% reduction in promoter activity as a result of deletion and mutation experiments performed in these regions [44]. The most striking change is that when the nucleotides −209/−210 are removed or mutated, the promoter activity is reduced by 6–30% relative to the natural promoter activity. A promoter library has been created with the changes made in these regions. Promoter activities in the generated library were ranging from 6% to 160% of the natural promoter activity. They characterized the library under the conditions of repression, derepression, and induction and have observed in silico the presence of different cis-effective regions. However, trans-acting proteins that bind to these cis-effective regions have not been experimentally determined [44].

Xuan et al. conducted a study to identify cis-acting regions of the *AOX1* promoter. For this aim the *AOX1* promoter region was divided into fragments, and deletion of the D fragment caused significant loss of transcriptional activity of the promoter. Furthermore, extract of cells grown under induction (methanol) and repression (glucose + methanol) conditions formed protein-DNA complexes with the D fragment. Xuan et al. concluded that the D fragment contains the sequences for binding of transcription regulatory proteins, and this fragment plays a role both in activation and repression [45]. The C fragment of Inan's study [43] and the D fragment of Xuan's study [45] correspond to the same region, and similar results were obtained from these two studies.

Lin-Cereghino et al. identified a protein that plays a role in methanol induction, and it was named methanol expression regulator 1 (Mxr1p) [46]. It was shown that transcription of genes in the methanol metabolism and transcription of other peroxisomal genes completely stops or significantly decreases in Δmxr1 strains obtained by deletion or inactivation of the *MXR1* gene. This case resulted in phenotypic effects in Δmxr1 strains like failure to grow in medium containing methanol or oleate as the sole carbon source and slow growth rate in glucose, glycerol, and ethanol. The DNA-binding region of Mxr1p is homologous with the DNA-binding region of Adr1p (alcohol dehydrogenase II regulator) of *S. cerevisiae*. Adr1p is expressed in the presence of glucose, but Mxr1p is produced constitutively in low levels. Mxr1p is found in the cytoplasm when cells were grown in glucose, and it migrates into the nucleus when cells were passed to methanol or oleate. In addition, Mxr1p localizes in the nucleus when cells grew in glycerol or ethanol, but the *AOX1* promoter was still repressed by these carbon sources. For this reason, it has been suggested that the expression of the *AOX1* gene has not been controlled only by the localization of Mxr1p [46].

Another positive regulator of the *AOX1* promoter is defined by Takagi et al. [47]. Positive regulator of methanol (Prm1) activates transcription from the methanol-inducible promoters even in the absence of methanol. *PRM1* gene deletion reduces the reporter phytase gene expression under the control of both *AOX1* and *DAS* promoters in the presence of methanol. Constitutive expression of Prm1p induces transcription with methanol-inducible promoters without the need for methanol in the growth medium.

Wang et al. identified a methanol-induced transcription factor 1 (Mit1) that regulates the expression of the *AOX1* gene [48]. According to a structural analysis of Mit1, it plays a critical role in repression of the *AOX1* promoter in glycerol medium. In addition, it strongly induces the *AOX1* promoter in methanol medium. Wang et al. reported that Mit1, Mxr1, and Prm1 are the positive regulators of *AOX1*, and they did not interact with each other. Mit1 and Prm1 bind to the *AOX1* promoter at different sites from Mxr1. The strength of binding of Mit1 and Prm1 to the *AOX1* promoter is affected by carbon sources. For example, weak, mild, and strong bindings are observed in glucose, glycerol, and methanol medium, respectively. They also concluded that Mit1, Mxr1, and Prm1 activate the *AOX1* promoter cooperatively. The data obtained from deletion study showed that deletion of any of them inhibits the transmission of the methanol induction signal to the *AOX1* promoter [48].

So far, three positive regulators of *AOX1* promoter in methanol induction have been identified. A study about the mechanism of catabolite repression of *P. pastoris* was conducted by Yılmaz [49]. This study showed that *MIG1* and *MIG2* genes play a role in catabolite repression of the *AOX1* gene. During cell growth in glycerol and glucose, no phenotypic effect was observed due to inactivation of *MIG1* and *MIG2* genes. Growth rates of *MIG1* knockout or *MIG1/MIG2* double knockout strains slightly increase in methanol medium. AOX enzyme activity was measured in all strains, but no activity was observed in glucose/methanol medium. However, in glycerol/methanol medium, 30% and 50% AOX activities were detected for Δmig2 and Δmig1Δmig2 strains, respectively. This is the first study suggesting that PpMig1 and PpMig2 have a role in a negative regulation mechanism of the *AOX1* promoter [49]. Similar results were found by Wang et al. [50]. They also reported that PpMig1 and PpMig2 proteins function in catabolite repression of the *AOX1* promoter.

Wang and coworkers identified another transcriptional repressor of the *AOX1* promoter, Nrg1, in glucose and glycerol medium [51]. In this study a growth defect has been observed in a Δnrg1 strain when growing in glucose, glycerol, and methanol. According to results of EMSA and DNase I footprinting assay, PpNrg1 binds to five sites including two binding sites of PpMxr1 of the *AOX1* promoter. It has been suggested that PpNrg1 competes with the PpMxr1 for binding to the *AOX1* promoter; thus it represses the activation of *AOX1*.

3 *GAP* Promoter

Glyceraldehyde 3-phosphate dehydrogenase (GAP) is an enzyme in glycolysis which is constitutively expressed. The *GAP* gene of *P. pastoris* was isolated by Waterham et al. with the colony hybridization technique [31]. The sequence of the *GAP* gene is 80% identical with the *S. cerevisiae TDH1* and *TDH3* genes. According to Southern blot results and genome sequences, *P. pastoris* has only one *GAP* gene encoded in its genome. The *GAP* gene is constitutively expressed in cells grown on glucose, glycerol, and methanol. In recombinant protein production under the control of pGAP, biomass increases, while protein is synthesized at the same time. The level of *GAP* promoter activity is two-thirds in glycerol medium and one-third in methanol medium when compared to glucose medium [31]. *AOX1* needs methanol induction to express protein. Methanol is a petrochemical product, and in some cases like production of food additives, its usage as a carbon source may not be desired. Additionally, handling large amounts of methanol for industrial applications requires extra safety precautions due to flammability and the risk of explosive mixtures with air. For this reason, the *GAP* promoter is an alternative for recombinant protein production in *P. pastoris.* The advantages of pGAP are there is no need for methanol for induction and there is no need to change the carbon source during growth. On the downside, this promoter is not useful for the production of a protein that is toxic to yeast cells.

A *GAP* promoter library was constructed by mutagenesis by Qin et al. [52]. They obtained pGAP variants with an activity ranging from 0.006- to 19.6-fold of the wild-type promoter activity. However, this study is insufficient in order to understanding of regulation mechanism of *GAP* promoter. Recently, a study was published related with the regulation of pGAP [53]. Firstly, putative transcription factor-binding sites (TFBS) and potential transcription factors (TF) of pGAP have been defined by in silico analysis. Carbon source affects the expression strength of the *GAP* promoter because it is responsible for the expression of glycolytic pathway enzymes. For this reason, TFs found in carbon source utilization, and their binding sites have a crucial importance in this study. pGAP variants have been constructed with the duplication and deletion of the TFBS and overexpression or deletion of TF genes related with the carbon source utilization. The expression of ten different pGAP variants has been tested in excess glucose, excess glycerol, limited glucose, and methanol medium by the production of enhanced green fluorescence protein (eGFP). Increase in eGFP production has been observed in three variants obtained by the deletion of binding sites of Mig1, Mig2, and Mig1/Mig2 in excess glucose. Similar eGFP production has been obtained with the wild-type promoter in limited glucose, excess glycerol, and methanol.

The other investigated TF was Rgt1p whose role in *P. pastoris* is unknown. The homolog of this TF induces the expression of the hexose transporter gene of *S. cerevisiae*. A variant has been obtained by duplication of the Rgt1p-binding site, leading to low-level expression of eGFP in limited and excess glucose. It has been suggested that this binding site may affect glucose induction [53].

There are four different binding sites of regulatory factor X (Rfx1p) in *P. pastoris* according to in silico analysis. A pGAP variant obtained by deleting these four binding sites showed an increase in eGFP expression in limited glucose and methanol, while there was no change in eGFP expression in excess glucose and glycerol.

CAT8 and *SIP4* genes encode transcription factors in *S. cerevisiae* that activate carbon source responsive elements. High expression levels in limited glucose medium were obtained in a variant with duplication of these TF-binding sites.

Gal4-*like* is the last evaluated TF for the regulation mechanism of pGAP. When the Gal4 binding region of *P. pastoris* is duplicated, an increase in expression of eGFP has been observed under the excess glucose, limited glucose, methanol, and excess glycerol conditions. It has been concluded that Gal4-*like* TF may be an activator for pGAP.

The effect of overexpression and knocking out of four different TFs such as Mig1-1, Mig1-2, Rfx1, and Gal4-*like* on the growth and expression of eGFP have also been evaluated. Overexpression of Mig1-1 and Rfx1 inhibits the growth of cells in all media tested. A 20–40% decrease of growth has been observed in Mig1-2 and Gal4-*like* overexpressing strains in excess glucose, glycerol, and methanol. According to the results obtained from overexpression of Gal4-*like* TF, eGFP production is in the range of 150% to 218% when compared to the wild-type strain in excess glucose and limited glucose medium, respectively. On the contrary, no significant increase or decrease was observed in growth and eGFP expression levels in the experiments done with Mig1-1, Mig1-2, and Rfx-1 knockout strains. Gal4-*like* TF knock-out increased the growth in excess glucose and glycerol, but it had no effect on the growth in limited glucose and methanol media. These results are consistent with the results obtained from Gal4-binding site duplicated strains. So it has been concluded that the Gal4-*like* TF may play a significant role in glucose metabolism of *P. pastoris*. In order to confirm these results, the Gal4-*like* TF has been overexpressed or deleted, and eGFP has been produced under the control of a *GAP* promoter variant with the Gal4-binding sites duplicated. While a 3.1-fold increase in eGFP expression is obtained by overexpression of the Gal4-*like* TF in limited glucose conditions, a decreased eGFP production is observed in the Gal4-*like* TF knockout strain under the all conditions tested. Therefore, all of these results showed that Gal4-*like* TF is an activator for pGAP regulation [53].

4 *ADH3* Promoter

P. pastoris can use ethanol as a carbon source like glucose, methanol, glycerol, sorbitol, mannitol, and others [37]. There is not much published about the ethanol metabolism of *P. pastoris*.

P. pastoris is a Crabtree-negative yeast, thus metabolizing glucose under aerobic conditions to CO_2, so that ethanol is not produced in the presence of high glucose in the medium. However, Inan and Meagher and Shon et al. have reported that *P. pastoris* produces small amount of ethanol as a by-product under aerobic conditions, and the produced ethanol affects recombinant protein production by repressing the pAOX1 [36, 54]. During recombinant protein production under the pGAP control, ethanol is produced when oxygen supply is limited. Unlike pAOX1, the produced ethanol does not repress pGAP, and an increased expression level was observed during pGAP controlled recombinant protein production in hypoxic conditions [55].

Alcohol dehydrogenase (ADH) genes are responsible for ethanol metabolism in *S. cerevisiae*. Two different alcohol dehydrogenase genes (ADH) of *P. pastoris*, namely, *ADH* and *ADH3*, have been annotated [22, 56, 57]. The functions of *ADH* and *ADH3* genes have been investigated by Karaoglan et al. [58]. Today, seven genes with putative ADH activity have been annotated (www.pichiagenome.org), of which *ADH3* described here is annotated as the homolog of *S. cerevisiae ADH2*. These genes have been disrupted in order to understand their role in ethanol metabolism of *P. pastoris*. When Δadh3 strains were grown on minimal glucose (MD) media, all glucose was consumed at the end of the 36 h. The wild-type strain X-33, however, did not use all glucose in the medium even until 90 h. The wild-type strain produced ethanol in the first 24 h and then utilized ethanol completely. Ethanol was also present in the Δadh3 strain at 24 h, but the produced ethanol was not consumed until the end of the fermentation. These results have been confirmed by Adh activity measurement assay. A 93% decrease in Adh activity has been observed in Δadh3 strain in MD media when compared to wild-type strain. According to these results, it was concluded that *ADH3* is responsible for ethanol utilization in *P. pastoris*. In order to determine the role of the *ADH* gene (gene name PP7435_Chr1-1573) in ethanol metabolism, it was disrupted, and ethanol productions of wild-type X-33, Δadh, Δadh3, and ΔadhΔadh3 strains were compared. No difference in ethanol production was observed between them, indicating that the *ADH* gene has no role in ethanol production or utilization in *P. pastoris* [58].

The same research group has also tested the *ADH3* promoter for recombinant protein production by comparing it with the commonly used *P. pastoris* promoters pAOX1 and pGAP [21]. The *Aspergillus niger* xylanase (*XylB*) gene was expressed under control of pADH3, pAOX1, and pGAP. In shake flask experiments, twofold lower xylanase enzyme activity was obtained from pADH3 controlled expression when compared to pAOX1 controlled expression. No difference was observed in enzyme activity between pGAP and pADH3 regulated enzyme production. However, in bioreactor scale the highest xylanase activity was measured in samples taken from pADH3 controlled xylanase production. Thus it can be concluded that the *ADH3* promoter can be used for recombinant protein production for large-scale applications [21]. After demonstrating that the *ADH3* promoter is an alternative strong promoter to the *AOX1* and *GAP* promoters, studies have oriented on the identification of regulatory DNA regions that play a role in the activity of the *ADH3* promoter. Upstream elements of pADH3 were investigated with deletion analyses by Erden Karaoglan [59].

5 Alternative Promoters

There are also different inducible and constitutive promoters in *P. pastoris*, but their usage is not as common as pAOX1 and pGAP. Dihydroxyacetone phosphate synthase (*DAS*) and alcohol oxidase 2 (*AOX2*) are examples of methanol-inducible alternative promoters of *P. pastoris*. pDAS is as strong as pAOX1 [19]. Two *DAS* genes are defined in *P. pastoris* [22, 56]. Studies indicate that Mxr1p and Prm1p are involved in the regulation mechanism of pDAS2 [60]. Another inducible promoter is formaldehyde dehydrogenase (*FLD1*). pFLD1 is induced by methanol or methylamine, and also the strength of pFLD1 is found similar to pAOX1 under the condition of methanol or methylamine induction [23]. The level of methanol induction of pFLD1 depends on the nitrogen source. This promoter can be used for recombinant protein production without methanol induction because of its induction with the simple and inexpensive nitrogen source methylamine [23, 61].

The promoter of the peroxisomal matrix protein gene *PEX8* is induced by methanol or oleic acid. Mxr1p binds to pPEX8 in five regions [60]. Menendez et al. have isolated the *P. pastoris* isocitrate lyase gene (*ICL1*) [24]. The *ICL1* promoter is repressed by glucose, while it is induced by the presence of ethanol or the absence of glucose. These promoters are not commonly used due to their low expression levels.

The *THI11* promoter regulates a gene involved in the biosynthesis of thiamine. Stadlmayr et al. have investigated the response of the promoter against thiamine [26]. The activity of the promoter was measured by eGFP expression on methanol and glucose. When same clones were grown on synthetic medium containing glucose, salts, and biotin, eGFP expression was completely inhibited by the addition of thiamine. Thus it was discovered that this promoter is regulated by the availability of thiamine in the growth medium [26]. Landes et al. investigated the regulation mechanism of thiamine concentrations on pTHI11 activity [27]. An extracellular thiamine concentration of 40 nM did not affect recombinant protein production, while 400 nM and 670 nM thiamine decreased the protein expression, and recombinant protein production was completely repressed by 4 and 40 μM thiamine. According to this study, pTHI11 activity can be controlled for a desired period of time by addition of a defined amount of thiamine into the culture medium [27].

Novel limited glucose inducible promoters have been identified by Prielhofer et al. [28]. Firstly, the gene expression under excess glycerol in batch and limited glucose in fed-batch, respectively, were determined by DNA microarray analysis. Based on the results of DNA microarray data, six novel promoters were identified and characterized in this study. Two novel promoters, namely, pG1 and pG6 reached the highest expression levels, and they were tightly repressed by excess glucose. In addition, the production of recombinant protein under the regulation of pG1 was twofold higher than under the control of pGAP. Prielhofer et al. suggested that pG1 is the promoter of a high-affinity glucose transporter (*GTH1*) gene [28].

Recently the rhamnose utilization pathway of *P. pastoris* was been investigated [25]. Four genes were determined to be involved in rhamnose metabolism of *P. pastoris*, and two of them have been studied in detailed, namely, *LRA3* and *LRA4*. When rhamnose is used as sole carbon source in the growth medium, the disruption of *LRA4* causes a growth defect of *P. pastoris* cells. The presence of glucose represses the expression of *LRA3* and *LRA4* genes, while rhamnose activates the expression of these two genes. The expression levels of *LRA3 and LRA4* promoters have been compared to expression level of the *GAP* promoter by using β-galactosidase as a reporter gene. pLRA3 and pLRA4 showed 75% and 22% expression levels at 2% rhamnose concentration, as compared to pGAP, respectively. It can be concluded that pLRA3 is a strong and pLRA4 is a weak rhamnose inducible promoter. Liu and coworkers suggested that *LRA3* can be useful for the production of food additives and therapeutic proteins because it is induced with rhamnose instead of methanol [25].

Inducible heterologous promoters are also used for recombinant protein production in *P. pastoris.* It has been shown that the *S. cerevisiae CUP1* promoter can actively work in *P. pastoris* and can be induced at desired levels by different Cu^{2+} concentrations [29]. In another study conducted by Kim et al., a copper inducible fungal laccase promoter was evaluated for recombinant laccase production in *P. pastoris* [30]. The *AOX1* promoter of *P. pastoris* was replaced with the laccase promoter, and the promoter activity of pPPLCC1 was compared to pAOX1 and pCUP1 by measuring the laccase activity. pAOX1-Lacc6 and pPPLCC1-Lacc6 cells were grown on BMDY agar plates containing 2,2-azino-bis(3-ethylbenzothiazoline-6-sulfonicacid) diammonium salt (ABTS) for activity staining. Laccase activity has been observed only in pPPLCC1-Lacc6 cells. Expression of the laccase gene was performed under control of both the *AOX1* and *PPLC1* promoters in synthetic medium (yeast nitrogen base, compete supplement mixture, zeocin with a suitable carbon source). Methanol and $CuSO_4$ induction was done for pAOX1-Lacc6 and pPPLCC1-Lacc6, respectively. Laccase activity began to decrease after reaching the maximum level at 96 h in the copper-induced clones, while increasing gradually in the methanol-induced clones after 96 h. Three different promoters, namely, pAOX1, pPPLCC1, and pCUP1, have been compared in BMDY medium. The maximum laccase activity (182 U/L) was observed in pPPLCC1 regulated expression. This study showed that pPPLCC1 is a new copper inducible promoter for *P. pastoris* [30].

YPT1 encodes a GTPase involved in secretion. pYPT1 is a weak constitutive promoter; it has a low expression level in the presence of glucose, methanol, or mannitol as a carbon source [32]. The translation elongation factor 1 (*TEF1*) promoter was clarified by Ahn et al. [33], comparing pGAP and pTEF1 by using a bacterial lipase gene. According to the results, pTEF1 was found to be as strong as pGAP, and its expression is constitutive. Stadlmayr et al. showed that pTEF1 has higher transcription levels than pGAP [26]. Another constitutive promoter, controlling the gene *GCW14*, which encodes the glycosylphosphatidylinositol of *P. pastoris*, has been identified in the past years [34]. The expression of *GCW14* gene reaches the highest level in glycerol media. When the carbon source shifted from glycerol to methanol, lower expression levels were obtained for pGCW14 compared to pAOX1. However, pGCW14 has a stronger promoter activity than pGAP and pTEF1 [34]. One negative and three positive regulatory regions of pGCW14 promoters of *P. pastoris* were been identified [62].

The 3-phosphoglycerate kinase gene (*PGK1*) of *P. pastoris* was identified by de Almeida et al. [35]. Expression levels controlled by pPGK1 were analyzed with α-amylase production. The highest expression value was obtained in cells grown on glucose when compared to cells grown on glycerol or methanol. de Almeida et al. suggest that pPGK1 is a constitutive promoter that can be used for recombinant protein production in *P. pastoris*.

References

1. Cregg JM, Tschopp JF, Stillman C, Siegel R, Akong M, Craig WS, Buckholz RG, Madden KR, Kellaris A, Davis GR, Smiley BL, Cruze J, Torregrossa R, Veliçelebi G, Thıll GP (1987) High-level expression and efficient assembly of hepatitis B surface antigen in the methylotrophic yeast, *Pichia pastoris*. Nat Biotechnol 5:479–485
2. Romanos M, Scorer CA, Clare JJ (1992) Foreign gene expression in yeast: a review. Yeast 8:423–488
3. Kjeldsen T, Frost-Petterson A, Hach M (1999) Secretory expression and characterization of insulin in *Pichia pastoris*. Biotechnol Appl Biochem 29:79–86
4. Wang Y, Liang ZH, Zhang YS, Yao SY, Xu YG, Tang YH, Zhu SQ, Cui DF, Feng YM (2001) Human insulin from a precursor overexpressed in the methylotrophic yeast *Pichia pastoris* and a simple procedure for purifying the expression product. Biotechnol Bioeng 73:74–79
5. Pais JM, Varas L, Valde´ s J, Cabello C, Rodriguez L, Mansur M (2003) Modeling of mini-proinsulin production in *Pichia pastoris* using the AOX promoter. Biotechnol Lett 25:251–255
6. Gurramkonda C, Polez S, Skoko N, Adnan A, Gabel T, Chugh D, Swaminathan S, Khanna N, Tisminetzky S, Rinas U (2010) Application of simple fed-batch technique to high-level secretory production of insulin precursor using *Pichia pastoris* with subsequent purification and conversion to human insulin. Microb Cell Factories 9(31):1–11
7. Raymond CK, Bukowski T, Holderman SD, Ching AFT, Vanaja E, Stamm MR (1998) Development of the methylotrophic yeast *Pichia methanolica* for the expression of the 65-kilodalton isoform of human glutamate decarboxylase. Yeast 14:11–23
8. Sakai Y, Rogi T, Takeuchi R, Kato N, Tani Y (1995) Expression of *Saccharomyces* adenylate kinase gene in *Candida boidinii* under the regulation of its alcohol oxidase promoter. Appl Microbiol Biotechnol 42:860–864
9. Sakai Y, Akiyama M, Kondoh H, Shibano Y, Kato N (1996) High-level secretion of fungal glucoamylase using the *Candida boidinii* gene expression system. Biochim Biophys Acta 1308:81–87
10. Weydemann U, Keup P, Piontek M, Strasser AWM, Schweden J, Gellissen G, Janowicz ZA (1995) High-level secretion hirudin by *Hansenula polymorpha*-authentic processing of three different preprohirudins. Appl Microbiol Biotechnol 44:377–385
11. Gellissen G, Melber K (1996) Methylotrophic yeast *Hansenula polymorpha* as production organism for recombinant pharmaceuticals. Drug Res 46:943–948
12. Hagenson MJ (1991) Production of recombinant proteins in the methylotrophic yeast *Pichia pastoris*. Bioprocess Technol 12:193–212
13. Lin-Cereghino J, Cregg JM (2000) Heterologous protein expression in the methylotrophic yeast *Pichia pastoris*. FEMS Microbiol Rev 24:45–66
14. Macauley-Patrick S, Fazenda ML, McNeil B, Harvey LM (2005) Heterologous protein production using the *Pichia pastoris* expression system. Yeast 22:249–270
15. Cregg JM, Vedvick TS, Raschke WC (1993) Recent advances in the expression of foreign genes in *Pichia pastoris*. Biotechnology 11:905–910
16. Ahmad M, Hirz M, Pichler H, Schwab H (2014) Protein expression in *P. pastoris*: recent achievements and perspectives for heterologous protein production. Appl Microbiol Biotechnol 98:5301–5317
17. Weinhandl K, Winkler M, Glieder A, Camattari A (2014) Carbon source dependent promoters in yeasts. Microb Cell Factories 13:5
18. Vogl T, Glieder A (2013) Regulation of *Pichia pastoris* promoters and its consequences for protein production. New Biotechnol 30 (4):386–404
19. Tschopp JF, Brust PF, Cregg JM, Stillman CA, Gingeras TR (1987) Expression of the lacZ

gene from two methanol-regulated promoters in *P. pastoris*. Nucleic Acids Res 15:3859–3876

20. Ellis SB, Brust PF, Koutz PJ, Waters AF, Harpold MM, Gingeras TR (1985) Isolation of alcohol oxidase and two other methanol regulatable genes from the yeast *Pichia pastoris*. Mol Cell Biol 5:1111–1121
21. Karaoglan M, Erden Karaoglan F, Inan M (2016) Comparison of *ADH3* promoter with commonly used promoters for recombinant protein production in *Pichia pastoris*. Protein Expr Purif 121:112–117
22. Küberl A, Schneider J, Thallinger GG, Ander I, Wibberg D, Hajek T, Jaenicke S, Brinkrolf K, Goesmann A, Szczepanowski R, Puhler A, Schwab H, Glieder A, Pichler H (2011) High-quality genome sequence of *Pichia pastoris* CBS7435. J Biotechnol 154:312–320
23. Shen S, Sulter G, Jeffries TW, Cregg JM (1998) A strong nitrogen source-regulated promoter for controlled expression of foreign genes in the yeast *Pichia pastoris*. Gene 216:93–102
24. Menendez J, Valdes I, Cabrera N (2003) The *ICL1* gene of *Pichia pastoris*, transcriptional regulation and use of its promoter. Yeast 20:1097–1108
25. Liu B, Zhang Y, Zhang X, Yan C, Zhang Y, Xu X, Zhang W (2016) Discovery of a rhamnose utilization pathway and rhamnose-inducible promoters in *Pichia pastoris*. Sci Rep 6:27352. https://doi.org/10.1038/srep27352
26. Stadlmayr G, Mecklenbraüker A, Rothmüller M, Maurer M, Sauer M, Mattanovich D, Gasser B (2010) Identification and characterisation of novel *Pichia pastoris* promoters for heterologous protein production. J Biotechnol 150:519–529
27. Landes N, Gasser B, Vorauer-Uhl K, Lhota G, Mattanovich D, Maurer M (2016) The vitamin-sensitive promoter P_{THI11} enables pre-defined autonomous induction of recombinant protein production in *Pichia pastoris*. Biotechnol Bioeng 113(12):2633–2643
28. Prielhofer R, Maurer M, Klein J, Wenger J, Kiziak C, Gasser B, Mattanovich D (2013) Induction without methanol: novel regulated promoters enable high-level expression in *Pichia pastoris*. Microb Cell Factories 12:5
29. Koller A, Valesco J, Subramani S (2000) The *CUP1* promoter of *Saccharomyces cerevisiae* is inducible by copper in *Pichia pastoris*. Yeast 16:651–656
30. Kim S-II, Ha B-S, Kim M-S, Park M, Ro H-S (2016) Evaluation of copper-inducible fungal laccase promoter in foreign gene expression in *Pichia pastoris*. Biotechnol Bioprocess Eng 21:53–59
31. Waterham HR, Digan ME, Koutz PJ, Lair SV, Cregg JM (1997) Isolation of the *Pichia pastoris* glyceraldehyde-3-phosphate dehydrogenase gene and regulation and use of its promoter. Gene 186:37–44
32. Sears IB, O'connor J, Rossanese OW, Glick BS (1998) A versatile set of vectors for constitutive and regulated gene expression in *Pichia pastoris*. Yeast 14:783–790
33. Ahn J, Hong J, Lee H, Park M, Lee E, Kim C, Choi E, Jung J (2007) Translation elongation factor 1-alpha gene from *Pichia pastoris*: molecular cloning, sequence, and use of its promoter. Appl Microbiol Biotechnol 74(3):601–608
34. Liang S, Zou C, Lin Y, Zhang X, Ye Y (2013) Identification and characterization of P GCW14: a novel, strong constitutive promoter of *Pichia pastoris*. Biotechnol Lett 35(11):1865–1871
35. de Almeida JRM, de Moraes LMP, Torres FAG (2005) Molecular characterization of the 3-phosphoglycerate kinase gene (*PGK1*) from the methylotrophic yeast *Pichia pastoris*. Yeast 22:725–737
36. Inan M, Meagher M (2001) The effect of ethanol and acetate on protein expression in *Pichia pastoris*. J Biosci Bioeng 92:337–341
37. Inan M, Meagher M (2001) Non-repressing carbon sources for alcohol oxidase (AOX1) promoter of *Pichia pastoris*. J Biosci Bioeng 92:585–589
38. Eggeling L, Sahm H (1980) Direct enzymatic assay for alcohol oxidase, alcohol dehydrogenase, and formaldehyde dehydrogenase in colonies of *Hansenula polymorpha*. Appl Environ Microbiol 39:268–269
39. Bruinenberg PG, Veenhuis M, Van Dijken JP, Duine JA, Harder W (1982) A quantitative analysis of selective inactivation of peroxisomal enzymes in the yeast, *Hansenula polymorpha*, by high-performance liquid chromatography. FEMS Microbiol Lett 15:45–50
40. Veenhuis M, Van Dijken JP, Harder W (1983) The significance of peroxisomes in the metabolism of one-carbon compounds in yeasts. Adv Microb Physiol 24:1–82
41. Cregg JM, Madden KR, Barringer KJ, Thill GP, Stillman CA (1989) Functional characterization of the two alcohol oxidase genes from the yeast *Pichia pastoris*. Mol Cell Biol 9:1316–1323
42. Nakagawa T, Mizumura T, Mukaiyama H, Miyaji T, Yurimoto H, Kato N (2002) Physiological role of the second alcohol oxidase gene

MOD2 in the methylotrophic growth of *Pichia methanolica*. Yeast 19:1067–1073

43. Inan M (2000) Studies on the alcohol oxidase (*AOX1*) promoter of *Pichia pastoris*. PhD thesis, University of Nebraska, 98 p
44. Hartner FS, Ruth C, Langeneegger D, Johnson SN, Hyka P, Lin-Cereghino GP, Lin-Cereghino J, Kovar K, Cregg JM, Glieder A (2008) Promoter library designed for fine-tuned gene expression in *Pichia pastoris*. Nucleic Acids Res 36(12):e76
45. Xuan Y, Zhou X, Zhang W, Zhang X, Song Z, Zhang Y (2009) An upstream activation sequence controls the expression of *AOX1* gene in *Pichia pastoris*. FEMS Yeast Res 9:1271–1282
46. Lin-Cereghino G, Godfrey L, Cruz BD, Johnson S, Khuongsathiene S, Tolstorukov I, Yan M, Lin-Cereghino J, Veenhuis M, Subramani S, Cregg JM (2006) Mxr1p, a key regulator of the methanol utilization pathway and peroxisomal genes in *Pichia pastoris*. Mol Cell Biol 26:883–897
47. Takagi S, Tsutsumi N, Terui Y, Kong XY, inventors; Novozymes A/S, Bagsvaerd (DK), assignee (2012) Method for methanol independent induction from methanol inducible promoters in Pichia. United States patent US 8,143,023
48. Wang X, Wang Q, Wang J, Bai P, Shi L, Shen W, Zhou M, Zhou X, Zhang Y, Cai M (2016) Mit1 transcription factor mediates methanol signaling and regulates the alcohol oxidase 1 (*AOX1*) promoter in *Pichia pastoris*. J Biol Chem 291(12):6245–6261
49. Yılmaz S (2012) Molecular and functional characterization of *Pichia pastoris MIG1* gene. MSc thesis, Akdeniz University, 100 p
50. Wang J, Wang X, Shi L, Qi F, Zhang P, Zhang Y, Zhou X, Song Z, Cai M (2017) Methanol-independent protein expression by *AOX1* promoter with *trans*-acting elements engineering and glucose-glycerol-shift induction in *Pichia pastoris*. Sci Rep 7:41850. https://doi.org/10.1038/srep41850
51. Wang X, Cai M, Shi L, Wang Q, Zhu J, Wang J, Zhou M, Zhou X, Zhang Y (2016) PpNrg1 is a transcriptional repressor for glucose and glycerol repression of *AOX1* promoter in methylotrophic yeast *Pichia pastoris*. Biotechnol Lett 38(2):291–298
52. Qin X, Qian J, Yao G, Zhuang Y, Zhang S, Chu J (2011) GAP promoter library for fine-tuning of gene expression in *Pichia pastoris*. Appl Environ Microbiol 77:3600–3608
53. Ata Ö, Prielhofer R, Gasser B, Mattanovich D, Çalık P (2017) Transcriptional engineering of the glyceraldehyde-3-phosphate dehydrogenase promoter for improved heterologous protein production in *Pichia pastoris*. Biotechnol Bioeng 114(10):2319–2327. https://doi.org/10.1002/bit.26363
54. Sohn SB, Graf AB, Kim TY, Gasser B, Maurer M, Ferrer P, Mattanovich D, Lee SY (2010) Genome-scale metabolic model of methylotrophic yeast *Pichia pastoris* and its use for in silico analysis of heterologous protein production. Biotechnol J 5:705–715
55. Baumann K, Maurer M, Dragosits M, Cos O, Ferrer P, Mattanovich D (2008) Hypoxic fed-batch cultivation of *Pichia pastoris* increases specific and volumetric productivity of recombinant proteins. Biotechnol Bioeng 100:177–183
56. de Schutter K, Lin YC, Tiels P, van Hecke A, Glinka S, Weber- Lehmann J, Rouze P, Van de Peer Y, Callewaert N (2009) Genome sequence of the recombinant protein production host *Pichia pastoris*. Nat Biotechnol 27:561–566
57. Mattanovich D, Graf A, Stadlmann J, Dragosits M, Redl A, Maurer M, Kleinheinz M, Sauer M, Altmann F, Gasser B (2009) Genome, secretome and glucose transport highlight unique features of the protein production host *Pichia pastoris*. Microb Cell Factories 2:8–29
58. Karaoglan M, Erden Karaoglan F, Inan M (2016) Functional analysis of alcohol dehydrogenase (ADH) genes in *Pichia pastoris*. Biotechnol Lett 38:463–469
59. Erden Karaoglan F (2016) Studies on *Pichia pastoris* alcohol dehydrogenase (*ADH3*) promoter. PhD thesis, Akdeniz University, 65 p
60. Kranthi BV, Kumar HRV, Rangarajan PN (2010) Identification of Mxr1p-binding sites in the promoters of genes encoding dihydroxyacetone synthase and peroxin 8 of the methylotrophic yeast *Pichia pastoris*. Yeast 27:705–711
61. Resina D, Cos O, Ferrer P, Valero F (2005) Developing high cell density fed-batch cultivation strategies for heterologous protein production in *Pichia pastoris* using the nitrogen source-regulated *FLD1* promoter. Biotechnol Bioeng 91:760–767
62. Zhang X, Zhang X, Liang S, Ye Y, Lin Y (2013) Key regulatory elements of a strong constitutive promoter, P GCW14, from *Pichia pastoris*. Biotechnol Lett 35(12):2113–2119

Chapter 4

Selection of the Optimal Yeast Host for the Synthesis of Recombinant Enzymes

Felix Bischoff, Martin Giersberg, Falko Matthes, Tobias Schwalenberg, Sebastian Worch, and Gotthard Kunze

Abstract

Yeasts, like *Arxula adeninivorans*, *Hansenula polymorpha*, *Pichia pastoris*, *Debaryomyces hansenii*, *Debaryomyces polymorphus*, *Schwanniomyces occidentalis*, *Yarrowia lipolytica*, and *Saccharomyces cerevisiae* are frequently used producers of recombinant enzymes, particularly when posttranslational modifications are mandatory to obtain full functionality. The wide-range transformation/expression platform presented in this chapter can be used to select the optimal yeast host for high-level synthesis of the desired enzyme with favorable biochemical properties. This platform is composed of a selection marker and up to four expression modules in a linearized cassette. Here we describe the protocols for the assembly as well as the transformation of yeast strains with the respective cassettes, screening of transformants, the isolation and biochemical characterization of the enzymes, and finally a simple fermentation strategy to achieve maximal yields of the chosen recombinant enzyme.

Key words Wide-range expression module, Gibson assembly, Yeast transformation, Transformant screening, Recombinant enzymes, Fermentation strategy

1 Introduction

A wide range of yeasts have been used as expression platforms during the last decades, as yeasts offer distinctive advantages over microbial and alternative eukaryotic systems combining safety aspects with efficient protein production from laboratory to industrial scale. However, no single system turns out to be optimal for all proteins. Secretion, processing, and glycosylation of the desired proteins are among the main factors that affect the choice of an expression system. While *Saccharomyces cerevisiae* has a long tradition as eukaryotic host for heterologous gene expression, certain drawbacks brought out the necessity to explore different hosts. For example, *S. cerevisiae* tends to hyperglycosylate proteins, and furthermore α-1,3-mannose bonds that are built to terminate

Brigitte Gasser and Diethard Mattanovich (eds.), *Recombinant Protein Production in Yeast*, Methods in Molecular Biology, vol. 1923, https://doi.org/10.1007/978-1-4939-9024-5_4,

N-linked carbohydrate chains contain the danger of allergenicity [1].

A vast number of alternative yeasts overcome these limitations and moreover demonstrate their inherent benefits, including *Kluyveromyces lactis* [2], *Yarrowia lipolytica* [3], and *Debaryomyces hansenii* [4] as well as the methylotrophic species *Pichia pastoris* [5] and *Hansenula polymorpha* [6] and the dimorphic yeast *Arxula adeninivorans* [7, 8].

A distinctive feature of *H. polymorpha* and *P. pastoris* is their capability to use methanol as sole energy and carbon source. Growth on methanol induces the expression of pathway-specific key enzymes such as alcohol oxidase, format dehydrogenase, and dihydroxyacetone synthase [9]. Accordingly, promoter elements derived from these genes can be exploited for gene expression depending on carbon source addition to the medium (e.g., *AOX1* and *AOX2* promoters in *P. pastoris*, *MOX* promoter in *H. polymorpha*). However, recent transcriptome analyses showed that methanol dissimilation genes in *H. polymorpha* including *MOX* and *FMD* are activated by derepression upon carbon source limitation and depletion rather than upon methanol induction [10]. Therefore, production processes in *H. polymorpha* can be activated by glycerol feed or the starvation of glucose, whereas promoter activity in *P. pastoris* is strictly dependent on the presence of methanol or methanol derivatives [5].

A. adeninivorans is an asexual, haploid yeast that has already been used in a wide range of applications, e.g., as microbial whole-cell biosensor [11], or for the secretion and accumulation of enzymes like cutinases that are of great biotechnological interest [12]. Beyond that, its special features like thermo- and osmotolerance and its capability to use a wide range of carbon and nitrogen sources make *A. adeninivorans* a favorable host for the production of recombinant proteins.

Selection systems that can be employed for heterologous gene expression in *A. adeninivorans* include the *Escherichia coli*-based *hph* gene conferring resistance to hygromycin B or the use of auxotrophic strains, e.g., incapable of producing leucine, isoleucine, tryptophan, or uracil, that can be complemented by the *A. adeninivorans*-derived *ALEU2*, *AILV1*, *ATRP1*, and *AURA3* genes, respectively. In all cases a 25S rDNA fragment from *A. adeninivorans* is used for integration of the transformation cassette into the yeast genome, resulting in the stable integration of up to ten cassettes into the ribosomal DNA [13]. The productivity of protein synthesis however strongly depends on the choice of the promoter-terminator system. For example, using the strong *Arxula*-derived *TEF1* promoter and *S. cerevisiae*-derived *PHO5* terminator, the expression of a *MF1-IL6* gene fusion led to the accumulation of up to 210 mg L^{-1} interleukin-6 on shaking flask scale [14], whereas the integration of genes for the production of polyhydroxyalkanoates using the same promoter-terminator system even led to the accumulation of 10.8 g L^{-1} of the desired poly (hydroxybutyrate-*co*-hydroxyvalerate) copolymer [15].

In the following chapter, we will introduce a straightforward protocol for the selection of the optimal yeast host for the synthesis of recombinant enzymes, including transgenic yeast strains of various species with a wide-range system and examples of screening assays in order to select the most productive transformants. Furthermore, we will give fast assays for estimating biochemical properties for the enzymes synthesized with each host to assess the differences due to folding and posttranslational modifications. Finally, a standardized fermentation procedure will be presented aiming at the production of a larger amount of the enzymes for further functional tests and trial product samples.

2 Materials

Prepare all solutions using ultrapure water and analytical reagents. Prepare and store all reagents at room temperature (20 °C) unless indicated otherwise.

2.1 Yeast Strains (Table 1)

2.2 Plasmids (Fig. 1)

2.3 Media (Table 2)

2.4 Buffers

McIlvaine buffer: *see* Table 3.

SBD-buffer:	PEG-bicine solution:	NaCl-bicine solution:
1 M sorbitol	40% PEG 1000	0.15 M NaCl
3% ethylene glycol	0.2 M bicine, pH 8.35	10 mM bicine, pH 8.35
5% DMSO		
10 mM bicine		
Autoclave		

3 Methods

3.1 Construction of Wide-Range Expression Modules

The Xplor® expression platform [4] suitable for several yeast species has been further developed through the years [24]. Besides the yeasts *A. adeninivorans*, *S. cerevisiae*, *D. hansenii*, *D. polymorphus*, *H. polymorpha*, and *P. pastoris*, the yeast *S. occidentalis* was also

Table 1
Yeast strains with their respective auxotrophy markers

Organism	Genotype	Auxotrophy	Ref.
A. adeninivorans LS3			[16]
A. adeninivorans G1211	[aleu2]	Leucine	[17]
A. adeninivorans G1212	[aleu2 atrp1::ALEU2]	Tryptophan	[18]
A. adeninivorans G1214	[aleu2 ura3::ALEU2]	Uracil	[18]
A. adeninivorans G1216	[aleu2 ade2::ALEU2]	Adenine	[19]
A. adeninivorans MS1001	[aleu2 atrp1::ALEU2 aura3::ALEU2]	Tryptophan Uracil	[18]
A. adeninivorans MS1004	[aleu2 aade2::ALEU2 aura3::ALEU2]	Adenine Uracil	[19]
A. adeninivorans MS1006	[aleu2 aade2::ALEU2 atrp1::ALEU2]	Tryptophan Adenine	[19]
H. polymorpha RB11	[odc1]	Uracil	[4]
P. pastoris GS115	[his4]	Histidine	[4]
D. hansenii H158			[4]
D. polymorphus H129			[4]
Schwanniomyces occidentalis ATCC 26077			[19]
S. cerevisiae SEY6210	[MATα leu2-3,112 ura3-52 his3-Δ200 trp1-Δ901 suc2-Δ9 lys2-801; GAL]	Leucine, uracil, histidine, lysine, tryptophan	[20]

successfully transformed [19]. The latest Xplor® version (Fig. 1) was designed for easy isothermal cloning [25] which is favorable in particular for one-step cloning of multiple fragments, e.g., for the generation of fusion genes. A modified protocol for the cloning procedure is given below. Furthermore, to broaden the range of available promoters, several putatively constitutive promoters, mostly from genes encoding transcription factors, ribosome or polymerase subunits, as well as glycolysis-related genes, were tested (*see* **Note 1**).

3.2 Gibson Assembly Protocol (Modified After [26])

1. Open the suitable Xplor® vector with the appropriate restriction enzyme between the desired promoter/terminator combination.
2. Use 0.2–1 pmol vector plasmid in a 0.2 mL PCR tube.

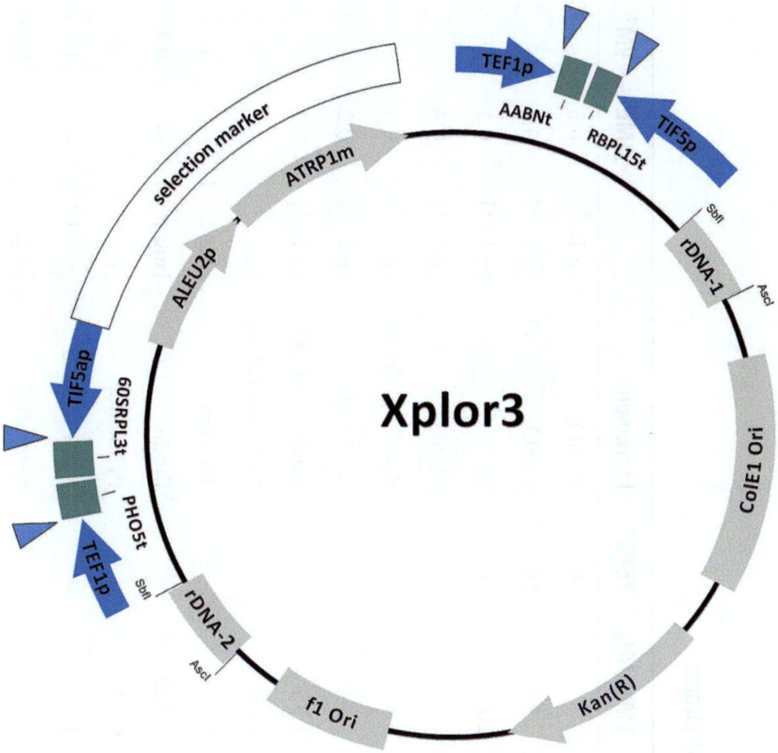

Fig. 1 Example of a recent Xplor® plasmid with four promoter/terminator combinations. Promoters are shown as blue arrows and terminators as boxes. The triangles mark cloning sites for genes to be expressed. Restriction sites for linearization of the expression construct prior to transformation are also shown [21]

3. Add the desired number of fragments (*see* **Note 2**).
4. Keep the total volume as small as possible, as little as 3 μL worked fine for us.
5. Add one volume of Gibson Assembly Master Mix (2×, NEB), and mix by pipetting.
6. Incubate in a gradient cycler for 1 h (*see* **Note 3**).
7. Use 4–6 μL for standard *E. coli* heat shock transformation.

3.3 Yeast Transformation

The DNA to be transformed has to contain a selection marker module to confer a particular resistance or to complement a specific auxotrophy of the yeast host cell. The Xplor® vector variations offer the possibility to convey hygromycin B resistance or to complement adenine, leucine, uracil, or tryptophan auxotrophy of a given yeast strain. Available yeast strains are listed in the Subheading 2.1.

3.3.1 Preparation of Competent Yeast Cells

1. Inoculate one colony in 10 mL YEPD medium (add 100 mg L^{-1} tryptophan, 40 mg L^{-1} adenine, or 40 mg L^{-1}

Table 2
Composition of the media which can be used for cultivation of different yeast strains

Ingredient		FM[a]_A	FM_B	FM_C	YMM[b]-NO_3	YMM-NH_4	YEPD[c]	Ingredient		ME[d]	VM[e]
Peptone	[g L^{-1}]	20	–	–	–	–	20	H_3BO_4	[g L^{-1}]	0.5	–
Yeast extract	[g L^{-1}]	10	–	–	–	–	10	$CuSO_4 \times 5H_2O$	[g L^{-1}]	0.1	–
Glucose	[g L^{-1}]	55	550	–	20	20	20	KI	[g L^{-1}]	0.1	–
$(NH_4)_2SO_4$	[g L^{-1}]	6.92	–	69.17	–	–	–	$MnSO_4$	[g L^{-1}]	0.4	–
$(NH_4)H_2PO_4$	[g L^{-1}]	–	–	–	–	5	–	$ZnSO_4 \times 7H_2O$	[g L^{-1}]	0.4	–
$NaNO_3$	[g L^{-1}]	–	–	–	3.7	–	–	Na_2MoO_4	[g L^{-1}]	0.2	–
KH_2PO_4	[g L^{-1}]	0.53	–	5.26	6.75	1.75	–	$CoCl_2 \times 6H_2O$	[g L^{-1}]	0.1	–
K_2HPO_4	[g L^{-1}]	0.22	–	2.25	1.75	1.75	–	Ca-Pantothenate	[mg L^{-1}]	–	40
$MgSO_4 \times 7H_2O$	[g L^{-1}]	1	–	10	1	1	–	Thiamin	[mg L^{-1}]	–	40
$FeCl_3 \times 6H_2O$	[g L^{-1}]	0.002	–	0.02	0.002	0.002	–	Biotin	[mg L^{-1}]	–	0.4
$Ca(NO_3)_2 \times 4H_2O$	[g L^{-1}]	0.02	–	0.2	0.02	0.02	–	Pyridoxine	[mg L^{-1}]	–	40
ME	[mL L^{-1}]	1	–	10	1	1	–	Nicotinic acid	[mg L^{-1}]	–	10
VM	[mL L^{-1}]	1	–	5	1	1	–	m-inositol	[mg L^{-1}]	–	400
Agar[f]	[g L^{-1}]	–	–	–	20	20	20				

[a]Fermentation medium
[b]Yeast minimal medium modified after [22]
[c]Yeast extract peptone dextrose
[d]Microelement solution
[e]Vitamin mix
[f]Add only for solid media

Table 3
Composition of McIlvaine buffer in the range from pH 2.2–8.0 [23]

	Composition ($g\ L^{-1}$)			$g_{KCl}\ L^{-1}$ for I of			Composition, $g\ L^{-1}$			$g_{KCl}\ L^{-1}$ for I of	
pH (25 °C)	$Na_2HPO_4 \times 12H_2O$	Citric acid $\times H_2O$	I^a [M]	1.0 M	0.5 M	pH (25 °C)	$Na_2HPO_4 \times 12H_2O$	Citric acid $\times H_2O$	I [M]	1.0 M	0.5 M
2.20	1.43	20.60	0.0108	74.50	37.20	5.20	38.40	9.75	0.2780	53.80	16.50
2.40	4.44	19.70	0.0245	72.70	35.40	5.40	40.00	9.29	0.3020	52.10	14.80
2.60	7.80	18.70	0.0410	71.50	34.20	5.60	41.50	8.72	0.3210	50.60	13.30
2.80	11.35	17.70	0.0592	70.20	32.90	5.80	43.30	8.32	0.3360	49.50	12.20
3.00	14.70	16.70	0.0771	68.70	31.40	6.00	45.20	7.74	0.3440	48.90	11.60
3.20	17.70	15.80	0.0934	67.60	30.30	6.20	47.50	7.12	0.3580	47.90	10.60
3.40	20.40	15.00	0.1120	66.20	28.90	6.40	49.60	6.47	0.3710	46.90	9.62
3.60	21.50	14.20	0.1280	64.90	27.60	6.60	52.10	5.72	0.3850	45.80	8.50
3.80	25.40	13.60	0.1420	64.00	26.70	6.80	55.40	4.79	0.3920	44.50	7.23
4.00	27.60	12.90	0.1570	62.80	25.50	7.00	58.90	3.70	0.4270	42.70	5.44
4.20	29.70	12.30	0.1730	61.70	24.40	7.20	62.30	2.74	0.4570	40.40	3.10
4.40	31.60	11.70	0.1900	60.40	23.10	7.40	65.00	1.91	0.4880	38.20	0.488
4.60	33.40	11.20	0.2100	58.90	21.60	7.60	67.20	1.35	0.5160	36.00	–
4.80	35.30	10.70	0.2320	57.20	19.90	7.80	68.60	0.89	0.5400	34.30	–
5.00	36.90	10.20	0.2560	55.50	18.20	8.00	69.60	0.59	0.5590	32.90	–

[a]Ionic strength

uracil if necessary) in a 100 mL Erlenmeyer flask, and incubate shaking (180 rpm) over night at 30 °C.

2. Dilute the pre-culture in 250 mL YEPD in a 1 L Erlenmeyer flask to an OD_{600} of 0.6–0.7.
3. Grow at 30 °C and 180 rpm to an OD_{600} of 1.5–1.7 (path length 10 mm).
4. Centrifuge 5 min at 3000 × *g* and at 4 °C.
5. Discard the supernatant.
6. Resuspend pellet in 0.5 initial volume (125 mL) ice-cold SBD buffer.
7. Centrifuge 5 min at 3000 × *g* and at 4 °C.
8. Discard supernatant.
9. Resuspend pellet in 1/50 initial volume ice-cold SBD buffer.
10. Prepare 200 μL aliquots in prechilled 1.5 mL tubes.
11. Freeze in liquid nitrogen and store at −80 °C until use.

3.3.2 Yeast Transformation

A modified protocol from [27] is used to transform the different yeast species. This protocol is easy in handling and usually results in a sufficient amount of transformants μg^{-1} DNA for standard screening procedures.

1. Heat carrier DNA (DNA from herring sperm, 5 mg mL^{-1}) for 10 min at 98 °C, and chill on ice.
2. Add 15 μL DNA (linearized plasmid or PCR product, 100–-500 ng μL^{-1}) + 15 μL carrier (5 mg mL^{-1}) to 200 μL frozen competent cells; do not mix.
3. Incubate the tubes at 37 °C for 5 min with gentle shaking.
4. Add 1 mL of PEG-bicine solution.
5. Mix once by inverting and flicking.
6. Incubate for 1 h at 30 °C without shaking (*see* **Note 4**).
7. Centrifuge for 2 min at 3000 × *g* and room temperature.
8. Discard the supernatant.
9. Resuspend pellet in 1 mL of NaCl-bicine solution.
10. Centrifuge for 2 min at 3000 × *g* at room temperature.
11. Discard the supernatant.
12. Resuspend pellet in 0.5 mL of NaCl-bicine solution.
13. Spread the cells on selective plates.
14. Incubate at 30 °C until colonies are formed (2–5 days).

3.4 Screening of Transformants Via PCR

1. Pick colonies and inoculate 500 μL YMM with 2% glucose as carbon source (+ antibiotics if required) in deep well plates (96-wells).

2. Incubate for 48 h at 30 °C and 180 rpm.
3. Transfer 20 μL cell suspension with a multichannel pipette into 96-well PCR plate.
4. Heat cells at 98 °C for 5 min.
5. Spin down cell debris.
6. Transfer 5 μL of supernatants (templates) in a clean 96-well PCR plate.
7. Add PCR reaction mix to 20 μL final volume.
8. Perform standard PCR and analyze on agarose gel.

3.5 Screening for Accumulation of Recombinant Enzymes

Choosing a screening method highly depends on the protein of choice as well as its modifications (e.g., tags). Non-specific methods include SDS-PAGE analysis. This method can be used in any case to estimate the amount of synthesized recombinant protein. For identification of the protein of interest, Western hybridization using a specific antibody should be subsequently performed. This antibody can detect either the recombinant protein or, if available, its tag (e.g., His-tag). For recombinant enzymes, a more direct and specific way for detection is to exploit its activity toward a certain substrate. For this purpose, the screening for enzymes in yeasts will be discussed in the following.

Like all eukaryotes, yeasts show differentiated subcellular localization for proteins. According to their putative signal peptides, enzymes can be intra- or extracellularly localized. They can even be transported to further subcellular compartments like endoplasmic reticulum (ER), mitochondria, or peroxisomes. Furthermore, localization on the extracellular cell wall is possible. Screening methods have to be chosen depending on those localization patterns (Fig. 2).

It is generally advisable to stick to one universal system for cultivation and strain maintenance. The potential positive transformants based on the screen described under Subheadings 3.4 and 3.5 should be transferred and kept on masterplates in 96-well format after their selection for faster colony picking via multi-pipettes.

1. Transfer colonies to 96-well deep well plates (*see* **Note 5**) with YEPD (*see* **Note 6**) using a multi-pipette.
2. Incubate for 24–48 h.
3. Extracellular enzyme:
   - Drop 3 μL on an agar plate containing a substrate of choice (Table 4).
   - Incubate for 24 h, and observe the reaction.
   - Spin down the remaining culture, and transfer the supernatant to a new 96-well deep well plate.

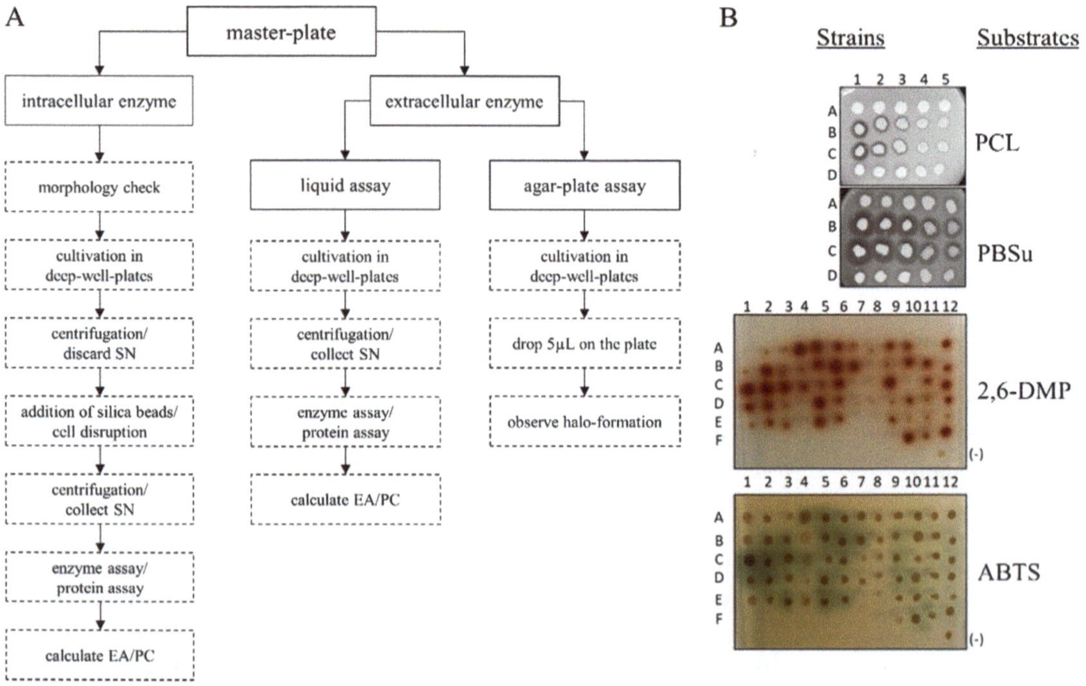

Fig. 2 General procedures for the screening of strains synthesizing recombinant enzymes (**a**). Examples for agar plate assays for the detection of extracellular cutinases with polycaprolactone (PCL) and polybutylene succinate (PBSu) and extracellular laccases with 2,6-dimethoxyphenol (2,6-DMP) or 2,2′-azino-bis(3-ethyl-benzothiazoline-6-sulfonic acid) (ABTS) as substrates (**b**) (*SN* supernatant, *EA* enzyme activity, *PC* protein concentration)

- Use between 10 and 50 μL for assaying the enzyme activity.
- Determine the protein concentration with Bradford assay [28].
- Calculate specific activity in U $mg_{protein}^{-1}$.

4. Intracellular enzyme:
   - Spin down the yeast cells.
   - Discard the supernatant.
   - Add 250 μL of silica beads and 250 μL of buffer of choice.
   - Disrupt the yeast cells with a bead mill.
   - Spin down the cell debris and silica beads.
   - Collect the supernatant in a new 96-well deep well plate.
   - Use between 10 and 50 μL for assaying the enzyme activity (Table 4).
   - Determine the protein concentration with Bradford assay [28].
   - Calculate the specific activity in U $mg_{protein}^{-1}$.

Table 4
Summary of assays which can be performed to detect certain enzymes

Enzymes	Liquid assay		Plate assay		Ref.
	Principle	Substrate	Principle	Substrate	
Oxidoreductases					
ADH	Detect formation/consumption of NAD(P)H at 340 nm	Alcohols, aldehydes, ketones, NAD(P), NAD(P)H	–	–	[29, 30]
Laccase	Detect formation of colored products	ABTS/ 2,6-DMP	Colored halo	ABTS/2,6-DMP/ guaiacol	[31–33]
Xanthine Oxidoreductase	Detect formation of uric acid at 293 nm	Hypoxanthine	–	–	Mod [34]
Urate oxidase	Detect consumption of uric acid at 293 nm	Uric acid	–	–	Mod [34]
Hydrolases					
Lipase	Detect *p*-nitrophenol at 348/405 nm	*p*-Nitrophenyl fatty acid esters	Clear zone	Tributyrin, triolein	[35]
Cutinase	Detect *p*-nitrophenol at 348/405 nm	*p*-Nitrophenyl fatty acid esters	Clear zone	Plastic nanoparticles, tributyrin	[12, 35, 36]
Tannase	Detect gallic acid via rhodanine at 520 nm	Methyl gallate	Clear zone	Tannic acid	[37, 38]
Phytase	Detect *p*-nitrophenol at 405 nm	*p*-Nitrophenyl phosphate	Clear zone	Calcium phytate, sodium phytate	[39]
Arabinanase	Detect monomers via DNSA or *p*-nitrophenol at 405 nm	Arabinans, *p*-nitrophenyl arabinofuranosides	Blue halo	Azurine-cross-linked arabinan	[31, 40, 41]
Glucanase/cellulase	Detect monomers via DNSA or *p*-nitrophenol at 405 nm	Glucans, *p*-nitrophenyl glucopyranosides	Blue halo	Cellulose azure	[31, 41, 42]

(continued)

Table 4
(continued)

Enzymes	Liquid assay		Plate assay		Ref.
	Principle	Substrate	Principle	Substrate	
Xylanase/xylosidase	Detect monomers via DNSA or *p*-nitrophenol at 405 nm	Xylans, *p*-nitrophenyl xylopyranosides	Blue halo	Xylan azure	[31, 41, 42]
Adenine deaminase	Detect formation of uric acid under addition of XOR at 293 nm	Adenine	–	–	Mod [34]
Guanine deaminase	Detect formation of uric acid under addition of XOR at 293 nm	Guanine	–	–	Mod [34]
Transferases					
Purine nucleoside phosphorylase	Detect formation of uric acid under addition of XOR at 293 nm	Inosine	–	–	[34]

3.6 Biochemical Characterization of Enzymes

The biochemical properties of the newly synthesized enzyme are a crucial factor for choosing one yeast host over the other. The weighing of different attributes depends on the purpose of the recombinant enzyme. It must be defined in the first place, whether, e.g., temperature stability or activity at certain pH values is the preferred property. Nevertheless, it is advisable to completely characterize the recombinant enzymes synthesized by different yeast host systems because the host's posttranslational modifications can have a vast influence on each of the parameters. For example, Giersberg et al. [43] showed those modifications are altering the thermostability of alcohol dehydrogenase synthesized with *S. cerevisiae*, *A. adeninivorans*, and *H. polymorpha*. The most important parameters are dependence on pH value, temperature, ionic strength, and cofactors as well as pH stability, temperature stability, substrate spectrum, and kinetic behavior.

1. Dependence on pH:
   - Prepare buffers of different pH. It is advisable to use wide-range buffer systems like McIlvaine buffer (Table 3) to exclude different influences of the buffer's components.
   - Adjust the ionic strength of the buffers by adding salts like NaCl or KCl (*see* **Note** 7). Therefore, calculate ionic strength at each pH, and fill it up to the lowest possible for all pH values.
   - Perform an activity assay using all prepared buffers (*see* **Note 8**).
   - Set pH with the highest activity to 100%, and correlate the others with this value.
2. pH stability:
   - Incubate the enzyme in all prepared buffers for 10 min, 30 min, 60 min, 120 min, 240 min, and 24 h.
   - Perform an activity assay directly after each incubation interval (*see* **Note 8**).
   - Set the non-incubated sample's activity to 100%, and correlate the others with this value.
   - Plot a decay curve for analysis.
3. Dependence on temperature:
   - Preincubate the reaction mixture without enzyme at different temperatures (*see* **Note 9**).
   - Add the enzyme, mix, and incubate for further reaction time in the thermocycler. The time of incubation depends on the enzyme concentration. Make sure that the reaction is still in the linear phase at the end of incubation time.

- Place the plates on ice, and transfer 100 μL to microtiter plate for endpoint measurement.

4. Temperature stability:
   - Prepare aliquots of all enzymes with similar protein concentrations.
   - Incubate them on ice, at 20, 30, 40, 50, 60, 70, 80, and 90 °C for 10, 30, 60, 120, and 240 min and 24 h.
   - Chill them down on ice directly after each interval.
   - Measure the residual activity as soon as possible.
   - Assign 100% to the activity after incubation on ice for 10 min, and correlate the others with this value.
   - Plot a decay curve for analysis.
5. Dependence on ionic strength:
   - Prepare buffer at an optimal pH (after performing assay 1).
   - Set the ionic strength using salts like NaCl or KCl.
   - Perform an activity assay.
   - Assign 100% to the highest activity, and correlate the others with this value.
6. Cofactor assay:
   - Prepare 10 mM stocks of $CaCl_2$, $CoCl_2$, $CoSO_4$, $CuCl_2$, $CuSO_4$, $FeCl_3$, $FeSO_4$, KCl, $MgCl_2$, $MgSO_4$, $MnCl_2$, $MnSO_4$, $NiCl_2$, $NiSO_4$, $ZnCl_3$, $ZnSO_4$, and EDTA.
   - Perform an activity assay with a final concentration of 1 mM and with water for control.
   - Set the control activity to 100%, and correlate the others with this value.

The results obtained from the biochemical characterization reveal the differences between the different yeast expression systems. At this point, the biochemical parameters of each tested enzyme have to be compared with the requirements of the enzyme's future application.

3.7 Fermentation Strategies

In order to come to a final decision, the productivity of the yeast system in question has to be tested in a standardized and reproducible fermentation procedure. This procedure is a growth-controlled fed-batch fermentation on a 5 L scale. The compositions of the media are shown in Table 2. The cultivation starts with 2 L of FM_A. After depletion of the initial glucose, the feeding procedure with FM_B and FM_C is started. The feed control is either realized via CO_2 concentration in the exhaust or via partial pressure of dissolved oxygen in the system (DO), kept on a constant level by controlling the stirrer speed (Fig. 3). Analyses of intracellular or extracellular enzymes are performed according to Subheading 3.5.

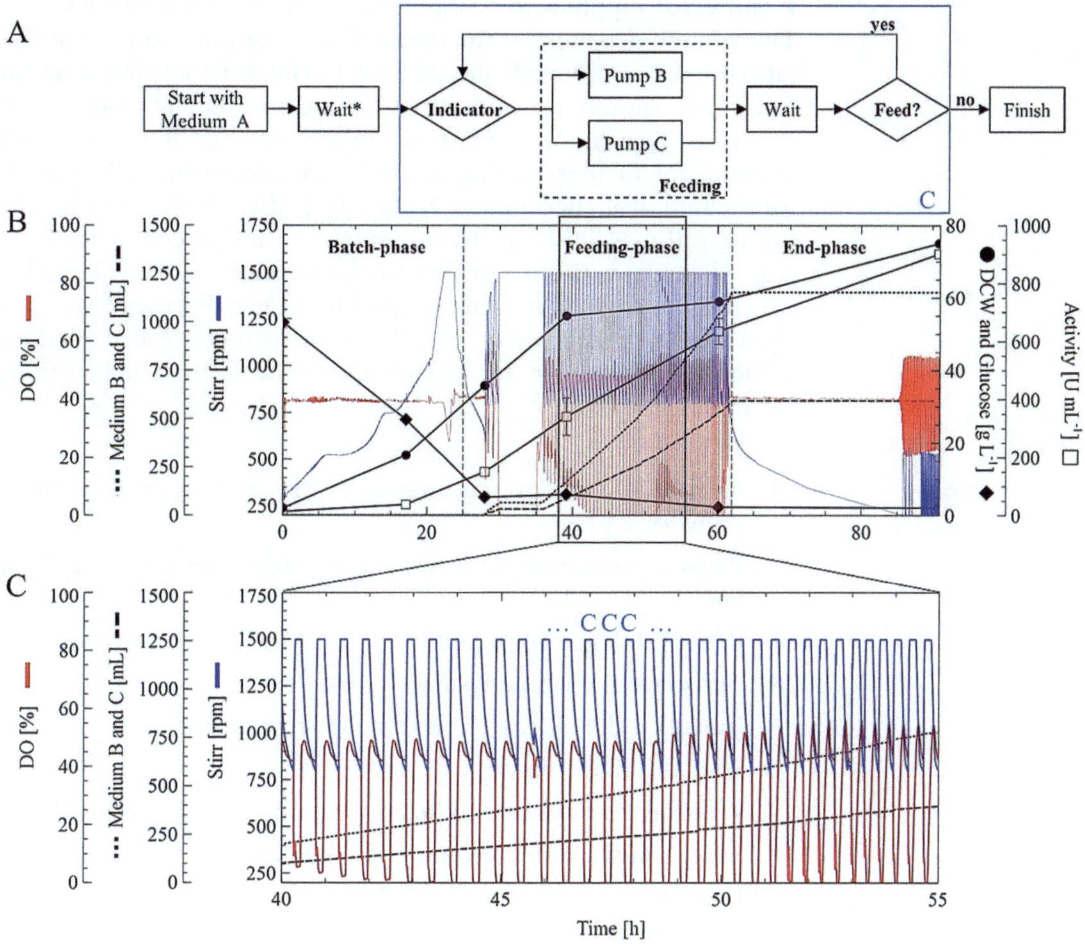

Fig. 3 Scheme of the general fermentation procedure for implementation in any laboratory fermentation control system (**a**). The * indicates a manual transition after the waiting phase for the end of the batch phase. Either stirrer speed, oxygen uptake rate, or carbon emission rate can be used as indicators for glucose depletion. The fermentation for the synthesis of Acut2-6hp with *A. adeninivorans* G1212/YRC-ACUT2-6H [12, 44] is shown in (**b**). A section between 40 and 55 h of fermentation time is given in (**c**). The feeding cycle (**c**) was started each time the stirrer speed went below 800 rpm. 1500 rpm was set as the maximum stirrer speed of the fermentation system

Additionally, the determination of the dry cell weight (DCW) and glucose concentration can be used for yield calculations of the formed products.

Scheme (a) of Fig. 3 can be implemented into any fermentation control system available in your lab (*see* **Note 10**). The first waiting phase has a manual transition into the feeding phase (*see* **Note 11**). After depletion of glucose at the end of the batch phase, the stirrer speed is dropping from 1500 rpm to 800 rpm due to the decreased oxygen uptake from the yeast cells. This is realized by the automated feeding algorithm, which is subsequently starting the pumps

B and C for supply with C source (FM_B) and N source (FM_C). The volume depends on the overall fermentation volume, but the ratio between both feeds should be 2:1. The short waiting phase of 5–10 min should provide enough time for stabilization of the system to prevent unwanted secondary feeding. If there is still medium for further feeding available, the algorithm will go back for the indicator checkup (*see* **Note 12**). If there is no medium left, the algorithm is finishing. The advantage of this procedure is a high glucose supply without reaching too high concentrations (Fig. 3b), which is important for Crabtree-positive yeasts. Furthermore, this algorithm is an easy way to achieve growth-controlled feeding without the need for knowledge about the growth behavior of each yeast species.

Procedure:

1. Prepare 200 mL pre-cultures in YEPD medium, and let them grow for 24 h.
2. Prepare your bioreactor system for sterilization with 2 L of FM_A (Table 2).
3. Prepare 1 L of FM_B and 0.5 L of FM_C in separate bottles.
4. Autoclave the whole system.
5. Inoculate the bioreactor with your pre-culture.
6. Start the algorithm.
7. Wait for the indicator to be at maximum level (*see* **Note 11**).
8. Force transit to the feeding phase manually.
9. Let the system run until FM_B and FM_C are empty (*see* **Note 12**).
10. Harvest cells and/or supernatant for analysis (*see* **Note 13**).

It is theoretically possible to find the most beneficial biochemical properties and the lowest enzyme yield within the same yeast host system. In this case it is advisable to choose a more suitable, thus stronger promoter for that specific yeast host (e.g., *MOX* promoter in *H. polymorpha*) in order to increase the productivity while maintaining the optimal biochemical characteristics. Here it should not only be taken into account that the performance of the promoter depends on the carbon sources that are used or may change after the diauxic shift during batch cultivation [45]. It is also important to bear in mind that promoter strength does not necessarily correlate with the amount of protein produced. In addition, increased synthesis levels can lead to misfolding or aggregation of the product, thus strongly impairing its enzymatic activity. Also, folding of smaller proteins can occur much faster than folding of multi-domain proteins where assistance by chaperones may be required [46]. However, the number of genomic integration events that are achieved using the Xplor® expression platform does not normally lead to critical local protein concentrations within the cell, especially not if proteins are designed to be secreted into the culture medium.

4 Notes

1. Several of the mentioned promoters worked well in the tests, and the *TEF1* promoter was the strongest constitutive promoter among them.
2. The manual recommends two to threefold excess of inserts, but we found five to tenfold excess (if possible) to work better.
3. We recommend a touchdown program (50 °C for 10 min, −2 °C per cycle, 5 cycles), especially when several fragments with different overlaps were used.
4. Invert the tube occasionally to float the cells.
5. The overall volume should not be higher than 500 μL to maintain a minimum of oxygen supply via surface movements at 300 rpm. For the same reason, the plate should be sealed with AeraSeal™ film (Sigma-Aldrich).
6. Use YEPD only in case of stable genomic integration of the expression cassette. Otherwise, stay with selective minimal medium, which has been used for plating after transformation.
7. NaCl and KCl are used due to the easy calculation of ionic strength. Basically, any salt can be used. It is important to know that some enzymes are inhibited by certain ions (e.g., laccases are inhibited by Cl^- ions).
8. For each pH value, one negative control (without addition of enzyme) is needed, in order to calculate the respective enzyme activity.
9. It is advisable to mix everything in a PCR plate and incubate in the thermocycler using the gradient function. Each cavity can be used for several enzymes at one temperature or for more repetitions with one enzyme.
10. Use either the batch control system delivered with your laboratory bioreactor or open-source software like MyOpenLab for implementation of the algorithm.
11. The transition should be performed after the indicator (rpm, OUR, CTR) has passed the corresponding threshold for the first time, e.g., passing 800 rpm after 20 h of the example fermentation depicted in Fig. 3b.
12. The feed volume can be either recorded by pump totalizers or by balancing the respective reservoir.

13. For bigger amounts of cell material, it is advisable to use a high-pressure homogenizer for proper disruption of the cell wall (intracellularly localized enzymes).

Acknowledgment

The work was supported by grants from the German Federal Ministry of Economics and Technology (Bundesministerium für Wirtschaft und Technologie, BMWi) within its program "Zentrales Innovationsprogramm Mittelstand, ZIM" (Grant Numbers KF2131620MD2, KF2131628SA4).

References

1. Böer E, Steinborn G, Kunze G, Gellissen G (2007) Yeast expression platforms. Appl Microbiol Biotechnol 77:513–523. https://doi.org/10.1007/s00253-007-1209-0
2. Gellissen G, Hollenberg CP (1997) Application of yeasts in gene expression studies: a comparison of *Saccharomyces cerevisiae*, *Hansenula polymorpha* and *Kluyveromyces lactis*—a review. Gene 190:87–97. https://doi.org/10.1016/S0378-1119(97)00020-6
3. Madzak C, Nicaud J-M, Gaillardin C (2005) Yarrowia lipolytica. In: Production of recombinant proteins: novel microbial and eukaryotic expression systems. Wiley-VCH Verlag GmbH & Co. KGaA, Weinheim, FRG, pp 163–189
4. Terentiev Y, Pico AH, Böer E et al (2004) A wide-range integrative yeast expression vector system based on *Arxula adeninivorans*-derived elements. J Ind Microbiol Biotechnol 31:223–228. https://doi.org/10.1007/s10295-004-0142-9
5. Ilgen C, Lin-Cereghino J, Cregg JM (2005) Pichia pastoris. In: Production of recombinant proteins. Wiley-VCH Verlag GmbH & Co. KGaA, Weinheim, FRG, pp 143–162
6. Gellissen G (2000) Heterologous protein production in methylotrophic yeasts. Appl Microbiol Biotechnol 54:741–750. https://doi.org/10.1007/s002530000464
7. Böer E, Gellissen G, Kunze G (2005) Arxula adeninivorans. In: Production recombinant proteins. Wiley-VCH Verlag GmbH & Co. KGaA, Weinheim, FRG, pp 89–110
8. Kunze G, Gaillardin C, Czernicka M et al (2014) The complete genome of *Blastobotrys (Arxula) adeninivorans* LS3—a yeast of biotechnological interest. Biotechnol Biofuels 7:66. https://doi.org/10.1186/1754-6834-7-66
9. Gellissen G, Janowicz ZA, Weydemann U et al (1992) High-level expression of foreign genes in *Hansenula polymorpha*. Biotechnol Adv 10:179–189. https://doi.org/10.1016/0734-9750(92)90002-Q
10. Stöckmann C, Scheidle M, Klee D et al (2009) Process development in *Hansenula polymorpha* and *Arxula adeninivorans*, a re-assessment. Microb Cell Factories 8:22. https://doi.org/10.1186/1475-2859-8-22
11. Chamas A, Pham HTM, Baronian K, Kunze G (2017) Biosensors based on yeast/fungal cells. In: Biotechnology of yeasts and filamentous fungi. Springer International Publishing, Cham, pp 351–371
12. Bischoff F, Litwińska K, Cordes A et al (2015) Three new cutinases from the yeast *Arxula adeninivorans* that are suitable for biotechnological applications. Appl Environ Microbiol 81:5497–5510. https://doi.org/10.1128/AEM.00894-15
13. Rösel H, Kunze G (1998) Integrative transformation of the dimorphic yeast *Arxula adeninivorans* LS3 based on hygromycin B resistance. Curr Genet 33:157–163. https://doi.org/10.1007/s002940050322
14. Böer E, Steinborn G, Matros A et al (2007) Production of interleukin-6 in *Arxula adeninivorans*, *Hansenula polymorpha* and *Saccharomyces cerevisiae* by applying the wide-range yeast vector (CoMed™) system to simultaneous comparative assessment. FEMS Yeast Res 7:1181–1187. https://doi.org/10.1111/j.1567-1364.2007.00254.x
15. Biernacki M, Marzec M, Roick T et al (2017) Enhancement of poly(3-hydroxybutyrate-co-3-hydroxyvalerate) accumulation in *Arxula adeninivorans* by stabilization of production.

Microb Cell Factories 16:144. https://doi.org/10.1186/s12934-017-0751-4

16. Kunze G, Kunze I (1994) Characterization of *Arxula adeninivorans* strains from different habitats. Antonie Van Leeuwenhoek 65:29–34. https://doi.org/10.1007/BF00878276
17. Samsonova IA, Kunze G, Bode R, Böttcher F (1996) A set of genetic markers for the chromosomes of the imperfect yeast *Arxula adeninivorans*. Yeast 12:1209–1217. https://doi.org/10.1002/(SICI)1097-0061(19960930)12:12<1209::AID-YEA12>3.0.CO;2-W
18. Böer E, Steinborn G, Florschütz K et al (2009) Arxula adeninivorans (Blastobotrys adeninivorans)—a dimorphic yeast of great biotechnological potential. In: Yeast biotechnology: diversity and applications. Springer Netherlands, Dordrecht, pp 615–634
19. Álvaro-Benito M, Fernández-Lobato M, Baronian K, Kunze G (2013) Assessment of *Schwanniomyces occidentalis* as a host for protein production using the wide-range Xplor®2 expression platform. Appl Microbiol Biotechnol 97:4443–4456. https://doi.org/10.1007/s00253-012-4527-9
20. Robinson JS, Klionsky DJ, Banta LM, Emr SD (1988) Protein sorting in *Saccharomyces cerevisiae*: isolation of mutants defective in the delivery and processing of multiple vacuolar hydrolases. Mol Cell Biol 8:4936–4948
21. Bischoff F, Chamas A, Litwińska K et al (2017) Applications of *Blastobotrys* (*Arxula*) *adeninivorans* in biotechnology. In: Yeast diversity in human Welfare. Springer Singapore, Singapore, pp 455–479
22. Tanaka A, Ohnishi N, Fukui S (1967) Studies on the formation of vitamins and their function in hydrocarbon fermentation. Production of vitamins and their function in hydrocarbon medium. J Ferment Technol 45:617–623
23. Elving PJ, Markowitz JM, Rosenthal I (1956) Preparation of buffer systems of constant ionic strength. Anal Chem 28:1179–1180. https://doi.org/10.1021/ac60115a034
24. Böer E, Piontek M, Kunze G (2009) Xplor® 2—an optimized transformation/expression system for recombinant protein production in the yeast *Arxula adeninivorans*. Appl Microbiol Biotechnol 84:583–594. https://doi.org/10.1007/s00253-009-2167-5
25. Gibson DG, Young L, Chuang R-Y et al (2009) Enzymatic assembly of DNA molecules up to several hundred kilobases. Nat Methods 6:343–345. https://doi.org/10.1038/nmeth.1318
26. Gibson DG, Smith HO, Hutchison CA et al (2010) Chemical synthesis of the mouse mitochondrial genome. Nat Methods 7:901–903. https://doi.org/10.1038/nmeth.1515
27. Dohmen RJ, Strasser AWM, Honer CB, Hollenberg CP (1991) An efficient transformation procedure enabling long-term storage of competent cells of various yeast genera. Yeast 7:691–692. https://doi.org/10.1002/yea.320070704
28. Bradford MM (1976) A rapid and sensitive method for the quantitation of microgram quantities of protein utilizing the principle of protein-dye binding. Anal Biochem 72:248–254. https://doi.org/10.1016/0003-2697(76)90527-3
29. Rauter M, Kasprzak J, Becker K et al (2014) ADH from *Rhodococcus ruber* expressed in *Arxula adeninivorans* for the synthesis of 1-(S)-phenylethanol. J Mol Catal B Enzym 104:8–16. https://doi.org/10.1016/j.molcatb.2014.02.015
30. Kasprzak J, Bischoff F, Rauter M et al (2016) Synthesis of 1-(S)-phenylethanol and ethyl (R)-4-chloro-3-hydroxybutanoate using recombinant *Rhodococcus erythropolis* alcohol dehydrogenase produced by two yeast species. Biochem Eng J 106:107–117. https://doi.org/10.1016/j.bej.2015.11.007
31. Pointing SB (1999) Qualitative methods for the determination of lignocellulolytic enzyme production by tropical fungi. Fungal Divers 2:17–33
32. Bleve G, Lezzi C, Mita G et al (2008) Molecular cloning and heterologous expression of a laccase gene from *Pleurotus eryngii* in free and immobilized *Saccharomyces cerevisiae* cells. Appl Microbiol Biotechnol 79:731–741. https://doi.org/10.1007/s00253-008-1479-1
33. Kiiskinen L-L, Ratto M, Kruus K (2004) Screening for novel laccase-producing microbes. J Appl Microbiol 97:640–646. https://doi.org/10.1111/j.1365-2672.2004.02348.x
34. Traut TW, Ropp PA, Poma A (1991) Purine nucleoside phosphorylase: allosteric regulation of a dissociating enzyme. Adv Exp Med Biol 309B:177–180
35. Glogauer A, Martini VP, Faoro H et al (2011) Identification and characterization of a new true lipase isolated through metagenomic approach. Microb Cell Factories 10:54. https://doi.org/10.1186/1475-2859-10-54
36. Halonen P, Reinikainen T, Nyyssölä A, Buchert J (2009) A high throughput profiling method for cutinolytic esterases. Enzym Microb

Technol 44:394–399. https://doi.org/10.1016/j.enzmictec.2008.12.012

37. Böer E, Breuer FS, Weniger M et al (2011) Large-scale production of tannase using the yeast *Arxula adeninivorans*. Appl Microbiol Biotechnol 92:105–114. https://doi.org/10.1007/s00253-011-3320-5
38. Bradoo S, Gupta R, Saxena RK (1996) Screening of extracellular tannase-producing fungi: development of a rapid and simple plate assay. J Gen Appl Microbiol 42:325–329. https://doi.org/10.2323/jgam.42.325
39. Bae H, Yanke L, Cheng K-J, Selinger L (1999) A novel staining method for detecting phytase activity. J Microbiol Methods 39:17–22. https://doi.org/10.1016/S0167-7012(99)00096-2
40. Shi H, Ding H, Huang Y et al (2014) Expression and characterization of a GH43 endo-arabinanase from *Thermotoga thermarum*. BMC Biotechnol 14:35. https://doi.org/10.1186/1472-6750-14-35
41. Miller GL (1959) Use of dinitrosalicylic acid reagent for determination of reducing sugar. Anal Chem 31:426–428. https://doi.org/10.1021/ac60147a030
42. Schiøtt M, De Fine Licht HH, Lange L, Boomsma JJ (2008) Towards a molecular understanding of symbiont function: identification of a fungal gene for the degradation of xylan in the fungus gardens of leaf-cutting ants. BMC Microbiol 8:40. https://doi.org/10.1186/1471-2180-8-40
43. Giersberg M, Degelmann A, Bode R et al (2012) Production of a thermostable alcohol dehydrogenase from *Rhodococcus ruber* in three different yeast species using the Xplor®2 transformation/expression platform. J Ind Microbiol Biotechnol 39:1385–1396. https://doi.org/10.1007/s10295-012-1134-9
44. Bischoff F (2017) Identifizierung und Charakterisierung von drei Cutinasen der Hefe *Arxula adeninivorans* LS3 und deren Einsatz zum Abbau von Polyestern. PhD Thesis. Ernst-Moritz-Arndt-Universität Greifswald
45. Peng B, Williams TC, Henry W et al (2015) Controlling heterologous gene expression in yeast cell factories on different carbon substrates and across the diauxic shift: a comparison of yeast promoter activities. Microb Cell Factories 14:91. https://doi.org/10.1186/s12934-015-0278-5
46. Liu Z, Tyo KEJ, Martínez JL et al (2012) Different expression systems for production of recombinant proteins in *Saccharomyces cerevisiae*. Biotechnol Bioeng 109(5):1259–1268. https://doi.org/10.1002/bit.24409

Chapter 5

Screening and Selection of Production Strains: Secretory Protein Expression and Analysis in *Hansenula polymorpha*

Su Jin Yoo, Hye Yun Moon, and Hyun Ah Kang

Abstract

The thermotolerant methylotrophic yeast *Hansenula polymorpha* has been used as a host for the high-level production of recombinant proteins from industrial enzymes to therapeutic proteins. Despite favorable characteristics of the *H. polymorpha*-based platform for application to heterologous gene expression, several problems and limitations, such as over-glycosylation and proteolytic degradation, can be encountered in the development of production strains for secretory proteins. Here, *H. polymorpha* genetic tools and host strains, developed for authentic processing and modification of secretory recombinant proteins, are introduced with the analytical protocols.

Key words *Hansenula polymorpha*, Protein expression, Multiple integration, Secretion, Glycosylation, Surface display

1 Introduction

Hansenula polymorpha is a thermotolerant methylotrophic yeast that can utilize methanol as sole energy and carbon source and grow at temperatures up to 50 °C. It has been a valuable model organism to study the mechanisms of peroxisomal biogenesis and degradation, regulation of methanol metabolism, nitrate assimilation, and stress response [1]. *H. polymorpha* has gained interest as a useful host system for the high-level production of recombinant proteins from industrial enzymes to therapeutic proteins [2]. The most representative *H. polymorpha*-derived therapeutic protein is the recombinant hepatitis B virus S antigen (HBsAg), which has been successfully commercialized with the brand names of HepaVax-Gene and AgB, respectively [3]. Other *H. polymorpha*-based recombinant proteins with the potential of pharmaceuticals are the leech *Hirudinaria manillensis* hirudin [4] and several human proteins including α_1-antitrypsin [5], insulin [6], IFNα-2a [7], serum albumin [8, 9], epidermal growth factor [10], and parathyroid hormone [11]. Several other proteins with application

Brigitte Gasser and Diethard Mattanovich (eds.), *Recombinant Protein Production in Yeast*, Methods in Molecular Biology, vol. 1923, https://doi.org/10.1007/978-1-4939-9024-5_5,

as food additives (hexose oxidase [12], phytase [13]) and as industrial enzymes (*Zymomonas mobilis* levansucrase [14] and *Aspergillus niger* glucose oxidase [15]) have been produced at high levels in *H. polymorpha*, supporting this methylotrophic yeast as a practical host for industrial production.

Moreover, peculiar physiological characteristics of *H. polymorpha*, such as resistance to high temperature, heavy metals, and oxidative stress, make this yeast attractive for several biotechnological purposes [16, 17]. So far, three representative *H. polymorpha* strains of independent origin, DL-1, CBS4732, and NCYC495, are widely used for basic and applied research. The CBS4732 and NCYC495 strains have been reclassified as *Ogataea polymorpha*, whereas the DL-1 strain is assigned to *Ogataea parapolymorpha*, respectively [18, 19]. The DL-1 strain has certain advantages in that it has a higher growth rate and adapts more quickly to culture media than the other parental strains; additionally, the DL-1 strain shows a higher frequency of homologous recombination than other strains [20, 21], facilitating genetic manipulation applicable for development of host strains.

Yeasts have been a preferred microbial host system for the secretory production of recombinant proteins. However, several limiting steps, such as different glycosylation processes and proteolytic degradation, are often encountered during secretory production of heterologous proteins in yeast. *H. polymorpha* does less hyperglycosylate secreted proteins and lacks the immunogenic α-1,3 terminal mannose epitope, which often causes a problem in heterologous protein production in the traditional yeast *Saccharomyces cerevisiae*. In this respect, *H. polymorpha* is advantageous over *S. cerevisiae* as a host for the production of humanized glycoproteins [15]. A series of host strains, including glyco-engineered strains to produce glycans with designed structures and protease-deficient strains to prevent aberrant proteolytic cleavage during secretion process, have been developed for authentic processing and modification of secretory recombinant proteins in *H. polymorpha* [22]. The glyco-engineered *H. polymorpha* strains were shown to produce human pentamannosyl *N*-glycans (Man_5-$GlcNAc_2$), core trimannosyl *N*-glycan (Man_3-$GlcNAc_2$), and human hybrid-type *N*-glycans with a terminal *N*-acetylglucosamine ($GlcNAc_1Man_5GlcNAc_2$ and $GlcNAc_1Man_3GlcNAc_2$, respectively) [23–25]. The *H. polymorpha* strains deleted in the genes coding for vacuolar proteases, such as proteinase A (*PEP4*) and carboxypeptidase Y (*PRC1*) [26], and in the genes encoding endoproteases localized in the secretory pathway, such as serine-type carboxypeptidase (*KEX1*) and yeast aspartyl proteases (*YPS1*, *YPS7*), were developed to avoid the overall degradation of recombinant proteins and particularly to prevent aberrant proteolytic cleavage of recombinant proteins during the secretion process [10, 11].

Besides secretory expression for release of recombinant proteins directly into the culture medium, glycosylphosphatidylinositol (GPI)-anchoring motifs have been exploited to develop a cell surface display system of industrial enzymes, such as carboxymethylcellulase from *Bacillus subtilis* and lipase B from *Candida antarctica* in *H. polymorpha*, which can be usefully applicable as a whole cell catalyst with the advantage of reusability [27, 28]. The recombinant *H. polymorpha* strains expressing surface-displayed *Aspergillus saitoi* α-1,2-mannosidase (MsdS), which was fused to C-terminal peptides of 40 amino acids derived from *H. polymorpha* GPI proteins, showed the potential not only as a whole cell catalyst for in vitro mannose trimming but also as a production host for recombinant proteins with designed glycans in vivo [29]. The availability of such versatile genetic tool boxes in *H. polymorpha* would expedite the development of this yeast as a cell factory for secretory production of recombinant proteins with various industrial applications.

The current challenge, which needs to be addressed to further improve *H. polymorpha* as a host for mass production of secretory proteins, is to increase the secretion capacity and the folding efficiency. Compared to animal cells, yeast cells appear to possess less number of chaperons and display limited extension of endoplasmic reticulum (ER) membrane, thus hindering the replacement of yeast as a practical host for industrial production of full-length antibodies. ER engineering based on the Hac1p-meidated unfolded protein response (UPR) might be a good strategy to develop host strains with enhanced secretion and folding capacity. A recent study on the gene regulatory networks required for protein secretion and modification in *H. polymorpha* identified a *HAC1* homolog of *H. polymorpha* (*HpHAC1*) as a master transcription factor regulating the UPR signal pathway, which is induced by the expression of an aberrant secreted protein or a misfolded heterologous protein [30]. Further studies of the UPR in *H. polymorpha* would generate more relevant information on the systematic manipulation of UPR to achieve successful exploitation of this yeast as an intelligent cell factory for the secretory production of correctly folded and processed recombinant proteins.

2 Materials

A variety of *H. polymorpha* genetic tools suitable for heterologous protein expression, including vectors, promoters, selection markers, secretion signals, and surface anchor motifs, have been developed in the DL-1 strain, which are also applicable to *H. polymorpha* CBS and NCYC strains (Tables 1–3). Several sets of *H. polymorpha* host strains, including multiple auxotrophic mutant strains, glyco-engineered strains, and protease-deficient strains, have been mostly developed in the background of DL-1 strain (Table 4).

Table 1
Promoters and selection markers for protein expression in *H. polymorpha* DL-1

	Elements	Characteristics	Regulation/source	Ref.
Promoters	*MOX1*	Strong inducible	Induced by methanol and glucose limitation	[8]
	SEO1	Moderate inducible	Induced by cadmium in concentration-dependent way	[31]
	GAP	Strong constitutive	–	[9]
	ACT1	Moderate constitutive	–	[32]
	MET3	Repressible	Repressed by cysteine in concentration-dependent way	[33]
	ScLEU2, ScURA3	Auxotrophic	Derived from *S. cerevisiae*	[34]
Markers	*LEU2, URA3, TRP3, ADE2, MET2*	Auxotrophic	Derived from *H. polymorpha*	[35]
	$G418^R$	Dominant	Under the control of the truncated *GAP* promoter sets	[36]
	Hyg^R	Dominant	Under the control of the truncated *ACT* promoter sets	[32]
	Zeo^R	Dominant	Under the control of the bacterial *EM7* and *ScTEF1* promoters	[37]

2.1 Promoters and Selection Markers for H. polymorpha

For heterologous gene expression in *H. polymorpha*, a range of homologous and heterologous promoters of different strengths is available. The *GAP* promoter of the *H. polymorpha* glyceraldehyde-3-phosphate dehydrogenase gene is suitable for strong and constitutive expression, whereas the *MOX* promoter derived from the *H. polymorpha* methanol oxidase gene is useful for carbon source-regulated expression. A novel repressible *MET3* promoter of the *H. polymorpha* ATP sulfurylase gene, which can be efficiently down-regulated by cysteine in a concentration-dependent way, has been recently developed as a useful molecular tool for regulated gene expression [33]. Besides a set of auxotrophic markers, several antibiotic resistance markers, such as the genes for resistance to G418, Hygromycin B, and Zeocin, are also available (Table 1).

2.2 Multiple Integration and Epitope Vectors

Multiple integration vectors of the DL-1 strain have been developed to harbor one of a set of several cloned sub-telomeric ARS sequences [36]. These vectors homologously integrate into a genomic counterpart and result in recombinant strains harboring single or multiple tandem repeated copies at the respective sub-telomeric genomic locus (*see* **Note 1**). Besides multiple integration systems based on complementation of auxotrophic mutations, rapid and copy-number-controlled selection systems have been developed

Table 2
Vectors for multiple integration and epitope tagging in *H. polymorpha* DL-1

Plasmid	Replication sequence	Selection markers	Epitope	Integration copy number	Promoter	Ref.
AMIpL1	HARS36	*HpLEU2*	–	1–5		[34]
AMIpLD1	HARS36	*HpLEU2-d*	–	~30–100	–	[34]
AMIpSL1	HARS36	*ScLEU2*		6–9	–	[34]
AMIpSU1	HARS36	*ScURA3*	–	30–50	–	[34]
pGLG61	TEL188	*HpLEU2*/ G418^R	–	1–50	–	[38]
pHACT90-HyL	TEL188	*HpLEU2*/ Hyg^R	–	1–25	–	[32]
pDTMOX	HARS36	*HpTRP1*	–	1–5	*MOX*	[35]
pDMMOX	HARS36	*HpMET2*	–	1–5	*MOX*	[35]
pDAMOX	HARS36	*HpADE2*	–	1–5	*MOX*	[35]
pHIGAZ-6HA	–	Zeo^{Ra}	HA[b]	1	*GAP*	[37]
pHIGAHT-4FLAG	–	Hyg^{Ra}	FLAG[c]	1	*GAP*	[37]
pHIGAZ-4FLAG	–	Zeo^{Ra}	FLAG[c]	1	*GAP*	[37]

[a]Zeo^R and Hyg^R are expressed under control of the bacterial *EM7* and *ScTEF1* promoter
[b]HA, human influenza hemagglutinin epitope
[c]FLAG, the sequence motif DYKDDDDK (where D = aspartic acid, Y = tyrosine, and K = lysine)

using antibiotic resistance markers as dominant selection markers, which allow selection of transformants on plates containing different concentrations of G418 or Hygromycin B (Table 2 and Fig. 1d).

2.3 Secretion Signals and Surface Anchor Motifs for H. polymorpha

A secretion signal is a prerequisite for a protein to be efficiently directed toward the secretion pathway. Several N-terminal signal sequences derived from various well-characterized heterologous secretory proteins, such as *Aspergillus niger* α-amylase, *Kluyveromyces marxianus* exoinulinase, *S. cerevisiae* mating factor α_1, and human serum albumin, have been exploited to optimize secretion of a recombinant protein in *H. polymorpha* DL-1 (Table 3). For a targeted expression of a recombinant protein at the cell surface of *H. polymorpha*, a set of C-terminal peptide-derived endogenous GPI-cell wall proteins has been identified and evaluated as useful motifs for surface display of enzymes with industrial potential (Table 3).

Table 3
Targeting elements for secretory expression in *H. polymorpha* DL-1

Source	Recombinant proteins	N- or C-terminal amino acid sequences	Ref.
Secretion signals			
K. marxianus exoinulinase	Human α_1-AT *Z. mobilis* LevU	MKLAYSLLLPLAGVSASVINYKRS	[5] [14]
S. cerevisiae mating factor α_1	Human EGF Human PTH	MRFPSIFTAVLFAASSALAAPVNTTTEDETAQIPAEAVIGYSDLEGDFDVAVLPFSNSTNNGLLFINTTIA SIAAKEEGVS	[10] [11]
A. niger α-amylase	*A. niger* GOD	MVAWWSLFLYGLQVAAPALA	[15]
Human serum albumin	Human serum albumin	MKWVTFISLLFLFSSAYS	[9]
A. saitoi aspergillopepsin I	*A. saitoi* msdS	MVVFSKTAALVLGLSTAVSAGST	[23]
Surface anchor motifs			
HpSed1	*B. subtilis* CMCase *A. niger* GOD	TAASSVAASSAPVISTAENAGAKVGAAGLAALAGAAAFLL	[27]
HpTip1	*A. saitoi* msdS *B. subtilis* CMCase	AGSSSAASNSSGHISTFSGAGAKLAVGAGAGIVGLAALLM	[29] [27]
HpCwp1	*A. niger* GOD *C. antarctica* LB	RSSAAASSSAKIATYEGAAAENVAKVGMGALLSGMAVLLM	[27] [28]
HpCcw14	*A. saitoi* msdS	VSSSSAAAPSSSIDVQVSAGAPALVANAAGLGLGVLIGLL	[29]
HpCrh1	*A. saitoi* msdS	GNSSSQSVSSTNGSESSHGFGSLNVFVSLVSATLFGLALI	[29]
HpOrf73	*A. saitoi* msdS	TASTASSSSAPAASIFEGAAGKINIGVGLISMVYGLICVL	[29]
HpCts2	*A. saitoi* msdS	YPDEMDGATASSYATQCFYTDIRVFLVIVLSWVFILVFIS	[29]
HpExg1	*A. saitoi* msdS	KYASVLSSDAAIVSSPLKNISKIEIIVLILTACALGLELL	[29]
HpOrf333	*A. saitoi* msdS	SSTVRNVTFQNAAGVSPDSSLSAVVSVVGFAFALHCFMFE	[29]
HpSps2	*A. saitoi* msdS	DGDHAKQTLARSGSSGSRPSGVGFAAVLLGFGVSWLAMQL	[29]
HpOrf518	*A. saitoi* msdS	PREDIPHPHVAAPVAEKSFIQKYWMYIVPPLIILMVMSPQ	[29]
HpOrf135	*A. saitoi* msdS	YDDEGTSTAASSSSAQANGVAPISGFAFGLTGIVLAALI	[29]

α_1-AT α_1-antitrypsin, *EGF* epidermal growth factor, *GOD* glucose oxidase, *LB* lipase B, *LevU* levansucrase, *MsdS* α-1,2-mannosidase, *PTH* parathyroid hormone

Table 4
Host strains based on *H. polymorpha* DL-1

Strain	Genotype	Phenotype	Ref.
Parental strain			
DL-1	NRRL-Y-7560/ATCC26012	Wild-type	[39]
Auxotrophic strains			
DL1-L	*leu2*	Leu$^-$	[20]
uDL10	*leu2 ura3*	Leu$^-$, Ura$^-$	[26]
DL1-LdU	*leu2 ura3Δ::lacZ*	Leu$^-$, Ura$^-$	[20]
DL1-LdT	*leu2 ura3Δ::lacZ trp1Δ::lacZ-URA3-lacZ*	Leu$^-$, Trp$^-$	[35]
DL1-LdTU	*leu2 ura3Δ::lacZ trp1Δ::lacZ*	Leu$^-$, Trp$^-$, Ura$^-$	[35]
DL1-LdM	*leu2 ura3Δ::lacZ met2Δ::lacZ-URA3-lacZ*	Leu$^-$, Met$^-$	[35]
DL1-LdMU	*leu2 ura3Δ::lacZ met2Δ::lacZ*	Leu$^-$, Ura$^-$, Met$^-$	[35]
DL1-LdTM	*leu2 ura3Δ::lacZ trp1Δ::lacZ met2Δ::lacZ-URA3-lacZ*	Leu$^-$, Trp$^-$, Met$^-$	[35]
DL1-LdTA	*leu2 ura3Δ::lacZ trp1Δ::lacZ-URA3-lacZ ade2Δ::tc-TRP1-tc*	Leu$^-$, Ade$^-$	[35]
DL1-LdTUA	*leu2 ura3Δ::lacZ trp1Δ::lacZ ade2Δ::tc-TRP1-tc*	Leu$^-$, Ura$^-$, Ade$^-$	[35]
DL1-LdTMU	*leu2 ura3Δ::lacZ trp1Δ::lacZ met2Δ::lacZ*	Leu$^-$, Ura$^-$, Trp$^-$, Met$^-$	[35]
DL1-LdTMA	*leu2 ura3Δ::lacZ trp1Δ::lacZ met2Δ::lacZ ade2Δ:: lacZ-URA3-lacZ*	Leu$^-$, Trp$^-$, Met$^-$, Ade$^-$	[35]
DL1-LdTMAU	*leu2 uraΔ3::lacZ trp1Δ::lacZ met2Δ::lacZ ade2Δ::lacZ*	Leu$^-$, Ura$^-$, Trp$^-$, Met$^-$, Ade$^-$	[35]
Protease-deficient strains			
uDLB11	*leu2 ura3 pep4Δ::lacZ*	Pep4$^-$	[26]
uDLB12	*leu2 ura3 prc1Δ::lacZ*	Prc1$^-$	KRIBB[a]
uDLB13	*leu2 ura3 kex1Δ::lacZ*	Kex1$^-$	[9]
uDLB14	*leu2 ura3 pep4Δ::lacZ prc1Δ::lacZ*	Pep4$^-$, Prc1$^-$	KRIBB[a]
uDLB15	*leu2 ura3 pep4Δ::lacZ kex1Δ::lacZ*	Pep4$^-$, Kex1$^-$	KRIBB[a]
uDLB16	*leu2 ura3 prc1Δ::lacZ kex1Δ::lacZ*	Prc1$^-$, Kex1$^-$	KRIBB[a]
uDLB17	*leu2 ura3 pep4Δ::lacZ prc1Δ::lacZ kex1Δ::lacZ*	Pep4$^-$, Prc1$^-$, Kex1$^-$	KRIBB[a]
uDLY1	*leu2 ura3 yps1Δ::lacZ*	Yps1$^-$	[11]
uDLY17	*leu2 ura3 yps1Δ::lacZ yps7Δ::lacZ*	Yps1$^-$, Yps7$^-$	[11]
Glyco-engineered strains			
DL1-g11	*leu2 ura3Δ::lacZ och1Δ::lacZ* pDTMOX-msdS	Human mannose-type *N*-glycans	[23]

(continued)

Table 4 (continued)

Strain	Genotype	Phenotype	Ref.
DL1-g16	*leu2 ura3Δ::lacZ och1Δ::lacZ alg3Δ::lacZ trp1Δ::lacZ* pDTMOX-msdS	Trimannose core *N*-glycans	[24]
DL1-g17Sm	*leu2 ura3Δ::lacZ och1Δ::lacZ alg3Δ::lacZ trp1Δ::lacZ* pDTMOX-msdS pHIMAZC7-SN9mHG1	Human hybrid-type *N*-glycans	[25]
DL1-g23	*leu2 ura3Δ::lacZ och1Δ::lacZ trp1Δ::lacZ* pDTG-msdS	Human mannose-type *N*-glycans	[25]
DL1-g24Sm	*leu2 ura3Δ::lacZ och1Δ::lacZ trp1Δ::lacZ* pDTG-msdS pHIMAZC7-SN9mHG1	Human hybrid-type *N*-glycans	[25]

[a] *KRIBB* Korea Research Institute of Bioscience and Biotechnology

2.4 Yeast Strains

Several host strains suitable for heterologous protein expression, including multiple auxotrophic mutants (*see* **Note 2**), protease-deficient strains, and glyco-engineered strains, have been constructed in the DL-1 strain mostly using gene disruption techniques based on homologous recombination (Table 4).

2.5 Culture Media

1. YPD medium: 1% bacto-yeast extract, 2% bacto-peptone, 2% glucose.
2. YPM medium: 1% bacto-yeast extract, 2% bacto-peptone, 2% methanol.
3. Synthetic complete (SC) medium: 0.67% yeast nitrogen base without amino acids, 2% glucose, all amino acid mixture supplemented with 30 mg/L each of uracil and adenine.
4. Synthetic minimal (SD) medium: 0.67% yeast nitrogen base without amino acids, 2% glucose.
5. *URA3* pop-out medium: YPD medium containing 0.5 g/L 5-fluoro-orotic acid (5-FOA; Thermo Scientific, USA), if necessary supplemented with 90 mg/L tryptophan and 60 mg/L uracil.
6. Selection medium for auxotrophic markers: SD medium supplemented with uracil (20 mg/mL), leucine (100 mg/mL), lysine (30 mg/mL), or histidine (20 mg/mL) in combination.
7. Selection medium for antibiotic resistance markers: YPD containing G418 (Invitrogen, USA), Zeocin (Invitrogen, USA), or Hygromycin B (Sigma-Aldrich, USA) in appropriate concentrations (*see* **Note 3**).

2.6 Fractionation of Cell Wall Proteins for Cell Surface Expression Analysis

1. Glass beads (425–600 μm; Sigma-Aldrich, USA).
2. A multi-bead beater (Precellys 24; Bertin Technologies, France).
3. Cell wall pellet washing solution: Solution A [1 mM phenylmethylsulfonyl fluoride (PMSF; Sigma-Aldrich, USA)], Solution B [5% (w/v) NaCl with 1 mM PMSF], Solution C [2% (w/v) NaCl with 1 mM PMSF], Solution D [1% NaCl with 1 mM PMSF].
4. SDS/β-ME or SDS/DTT extraction solution: 50 mM Tris–HCl, pH 7.8, 2% (w/v) sodium dodecyl sulfate (SDS), 100 mM sodium ethylenediaminetetraacetic acid (EDTA), and 40 mM β-mercaptoethanol (β-ME) or 10 mM dithiothreitol (DTT).
5. TCA precipitation: trichloroacetic acid (Sigma-Aldrich, USA).
6. HF-pyridine (hydrogen fluoride-pyridine, Sigma-Aldrich, USA).
7. CBB staining solution: 0.1% Coomassie Brilliant Blue R-250 (Bio-Rad, USA), 50% methanol, and 10% glacial acetic acid.
8. Western blotting reagents: anti-FLAG monoclonal antibody (Sigma-Aldrich, USA), AP-conjugated substrate kit (Bio-Rad, USA).

2.7 High-Throughput N-Glycan Analysis Using DNA Sequencer

1. APTS labeling reagent: 8-aminopyrene-1,3,6-trisulfonic acid (Molecular Probes, USA).
2. Purification column: Carbograph™ graphitized carbon resin (Alltech, USA), packed in 96-well plates of Multi-Screen-HV (Millipore, USA).
3. Washing buffer: 80% (v/v) acetonitrile containing 0.1% trifluoroacetic acid (TFA; Sigma-Aldrich, USA).
4. Elution buffer: 25% acetonitrile containing 0.075% TFA.
5. Glycan-processing enzymes: PNGase F (New England Biolabs, USA), α-1,2 mannosidase (Prozyme, USA), β-*N*-hexosaminidase (Prozyme, USA), α-1,3 mannosidase (New England Biolabs, USA).
6. Internal standard: ROX-labeled GeneScan 500 standard mixture (Applied Biosystems, USA).
7. Glycan profiling instrument: ABI 3130 sequencer (Applied Biosystems, USA).

3 Methods

3.1 Rapid and Copy-Number-Controlled Selection of *H. polymorpha* Transformants

1. Transform *H. polymorpha* cells with minor modification of the DMSO-enhanced whole cell yeast transformation [40]. The minor modifications include the increase of the volume of PEG4000 solution (50% PEG4000 in lithium acetate solution), from 280 to 600 μL, and DMSO, from 43 to 70 μL, along with the extension of heat shock time at 42 °C from 5 to 15 min.
2. Select *H. polymorpha* transformants obtained with pGLG or pHACT-HyL vectors on synthetic complete medium lacking leucine (SC-LEU). Macroscopic colonies typically appear after 4–5 days of incubation at 37 °C (Fig. 1a, b).
3. For the stabilization of transformants, pool all LEU+ transformants with YPD broth and inoculate into a 500-mL baffled shaker flask containing 50 mL of YPD medium. After 24 h, transfer 1 mL of culture broth to 50 mL of fresh YPD medium, and repeat this procedure until 50 generations.

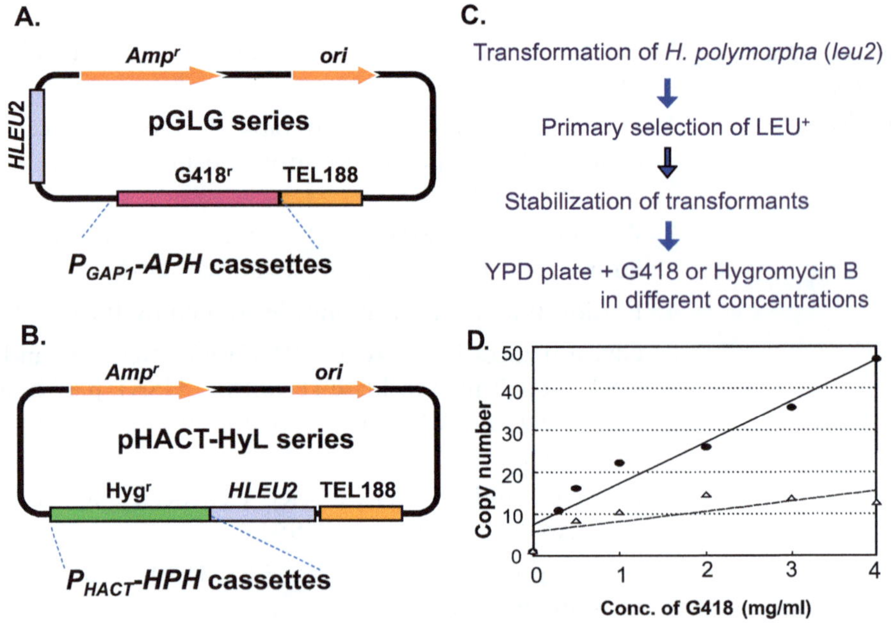

Fig. 1 Copy-number-controlled gene integration in *H. polymorpha* DL-1. (**a**) Vector pGLG series carry, as a selective marker, a variant of the G418 resistance cassette ($G418^R$) composed of the *APH* gene under the control of the truncated *GAP* promoter in different lengths. (**b**) Vector pHACT-HyL series carry, as a selective marker, a variant of the Hygromycin B resistance cassette (Hyg^R) composed of the *HPH* gene under the control of the truncated *ACT* promoter in different lengths. (**c**) Selection procedure of multiple integrants based on G418 and Hygromycin B resistance. (**d**) Copy numbers of pGLG vectors, pGLG578 (Δ) and pGLG 61 (●), in different G418 concentrations

4. Then plate the cells on minimal plates to select out transformants that have lost the vector in the episomal state during cultivation in YPD medium (*see* **Note 4**).
5. After complete stabilization, plate approximately 1×10^5 cells grown in YPD broth onto a series of YPD plates containing G418 or Hygromycin B in different concentrations (Fig. 1c, d).
6. Due to the strong correlation between antibiotic resistance and integration copy number, construction of transformants harboring different copy numbers ranging from 1 to 50 can be easily achieved (*see* **Note 5**).

3.2 Co-Transformation of H. polymorpha with Multiple ARS Vectors

1. Co-transform the multiple auxotrophic mutant strains simultaneously with several ARS vectors with different selection markers using the DMSO-enhanced whole cell yeast transformation (*see* **Note 6**).
2. Select *H. polymorpha* transformants on synthetic complete media lacking appropriate nutrients, such as SC-ADE-TRP-MET-LEU, or synthetic minimal SD media supplemented with appropriate nutrients, such as SD + URA.
3. Analyze the expression of the recombinant protein from each of vectors. For example, the targeted expression of the epitope-tagged recombinant *A. saitoi* MsdS proteins at different subcellular compartments of the secretory pathway can be analyzed by Western blotting (Fig. 2).

3.3 Localization Analysis of Surface-Displayed GPI-Motif Fusion Proteins

1. Cultivate the *H. polymorpha* transformants, harboring a construct containing a GPI-motif fusion protein under control of the *MOX* promoter, in YPM medium for 48 h.
2. Harvest cells by centrifugation, wash cell pellets with cold distilled water and subsequently with 10 mM Tris–HCl (pH 7.5), and break cells with glass beads using a multi-bead beater.
3. Obtain the crude cell lysate and the cell wall pellet by centrifugation at $1000 \times g$ for 5 min.
4. To remove non-covalently linked proteins and the intracellular crude lysate, wash the cell walls with washing solution A (1 mM PMSF), washing solution B [5% (w/v) NaCl with 1 mM PMSF], washing solution C [2% (w/v) NaCl with 1 mM PMSF], and washing solution D [1% NaCl with 1 mM PMSF], and wash again with solution A followed by centrifugation at $1000 \times g$ for 5 min.
5. Incubate twice the cell walls with the SDS/β-ME or SDS/DTT extraction solution at 100 °C for 5 min, and precipitate the solution by TCA (*see* **Note 7**).

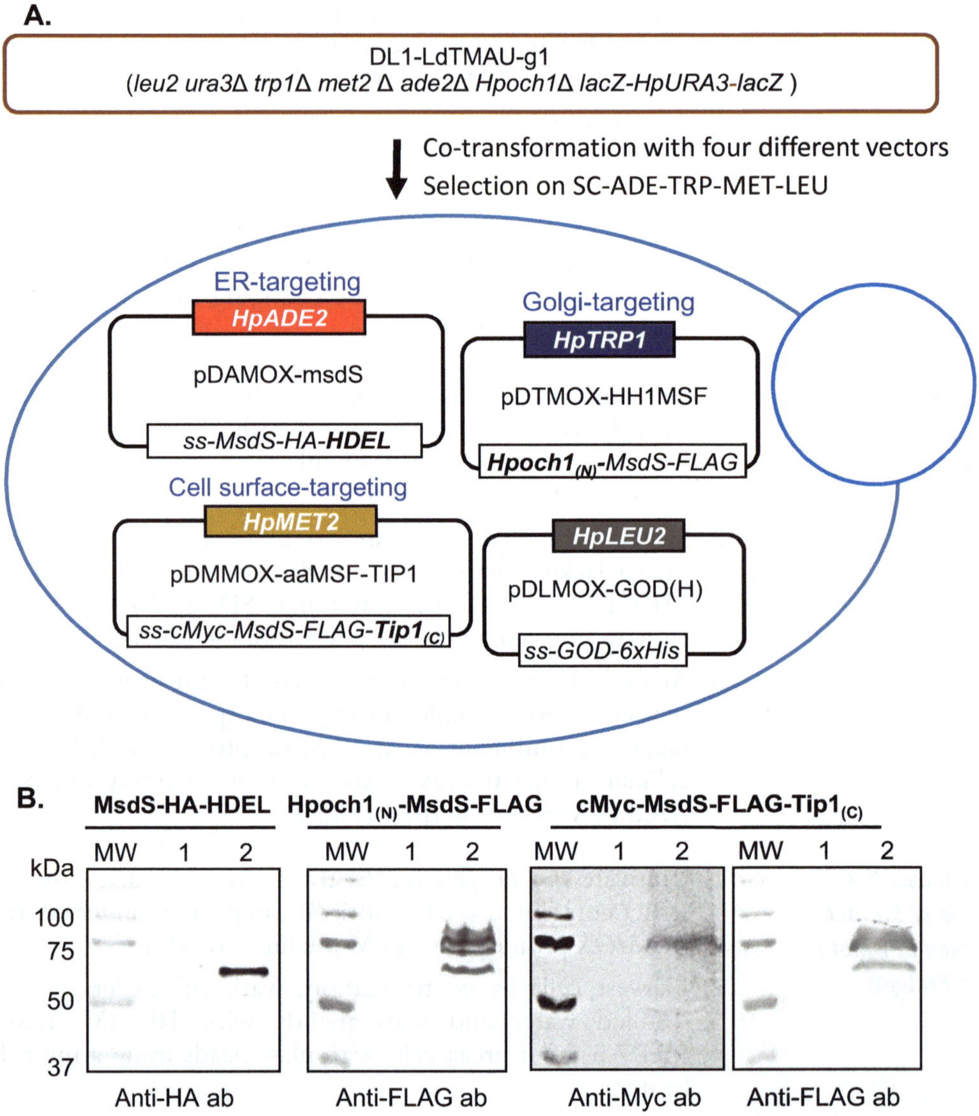

Fig. 2 Co-transformation of *H. polymorpha* with four different ARS vectors. (**a**) Diagrams for the ER-, Golgi-, and surface-targeted *Aspergillus saitoi* α-1,2 mannosidase (MsdS) expression vectors, pDAMOX-msdS, pDTMOX-HH1MSF, and pDMMOX-aaMSF-TIP1, and with another expression vector, pDLMOX-GOD(H), encoding recombinant glucose oxidase (rGOD) as a reporter glycoprotein. For targeted expression of MsdS at different subcellular compartments of the secretory pathway, four ARS vectors were simultaneously co-transformed, and transformants were selected on SC-ADE-TRP-MET-LEU. (**b**) Expression analysis of MsdS localized at the ER, Golgi, and cell surface. The DL1-LdTMAU-g1 strain (lane 1) and its transformants harboring pDAMOX-msdS, pDTMOX-HH1MSF, pDMMOX-aaMSF-TIP1, and pDLMOX-GOD(H) (lane 2) were incubated in YPM medium at 37 °C for 48 h. The total cell lysates and cell wall fraction were loaded on 10% SDS-polyacrylamide gel and analyzed by Western blotting with anti-HA, anti-FLAG, and anti-cMyc antibodies to detect the ER-, Golgi-, and surface-targeted MsdS expression, respectively

6. Wash SDS-extracted cell walls three times with distilled water, and then freeze-dry.
7. Release GPI-cell wall proteins by treating the cell walls with 800 μL HF-pyridine at 0 °C for 3 h.
8. Freeze-dry HF-pyridine-extractable cell wall proteins after dialysis against distilled water.
9. Separate crude lysate, SDS- and HF-pyridine-extractable proteins using 10% SDS-PAGE, stained with CBB R-250.
10. Detect GPI-motif fusion proteins with FLAG tagging by immunoblotting with an anti-FLAG monoclonal antibody, followed by an AP-conjugated substrate kit (Fig. 3).

3.4 High-Throughput N-Glycan Structure Analysis of Yeast Cell Wall Proteins

1. Prepare glycoprotein samples purified from the culture supernatants, or extract yeast cell wall proteins (CWPs) by a slightly modified method using hot citrate buffer (20 mM sodium citrate buffer, pH 7.0) [41].
2. Load the purified glycoproteins or CWP samples on the 96-well plates (Fig. 4a).
3. Add PNGase F reaction mixtures (100 U in 20 μL of 10 mM Tris-acetate [pH 8.3]) to release *N*-glycans from the purified glycoproteins or CWPs.
4. Purify the released *N*-glycans using nonporous graphitized carbon (GC) resin, packed in 96-well plates (*see* **Note 8**).
5. Elute *N*-glycans three times via the addition of 200 μL elution buffer and centrifugation at 750 × *g*.
6. Label the eluted *N*-glycans with the fluorophore APTS by reductive amination.
7. Transfer labeled *N*-glycans to Sephadex G10 packed plate for cleanup and collection.
8. After drying the collected elutes, add appropriate glycan-processing enzymes, such as α-1,2 mannosidase and β-*N*-hexosaminidase, if exoglycosidase treatment is required.
9. Dry and reconstitute the incubated samples in 10 μL of water containing Liz-labeled GeneScan standard.
10. Analyze the glycan profile using capillary electrophoresis equipped with an ABI3130 sequencer (Fig. 4b) (*see* **Note 9**).

4 Notes

1. One of the unique features of *H. polymorpha* is its high frequency of multiple integration of vector, even in circular form with autonomous replication sequence (ARS), into the host chromosomes. Despite the presence of *H. polymorpha*-derived

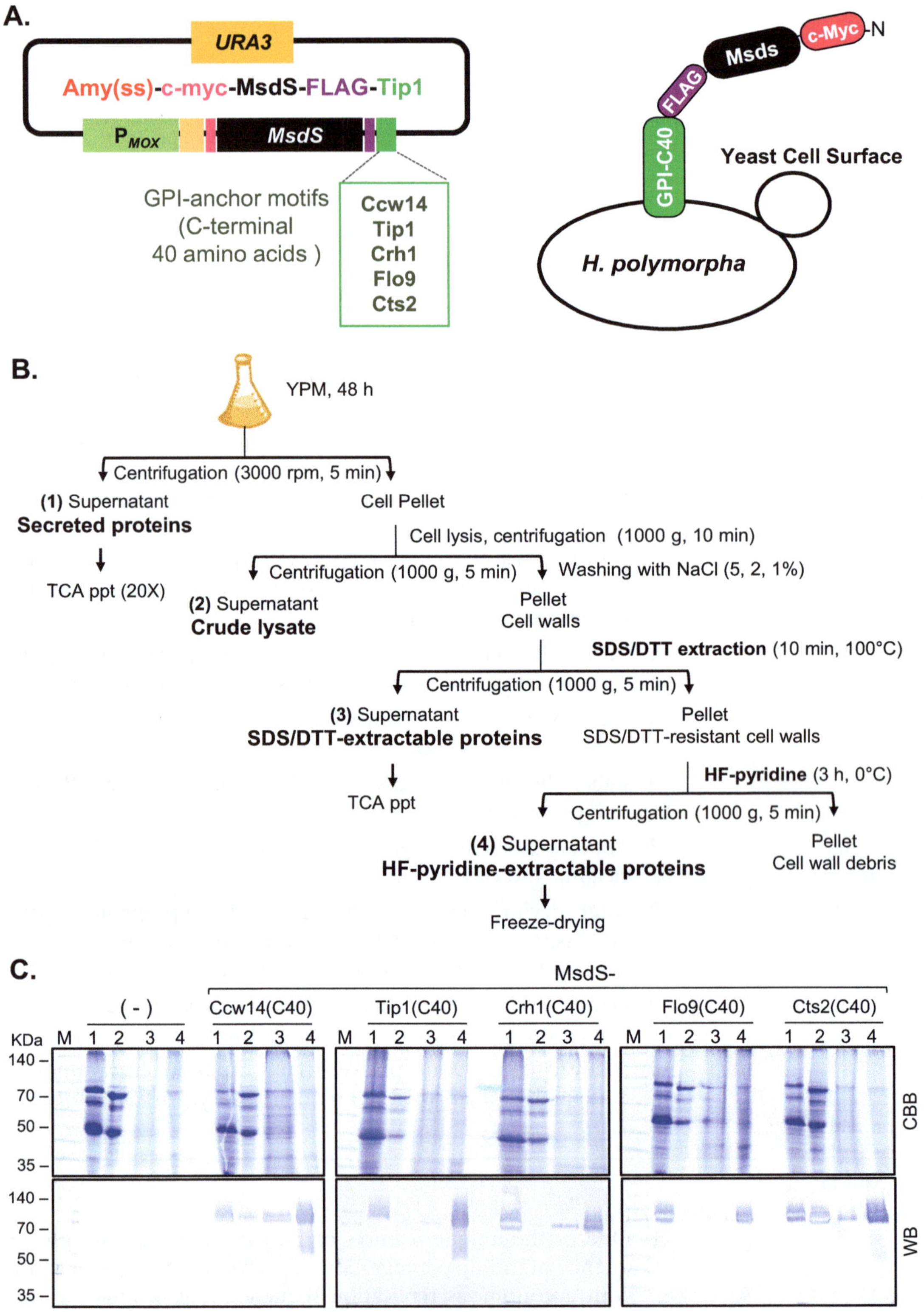

Fig. 3 Surface display expression using *H. polymorpha* GPI-anchoring motifs and localization analysis. (**a**) The yeast surface display system was depicted for expression of the reporter protein, *A. saitoi* α-1,2 mannosidase

ARS (HARS) sequences, circular plasmids are mostly integrated into the host chromosomes and do not exist in an episomal fate in this yeast.

2. All *H. polymorpha* auxotrophic mutant strains, except for the *HpTRP1* deletion mutant (*Hptrp1Δ*) strains, were routinely grown on YPD medium, SC medium, or SD medium, with appropriate nutritional supplements at 37 °C. The *Hptrp1Δ* mutants required additional tryptophan supplementation for their full growth, even on complex media such as YPD. The *Hptrp1Δ* mutant strains were maintained in YPD supplemented with 90 mg/L tryptophan (YPD + 3 × T) or SC medium supplemented with 90 mg/L tryptophan (SC + 3 × T), which is threefold higher than the normal tryptophan concentration used for *Yarrowia lipolytica* (30 mg/L) and more than fourfold higher than that used for *S. cerevisiae* (20 mg/L).
3. When employing the vectors based on Zeo^R and Hyg^R, in which the antibiotic resistance genes are expressed under control of the bacterial *EM7* and *S. cerevisiae TEF1* promoters, *Escherichia coli* DH5α used for plasmid propagation is cultured in LB medium (1% bacto-tryptone, 0.5% bacto-yeast extract, 1% sodium chloride) with ampicillin (100 μg/mL) or 2× YT medium (1.6% bacto-tryptone, 1% bacto-yeast extract, 0.5% sodium chloride) with Zeocin™ (25 μg/mL) or Hygromycin B (50 μg/mL) at 37 °C.
4. Complete stabilization of transformants bearing the *LEU2* vector integrated into the genome can be monitored by spreading cells onto minimal selective plates and checking for an even growth rate, judging from the sizes of colonies. The stability in the LEU^+ phenotype of the individual stabilized colony can be also further confirmed by replica plating of cells on minimal and complex media.
5. Notably, a high number of integrated copies are not always a prerequisite for high-level protein production, especially in the case of secretory production. Thus, using controllable multiple integration vectors would be a good strategy to introduce optimal copy numbers for high-level expression of recombinant proteins.

Fig. 3 (continued) (MsdS) fused to the C-terminal 40 amino acids of putative GPI proteins of *H. polymorpha*. (**b**) Fractionation scheme of cell wall proteins. (**c**) Culture supernatant (lane 1, 20 μL of 20 × concentrate by TCA precipitation), crude lysate (lane 2, 20 μg of the soluble fraction of crude cell lysates), SDS/β-ME-extractable plasma membrane proteins (lane 3), and HF-pyridine-extractable cell membrane proteins (lane 4), which were prepared from *H. polymorpha* strains expressing MsdS-Ccw14, -Tip1, -Crh1, -Flo1, and -Cts2, were analyzed by CBB staining (top) and immunoblotting with an anti-FLAG antibody (bottom)

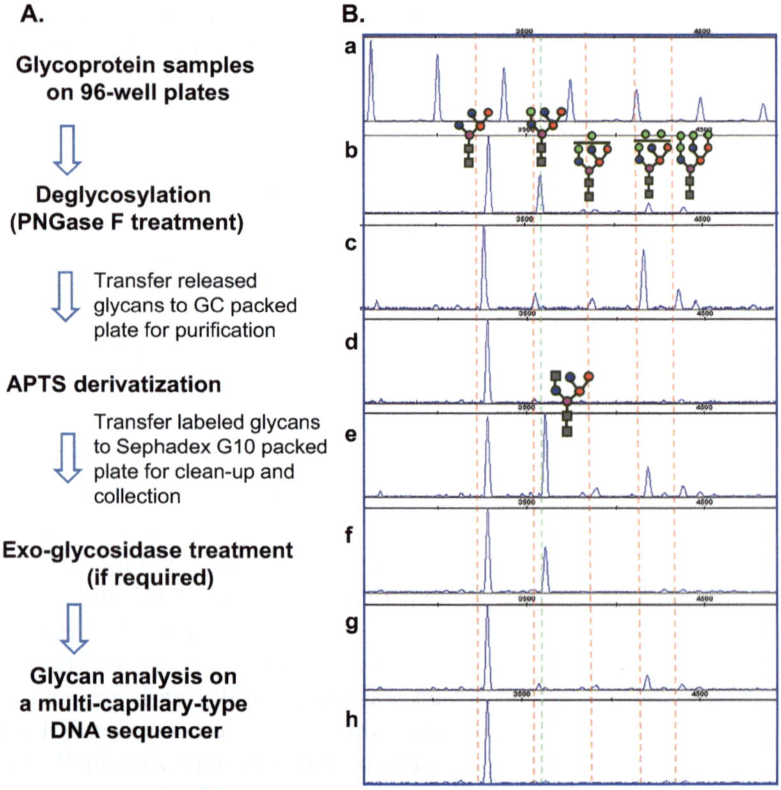

Fig. 4 High-throughput *N*-glycan analysis of glyco-engineered *H. polymorpha* strains. (**a**) Workflow scheme for high-throughput *N*-glycan analysis using DNA sequencer. (**b**) ATPS-labeled glycans of oligomaltose hydrolysate (*a*) and bovine RNase B (*b*) that serve as a size reference. ATPS-labeled *N*-glycans released from the CWPs of the *Hpoch1Δ* strain expressing MsdS (strain DL1-g23) (*c*) and treated with α-1,2 mannosidase (*d*). ATPS-labeled *N*-glycans from the CWPs of the *Hpoch1Δ* strain expressing MsdS and human β-1,2-*N*-acetylglucosaminyl-transferase I (strain DL1-g24Sm) (*e*), treated with α-1,2 mannosidase (*f*), treated with β-*N*-hexosaminidase (*g*), and treated with β-*N*-hexosaminidase and subsequently with α-1,2 mannosidase (*h*)

6. Since *H. polymorpha* shows high frequency of multiple integration of vector, even in circular form with ARS, *H. polymorpha* expression vectors with different selective markers could be simultaneously co-transformed in the same background strain with multiple auxotrophic markers. We were able to obtain the expected transformants, acquiring multiple selection markers at high frequency, from a single co-transformation with up to four vectors, using the quintuple auxotrophic DL1-LdTMAU strain as a host [35]. The feasibility of co-transformation will surely shorten the time and effort required for multiple introductions of several heterologous genes into *H. polymorpha*.

7. Non-covalently linked cell wall proteins (CWPs) can be extracted with SDS/β-ME from the membrane of well-washed cell pellet, whereas the GPI-anchored CWPs can be obtained from SDS/β-ME-resistant cell walls by treatment of HF-pyridine, which is known to specifically cleave the phosphodiester bonds linking GPI-anchored CWPs to β-1,6-glucan chains [42]. Therefore, SDS/β-ME and the successive HF-pyridine extraction of the cell wall fraction samples is expected to discriminate the localization of the GPI proteins at the cell plasma membrane (PM) or cell wall, as GPI-PM proteins or GPI-cell wall proteins, respectively.
8. After packing of graphitized carbon resin in 96 wells, the resin should be pre-activated by five cycles of washing with 200 μL of washing buffer and 200 μL of distilled water. Buffer and water are removed from the wells via the application of a vacuum under the plate [43].
9. The APTS-labeled glycans are examined with an ABI 3130 DNA sequencer by using a 12% polyacrylamide gel and a buffer containing 89 mM Tris, 89 mM borate, and 2.2 mM EDTA. On each gel, *N*-glycans of bovine RNase B and a maltodextrose ladder are included as references. Glycans are detected at the end of the gel with an Ar laser, and data analysis is performed by using the GeneScan 3.1 software as described previously [44]. This technique allowed us high-throughput glycan analysis and screening for glyco-engineered *H. polymorpha* host strains designed for production of recombinant proteins with a human-compatible glycan [25].

Acknowledgment

The work was supported by the grant No. NRF-2013M3A6-A8073554 (Global Frontier Program for the Intelligent Synthetic Biology) from the National Research Foundation of Korea (NRF).

References

1. van der Klei IJ, Yurimoto H, Sakai Y, Veenhuis M (2006) The significance of peroxisomes in methanol metabolism in methylotrophic yeast. Biochim Biophys Acta 1763(12):1453–1462
2. Kang HA, Gellissen G (2005) *Hansenula polymorpha*. In: Gellissen G (ed) Production of recombinant proteins-novel microbial and eukaryotic expression systems. Wiley-VCH, Weinheim, p 111
3. Seo HS, Park JS, Han KY, Bae KD, Ahn SJ, Kang HA et al (2008) Analysis and characterization of hepatitis B vaccine particles synthesized from *Hansenula polymorpha*. Vaccine 26(33):4138–4144
4. Weydemann U, Keup P, Piontek M, Strasser AW, Schweden J, Gellissen G et al (1995) High-level secretion of hirudin by *Hansenula polymorpha*—authentic processing of three different preprohirudins. Appl Microbiol Biotechnol 44(3-4):377–385
5. Kang HA, Sohn JH, Choi ES, Chung BH, Yu MH, Rhee SK (1998) Glycosylation of human alpha 1-antitrypsin in *Saccharomyces cerevisiae*

and methylotrophic yeasts. Yeast 14 (4):371–381

6. Melmer G (2005) Biopharmaceuticals and the industrial environment. In: Gellissen G (ed) Production of recombinant proteins—novel microbial and eukaryotic expression systems. Wiley-VCH, Weinheim, pp 361–383
7. Degelmann A, Müller F, Sieber H, Jenzelewski V, Suckow M, Strasser AW et al (2002) Strain and process development for the production of human cytokines in *Hansenula polymorpha*. EMS Yeast Res 2 (3):349–361
8. Kang HA, Kang W, Hong WK, Kim MW, Kim JY, Sohn JH et al (2001) Development of expression systems for the production of recombinant human serum albumin using the *MOX* promoter in *Hansenula polymorpha* DL-1. Biotechnol Bioeng 76(2):175–185
9. Heo JH, Hong WK, Cho EY, Kim MW, Kim JY, Kim CH et al (2003) Properties of the *Hansenula polymorpha*-derived constitutive *GAP* promoter, assessed using an HSA reporter gene. FEMS Yeast Res 4(2):175–184
10. Heo JH, Won HS, Kang HA, Rhee SK, Chung BH (2002) Purification of recombinant human epidermal growth factor secreted from the methylotrophic yeast *Hansenula polymorpha*. Protein Expr Purif 24(1):117–122
11. Sohn MJ, Oh DB, Kim EJ, Cheon SA, Kwon O, Kim JY et al (2012) *HpYPS1* and *HpYPS7* encode functional aspartyl proteases localized at the cell surface in the thermotolerant methylotrophic yeast *Hansenula polymorpha*. Yeast 29(1):1–16
12. Cook MW, Thygesen HV (2003) Safety evaluation of a hexose oxidase expressed in *Hansenula polymorpha*. Food Chem Toxicol 41 (4):523–529
13. Mayer AF, Hellmuth K, Schlieker H, Lopez-Ulibarri R, Oertel S, Dahlems U et al (1999) An expression system matures: a highly efficient and cost-effective process for phytase production by recombinant strains of *Hansenula polymorpha*. Biotechnol Bioeng 63(3):373–381
14. Park BS, Vladimir A, Kim CH, Rhee SK, Kang HA (2004) Secretory production of *Zymomonas mobilis* levansucrase by the methylotrophic yeast *Hansenula polymorpha*. Enzym Microb Technol 34(2):132–138
15. Kim MW, Rhee SK, Kim JY, Shimma Y, Chiba Y, Jigami Y et al (2004) Characterization of *N*-linked oligosaccharides assembled on secretory recombinant glucose oxidase and cell wall mannoproteins from the methylotrophic yeast *Hansenula polymorpha*. Glycobiology 14(3):243–251
16. Blazhenko OV, Zimmermann M, Kang HA, Bartosz G, Penninckx MJ, Ubiyvovk VM et al (2006) Accumulation of cadmium ions in the methylotrophic yeast *Hansenula polymorpha*. Biometals 19(6):593–599
17. Dmytruk OV, Dmytruk KV, Abbas CA, Voronovsky AY, Sibirny AA (2008) Engineering of xylose reductase and overexpression of xylitol dehydrogenase and xylulokinase improves xylose alcoholic fermentation in the thermotolerant yeast *Hansenula polymorpha*. Microb Cell Factories 7:21
18. Suh SO, Zhou JJ (2010) Methylotrophic yeasts near *Ogataea* (*Hansenula*) *polymorpha*: a proposal of *Ogataea angusta* comb. nov. and *Candida parapolymorpha* sp. nov. FEMS Yeast Res 10(5):631–638
19. Kurtzman CP (2011) A new methanol assimilating yeast, *Ogataea parapolymorpha*, the ascosporic state of *Candida parapolymorpha*. Antonie Van Leeuwenhoek 100(3):455–462
20. Kang HA, Sohn JH, Agaphonov MO, Choi ES, Ter-Avanesyan MD, Rhee SK (2002) Development of expression systems for the production of recombinant proteins in *Hansenula polymorpha* DL-1. In: Gellissen G (ed) *Hansenula polymorpha*: biology and applications. Wiley-VCH, Weinheim, pp 124–146
21. Lahtchev KL, Semenova VD, Tolstorukov II, van der Klei I, Veenhuis M (2002) Isolation and properties of genetically defined strains of the methylotrophic yeast *Hansenula polymorpha* CBS4732. Arch Microbiol 177 (2):150–158
22. Kim H, Yoo SJ, Kang HA (2015) Yeast synthetic biology for the production of recombinant therapeutic proteins. FEMS Yeast Res 15 (1):1–16
23. Kim MW, Kim EJ, Kim JY, Park JS, Oh DB, Shimma Y et al (2006) Functional characterization of the *Hansenula polymorpha HOC1*, *OCH1*, and *OCR1* genes as members of the yeast *OCH1* mannosyltransferase family involved in protein glycosylation. J Biol Chem 281(10):6261–6272
24. Oh DB, Park JS, Kim MW, Cheon SA, Kim EJ, Moon HY et al (2008) Glycoengineering of the methylotrophic yeast *Hansenula polymorpha* for the production of glycoproteins with trimannosyl core *N*-glycan by blocking core oligosaccharide assembly. Biotechnol J 3 (5):659–668
25. Cheon SA, Kim H, Oh DB, Kwon O, Kang HA (2012) Remodeling of the glycosylation pathway in the methylotrophic yeast *Hansenula polymorpha* to produce human hybrid-type *N*-glycans. J Microbiol 50(2):341–348

26. Bae JH, Sohn JH, Rhee SK, Choi ES (2005) Cloning and characterization of the *Hansenula polymorpha PEP4* gene encoding proteinase A. Yeast 22(1):13–19
27. Kim SY, Sohn JH, Pyun YR, Choi ES (2002) A cell surface display system using novel GPI-anchored proteins in *Hansenula polymorpha*. Yeast 19(13):1153–1163
28. Kim SY, Sohn JH, Pyun YR, Yang IS, Kim KH, Choi ES (2007) *In vitro* evolution of lipase B from *Candida antarctica* using surface display in *Hansenula polymorpha*. J Microbiol Biotechnol 17(8):1308–1315
29. Cheon SA, Jung J, Choo JH, Oh DB, Kang HA (2014) Characterization of putative glycosylphosphatidylinositol-anchoring motifs for surface display in the methylotrophic yeast *Hansenula polymorpha*. Biotechnol Lett 36 (10):2085–2094
30. Moon HY, Cheon SA, Kim H, Agaphonov MO, Kwon O, Oh DB et al (2015) *Hansenula polymorpha* Hac1p is critical to protein *N*-glycosylation activity modulation, as revealed by functional and transcriptomic analyses. Appl Environ Microbiol 81(20):6982–6993
31. Park JN, Sohn MJ, Oh DB, Kwon O, Rhee SK, Hur CG et al (2007) Identification of the cadmium-inducible *Hansenula polymorpha SEO1* gene promoter by transcriptome analysis and its application to whole-cell heavy-metal detection systems. Appl Environ Microbiol 73 (19):5990–6000
32. Kang HA, Hong WK, Sohn JH, Choi ES, Rhee SK (2001) Molecular characterization of the actin-encoding gene and the use of its promoter for a dominant selection system in the methylotrophic yeast *Hansenula polymorpha*. Appl Microbiol Biotechnol 55(6):734–741
33. Yoo SJ, Chung SY, Lee DJ, Kim H, Cheon SA, Kang HA (2015) Use of the cysteine-repressible *HpMET3* promoter as a novel tool to regulate gene expression in *Hansenula polymorpha*. Biotechnol Lett 37(11):2237–2245
34. Agaphonov MO, Trushkina PM, Sohn JH, Choi ES, Rhee SK, Ter-Avanesyan MD (1999) Vectors for rapid selection of integrants with different plasmid copy numbers in the yeast *Hansenula polymorpha* DL1. Yeast 15 (7):541–551
35. Cheon SA, Choo J, Ubiyvovk VM, Park JN, Kim MW, Oh DB et al (2009) New selectable host-marker systems for multiple genetic manipulations based on *TRP1*, *MET2* and *ADE2* in the methylotrophic yeast *Hansenula polymorpha*. Yeast 26(9):507–521
36. Sohn JH, Choi ES, Kang HA, Rhee JS, Agaphonov MO, Ter-Avanesyan MD et al (1999) A dominant selection system designed for copy-number-controlled gene integration in *Hansenula polymorpha* DL-1. Appl Microbiol Biotechnol 51(6):800–807
37. Kim H, Moon HY, Lee DJ, Cheon SA, Yoo SJ, Park JN et al (2013) Functional and molecular characterization of novel *Hansenula polymorpha* genes, *HpPMT5* and *HpPMT6*, encoding protein *O*-mannosyltransferases. Fungal Genet Biol 58-59:10–24
38. Sohn JH, Choi ES, Kang HA, Rhee JS, Rhee SK (1999) A family of telomere-associated autonomously replicating sequences and their functions in targeted recombination in *Hansenula polymorpha* DL-1. J Bacteriol 181 (3):1005–1013
39. Levine DW, Cooney CL (1973) Isolation and characterization of a thermotolerant methanol-utilizing yeast. Appl Microbiol 26(6):982–990
40. Hill J, Donald KA, Griffiths DE (1991) DMSO-enhanced whole cell yeast transformation. Nucleic Acids Res 19(20):5791
41. Thak EJ, Kim J, Lee DJ, Kim JY, Kang HA (2018) Structural analysis of *N*-/*O*-glycans assembled on proteins in yeasts. J Microbiol 56(1):11–23
42. de Groot PW, de Boer AD, Cunningham J, Dekker HL, de Jong L, Hellingwerf KJ et al (2004) Proteomic analysis of *Candida albicans* cell walls reveals covalently bound carbohydrate-active enzymes and adhesins. Eukaryot Cell 3(4):955–965
43. Lee KJ, Jung JH, Lee JM, So Y, Kwon O, Callewaert N et al (2009) High-throughput quantitative analysis of plant *N*-glycan using a DNA sequencer. Biochem Biophys Res Commun 380(2):223–229
44. Callewaert N, Geysens S, Molemans F, Contreras R (2001) Ultrasensitive profiling and sequencing of *N*-linked oligosaccharides using standard DNA-sequencing equipment. Glycobiology 11(4):275–281

Chapter 6

Selection of Heterologous Protein-Producing Strains in *Yarrowia lipolytica*

Paul Soudier, Macarena Larroude, Ewelina Celińska, Tristan Rossignol, and Jean-Marc Nicaud

Abstract

Yarrowia lipolytica has emerged as an alternative expression system for heterologous protein production and enzyme evolution. Several different expression systems dedicated for this species have been developed, ranging from the simple cloning of expression vectors to recently developed high-throughput methodologies using efficient cloning and assembly such as Gateway and Golden Gate strategies. The latter strategies, due to their modular character, enable multiple vector construction and the construction of expression cassettes containing different genes or a gene under different promoters of various strengths.

Here, we present the Golden Gate cloning strategy for the construction of multiple expression cassettes, the transformation into *Y. lipolytica*, and the selection of efficient enzyme-producing strains using an insect alpha-amylase as a reporter detected via a thermal cycler-based microassay.

Key words *Yarrowia lipolytica*, Heterologous protein, Secretion, Golden Gate, Gateway, Gene assembly, Inducible promoter, Targeting sequence, High-throughput techniques

1 Introduction

Yarrowia lipolytica has emerged as an attractive alternative host for heterologous protein production, thanks to its ability to efficiently secrete proteins due to co-translational translocation of a polypeptide, performing low level of glycosylation and having a GRAS status. To have a detailed overview of heterologous protein expression and secretion in *Y. lipolytica*, see the recent reviews of Madzak and Beckerich [1, 2]. We will here briefly summarize the evolution of the current tools for protein production in *Y. lipolytica*, the methods and their specific application, and the more recent achievements using high-throughput techniques.

The expression system employing *Y. lipolytica* was initially based mainly on integrative vectors, since replicative plasmids are maintained in low copy number and tend to be relatively unstable in the cells. The first plasmids were based mainly on vectors carrying

Brigitte Gasser and Diethard Mattanovich (eds.), *Recombinant Protein Production in Yeast*, Methods in Molecular Biology, vol. 1923, https://doi.org/10.1007/978-1-4939-9024-5_6, © Springer Science+Business Media, LLC, part of Springer Nature 2019

specific restriction sites for the cloning of the gene of interest. Then, vectors for mono- and multi-copy integration were developed using the defective ura3d4 marker allowing to have up to more than 30 integrated copies of the vector within the genome [3–6]. Further developments covered the construction of auto-cloning vectors based on vectors from which the expression cassette could be released by enzymatic digestion in order to eliminate the bacterial backbone prior to transformation [4].

Various vectors were designed for directing integration at specific sites in the genome. The strategies relied on either insertion at specific *loci* introduced into defined recipient strains or into repeated genomic sequences such as the rDNA region or the long terminal repeat sequence of *Y. lipolytica* retrotransposon *Yl*TY1 called ZETA region (714 bp).

For specific *loci* integration, the main strains used are Po1g and JMY1212. Strain Po1g (CLIB725) *MatA*, *leu2-270*, *ura3-302::URA3-xpr2-322*, *axp1-2* [7], a leucine auxotroph which bears a pBR322 docking platform in the genome, is included in the *Y. lipolytica* expression kit (Yeastern Biotech). This strain is deleted for the two main extracellular proteases: the alkaline extracellular protease (AEP) and the acid extracellular protease (AXP). Another strain widely used for enzyme evolution is strain JMY1212 *MatA*, *leu2-270*, *ura3-LEU2-ZETA*, *xpr2-322*, *lip2Δ*, *lip7Δ*, *lip8Δ*, deleted for the AEP and for the three major extracellular lipases Lip2, Lip7, and Lip8, which contains a ZETA-LEU2 platform at the *URA3* locus, Ura^- Leu+ [8].

For non-specific *loci* integration, the expression cassette carries either a rDNA fragment or the ZETA motif as targeting sequences [9]. The American wild-type strain YB423 and its inbreeding derivative strain E150 [10] contain a large number of *Yl*TY1 and ZETA solo sequences. Therefore integration of a given expression cassette containing the zeta motif will occur at different *loci*. In contrast, *Yl*TY1 and ZETA solo sequences were not detected in the wild-type German strain H222 and in the wild-type French strain W29 and its derivative Po1d [9]. In such zeta-free strains, integration of the expression cassette occurred at random in the genome [4].

Along with the design of new expression systems dedicated for *Y. lipolytica*, new promoter sequences have been developed. For many years, the constitutive pTEF promoter of the translation elongation factor-1α isolated by Müller in 1998 [11] was considered to be the strongest known promoter operating in *Y. lipolytica* cells. Then, the first hybrid promoter bearing four tandem copies of the upstream activating sequences UAS1B of the *XPR2* gene was developed by Madzak and colleagues in 2000, giving rise to the hp4d promoter presenting a very strong phase-dependent expression [7]. Since then, other hybrid promoters of semi-constitutive character have been developed.

Nevertheless, truly potent inducible promoters remained to be established. The only native inducible promoters that have been developed are based on promoters induced by hydrophobic substrates such as alkanes and lipids. These promoters mainly originate from the peroxisomal acyl-CoA oxidase 2 (*POX2*), the peroxisomal 3-ketoacyl-thiolase (*POT1*), the extracellular lipase Lip2 (*LIP2*), and the cytochrome P450 oxidase (*ALK1*) [12] genes.

Recently a new set of inducible hybrid promoters, activated by erythritol, has been developed [13]. They are based on the upstream activating sequence UAS1eyk of the *EYK1* gene coding for erythrulose kinase, involved in the utilization of erythritol as carbon source [14].

The main strains and expression strategies, used for different protein production purposes and projects, are depicted in Fig. 1.

Here we will describe the procedure for the cloning steps using the Golden Gate strategy, the transformation method for introducing the expression cassettes into *Y. lipolytica* cells, and exemplify this with the expression of the insect *Sitophilus oryzae* starch-digesting alpha-amylase, including the selection strategy to isolate the best producers.

2 Materials

1. Apparatus.
2. PCR apparatus (thermocycler).
3. Incubator with a rotary shaker at 28 °C.
4. Photometer at 600 nm.
5. Microcentrifuge and centrifuge at 4 °C.
6. Baffled Erlenmeyer flasks (250 and 500 mL).
7. Microtiter plate (MTP) reader (operating at 600 nm and 580 nm).
8. Sterile MTP (for the strain cultivation).
9. PCR 96-well plates (for the activity test).

2.1 Strains

1. *E. coli* strain

   Escherichia coli DH5α strain will be used for transformation and amplification of recombinant plasmid DNA. Cells will be grown at 37 °C in Luria broth (LB) medium supplemented with kanamycin (50 μg/mL) or ampicillin (100 μg/mL) for isolation of plasmids.
2. *Y. lipolytica* strain

   The French wild-type W29 derivative JMY1212 *MatA*, *leu2-270*, *ura3-LEU2-ZETA*, *xpr2-322*, *lip2Δ*, *lip7Δ*, *lip8Δ* was selected for these studies due to its inability to produce

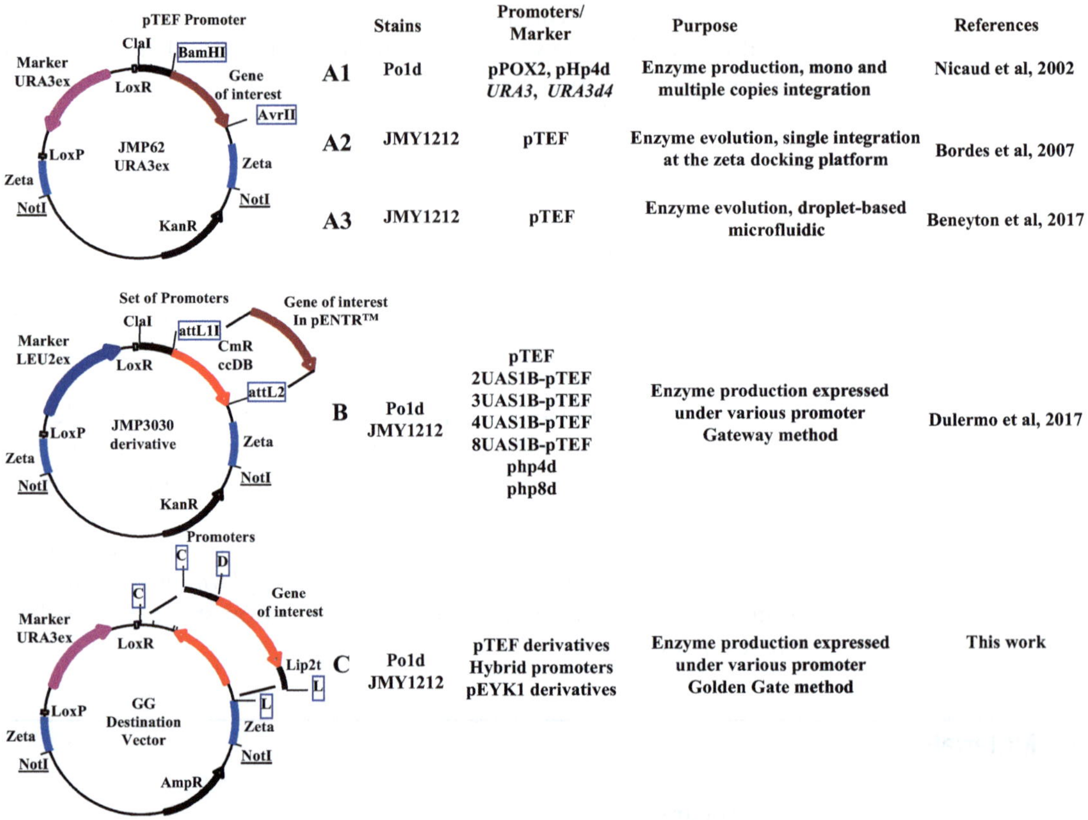

Fig. 1 Expression strategies for protein expression and enzyme evolution in *Y. lipolytica*. Schematic plasmid map and expression system for single or multiple integration for classical enzyme production (**A1**), for single enzyme comparison or enzyme evolution (**A2**) or for high-throughput clone analysis using microfluidic selection screening (**A3**). New high-throughput expression system using either Gateway system (**B**) or the Golden Gate methodologies (**C**) in order to select the best enzyme producer. Cloning of the gene of interest could be performed by enzyme digestion of the donor plasmid by, for example, *Bam*H1 and *Avr*II (**A**), by transferring in vitro the gene of interest previously cloned into the Gateway vector pENTR™/D-TOPO using LR Clonase^{R} (**B**) or by Golden Gate assembly using the Golden Gate destination vector GGDV002 together with the BioBricks carrying the promoter region and the gene of interest (**C**) as depicted also in Fig. 2. Markers *URA3*ex and *LEU2*ex are used for selection of transformants in *Y. lipolytica* and AmpR and KanR for selection in *E. coli*. The zeta region is used for integration in *Y. lipolytica* genome

protease and lipase activities and also because it presents a docking zeta platform for the insertion of the expression cassettes.

List of the strains constructed

GGY_1 GGA-SoAMY1, JMY1212 + pTEF1-SoAMY.

GGY_2 GGA-SoAMY2, JMY1212 + p8UAS1-TEF1-SoAMY.

GGY_5 GGA-SoAMY5, JMY1212 + pEYK3AB-SoAMY.

3 Expression of an Efficient Starch-Digesting Alpha-Amylase Gene from the Insect *Sitophilus oryzae*

3.1 Construction of Expression Cassettes Using a High-Throughput Golden Gate Strategy

Recently, the high-throughput Golden Gate cloning system [15] was adopted for pathway engineering in the yeast *Y. lipolytica* [16]. This system allowed to assemble one, two, or three transcription units, each containing a promoter, a gene of interest, and a terminator, together with a selection marker for selection of transformants in *Y. lipolytica* and a targeting sequence for the integration of the expression cassette. This system is based on the assembly of DNA fragments (BioBricks) carrying *Bsa*I sites which generates specific overhang sequences defined by a Roman letter A to M according to Celinska et al. [16]. Here, we use this strategy in order to construct several expression cassettes with the SoAMY expressed under various promoters of different strengths and presenting different expression profiles. SoAMY is an efficient starch-digesting alpha-amylase gene from the insect *S. oryzae* [17–19] that can be functionally expressed in *Y. lipolytica* strains [18] (Fig. 2).

1. The Golden Gate destination vector GGDV002 carrying the zeta region for integration into the *Y. lipolytica* genome, the *URA3ex* excisable marker for selection of transformants in

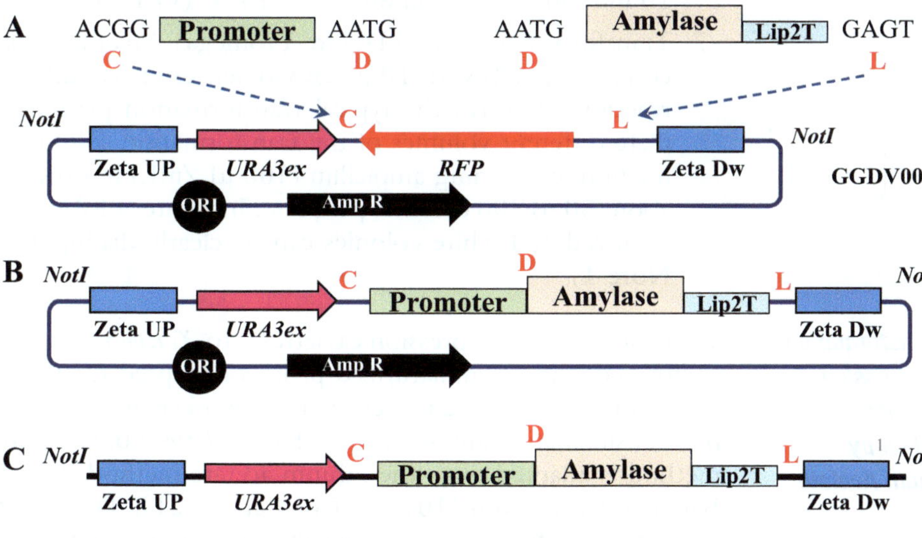

Fig. 2 Construction of expression cassettes bearing *SoAMY* gene, expressed under various promoters of different strengths and expression profiles. (**a**) Vectors and fragments assembly: fragment Bsa1-C-D carrying various promoters (*Bsa*I-C-Promoters-D-*Bsa*I) were assembled with the BsaI-D-L fragment carrying the amylase gene and the Lip2 terminator (*Bsa*I-D-Amylase-Lip2t-L-*Bsa*I) and with the destination vector GGDV002 carrying *URA3*ex selection marker and the zeta upstream (up) and downstream (dw) regions and the bacterial region containing the Ori and the ampicillin resistance gene (AmpR). (**b**) Schematic representation of the assembled vectors. (**c**) Schematic representation of the expression cassette obtained by *Not*I digestion for transformation into *Y. lipolytica*

Y. lipolytica, the ampicillin marker for the selection of *E. coli* transformants, and the chromophore red fluorescent protein (RFP) resulting in red *E. coli* colonies to differentiate between clones that contain or not the assembly.

2. Golden Gate donor vectors carrying promoter bricks with the flanking overhang *Bsa*I C (ACGG) and *Bsa*I D (AATG) restriction sites.
3. Golden Gate donor vectors GGE0158 carrying the SoAMY *Sitophilus oryzae* alpha-amylase gene with the Lip2 terminator of the *Y. lipolytica LIP2* gene coding for the extracellular lipase Lip2p with the flanking overhang *Bsa*I D (AATG) and *Bsa*I L (GAGT) restriction sites cloned into pCRBlunt-TOPO vector (Invitrogen) (*see* **Note 1**).
4. Mix equimolar concentrations (3.3 p mol/μL) of the GG fragments (in the destination vector) and the donor vectors in a PCR tube (*see* **Note 2**).
5. Add 0.5 μL of T4 DNA ligase (New England Biolabs NEB, 400 U/μL), 1 μL of *Bsa*I (NEB, 10 U/μL), and 2 μL of NEB Cutsmart R buffer, and complete to 15 with ddH_2O.
6. Place the PCR tube (0.2 mL from Thermo Scientific) in the PCR apparatus (Eppendorf, Mastercycler Gradient), and run the following cycles: (5 min 37 °C; 5 min 16 °C) × 50 cycles; 10 min 60 °C; 10 min 80 °C; ∞ 15 °C (*see* **Note 3**).
7. Transform *E. coli* strain DH5α (Promega). Mix 80 μL of *E. coli* competent cells with 10 μL of Golden Gate Assembly (GGA) reaction. Perform the typical transformation procedure, and plate different volumes of the transformation solution on LB medium containing ampicillin (100 μL/mL) in order to have about 50 to 200 colonies per plate. Incubate at 37 °C for 1 day until red and white colonies can be clearly distinguished (*see* **Note 4**).

3.2 Transformation of the Expression Cassette into *Y. lipolytica* by the Lithium Acetate Method

Transformation of expression cassettes into *Y. lipolytica* strains was performed using a miniaturized protocol adapted for the 96-well plate format, according to Leplat and coworkers for high-throughput transformation and screening of *Y. lipolytica* libraries [20]. This method is adapted from the lithium acetate method described by Barth and Gaillardin [10] and Chen et al. [21]. The transformations were performed in 96-well PCR plates, and incubation was performed in a thermocycler.

1. Lithium acetate solution. LiAC 0.1 M pH 6, dissolve 1.02 g of LiAc in 100 mL, add acetic acid to bring the pH to 6, and sterilize by filtration. Keep at room temperature.
2. PEG 4000 solution. PEG 4000 at 40%, dissolve 20 g of PEG 4000 (VWR) in 50 mL of LiAC 0.1 M pH 6, and sterilized by filtration. Keep at −20 °C.

3. Carrier DNA solution. Dissolve carrier DNA in TE (Tris50mM Tris–HCl, 5 mM EDTA, pH 8) at 5 mg/mL (*see* **Note 5**).
4. Streak strain to be transformed on YPD plates and grown for 24 h at 28 °C.
5. For 100 single transformations, 15 loopfuls of cells (10 μL loops) were scraped from the agar surface, washed in 15 mL TE buffer, and centrifuged at 3000 rpm (604 × *g*) at room temperature for 5 min (around 5 × 10^8 cells/mL). Cells were resuspended in 10 mL of 0.1 M LiAC, pH 6 and incubated 1 h at 28 °C in order to obtain competent cells.
6. Cells were centrifuged at 2000 rpm (268 × *g*) at room temperature for 2 min and resuspended in 1 mL of 0.1 M LiAC pH 6.
7. Then, 10 μL of cells was transferred into each well of a 96-well PCR plates, gently mixed with 3 μL of linearized DNA vector (~50 ng) and 1 μL of carrier DNA (5 mg/mL), and incubated for 15 min at 28 °C (*see* **Note 6**).
8. Following this, 66 μL of reaction buffer (40% PEG, 0.1 M LiAC pH 6) was added to each transformation and mixed gently. The mix was incubated 1 h at 28 °C followed by a 10-min heat shock at 39 °C.
9. The cells were platted on an appropriate selective solid minimal YNB medium.
10. Transformants appear after 2 days and should be purified by reisolating single colonies onto new YNB plates. Two to six purified independent transformants were kept for subsequent enzyme and protein production analysis.

4 Selection of Positive Clones: Comparison of Alpha-Amylase Production in the Transformant Strains

Currently, due to introduction of efficient cloning techniques, such as Gateway or Golden Gate, as well as high-capacity cloning, as mentioned above in PCR 96-well plates, the need for high-throughput screening protocols is even more pronounced. Different methodologies could be used for the comparison of enzyme production in the transformant strains depending on the promoter strength and on the characteristics of the protein under study as well as equipment availability [22, 23].

4.1 Selection of Positive Clones by Drop Test

One of the most straightforward methods is adopting one of the classical plate tests—relying on qualitative investigation of some biochemical (lipolytic, amylolytic) trait of the transformant strain. In such a strategy, the culture suspensions are spotted on test plates, and the plates are incubated for several days at 28 °C. The size of the clear zone around the colonies reflects the enzyme production (*see* **Note 7**).

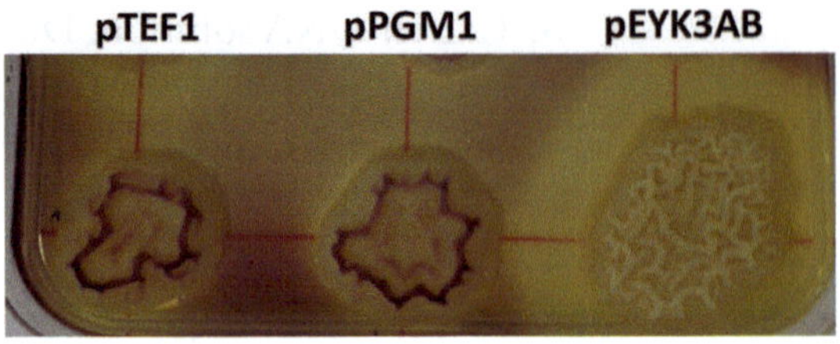

Fig. 3 Starch degradation by *Y. lipolytica* strains expressing amylase. Dilution of cell suspension was dropped on YPDOLS plates [YPDOL agar (yeast extract 10 g/L, peptone 10 g/L, dextrose 10 g/L, erythritol 10 g/L) and top agar of YPDOL containing wheat raw starch]. The strains able to clarify starch were distinguished by the clear zone around the colonies

For lipase, YNBT plates could be used for the analysis of overexpression of lipase as described previously by Pignede et al. [4] (*see* **Note 8**).

For amylase, starch plates could be used as described previously [24, 25] (*see* **Note 9**, Fig. 3).

One could compare also activities during growth and analyze enzyme production by analyzing protein production by gel electrophoresis by SDS-PAGE analysis or by measuring enzyme activities depending on medium and promoter strength (for more details and the choice of the media, *see* [13, 25]).

4.2 Selection of Positive Clones by Protein Electrophoresis Separation

Another important step in the process of selecting positive transformant strains is verifying if the protein under study is expressed and secreted into the culture medium (as preferred during the protein overexpression). A classical method involves running a protein electrophoresis protocol. It will additionally give information about eventual over glycosylation of the protein in the heterologous expression system, when a significant upward band shift is observed. Protein electrophoresis may be also extremely helpful if the initial screening tests using plate drop assays fail to indicate positive transformants, which may result from (1) lack of expression at all or (2) expression of an inactive enzymatic protein, i.e., when a specific band in the protein electrophoretic profile is detectable while no activity was recorded on the plate drop test.

1. Prepare YNBD and YNBOL minimal media.
2. Inoculate flask at OD600 of 0.5 with an overnight culture (50 mL in 250 mL Erlenmeyer flask). Cultures were incubated for 72 h at 28 °C at 150 rpm.
3. Cultures were sampled at different time points (1 mL).
4. Concentrate the supernatant ten times on Amicon Ultra-0.5 10K centrifugal filters (Merck Millipore Ltd., Ireland).
5. Load the equivalent of 15–150 μL of supernatant on SDS-PAGE (Fig. 4).

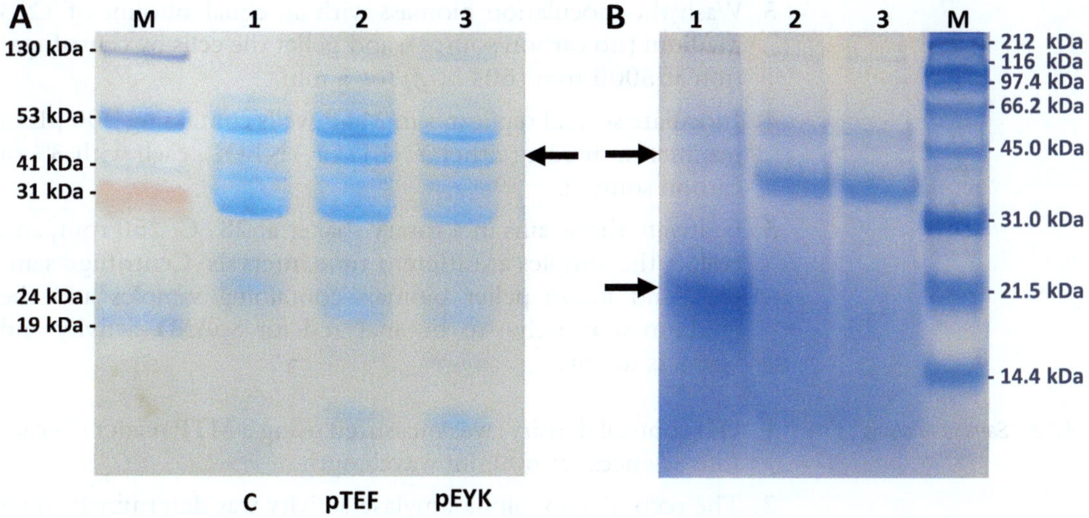

Fig. 4 Analysis of protein production by *Y. lipolytica* transformants. (**a**) Production of amylase. (**b**) Production of two heterologous lipases. Panel (**a**) Production of amylase by strains expression the AMY gene under the constitutive pTEF and the erythritol-inducible pEYK3AB promoters after 72 h of culture in YPD, molecular marker prism (M), control (line 1, C), pTEF-AMY (line 2, pTEF), pEYK3AB-AMY (line 3, pEYK). Proteins were separated on a novex 10–20% Tris-glycine. Equivalent of 150 μL of supernatant was loaded per lines. Panel (**b**) Production of the codon-optimized lipase genes under the promoter pTEF in YNBD3OL1, a 20 kDa lipase (line 1) and a 38 kDa lipase [two independent clones (lines 2 and 3)]. The wide-range protein molecular weight marker (M). Proteins were separated on a novex 10–20% Tris-glycine. Equivalent of 20 μL of supernatant were loaded

4.3 Selection of Positive Clones by Enzymatic Assays: Microassays—Rapid Screening of Recombinant Strains Rendering Quantitative Data

As discussed above, initial screening by plate drop test or protein electrophoresis results in selection of positive recombinant strains. The following step requires quantitative description of the production capacity of the obtained transformants. To this end, we adopted classical biochemical test relying on iodine staining of starch helices and determination of the staining degree after treatment with an enzymatic preparation (here: culture media containing secreted protein). For a detailed protocol of the microassay, *see* [26].

4.3.1 Microassay: Starch-Iodine Test

1. Prepare production media YNBD and YNBOL with 3% concentration of the respective carbon source (glucose or erythritol) (*see* **Note 10**). Prepare a standard medium for inoculation culture—YNBD with 1% carbon source. Prepare YNB medium devoid of carbon source—required for washing the cells.
2. Prepare inoculation cultures—inoculate 4 mL of YNBD (1%) in a test tube with a loopful of the strain to analyze. Cultivate in leaned position over 20 h in 28 °C at 250 rpm.

3. Wash the inoculation biomass with an equal amount of YNB medium (no carbon source), and pellet the cells by centrifugation at 3000 rpm (604 × *g*) for 5 min.
4. Inoculate several replicates in MTP wells containing 200 μL of production media (either YNBD or YNBOL, each with 3% of carbon source).
5. Cultivate the strains in a rotary shaker at 28 °C, 250 rpm, and collect the samples at different time intervals. Centrifuge samples, and freeze pellet biomass-containing samples and the medium separately—to be analyzed for SoAMY activity and biomass density.

4.3.2 Sample Analysis

1. OD (optical density) was measured using a MTP reader (Tecan, Life Sciences) at 600 nm wavelength.
2. The recombinant alpha-amylase activity was determined using a protocol developed by Borkowska et al. [26], relying on modification of a standard starch-iodine test, adapted to MTPs and thermocycler format.
   (a) Prepare the substrate solution—starch in 0.1 M acetate buffer of pH 5.0.
   (b) Dispense the solution into PCR 96-well plate, each containing two parts of the substrate (to be adjusted individually).
   (c) Conduct the assay by adding stepwise an equal amount of the sample into the PCR plate wells, incubate at 40 °C, and terminate the reaction by adding one part HCl 0.1 M. The time of the reaction termination should be individually determined from the set of analyzed samples.
   (d) Stain the remaining starch by adding iodine solution (I2 in KI standard Lugol solution).
   (e) Read the plates immediately using an MTP reader (*see* **Note 11**).
   (f) Calculate the amount of the remaining starch-iodine complexes, not digested by the enzymatic preparation (*see* **Note 12**).
   Following the procedure described above, we obtained the following time-production profiles by the recombinant *Y. lipolytica* strains bearing the insect alpha-amylase gene (SoAMY) (Fig. 5a, b).

As shown in Fig. 5a, b, depending on the promoter used, the profile of the SoAMY activity differed significantly. The profiles represented by strains GGY_1 and GGY_2, bearing pTEF and p8UASpTEF promoters, respectively, follow typical course for these regulatory elements described in previous reports

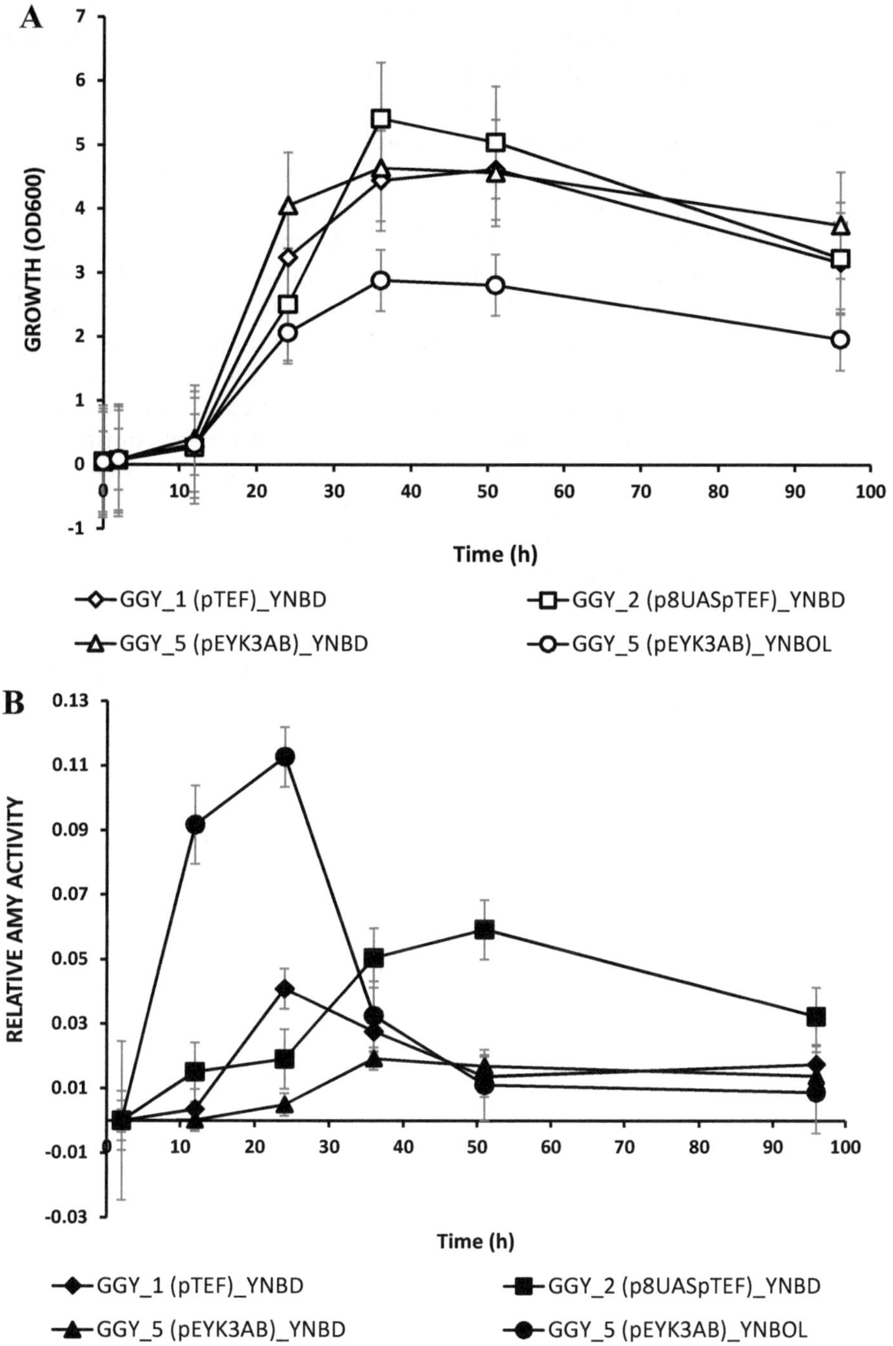

Fig. 5 Kinetics growth (**a**) and of the recombinant SoAMY enzyme activity in the culture medium (**b**) of the recombinant *Y. lipolytica* strains bearing the heterologous gene under the control of different promoters: GGY_1 (pTEF; diamonds), GGY_2 (p8UASpTEF; squares), and GGY_5 (pEYK1); the strains GGY_5 were cultured either in YNBD (3%; triangles) or YNBOL (3%; circle) to demonstrate inducible character of the pEYK1 promoter. X axis, growth [hours]; Y axis, optical density at 600 nm wavelength units (open symbols, **a**) and relative amylolytic activity units (close symbols, **b**). YNBD, YNB medium supplemented with 3% glucose;

[13, 25]. Briefly, pTEF promoter is a constitutive promoter, active in metabolically active (protein-producing) cells. The hybrid promoter—p8UASpTEF equipped with eight subsequent UAS1 motives—is most active during stationary phase of growth, which can be observed upon comparison of Fig. 5a, b, depicting growth curve of the strains. Also, the hybrid promoter is characterized by a higher activity, as discussed in the introduction to this chapter. In this specific case of SoAMY enzyme, it is important to explain the decrease in the activity of the amylase observed at the end of the cultures. This phenomenon results from changes in the pH value of the culture medium, normally observed in *Y. lipolytica* batch cultures without continuous pH regulation (*see* [17]). Nevertheless, from the curves showed in Fig. 5a, it can be inferred that the lower drop in the SoAMY activity is correlated with continuous activity of the promoter under study. Consequently it can be concluded that pEYK3AB promoter was highly active within the first 24 h of culturing—as long as the inducer (erythritol) was contained in the culture medium at relatively high concentrations (as demonstrated in [13]). Upon consumption of the inducer—the activity of the promoter ceased, which was concomitant with a drop in its activity (as the protein could not be accumulated in its active form in the medium due to changes in the pH value). From Fig. 5b, a slight induction of the pEYK3AB in the glucose-containing medium was described upon first characterization of this regulatory sequence. In this specific case, high activity of the pEYK3AB promoter in the erythritol-containing medium within the first 24 h of culturing importantly compromised growth of the recombinant GGY_5 strains, as shown in Fig. 5b. In contrast, delayed activity of the hybrid p8UASpTEF promoter enabled higher biomass buildup in the GGY_2 strains (Fig. 5b). Similar tendency can be observed for pTEF- and pEYK3AB-bearing strains.

5 Notes

1. The donor fragments can be a PCR product or digestion from vectors carrying the cloned PCR fragment. The donor vectors must carry a different *E. coli* selection marker. We typically used pCRBlunt-TOPO vector carrying the kanamycin resistance gene (KanR). Typically to amplify/clone the gene of interest, the forward primer has the following sequence:

Fig. 5 (continued) YNBOL, YNB medium supplemented with 3% erythritol. The decrease in the activity of the SoAMY enzyme observed at the end of the cultures results from changes in the pH value of the culturing medium, normally observed in *Y. lipolytica* batch cultures without continuous pH regulation. Each curve represents mean from at least two independent batch cultures, each analyzed in technical duplicate ±SD

AMY forward: gg*GGTCTC*t**A** **ATG** AAGGTGCTCGCCCTGC which contains the *Bsa*I-D site in bold and the beginning of the SoAMY sequence.

The common reverse primer contained within the Lip2 terminator, cc*GGTCTC*t**ACTC**CGATTTGTCTTAGAGGAA, which contains the *Bsa*I-L site in bold and the complementary sequence of the LIP2 terminator. The template used for the PCR is a GGA assembly (GGE0158) containing the following assembly pTEF-SoMAY-Lip2ter.

To initially amplify the SoAMY gene, the AMY reverse primer gg*GGTCTC*t**GTAC** CTA GTGGTGGTGGTGG was used, containing the *Bsa*I-K site in bold and the complementary sequence of the SoAMY gene; the stop codon TAG is boxed (Celinska et al., to be published).

Underlined are regions containing the *Bsa*I site and the cleavage site, in italic the *Bsa*I site, in bold the specific overhang for assembly, and then the gene specific sequence with the ATG or the stop boxed (5′ to 3′ sequences).

2. Using high-quality purified PCR fragment or DNA greatly enhances the assembly success.
3. We recommend assembling all reaction components on ice and quickly transferring the reaction to a thermocycler preheated at 37 °C.
4. We recommend to plate different quantities of the transformation mixture in order to have well-separated colonies. This will greatly facilitate to have pure white GGA positive clones and adjust accordingly the destination vector concentration in the GGA reaction and the volume of transformation reaction plated depending on transformation frequencies and the percentage of positive white clones.
5. It is important to have a good carrier DNA with 500 bp fragments. Carrier DNA could be, for example, from Clontech (Yeastmaker carrier DNA) or from Sigma (single stranded from salmon testes).
6. Some protocols include the addition of 1 μL of DTT (2 M) in the transformation reaction. However we have clearly showed that the DTT addition increases the transformation frequency but also increases the integration of the expression cassette at random in the genome. This also increases the number of clones containing several copies of the expression cassette. This could be added if one wants to have a higher expression; however, this will impaired the comparison of promoter efficiency.
7. Agar plates contains 0.67% if yeast nitrogen base w/o amino acids (Difco) and 1% of carbon source. Typically, 50 mL is required for a square Petri dish (120 × 120 × 16 mm) and

20 mL for circular Petri dish (90 × 16 mm). The substrates could be added either directly into the agar medium or added only as a top agar (usually 15 mL and 5 mL for square or round plates, respectively). Cells were centrifuged and resuspended at an OD600 of 1 and 3 μL spotted on the plate.

8. YNBT: Tributyrin is added in a 1% final concentration. A 20-fold concentrated tributyrin emulsion (20% tributyrin, 0.5% Tween 40) sonicated three times for 1 min was added.
9. Agar minimal media (YNB) or rich media (YPD) were prepared as required with glucose (D) or erythritol (OL) at 1% final concentration. Wheat raw starch (Sigma-Aldrich) was added at 0.2% as top agar. Starch degradation is revealed by placing the agar plate above a dish containing iodine crystals. Coloration appeared within about 10 min.
10. The enzyme under study in this example is susceptible to significant changes in pH (*see* [17]. Therefore it was important to buffer the medium with 100 mM (final concentration) phosphate buffer of pH 5.7. Remember to consider individual properties of your enzyme—like pH and temperature range.
11. Importantly, the staining values of the starch-iodine complexes decrease after 10 min after adding the iodine and is subjected to photo-bleaching afterward. Thus it is important to make the measurement within this time.
12. To detect any interfering agents contained in the culture media which may adversely affect the readouts, it is important to simultaneously run negative controls. In this case, a negative control is run for each individual reaction by adding the stopping agent first (HCl), prior to the sample addition. The amount of residual starch-iodine complexes is calculated by subtracting the staining value of the sample-treated substrate, from its negative control (rendering relative SoAMY activity, as shown in Fig. 5a).

Acknowledgment

The experiments conducted by EC were financially supported by the Polish Ministry of Sciences and Higher Education (Project No IP2015 011074).

References

1. Madzak C (2015) *Yarrowia lipolytica*: recent achievements in heterologous protein expression and pathway engineering. Appl Microbiol Biotechnol 99:4559–4577. https://doi.org/10.1007/s00253-015-6624-z
2. Madzak C, Beckerich J-M (2013) Heterologous protein expression and secretion in *Yarrowia lipolytica*. In: Barth G (ed) Yarrowia lipolytica: biotechnological applications. Springer Berlin Heidelberg, Berlin,

Heidelberg, pp 1–76. https://doi.org/10.1007/978-3-642-38583-4_1

3. Le Dall MT, Nicaud JM, Gaillardin C (1994) Multiple-copy integration in the yeast *Yarrowia lipolytica*. Curr Genet 26(1):38–44
4. Pignede G, Wang H-J, Fudalej F, Seman M, Gaillardin C, Nicaud J-M (2000) Autocloning and amplification of *LIP2* in *Yarrowia lipolytica*. Appl Environ Microbiol 66 (8):3283–3289. https://doi.org/10.1128/aem.66.8.3283-3289.2000
5. Nicaud J-M, Madzak C, van den Broek P, Gysler C, Duboc P, Niederberger P, Gaillardin C (2002) Protein expression and secretion in the yeast *Yarrowia lipolytica*. FEMS Yeast Res 2 (3):371–379. https://doi.org/10.1111/j.1567-1364.2002.tb00106.x
6. Ogrydziak DM, Nicaud JM (2012) Characterization of *Yarrowia lipolytica XPR2* multi-copy strains over-producing alkaline extracellular protease—a system for rapidly increasing secretory pathway cargo loads. FEMS Yeast Res 12 (8):938–948. https://doi.org/10.1111/j.1567-1364.2012.00846.x
7. Madzak C, Treton B, Blanchin-Roland S (2000) Strong hybrid promoters and integrative expression/secretion vectors for quasi-constitutive expression of heterologous proteins in the yeast *Yarrowia lipolytica*. J Mol Microbiol Biotechnol 2(2):207–216
8. Bordes F, Fudalej F, Dossat V, Nicaud JM, Marty A (2007) A new recombinant protein expression system for high-throughput screening in the yeast *Yarrowia lipolytica*. J Microbiol Methods 70(3):493–502
9. Juretzek T, Le Dall M-T, Mauersberger S, Gaillardin C, Barth G, Nicaud J-M (2001) Vectors for gene expression and amplification in the yeast *Yarrowia lipolytica*. Yeast 18 (2):97–113. https://doi.org/10.1002/1097-0061(20010130)18:2<97::aid-yea652>3.0.co;2-u
10. Barth G, Gaillardin C (1996) The dimorphic fungus *Yarrowia lipolytica*. In: Wolf K (ed) Genetics, biochemistry and molecular biology of non conventional yeasts. Springer Verlag, Heidelberg, pp 313–388
11. Müller S, Sandal T, Kamp-Hansen P, Dalbøge H (1998) Comparison of expression systems in the yeasts *Saccharomyces cerevisiae*, *Hansenula polymorpha*, *Klyveromyces lactis*, *Schizosaccharomyces pombe* and *Yarrowia lipolytica*. Cloning of two novel promoters from *Yarrowia lipolytica*. Yeast 14(14):1267–1283. https://doi.org/10.1002/(sici)1097-0061(199810)14:14<1267::aid-yea327>3.0.co;2-2
12. Sumita T, Iida T, Yamagami S, Horiuchi H, Takagi M, Ohta A (2002) *YlALK1* encoding the cytochrome P450ALK1 in *Yarrowia lipolytica* is transcriptionally induced by n-alkane through two distinct cis-elements on its promoter. Biochem Biophys Res Commun 294(5):1071–1078. https://doi.org/10.1016/S0006-291X(02)00607-1
13. Trassaert M, Vandermies M, Carly F, Denies O, Thomas S, Fickers P, Nicaud J-M (2017) New inducible promoter for gene expression and synthetic biology in *Yarrowia lipolytica*. Microb Cell Factories 16:141. https://doi.org/10.1186/s12934-017-0755-0
14. Carly F, Vandermies M, Telek S, Steels S, Thomas S, Nicaud J-M, Fickers P (2017) Enhancing erythritol productivity in *Yarrowia lipolytica* using metabolic engineering. Metab Eng 42:19–24. https://doi.org/10.1016/j.ymben.2017.05.002
15. Engler C, Gruetzner R, Kandzia R, Marillonnet S (2009) Golden Gate shuffling: a one-pot DNA shuffling method based on type IIs restriction enzymes. PLoS One 4(5):e5553. https://doi.org/10.1371/journal.pone.0005553
16. Celinska E, Ledesma-Amaro R, Larroude M, Rossignol T, Pauthenier C, Nicaud JM (2017) Golden Gate Assembly system dedicated to complex pathway manipulation in *Yarrowia lipolytica*. Microb Biotechnol 10(2):450–455. https://doi.org/10.1111/1751-7915.12605
17. Celińska E, Borkowska M, Białas W (2016) Evaluation of heterologous alpha-amylase production in two expression platforms dedicated for *Yarrowia lipolytica*—commercial Po1g-pYLSC (php4d) and custom-made A18-pYL-TEF (pTEF). Yeast 33(5):165–181. https://doi.org/10.1002/yea.3149
18. Celińska E, Borkowska M, Białas W (2016) Evaluation of a recombinant insect-derived amylase performance in simultaneous saccharification and fermentation process with industrial yeasts. Appl Microbiol Biotechnol 100 (6):2693–2707. https://doi.org/10.1007/s00253-015-7098-8
19. Celińska E, Białas W, Borkowska M, Grajek W (2015) Cloning, expression, and purification of insect (*Sitophilus oryzae*) alpha-amylase, able to digest granular starch, in *Yarrowia lipolytica* host. Appl Microbiol Biotechnol 99 (6):2727–2739. https://doi.org/10.1007/s00253-014-6314-2
20. Leplat C, Nicaud J-M, Rossignol T (2015) High-throughput transformation method for *Yarrowia lipolytica* mutant library screening. FEMS Yeast Res 15(6). https://doi.org/10.1093/femsyr/fov052
21. Chen DC, Beckerich JM, Gaillardin C (1997) One-step transformation of the dimorphic yeast *Yarrowia lipolytica*. Appl Microbiol

Biotechnol 48(2):232–235. https://doi.org/10.1007/s002530051043

22. Back A, Rossignol T, Krier F, Nicaud J-M, Dhulster P (2016) High-throughput fermentation screening for the yeast *Yarrowia lipolytica* with real-time monitoring of biomass and lipid production. Microb Cell Factories 15(1):147. https://doi.org/10.1186/s12934-016-0546-z
23. Beneyton T, Thomas S, Griffiths AD, Nicaud J-M, Drevelle A, Rossignol T (2017) Droplet-based microfluidic high-throughput screening of heterologous enzymes secreted by the yeast *Yarrowia lipolytica*. Microb Cell Factories 16(1):18. https://doi.org/10.1186/s12934-017-0629-5
24. Ledesma-Amaro R, Dulermo T, Nicaud J (2015) Engineering *Yarrowia lipolytica* to produce biodiesel from raw starch. Biotechnol Biofuels 8(1):148
25. Dulermo R, Brunel F, Dulermo T, Ledesma-Amaro R, Vion J, Trassaert M, Thomas S, Nicaud J-M, Leplat C (2017) Using a vector pool containing variable-strength promoters to optimize protein production in *Yarrowia lipolytica*. Microb Cell Factories 16(1):31. https://doi.org/10.1186/s12934-017-0647-3
26. Borkowska M, Białas W, Kubiak M, Celinska E (2017) Micro-assays for determination of amylolytic activity using a thermalcycler: miniaturization of Somogyi-Nelson and Starch-Iodine Tests. To be published

Chapter 7

High-Throughput Screening and Selection of *Pichia pastoris* Strains

Roland Weis

Abstract

Clone screening procedures for *Pichia pastoris* expression strain comparison rely on the availability of a cultivation environment that ensures equal growth and production capabilities for all assessed transformants. As clonal variation in such experiments is caused by diverging numbers and possibly also genomic locations of integrated (linearized) expression constructs, the productivity assessment of a larger number of strains is mandatory for selecting a set of strains for follow-up bioreactor cultivations in order to define the best-producing clone. Microscale cultivation provides the means to reliably compare growth and productivity of a large number of transformants and by that narrows down the amount of selected strains for scaling up.

Key words Small-scale cultivation, Deep-well plates, High-throughput, Protein production, Screening

1 Introduction

The methylotrophic yeast *Pichia pastoris* is an established (FDA and EMA approved), safe (GRAS), and highly competitive expression host with strong and effective secretory capacities often resulting in higher g/L levels of recombinant protein in the culture supernatant while retaining most endogenous proteins within the cells. A simplified downstream procedure is thus enabled by delivering a secreted raw product with a high purity in the culture supernatant. Combining the advantages of prokaryotes and mammalian cells, the ease of genetic manipulation and fast growth on cheap media paired with eukaryotic features like a subcellular protein processing system required for post-translational modifications render the *Pichia*-expression system as "best of both worlds."

Unlike protein production in *Escherichia coli*, cell lysis and time-consuming isolation from a crude lysate are not required to obtain the target protein in pure form. Also, in contrast to expensive, laborious, and intricate mammalian cell line generation often

Brigitte Gasser and Diethard Mattanovich (eds.), *Recombinant Protein Production in Yeast*, Methods in Molecular Biology, vol. 1923, https://doi.org/10.1007/978-1-4939-9024-5_7, © Springer Science+Business Media, LLC, part of Springer Nature 2019

associated with inherent genetic instabilities, establishing stable recombinant *P. pastoris* cells is a straightforward and time-saving process. Furthermore, the absence of contamination with endotoxins or viruses alleviates safety issues and further simplifies downstream processing in bioproduction with *P. pastoris*.

Production of recombinant proteins in intracellular or secreted form by *P. pastoris* follows either an inducible expression regime for the target gene(s) (usually by methanol-inducible promoters), a constitutive expression path, or a strategy relying on expression of the respective genes upon derepressive abundance of a given carbon (and energy) source [1–3]. The common feature of all production protocols is an initial growth phase on glucose or glycerol aiming at biomass accumulation prior to production (although any strict constitutive expression would happen already in this phase as well).

For a comparative clone screening experiment, it is imperative to ensure that all assessed strains are physiologically equally fit for heterologous protein production, otherwise a bias toward certain strain populations is introduced favoring, e.g., faster growth over actual productivity of the target protein(s). Therefore, the baseline prior to the decisive cultivation and production screening needs to be as comparable as possible. Here, reliable cultivation protocols using 24-deep-well plates with some degree of process control and good comparability between individual wells and inoculated strains are published [4].

In the described protocol for cultivation and screening of secreted expression in 96-deep-well plates [5], methods are compiled that allow for highly comparable productivity assessment of a large number of clones using minimal media for methanol-induced expression. This media formulation mimics cultivation conditions in scale-up bioreactor processes and therefore maximizes the transferability of selected expression strains from the small-scale experiments with minimal effort in labor and time. Moreover, screening tools for rapid evaluation of strain productivity are enclosed.

2 Materials

2.1 Media

Prepare all media components using ultrapure water and analytical grade reagents. Always admix denoted final media freshly.

2.1.1 Stock Solutions for Minimal Media

10× YNB: dissolve 134 g of yeast nitrogen base with ammonium sulfate and without amino acids in 1000 mL ultrapure water by heating, filter sterilize, and store at 4 °C (shelf life 1 year).

500× biotin: dissolve 20 mg biotin in 100 mL ultrapure water, filter sterilize, and store at 4 °C (shelf life 1 year).

10× D: dissolve 100 g of a-D(+)-glucose monohydrate in 1000 mL ultrapure water, autoclave (standard procedure), and store at ambient temperature (shelf life 1 year).

1 M sodium-phosphate buffer pH 6.0: prepare stock solutions of $NaH_2PO_4 \times H_2O$ (138 g/L) and Na_2HPO_4 (142 g/L), and mix 877 mL $NaH_2PO_4 \times H_2O$—stock with 123 mL Na_2HPO_4. Adjust pH using phosphoric acid or NaOH, as required.

2.1.2 Media Composition

Buffered Minimal Medium Supplemented with 1% Glucose BMD1

Admix 100 mL of *1 M sodium-phosphate buffer pH 6.0* (*see* **Note 1**), 100 mL of *10× D*, 100 mL of *10× YNB*, and 2 mL of *500× biotin* with 698 mL ultrapure water per liter of BMD1 medium.

Buffered Minimal Medium Supplemented with 1% Methanol (BMM2)

Admix 100 mL of *1 M sodium-phosphate buffer pH 6.0*, 10 mL of absolute methanol, 100 mL of *10× YNB*, and 2 mL of *500× biotin* with 698 mL ultrapure water per liter of BMM2 medium.

Buffered Minimal Medium Supplemented with 5% Methanol (BMM10)

Admix 100 mL of *1 M sodium-phosphate buffer pH 6.0*, 50 mL of absolute methanol, 100 mL of *10× YNB*, and 2 mL of *500× biotin* with 698 mL ultrapure water per liter of BMM10 medium.

Valid for all minimal media: In case the strain background to be analyzed is auxotroph for one or more amino acids also after transformation, the appropriate amount of the necessary amino acid must be supplied accordingly (*see* **Note 2**).

2.2 Hardware

96-deep-well plate(s) with square bottom and cover (*see* **Note 3**).

Sterile toothpicks.

Multichannel pipettes or pipetting robot.

INFORS HT Multitron (or comparable shaking device).

Microfluidic capillary electrophoresis apparatus (e.g., Perkin Elmer LabChip GX Touch, GXII, or comparable device) or classical devices for SDS-PAGE analysis.

Centrifugation device for microplates and deep-well plates.

Spectrophotometer capable of handling 96-well plates.

PCR and microtiter plates compatible with machines used.

3 Methods

As a starting point prior to cultivation and screening, a short evaluation of methods for transformation of *P. pastoris* and selection of transformed strains is provided.

A variety of developed transformation techniques to safely introduce the expression cassettes into the genome is available [*Pichia* protocols www.thermofisher.com; 6], with clear descriptions of expectations for the number and locus of integration events [7]. Genomic integration, ideally into a targeted locus, is still the predominantly applied strategy, despite the fact that recombinant protein production can also be performed using episomal replicative plasmids [8]. However, episomal stability as well as unwanted genomic integration events of this plasmid was not shown by the authors under cultivation conditions other than microscale, e.g., in bioreactors, which limits the potential use of this technology in an industrial setting up to now.

Moreover, methodologies to increase the copy number by the transformation/integration strategy are published [9–12]. As nicely presented [10–12], an increased copy number not always correlates with elevated production of the target protein but rather seems to depend on the individual protein as well as the production regime, i.e., intracellular versus secreted. Additionally, these reports also highlight potential instabilities of multi-copy strains during cultivation caused by recombination events.

A yet undiscussed pitfall of multi-copy strains in an industrial setting is the necessity to repeatedly prove that the selected expression strain does not undergo (significant, e.g., loss of expression cassettes) genetic modifications during the production runs, which represents a highly demanding effort for strains with elevated number of integrated expression cassettes.

Although no valued, defined transformation/selection strategy can be proposed for every user, every project constellation, and every target protein, a general recommendation for a course of action is given below.

3.1 Transformation

Irrespective of the method used, it is advised to transform a linearized plasmid, aiming for targeted integration into the genome, in two concentrations, being "lower range 200–500 ng" and "higher range 1–3 μg." In doing so, a small but present bias is given to a higher propensity of low copy number of integrations for the "lower range" and elevated copy numbers for the "higher range" (*see* **Note 4**).

3.2 Selection

Increasing concentrations of selective agent (only valid for antibiotic selection, not complementation strategies in auxotrophic strains, of course [7]) are thought to set a bias toward higher copy numbers, simply due to the fact that the counter-acting gene needs to be transcribed to sufficient quantities for the cells to survive high selection pressure.

On the other hand, applying a "normal" range of selection pressure does also allow for an increased integration number of expression cassettes. If the aim of the screening project was the identification of the ultimate production strain which is also manageable in industrial settings regarding analysis requirements (*see* above), it is advised to select on only slightly increased antibiotic concentrations (in the range of two to threefold higher than proposed in the *Pichia* protocols, i.e., 200–300 μg/mL of zeocin or 750–1000 μg/mL of Geneticin). This will trigger the generation of strains both with low and higher copy numbers and lead to a sufficient genetic diversity to deliver better and worse performance.

3.3 Liquid Handling and Cultivation

In order to ensure a quick handling time for liquid transfer, it appears mandatory to use multichannel pipets or a suitable automated liquid handling device.

It is required that the shaking device is capable of:

- Efficient humidification to prevent a loss of volume by evaporation.
- Efficient temperature control, i.e., cooling the chamber.
- High rotation speed (a minimum of 280 rpm at a diameter of 50 mm).
- Adequate fixing table for deep-well plates.

The distinct minimal media composition paired with the cultivation protocol aims at highest comparability of growth and potential productivity performance at all positions on one 96-well plate, as well as across individual plates (*see* **Note 5**). Furthermore, handling steps are minimized in order to allow for rapid parallel plate processing.

The long initial phase of 60–65 h in batch medium serves as time buffer to ensure that all colonies consumed the supplied carbon source (glucose in this description). The limiting glucose amount in this special medium formulation secures physiological competence for heterologous protein expression and also for those cultures that utilize glucose faster than others and consequently rest longer in the stationary phase before being induced by methanol. Additionally, this procedure and the media formulation lead to reduced fractions of necrotic and apoptotic cells as compared to other media compositions [5].

3.3.1 Basic Protocol in 96-Well Plates: Methanol-Induced Expression

The cultivation protocol in 96-deep-well plates using methanol induction for a mut^{s}-strain background (*see* **Note 6**) is schematically depicted in Fig. 1 with a time frame of 5½ days (~134–140 h).

Methanol induction is carried out for a total of 72 h, thereby maximizing the time frame of the productive phase of cultivation (*see* **Note 7**). The inducer is supplied four times in a buffer formulation aiming at compatibility to standard multichannel liquid handling tools for high throughput.

The methanol-induced protocol description spans the range of Friday afternoon until Thursday morning in the next week with five manipulation steps before harvest. Note that the time span between

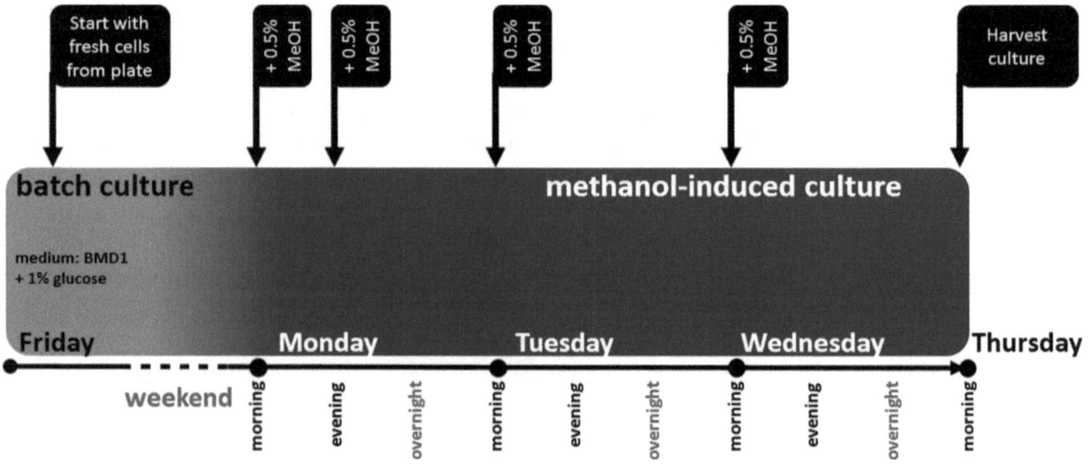

Fig. 1 Clone screening protocol for methanol-driven recombinant production in 96-well format for mut^{S}-background

"morning" and "evening" should be at least 8 h. A person without experience in this particular handling can easily handle 24–36 plates per week, given that the necessary instruments are available (liquid handling device, shaker). One always needs to carry out handling steps very carefully and aseptically, even though only minimal media are applied in this protocol.

Deviations between biological replicates might occur mainly among different plates, hence "normalizer clones" and/or "mock strains" are advised on each separate plate (Fig. 2a) (*see* **Note 8**). For clone screening applications, i.e., the initial and first liquid cultivation of freshly prepared transformants, it is not necessarily required to run these "normalize clones" on each plate because a rescreening of selected clones from this initial round is advised in order to validate results obtained from this first liquid cultivation (Fig. 2b).

Step 1: Inoculation of culture with transformants from plates, Day 1 (Friday), afternoon.

Inoculate 0.25 mL BMD1 medium with cells from a transformation plate (*see* **Note 9**) or a master plate (*see* **Note 10**), and incubate at 28 °C at 260–340 rpm and 80% humidity.

Step 2: Methanol induced culture, Day 4 (Monday), morning.

Add 0.25 mL of BMM2 into each well (*see* **Note 11**) and incubate at max. 28 °C (*see* **Note 12**) at 260–340 rpm.

Step 3: Methanol induction, Day 4 (Monday), evening.

Add 0.05 mL of BMM10 into each well and incubate at max. 28 °C at 260–340 rpm.

Step 4: Methanol induction, Day 5 (Tuesday), morning.

Add 0.05 mL of BMM10 into each well and incubate at max. 28 °C at 260–340 rpm.

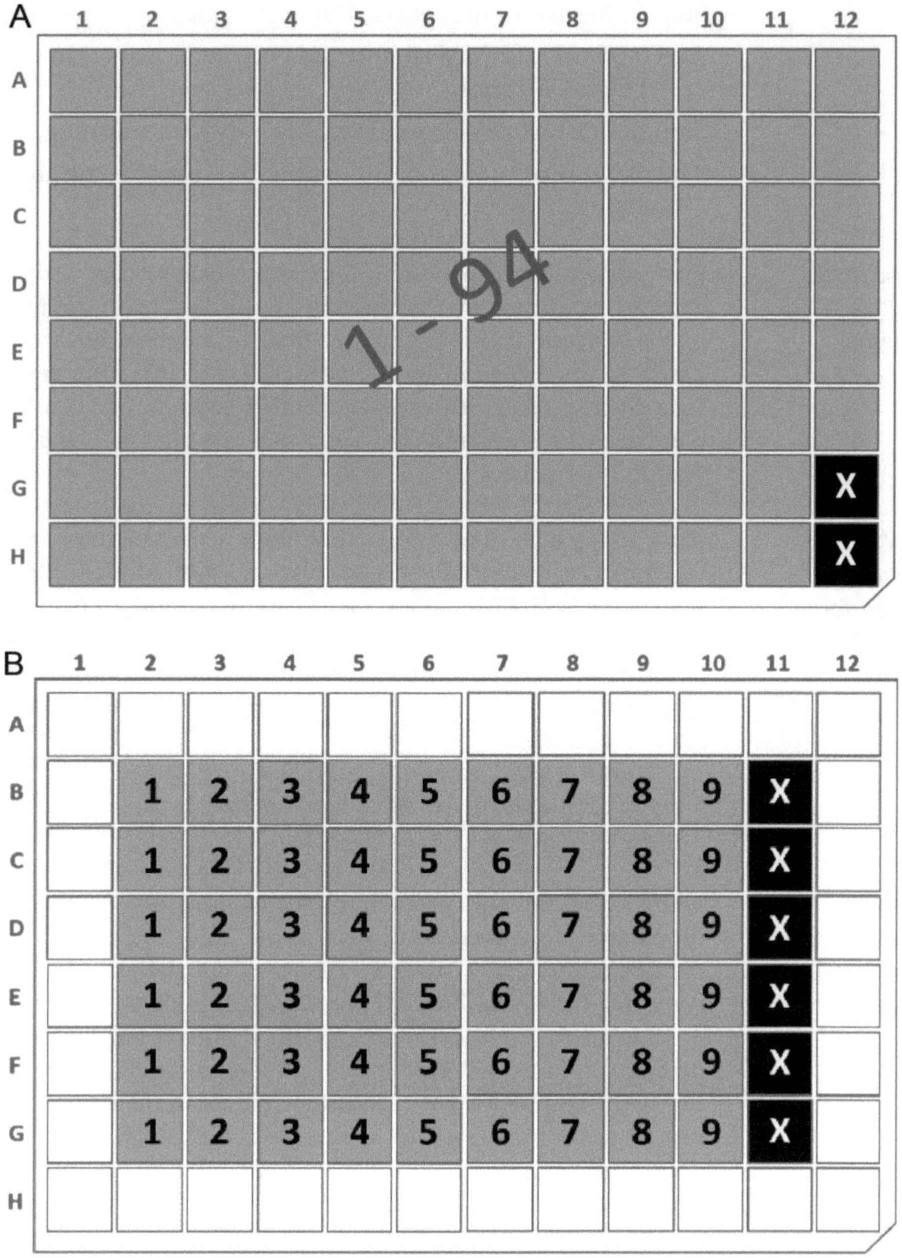

Fig. 2 Plate layout for 96-deep-well plate cultivations for initial clone screening and rescreening: (**a**) Recommended layout for regular production clone screening: individual clones are cultivated and analyzed in comparison to biological replicates of a "normalizer clone" and/or a "mock strain" X (wells G and H 12, not necessarily performed for each plate). (**b**) Layout for rescreening of selected strains from the initial screening round or for parameter check with respect to media formulation, pH, buffer strength, and the like; a potential evaporation in perimeter wells by, e.g., insufficient humidification is prohibited by using well B2 to G11, only; a normalizer strain or any other *Pichia* strain should be inoculated into the perimeter wells, nevertheless, in order to have all wells filled with cell material

Step 5: Methanol induction, Day 6 (Wednesday), morning.

Add 0.05 mL of BMM10 into each well and incubate at max. 28 °C at 260–340 rpm.

Step 6: Harvest of culture, Day 7 (Thursday), morning.

After ~72 h of methanol induction, harvest the culture (*see* next section: Harvest).

3.3.2 Basic Protocol in 96-Well Plates: Glucose-Based Expression

For glucose-based production systems (using other promoters for expression of the target gene), an established protocol features the repeated feeding of 1% glucose (final concentration) at Day 4 and Day 5 in the morning and harvest at Day 6. It is important to note that glucose-fed cultures have a higher tendency of acidification as compared to methanol-fed processes, so that an efficient buffer system is mandatory.

3.4 Harvest of Supernatant for Secreted Target Proteins

Prior to separation of culture supernatant and cell mass, a minimum of 10 μL per well of all wells per plate are diluted 40-fold in water in a separate microtiter plate (*see* **Note 13**), and optical density at 600_{nm} is measured in an appropriate spectrophotometer. This determination serves the calibration of detected productivity with approximate biomass levels and reveals strains with apparent growth deficits.

Culture supernatant is collected into stock microtiter plates after centrifugation of all deep-well plates for 10 min at min. 2500 × *g* and 4–8 °C.

3.5 Analysis of Culture Supernatant for Target Protein Abundance

Depending on the screening size as measure of transformants inoculated for microscale cultivation and therefore number of supernatants to be analyzed, either a direct analysis of all generated supernatants is feasible or a pre-selection method is required.

Using the methodology described above, hardly any homologous protein is secreted to levels above 1 mg/L by the most regularly employed basic strains used for transformation, mostly due to the minimal media applied for cultivation. Hence, all easily detectable proteins in the supernatant refer to the target protein (s) to be secreted.

As word of caution, it needs to be mentioned that co-overexpression of helper factors aiding in production/folding/ secretion of the target protein(s), e.g., chaperones, might result in co-secretion due to cellular stress during recombinant protein production. These proteins would of course negatively influence a pre-selection method that is NOT associated with a direct detection of the amount of target protein.

Thus, a verification of pre-selection results obtained is mandatory when using, e.g., total protein content (*see* **Note 14**) or ELISA (*see* **Note 15**) as means of determination.

As "final" evaluation of actual productivity, and most importantly the identification of significant differences in productivity of certain strains over others, several methods are available.

Western blot analysis after electrophoretic separation of proteins offers a precise assignment of the target protein and potentially presents processing products, in case the antibody recognizes these fragments. Of course, limited sample throughput and required time render this method to be valuable for a rescreening, rather.

SDS-PAGE analysis delivers rather sound data for the determination of superior producers provided that the target protein(s) are secreted to levels high enough for visualization with conventional staining. As opposed to the clearly limited detection capabilities of Coomassie Blue staining, more sensitive methods, e.g., silver staining, are available but come at the expense of sample throughput and time.

Microfluidic capillary electrophoresis (mCE) might relieve the described shortcomings of SDS-PAGE regarding throughput and time and, on top of that, offers automatable (at least) semiquantitative analysis for rapid identification of supernatants with most desired characteristics: product quantity and (if suitable/possible) product quality concerning proteolytic processing.

The primary output of a 96- or 384-well mCE-run can be displayed as SDS-PAGE-like graphic (Fig. 3) where automatable selection of a desired combination (e.g., strongest signal of full-length target protein paired with lowest signals for processing fragments or lowest abundance of general by-products) simplifies the identification of valuable supernatants from a large screening. The analysis output as electropherogram, however, better visualizes differences in productivity (Fig. 4).

The much higher peak resolution compared to SDS-PAGE analysis enables an identification of subtle (micro-) heterogeneities of target protein signals which potentially would effect a huge impact on product quality (Fig. 5). Supernatants displaying such characteristics can therefore be excluded from closer analysis, and the originating strain would not be selected for rescreening and bioreactor cultivation, eventually.

3.6 Selection of Strains for Rescreening

3.6.1 Monoclonality/Cultivation Quality

Ideally, five to ten best-performing strains with respect to target protein quantity and quality (if applicable) as well as at least one strain showing average production are selected for rescreening. After restreak from the selected wells onto agar plates (*see* **Note 16**) and cultivation of single colonies per strain in individual wells, the comparative analysis of all supernatants per strain per well will reveal the following important parameters: monoclonality/cultivation quality.

In case a deviation in productivity above 15% is determined, either the cultivation protocol applied is not delivering an equal

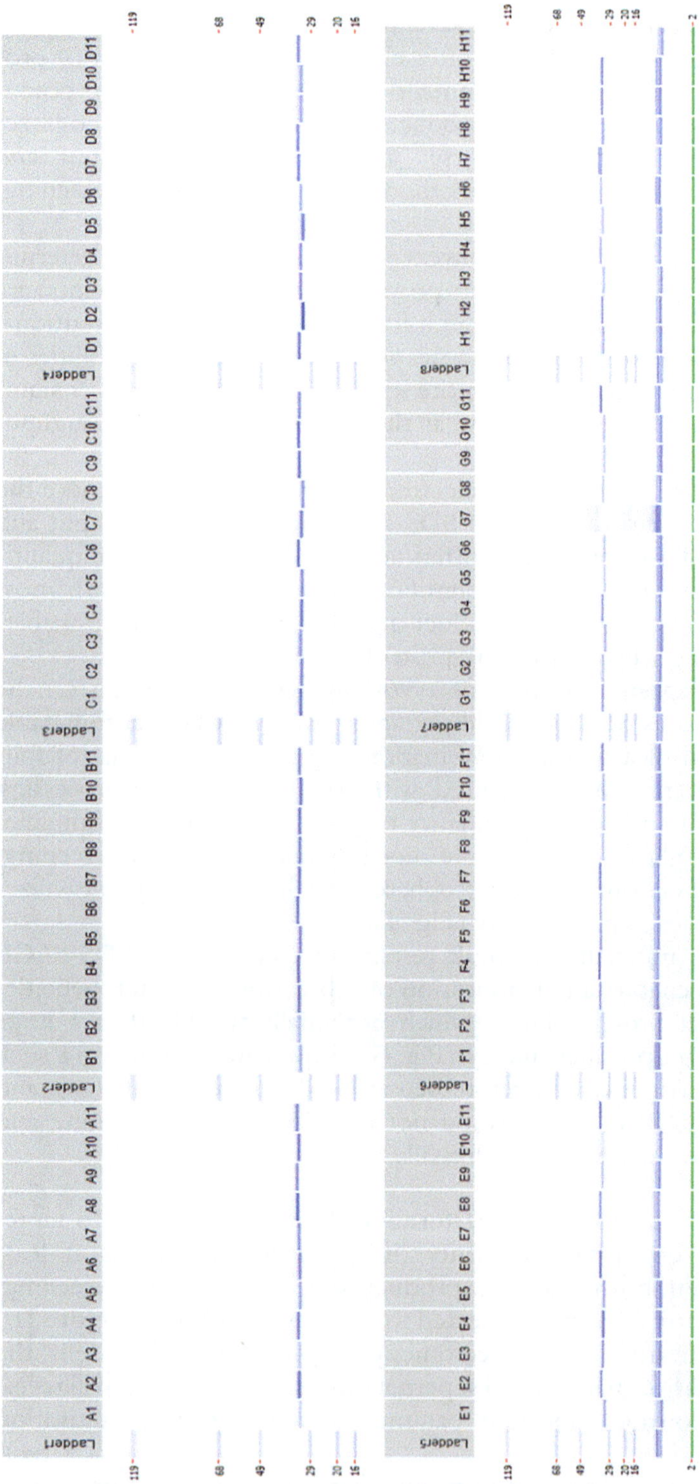

Fig. 3 SDS-PAGE-like image of a 96-well plate analyzed by mCE

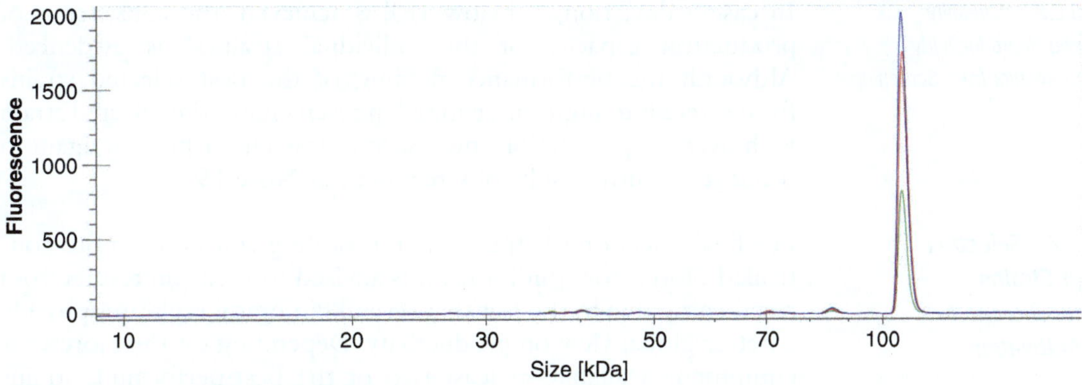

Fig. 4 Electropherogram overlay of individual supernatants, with top performers given in blue, red, and brown lines and an average performer in green line

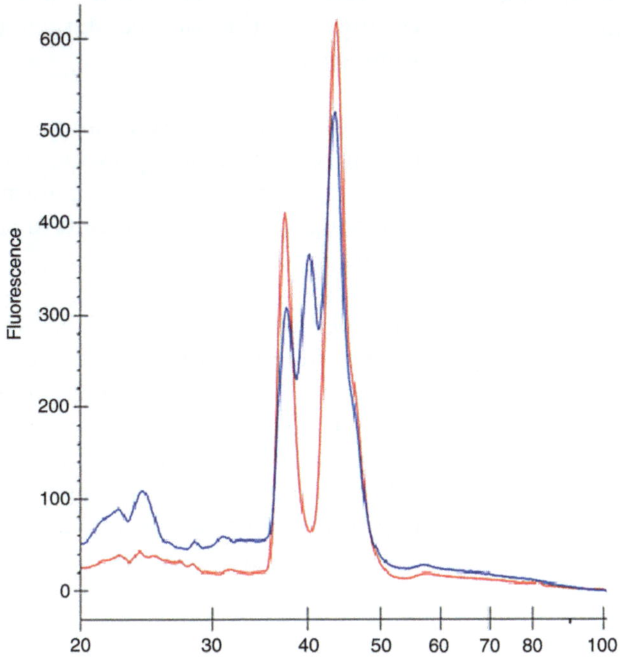

Fig. 5 Electropherogram overlay of individual supernatants. Microheterogeneity delivering three peaks (blue) as opposed to two peaks (red)

environment to all assessed wells (*see* **Note 17**) or the originally inoculated colony from transformation plate to the well of the -deep-well plate in screening was not of monoclonal origin (*see* **Note 18**). For the second case, a repeated restreak to single-colony level from the well of the best-performing supernatant followed by another rescreening needs to be done.

3.6.2 Reliability and Reproducibility of Results from Screening

In case a deviation of below 15% is achieved, the consistency of production capacity of the individual strain(s) is evidenced. Although the performance ranking of the best selected strains from screening might be changed in rescreening, the selected strain with average production must secrete less target protein again in order to confirm results of screening (*see* **Note 19**).

3.7 Selection of Strains for Bioreactor Cultivation

For final selection of strains in an upscaling run under more controlled bioreactor conditions, it is advised to focus on results from rescreening due to the fact that several biological replicates provide a better global view on productivity. Depending on the bioreactor equipment available, at least two of the best-performing strains should be assessed (*see* **Note 20**).

3.8 Proposed Strain Screening Experiment Dimension

Assuming the transformation of a linearized plasmid with one dedicated promoter driving the expression of one dedicated gene coding for the desired target protein, several strategies are suggested.

3.8.1 Feasibility Screening

The cultivation and screening of 12–24 transformants will give a definitive answer of expressibility of the gene/protein of interest. This result might serve as decision point to proceed with a larger screening upon positive results or stop the project due to a negative result.

3.8.2 Extended Screening for Final Production Host: Single Transformation Construct

In order to best cover the introduced variability of integration events, iterative rounds of screening starting with 200 transformants and potentially going up to 1000 transformants can be assessed. In the largest cultivation/screening endeavor, there of course will be genetically identical strains present, but on the other hand, virtually every genetic constellation with respect to copy number variations, integration loci (*see* **Note 21**), and expression cassette architectures (e.g., head-to-tail orientation, also multiple times) is included.

3.8.3 Extended Screening for Final Production Host: Several Transformation Constructs

An experiment with more than one transformation construct, e.g., the co-transformation of a plasmid pool carrying different promoters or promoter variants driving the expression of the target gene as well as a plasmid pool with different promoters or promoter variants for the production of a helper protein (like a chaperone), generates such a huge number of potential genetic constellations that a comprehensive coverage of all events would result in a screening size beyond the means of every laboratory. From experience, a project with a cultivation/screening size of between 6000 and 10,000 transformants delivered satisfying results (*see* **Note 22**).

4 Notes

1. Any other buffer composition can be applied, with the prerequisite that it is suitable for pH ranges of 4 to above 6 (most frequently used range is 5–6). As word of caution, it must be said that the applicability of the protocol must be newly assessed when changing the buffer system. As example, potassium-containing buffers are not compatible with the microfluidic capillary electrophoresis systems from Perkin Elmer.
2. Appropriate concentrations for amino acid supplementations can be found in the "Invitrogen Pichia Protocols" via www.thermofisher.com.
3. Best results were obtained with plates from Bel-Art Biosciences (https://www.belart.com/bel-art-f37860-0001-sterile-polypropylene-96-deep-well-plate-pack-of-24.html, available in Europe via VWR) (Fig. 6).
4. An absolute bias for copy number of integration events cannot be forced using the described procedure, but empirical, unpublished investigation yielded a higher number of transformants with 1–2 copies integrated for the "lower range" and 3+ copies for the "higher range" (both on agar plates with identical selection pressure). Eventually, microscale screening will

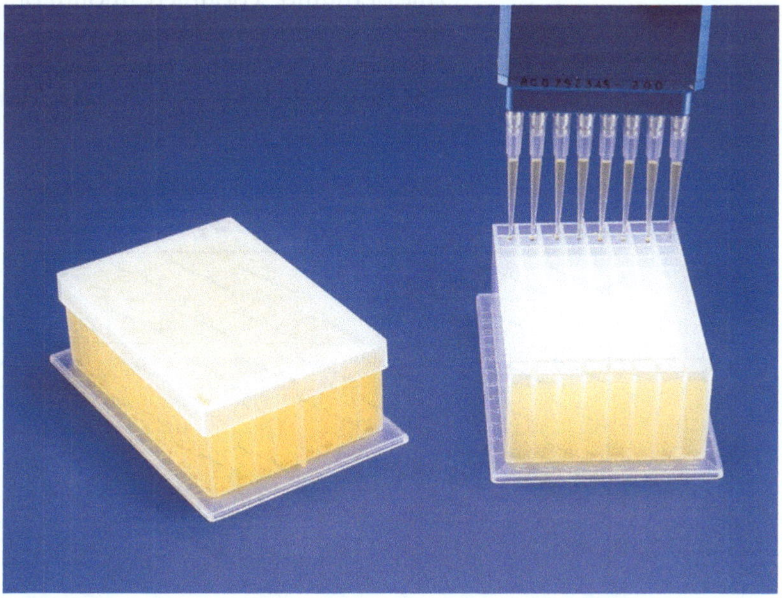

Fig. 6 96-deep-well plates with flat bottom, square-shaped wells, and cover (© Bel-Art SP Scienceware)

identify the best-performing strain WITHOUT knowledge about the actual copy number, anyway.

5. The comparability of growth and potential productivity performance at all positions on one 96-well plate, as well as across individual plates, can be best assessed as follows:

   Inoculate two entire deep-well plates with single colonies of one selected strain producing, e.g., an enzyme where the activity directly correlates with productivity.
   Place the deep-well-plates at opposite positions of the table in the shaking device, and run cultivation.
   Measure optical densities of cultures from all wells of both plates: if the deviation in optical densities across one individual plate AND across both plates is lower than a maximum of 10%, comparable growth is ensured.
   Measure activity of the secreted enzyme of culture supernatants from all wells of both plates: if the deviation in optical densities across one individual plate AND across both plates is lower than a maximum of 10%, comparable productivity is ensured.
   Compare the obtained results to the example provided in Fig. 7, showing an activity assay for a lipase-secreting strain after the procedure described above.

6. Using the strain background mut$^+$, considerably higher methanol uptake rates can be reached, accompanied by a higher growth rate (and sometimes also increased productivity, being target protein-dependent of course); hence, also higher concentrations of methanol need to be added for induction. As rule of thumb, a doubled amount of methanol in the induction media should be administered at higher feeding frequency: following the time-scheme of Fig. 1, induce with 50 μL BMM10 ADDITIONALLY on Monday noon (or 4 h after

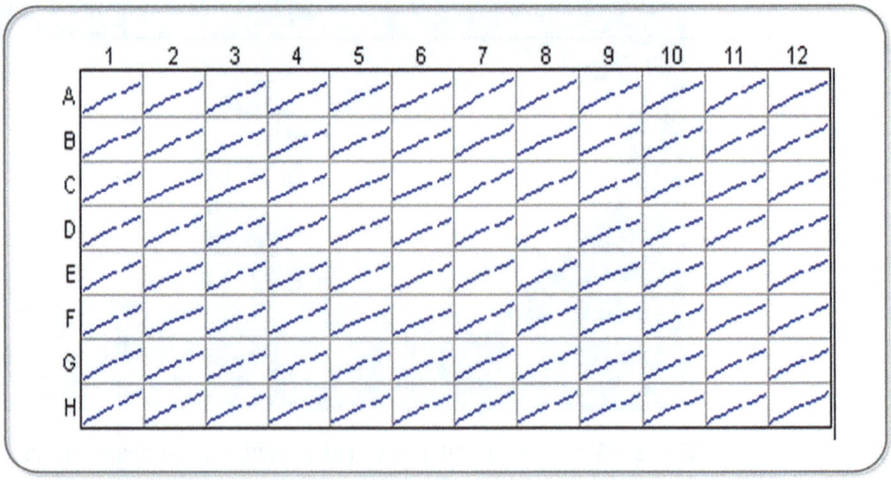

Fig. 7 Results of spectrophotometric activity assay

initial induction with BMM2 at the earliest) and Tuesday and Wednesday evening. The ideal procedure needs to be fixed empirically, of course.

7. Due to the faster growth of mut$^+$-strains, it might also be possible to stop cultivation on day 6 (Wednesday morning in Fig. 1) and assess productivity.
8. Such “normalizer clones” can be, e.g., the best-performing clone from previous screenings or a randomly chosen clone from one transformation round that has not yet been tested in liquid culture, or any other *P. pastoris* strain producing a target protein (also different from the one tested in this cultivation) which can be easily assayed. Including such a “normalizer clone” facilitates the assessment of potentially occurring systemic differences between individual deep-well plates, be it through unequal treatment during cultivation, problematic differences of oxygen supply or humidity between certain positions on the shaker, subtle material alterations of deep-well plates caused by repeated usage or aging, or any other reason.

   Parallel cultivation of untransformed mock strain(s) would provide an authentic blank supernatant for comparison.
9. A transformation plate means to be the selection agar plate from the initial transformation (not older than ~2 weeks). Inoculate with the approximate equal cell mass per colony on transformation plates: as it is impossible to estimate the exact cell mass to apply to the well, one should strive for avoiding the transfer of the entire cell mass of the largest colonies on the plate while taking only a fraction of small colonies on the plate. Ideally, a well-balanced cell mass per transferable colony will do. Additionally, it is of utmost importance to pick only isolated, morphologically well-defined colonies from transformation plates in order to avoid a mixture of strains in one well.
10. A master plate represents an agar plate with cell mass originating from a restreak from a colony on a transformation plate (not older than ~2 weeks). Inoculate with the approximate equal cell mass per colony on master plates: as it is impossible to estimate the exact cell mass to apply to the well, one should strive for avoiding the transfer of the entire cell mass of the largest colonies on the plate while taking only a fraction of small colonies on the plate. Ideally, a well-balanced cell mass per transferable colony will do.
11. Without proper liquid handling tools, this step is extremely time-consuming when cultivating more than two to three plates (true for all methanol induction steps).
12. As growth on methanol (especially for mut^s strains) is limited anyhow, a reduction of the incubation temperature to 25 °C or even as low as 20 °C is not detrimental to the cell’s fitness but

may well positively influence your screening results (valid for the whole production phase).

13. Volumes lower than 10 μL might be inadequately pipetted, so for a 40-fold dilution, it is proposed to admix 10 μL of broth with 90 μL of *dd*H_2O in a first step and 50 μL of this first dilution with 150 μL of *dd*H_2O in a second step.
14. Assuming the production of a fusion protein prone to be proteolytically processed to a certain extent, a high total protein content would not necessarily correlate with high levels of unprocessed, full-length target protein.
15. Assuming the production of a target protein suffers from considerable proteolytic processing, a high readout of target protein bound to the antibody in ELISA analysis would not necessarily correlate with high levels of unprocessed, full-length target protein.
16. Aseptical transfer of cell material from the selected wells using a sterile device (inoculating loop or toothpick) onto agar plates without or with low selection pressure (i.e., 30–50% of the proposed concentration of the antibiotic) is followed by dilution streak to definitely obtain single colonies. This "master plate" serves as starting material for rescreening cultivation as well as for subsequent cell banking (*see* **Note 19**).
17. This would normally be readily observable by different final culture volumes in the individual wells.
18. Referring to **Note 8**, the described problem of polyclonality can be prevented by picking isolated, morphologically well-defined colonies from transformation plates, only.
19. As soon as the final decision for candidate strains is drawn, the cellular material on the "master plate" (*see* **Note 16**) serves as inoculate for preparation of research cell bank, i.e., glycerol stocks ("Invitrogen Pichia Protocols" via www.thermofisher.com). Usually, at least two glycerol stocks are prepared from all rescreening candidates, and ultimately more cell banks are generated from final production strains after evaluation or results from bioreactor cultivation.
20. Numerous in-house expression projects provided evidence that the performance ranking from microscale rescreening is transferable to bioreactor scale. Still, there might be product-specific bottlenecks in individual selected strains that are detected only upon bioreactor cultivation but had no negative effect in deep-well plate cultivation. Hence, assessment of a higher number of strain candidates in upscaling experiments tremendously increases the chances to find the best-possible high-level production strain.
21. Even though the linearization of the plasmid normally provides a rather long homologous sequence for recombination into the

desired genomic locus (depending on the plasmid), *P. pastoris* also features nonhomologous recombination events. Disregarding the amount of additional work, this might generate for characterization of a final high-level production strain for regulatory purposes; such integration events might beneficially influence productivity to significant levels.

22. "Satisfying results" in this context reflect the range of productivity from the average level of the entire performance landscape to the best expression strains. A fold change in productivity of two to three is a regularly observed result, while above fivefold enhanced production is also encountered frequently.

References

1. Weinhandl K, Winkler M, Glieder A, Camattari A (2014) Carbon source dependent promoters in yeasts. Microb Cell Factories 13:5. https://doi.org/10.1186/1475-2859-13-5
2. Hartner FS, Ruth C, Langenegger D, Johnson SN, Hyka P, Lin-Cereghino GP, Lin-Cereghino J, Kovar K, Cregg JM, Glieder A (2008) Promoter library designed for fine-tuned gene expression in *Pichia pastoris*. Nucleic Acids Res 36(12):e76. https://doi.org/10.1093/nar/gkn369
3. Prielhofer R, Maurer M, Klein J, Wenger J, Kiziak C, Gasser B, Mattanovich D (2013) Induction without methanol: novel regulated promoters enable high level expression in *Pichia pastoris*. Microb Cell Factories 12:5. https://doi.org/10.1186/1475-2859-12-5
4. Marsalek L, Gruber C, Altmann F, Aleschko M, Mattanovich D, Gasser B, Puxbaum V (2017) Disruption of genes involved in CORVET complex leads to enhanced secretion of heterologous carboxylesterase only in protease deficient *Pichia pastoris*. Biotechnol J 12(5). https://doi.org/10.1002/biot.201600584
5. Weis R, Luiten R, Skranc W, Schwab H, Wubbolts M, Glieder A (2004) Reliable high-throughput screening with *Pichia pastoris* by limiting yeast cell death phenomena. FEMS Yeast Res 5:179–189. https://doi.org/10.1016/j.femsyr.2004.06.016
6. Lin-Cereghino J, Wong WW, Xiong S, Giang W, Luong LT, Vu J, Johnson SD, Lin-Cereghino GP (2005) Condensed protocol for competent cell preparation and transformation of the methylotrophic yeast *Pichia pastoris*. BioTechniques 38(1):44–48
7. Ahmad M, Hirz M, Pichler H, Schwab H (2014) Protein expression in *Pichia pastoris*: recent achievements and perspectives for heterologous protein production. Appl Microbiol Biotechnol 98:5301–5317. https://doi.org/10.1007/s00253-014-5732-5
8. Camattari A, Goh A, Yee Yip L, Hee Meng Tan A, Wai Ng S, Tran A, Liu G, Liachko I, Dunham MJ, Rancati G (2016) Characterization of a panARS-based episomal vector in the methylotrophic yeast *Pichia pastoris* for recombinant protein production and synthetic biology applications. Microb Cell Factories 15:139. https://doi.org/10.1186/s12934-016-0540-5
9. Sunga AJ, Tolstorukov I, Cregg JM (2008) Posttransformational vector amplification in the yeast *Pichia pastoris*. FEMS Yeast Res 8:870–876. https://doi.org/10.1111/j.1567-1364.2008.00410.x
10. Aw R, Polizzi KM (2013) Can too many copies spoil the broth? Microb Cell Factories 12:128. https://doi.org/10.1186/1475-2859-12-128
11. Marx H, Mecklenbräuer A, Gasser B, Sauer M, Mattanovich D (2009) Directed gene copy number amplification in *Pichia pastoris* by vector integration into the ribosomal DNA locus. FEMS Yeast Res 9:1260–1270. https://doi.org/10.1111/j1567-1364.2009.00561.x
12. Zhu T, Guo M, Sun C, Qian J, Zhuang Y, Chu J, Zhang S (2009) A systematical investigation on the genetic stability of multi-copy *Pichia pastoris* strains. Biotechnol Lett 31:679–684. https://doi.org/10.1007/s10529-009-9917-4

Chapter 8

Analyzing Recombinant Protein Production in *Pichia pastoris* with Targeted Proteomics

Roslyn M. Bill, Annegret Ulke-Lemée, Stephanie P. Cartwright, Rena Far, Jay Kim, and Justin A. MacDonald

Abstract

New mass spectrometry approaches enable antibody-independent tracking of protein production. Herein, we outline an antibody-independent mass spectrometry method for tracking recombinant protein production in the methylotrophic yeast *Pichia pastoris* system.

Key words Multiple reaction monitoring, Mass spectrometry, MRM-MS, Yeast, Recombinant protein production, ZIPK, DAPK3

1 Introduction

New targeted proteomic approaches to quantitatively examine protein levels have emerged [1, 2]. Multiple reaction monitoring-mass spectrometry (MRM-MS) can identify specific targets (e.g., protein, peptide, and small molecules) in complex samples with high sensitivity and selectivity. For the specific identification of proteins, MRM-MS relies on the detection of a prototypic peptide derived from the target protein. MRM-MS has the advantage over antibody-based methods in that it is only dependent on knowledge of tryptic peptide masses that can be easily derived from the protein sequence. Importantly, MRM-MS assays can be multiplexed, and the simultaneous quantification of dozens of different proteins in a complex sample has been previously reported [3]. Finally, it is possible to quantify protein levels; in this case, specific peptides can be used as stoichiometric representatives of the protein from which they are cleaved and quantified against a spiked internal standard (e.g., a *s*table *i*sotope-labeled *s*ynthetic peptide, SIS peptide) to yield protein concentration [1, 2].

The efficient production of recombinant proteins is of critical importance for advancing our understanding of protein structure-

Brigitte Gasser and Diethard Mattanovich (eds.), *Recombinant Protein Production in Yeast*, Methods in Molecular Biology, vol. 1923, https://doi.org/10.1007/978-1-4939-9024-5_8, © Springer Science+Business Media, LLC, part of Springer Nature 2019

function relationships as well as in drug development. Recombinant protein production is often examined in whole cell lysates with immunoreactive detection reagents; however, in many cases appropriate bioreagents (i.e., sensitive and specific antibodies) are not available to provide efficient tracking of protein synthesis. As an alternative to antibody-based detection, we have established an MRM-MS assay protocol which provides a convenient means for analyzing recombinant protein production in the methylotrophic yeast *Pichia pastoris* expression system. MRM-MS assessments of yeast whole cell lysates can provide information on yields with ratios of recombinant protein versus an internal standard yeast protein, recombinant protein quality and likelihood of full-length protein production, and temporal dynamics of recombinant protein production. The following describes the use of the MRM-MS assay protocol using the recombinant ZIPK-expressing *P. pastoris* X-33 strain as an example.

2 Materials

2.1 Growth of P. pastoris Cultures and Induction of Protein Production

1. Shaking incubator with adjustable thermostat, capable of heating to 30 °C and 250 rpm shaking.
2. pH meter.
3. Microcentrifuge for 1.5 mL microfuge tubes.
4. Disposable vacuum filter/storage bottle (0.2 μm, 500 mL receiver flask).
5. VIS spectrophotometer (600 nm) and polystyrene cuvettes (1 mL).
6. Methanol (5% v/v): Filter sterilize and store at room temperature.
7. Biotin (0.02% w/v): Dissolve 10 mg biotin in 50 mL dH_2O.-Filter sterilize and store at 4 °C.
8. Glycerol (10% v/v): Mix 50 mL analytical grade glycerol with 450 mL dH_2O. Filter sterilize and store at room temperature.
9. Potassium phosphate pH 6.0 (1 M): 132 mL 1 M K_2HPO_4 and 868 mL 1 M KH_2PO_4 mixed together. Adjust pH to 6.0 with phosphoric acid or KOH.
10. Yeast nitrogen base with ammonium sulfate without amino acids (YNB): Dissolve 33.5 g in 250 mL dH_2O. Warm solution gently to dissolve completely. Filter sterilize and store at 4 °C.
11. Buffered glycerol-complex medium (BMGY): 1% (w/v) yeast extract, 2% (w/v) peptone, 100 mM potassium phosphate pH 6.0, 1.34% YNB, 4×10^{-5}% (w/v) biotin, 1% (v/v) glycerol.

12. Buffered methanol-complex medium (BMMY): 1% (w/v) yeast extract, 2% (w/v) peptone, 100 mM potassium phosphate pH 6.0, 1.34% YNB, 4×10^{-5}% (w/v) biotin, 0.5% (v/v) methanol.
13. YPD: 1% (w/v) yeast extract, 2% (w/v) peptone, 2% (w/v) dextrose. Mix dry ingredients minus dextrose with dH_2O to 900 mL. Autoclave. Add 100 mL of 20% (w/v) filter-sterilized dextrose solution, mix thoroughly, and store at 4 °C.
14. Zeocin: 100 mg/mL solution in dH_2O. Filter sterilize, and store solution at −20 °C. Protect from light.
15. YPDZ agar: YPD plus 2% (w/v) agar. Mix dry ingredients minus dextrose with dH2O to 180 mL. Autoclave and cool to ~50 °C. Add 20 mL of 20% (w/v) filter-sterilized dextrose solution. Add 0.1 mL of 100 mg/mL zeocin, mix thoroughly, and pour into Petri dishes.
16. Generated with previously described procedures [4], a *P. pastoris* X33 strain transformed by electroporation with human zipper-interacting protein kinase (ZIPK/DAPK3: UniProtKB ID O43293, 454 amino acids) in pPICZ-A plasmid (*see* **Note 1**).

2.2 Preparation of Yeast Whole Cell Extracts

1. 0.815 M NaOH/0.74% (v/v) β-mercaptoethanol solution.
2. Ice bucket.
3. Desktop vortexer.
4. Refrigerated desktop mini-centrifuge.
5. VIS spectrophotometer and polystyrene cuvettes (1 mL).
6. Trichloroacetic acid (TCA, 100% w/v).
7. Acetone (ice-cold, stored at −20 °C).
8. ABC: 100 mM ammonium biocarbonate (ABC), pH 8. Make fresh by dissolving 0.078 g ammonium bicarbonate in 10 mL pure dH_2O (*see* **Note 2**).
9. ABC + urea: 50 mM ammonium bicarbonate (ABC), pH 8 containing 6 M urea.
10. Bradford protein assay.

2.3 Development of Skyline Analysis Criteria for MRM-MS Application

1. Skyline v3.7 program for creating and analyzing targeted proteomics experiments [5, 6]. This program requires Microsoft Windows or Linux and does not operate with Apple OS. Downloadable from https://skyline.ms/project/home/software/Skyline/begin.view.
2. Microsoft Excel or similar spreadsheet software program.

2.4 Tryptic Digestion of Protein Extracts

1. His-tagged ZIPK protein (1 μg) produced in HEK293 cells and purified using NiNTA-Agarose [7] (*see* **Note 3**).
2. Yeast whole cell protein extracts.
3. ABC: 50 mM ammonium bicarbonate, pH 8. Make fresh by dissolving 0.039 g ammonium bicarbonate in 10 mL pure H_2O.
4. IAA: 200 mM stock iodoacetamide. Make fresh by dissolving 0.037 g iodoacetamide in 1 mL of 50 mM ABC. Store in the dark.
5. DTT: 1 M stock dithiothreitol, dissolve 0.154 g in 1 mL pure H_2O (*see* **Note 4**).
6. Mass spectrometry grade trypsin: 0.5 mg/mL (Promega). Solution made according to the manufacturer's manual (*see* **Note 5**).
7. Heating block for 1.5 mL tubes. Constant temperature settings of 50 °C and 37 °C.
8. Microcentrifuge for 1.5 mL tubes, capable of reaching $> 15{,}000 \times g$.
9. TFA, trifluoroacetic acid, HPLC-grade.
10. ACN, acetonitrile, HPLC-grade.
11. 0.5% (v/v) TFA/50% (v/v) ACN: make up fresh by mixing 10 μL TFA, 100 μL ACN, and 90 μL pure H_2O.
12. Autosample vials (*see* **Note 6**).

2.5 Profiling Protein Expression with MRM-MS

1. HPLC, Dionex Ultimate 3000, running Chromeleon Express, and Dionex Chromatography MS Link or similar HPLC.
   (a) Trap column: for peptide concentration and desalting with capacity for up to 200 μg peptide (*see* **Note 7**).
   (b) Precolumn: same bead type and same or smaller size as the separation column (*see* **Note 8**).
   (c) Separation column: for providing analytical separation of tryptic peptides (*see* **Note 9**).
2. Mass spectrometer, QTRAP (ABSciex QTrap4500, running Analyst software), or similar tandem mass spectrometer.
3. Organic running buffer: HPLC-grade ACN with 0.5% (v/v) formic acid.
4. Aqueous running buffer: HPLC-grade H_2O with 0.5% (v/v) formic acid.
5. Skyline v3.7 (*see* above).

3 Methods

3.1 Growth of P. pastoris Cultures and Induction of Protein Production

The following steps should be viewed as guidelines and are presented as an initial starting point for protein production. Additional methods may be obtained in other detailed protocol descriptions [4]. Optimization of the growth conditions may be required for a particular recombinant protein. Shake flasks are employed due to ease of use; they are available in a variety of sizes (50–2000 mL). If a number of cultures are to be analyzed, it is possible to use smaller volumes and 24-well deep-well plates or test tubes. Be certain that the medium is well-aerated by increasing the rate of shaking. Untransformed X-33 can be used as a control for background production of endogenous yeast proteins.

1. Using sterile techniques, prepare a streak plate using the yeast glycerol stock (e.g. pPICZ-hsZIPK X-33 *P. pastoris*) onto a YPDZ plate, and grow at 30 °C for approximately 3 days or until distinct colonies are apparent (*see* **Note 10**).
2. Using a single colony (selected from YPDZ plate), inoculate 10 mL of YPD in a 100 mL shake flask. Grow at 30 °C in a shaking incubator (250–300 rpm) overnight (*see* **Note 11**).
3. Measure OD_{600}. Transfer 25 OD of yeast cells to a clean conical tube. Harvest the cells by centrifuging at 2500 × *g* for 10 min at room temperature. Decant supernatant and resuspend cell pellet to give a final OD_{600} of 1.0 in 25 mL of BMGY.
4. Transfer to a 100 mL shake flask. Grow at 30 °C in a shaking incubator (250–300 rpm) overnight. Culture should reach an OD_{600} ~5 (approximately 16–18 h), signifying the yeast are in log-phase growth.
5. Harvest the cells by centrifuging at 2500 × *g* for 10 min at room temperature. Decant supernatant, and resuspend the cell pellet to give a final OD_{600} of 1.0 in 50 mL of BMMY to induce expression. Place culture in a 250 mL shake flask. Return to 30 °C incubator to continue growth with shaking as before.
6. Add methanol to give a final concentration of 0.5% (v/v) methanol every 24 h to maintain induction (e.g., for 50 mL BMMY, add 0.25 mL of 100% methanol).
7. At each of the times indicated, remove 1 mL of yeast culture using sterile techniques and transfer to a spectrophotometer cuvette. Measure OD_{600} (*see* **Note 12**). Also transfer 1 mL of the culture to a 1.5 mL microcentrifuge tube. These samples will be used to analyze expression levels and determine the optimal time post-induction to harvest. Suggested time points (h): 0, 6, 12, 24, 36, 48 (2 days), 60, 72 (3 days), 84, and 96 (4 days).
8. Centrifuge 1 mL sample at 15,000 × *g* in a tabletop microcentrifuge for 5 min at room temperature.

For intracellular proteins, decant the supernatant, and store the yeast cell pellets at −80 °C until ready to assay.

For secreted proteins, transfer the supernatant to a separate 1.5 mL tube. Store at −80 °C until ready to assay. Freeze quickly in liquid N_2 or dry ice/alcohol bath.

3.2 Preparation of Yeast Whole Cell Extracts

Prepare protein extracts from whole yeast cell pellets following an alkaline-lysis protocol as reported previously [8].

1. Thaw yeast cell pellets on ice.
2. Add 0.815 M NaOH/0.74% BME (0.9 mL) to each 1.5 mL microcentrifuge tube containing yeast pellets. Ensure the complete disruption of the yeast cell pellet.
3. Incubate on ice for 5 min; vortex with maximum agitation for 1 min and then return tubes to the ice bath. Incubate another 5 min on ice.
4. Add 100% (w/v) TCA (0.1 mL) to give final [TCA] = 10% (w/v). Incubate at −20 °C for 30 min to precipitate protein.
5. Centrifuge: 15,000 × *g*, 20 min, 4 °C.
6. Wash protein pellet with 1 mL ice-cold (−20 °C) acetone. Vortex repeatedly (30 s) with incubation at −20 °C for an additional 10 min.
7. Centrifuge: 15,000 × *g*, 20 min, 4 °C.
8. Resuspend precipitated protein in 50 mM ABC, 6 M urea (0.3 mL). Centrifuge at 2500 × *g* to pellet debris. Retain supernatant and measure protein concentration by Bradford protein assay.
9. Store protein samples at −20 °C until use.

3.3 Development of Skyline Analysis Criteria for MRM-MS Application

1. Import the entire human ZIPK protein sequence into Skyline, as verified by sequencing the vector utilized in Subheading 3.1 (*see* **Note 13**).
2. For peptide settings, select "Trypsin [KR/P]" as enzyme with one missed cleavage allowed. Filter peptides to include those with 6–25 amino acids, exclude peptides that originate from the N-terminus, and auto-select all matching peptides.
3. Allow the structural modifications "Carbamidomethyl Cysteine" and "Oxidation (M)" (*see* **Note 14**).
4. Select transitions by using monoisotopic precursor and product ion masses, with collision energy and declustering potentials adjusted for the tandem MS instrument used (*see* **Note 15**).
5. Filter and auto-select all matching "Precursors with two charges," "Ion charge of one" and b- and y-type ions in a *m/z* range of 150–1500. Select precursors from "$m/z >$ product ion" to four ions (*see* **Note 16**).

6. Paste the control peptides derived from *P. pastoris* glyceraldehyde-3-phosphate dehydrogenase (G3P_PICPA) and beta-tubulin (ANZ77882.1) sequences into Skyline (*see* **Note 17**).
   (a) G3P: TASGNIIPSSTGAAK, precursor charge = 2+. Select transitions: y11, y10, y9, y8.
   (b) Tubulin: LAVNLVPFPR, precursor charge = 2+. Select transitions: y7, y6, y5, y4.
7. Select SSRCalc 3.0, 100 Å retention time predictor, with a slope determined based on a previous run with known peptides, using an identical HPLC program (*see* **Note 18**).
8. Export the scheduled transition list from Skyline.
9. Open the list in MS Excel, and manually enter the entrance potential = 10 and the cell exit potential = 17 into the resulting transition list (*see* Table 1 as an example of transitions employed for an MRM-MS run with ZIPK as the target).

3.4 Tryptic Digestion of Protein Samples

1. Prepare samples for tryptic digestion by diluting 5 μg of total protein (e.g., obtained from whole yeast cell lysate, secreted products in the supernatant, or purified ZIPK) with 50 mM ABC to a final volume of 50 μL (*see* **Note 19**).
2. Add 5 mM DTT (0.25 μL from a 1 M stock) and incubate for 30 min at 50 °C (*see* **Note 20**).
3. Alkylate the protein by adding 3.75 μL iodoacetamide to a final concentration of 15 mM. Incubate for 30 min in the dark at room temperature.
4. Residual iodoacetamide is quenched with a second addition of 5 mM DTT and incubated again for 30 min at 50 °C.
5. Digest protein extract with trypsin (1:50 w/w ratio for whole cell lysates or 1:100 w/w ratio for purified material) overnight at 37 °C (approximately 16 h) (*see* **Note 21**).
6. The tryptic peptide solution is acidified with 0.5% (v/v) trifluoroacetic acid and 5% (v/v) acetonitrile by adding 5.5 μL TFA/ACN (5% TFA (v/v), 50% (v/v) ACN in H_2O) (*see* **Note 22**).
7. Remove particulate matter by centrifugation (15,000 × *g*, 10 min, 22 °C) and carefully transfer the supernatant to a new tube.
8. Repeat the centrifugation step and transfer the peptide mixture to HPLC-autosampler vials.

3.5 Profiling Protein Expression with MRM-MS

1. Create an Analyst program using the predicted transition list from Subheading 3.3 by copy-pasting the entire transition list from Microsoft Excel into a new MRM-Analyst method (*see* **Note 23**).

Table 1
List of ion transitions used for ZIPK, G3P, and β-tubulin in the described MRM-MS experiments

Q1	Q3	RT	Tryptic peptide sequence	DP	EP	CE	CXP
1327.587	1305.680	40.0	hsZIPK.QEDVEDHYEM[Oxi] GEELGSGQFAIVR.+2y12	128	10	65	17
1327.587	1176.637	40.0	hsZIPK.QEDVEDHYEM[Oxi] GEELGSGQFAIVR.+2y11	128	10	65	17
1327.587	1047.595	40.0	hsZIPK.QEDVEDHYEM[Oxi] GEELGSGQFAIVR.+2y10	128	10	65	17
1327.587	934.510	40.0	hsZIPK.QEDVEDHYEM[Oxi] GEELGSGQFAIVR.+2y9	128	10	65	17
1327.587	877.489	40.0	hsZIPK.QEDVEDHYEM[Oxi] GEELGSGQFAIVR.+2y8	128	10	65	17
1391.635	1433.775	38.9	hsZIPK.QEDVEDHYEM[Oxi] GEELGSGQFAIVRK.+2y13	133	10	68	17
1391.635	1304.732	38.9	hsZIPK.QEDVEDHYEM[Oxi] GEELGSGQFAIVRK.+2y12	133	10	68	17
1391.635	1175.690	38.9	hsZIPK.QEDVEDHYEM[Oxi] GEELGSGQFAIVRK.+2y11	133	10	68	17
1391.635	1062.605	38.9	hsZIPK.QEDVEDHYEM[Oxi] GEELGSGQFAIVRK.+2y10	133	10	68	17
309.682	490.298	15.6	hsZIPK.QKGTGK.+2y5	54	10	14	17
309.682	362.203	15.6	hsZIPK.QKGTGK.+2y4	54	10	14	17
309.682	305.182	15.6	hsZIPK.QKGTGK.+2y3	54	10	14	17
309.682	204.134	15.6	hsZIPK.QKGTGK.+2y2	54	10	14	17
309.682	257.161	15.6	hsZIPK.QKGTGK.+2b2	54	10	14	17
462.743	867.457	18.3	hsZIPK.GTGKEYAAK.+2y8	65	10	22	17
462.743	766.409	18.3	hsZIPK.GTGKEYAAK.+2y7	65	10	22	17
462.743	709.388	18.3	hsZIPK.GTGKEYAAK.+2y6	65	10	22	17
462.743	581.293	18.3	hsZIPK.GTGKEYAAK.+2y5	65	10	22	17
462.743	452.250	18.3	hsZIPK.GTGKEYAAK.+2y4	65	10	22	17
485.274	840.498	29.6	hsZIPK.EYAAKFIK.+2y7	67	10	23	17
485.274	677.434	29.6	hsZIPK.EYAAKFIK.+2y6	67	10	23	17
485.274	606.397	29.6	hsZIPK.EYAAKFIK.+2y5	67	10	23	17
485.274	535.360	29.6	hsZIPK.EYAAKFIK.+2y4	67	10	23	17
485.274	293.113	29.6	hsZIPK.EYAAKFIK.+2b2	67	10	23	17
353.204	549.299	18.0	hsZIPK.RLSSSR.+2y5	57	10	17	17

(continued)

Table 1
(continued)

Q1	Q3	RT	Tryptic peptide sequence	DP	EP	CE	CXP
353.204	436.215	18.0	hsZIPK.RLSSSR.+2y4	57	10	17	17
353.204	349.183	18.0	hsZIPK.RLSSSR.+2y3	57	10	17	17
353.204	262.151	18.0	hsZIPK.RLSSSR.+2y2	57	10	17	17
353.204	270.192	18.0	hsZIPK.RLSSSR.+2b2	57	10	17	17
353.204	592.316	17.2	hsZIPK.LSSSRR.+2y5	57	10	17	17
353.204	505.284	17.2	hsZIPK.LSSSRR.+2y4	57	10	17	17
353.204	418.252	17.2	hsZIPK.LSSSRR.+2y3	57	10	17	17
353.204	331.220	17.2	hsZIPK.LSSSRR.+2y2	57	10	17	17
537.781	918.464	21.7	hsZIPK.GVSREEIER.+2y7	70	10	26	17
537.781	831.432	21.7	hsZIPK.GVSREEIER.+2y6	70	10	26	17
537.781	675.331	21.7	hsZIPK.GVSREEIER.+2y5	70	10	26	17
537.781	546.288	21.7	hsZIPK.GVSREEIER.+2y4	70	10	26	17
537.781	157.097	21.7	hsZIPK.GVSREEIER.+2b2	70	10	26	17
372.224	614.398	26.1	hsZIPK.EVNILR.+2y5	58	10	17	17
372.224	515.330	26.1	hsZIPK.EVNILR.+2y4	58	10	17	17
372.224	401.287	26.1	hsZIPK.EVNILR.+2y3	58	10	17	17
372.224	456.245	26.1	hsZIPK.EVNILR.+2b4	58	10	17	17
372.224	569.329	26.1	hsZIPK.EVNILR.+2b5	58	10	17	17
1045.063	1456.779	41.0	hsZIPK.EIRHPNIITLHDIFENK.+2y12	107	10	51	17
1045.063	1342.737	41.0	hsZIPK.EIRHPNIITLHDIFENK.+2y11	107	10	51	17
1045.063	1229.652	41.0	hsZIPK.EIRHPNIITLHDIFENK.+2y10	107	10	51	17
1045.063	1116.568	41.0	hsZIPK.EIRHPNIITLHDIFENK.+2y9	107	10	51	17
845.949	1229.652	40.4	hsZIPK.HPNIITLHDIFENK.+2y10	93	10	41	17
845.949	1116.568	40.4	hsZIPK.HPNIITLHDIFENK.+2y9	93	10	41	17
845.949	1015.521	40.4	hsZIPK.HPNIITLHDIFENK.+2y8	93	10	41	17
845.949	902.437	40.4	hsZIPK.HPNIITLHDIFENK.+2y7	93	10	41	17
845.949	235.119	40.4	hsZIPK.HPNIITLHDIFENK.+2b2	93	10	41	17
755.867	1181.568	35.9	hsZIPK.ESLTEDEATQFLK.+2y10	86	10	37	17
755.867	1080.521	35.9	hsZIPK.ESLTEDEATQFLK.+2y9	86	10	37	17
755.867	951.478	35.9	hsZIPK.ESLTEDEATQFLK.+2y8	86	10	37	17
755.867	836.451	35.9	hsZIPK.ESLTEDEATQFLK.+2y7	86	10	37	17

(continued)

Table 1
(continued)

Q1	Q3	RT	Tryptic peptide sequence	DP	EP	CE	CXP
755.867	636.372	35.9	hsZIPK.ESLTEDEATQFLK.+2y5	86	10	37	17
755.867	217.082	35.9	hsZIPK.ESLTEDEATQFLK.+2b2	86	10	37	17
705.380	1055.527	32.4	hsZIPK.QILDGVHYLHSK.+2y9	83	10	34	17
705.380	940.500	32.4	hsZIPK.QILDGVHYLHSK.+2y8	83	10	34	17
705.380	883.478	32.4	hsZIPK.QILDGVHYLHSK.+2y7	83	10	34	17
705.380	784.410	32.4	hsZIPK.QILDGVHYLHSK.+2y6	83	10	34	17
705.380	242.150	32.4	hsZIPK.QILDGVHYLHSK.+2b2	83	10	34	17
783.431	1096.601	32.3	hsZIPK.QILDGVHYLHSKR.+2y9	88	10	38	17
783.431	1039.580	32.3	hsZIPK.QILDGVHYLHSKR.+2y8	88	10	38	17
783.431	940.511	32.3	hsZIPK.QILDGVHYLHSKR.+2y7	88	10	38	17
783.431	803.452	32.3	hsZIPK.QILDGVHYLHSKR.+2y6	88	10	38	17
957.014	1329.745	41.3	hsZIPK.IAHFDLKPENIM[Oxi]LLDK.+2y11	101	10	47	17
957.014	1216.661	41.3	hsZIPK.IAHFDLKPENIM[Oxi]LLDK.+2y10	101	10	47	17
957.014	1088.566	41.3	hsZIPK.IAHFDLKPENIM[Oxi]LLDK.+2y9	101	10	47	17
957.014	991.513	41.3	hsZIPK.IAHFDLKPENIM[Oxi]LLDK.+2y8	101	10	47	17
957.014	185.128	41.3	hsZIPK.IAHFDLKPENIM[Oxi]LLDK.+2b2	101	10	47	17
348.693	582.336	18.7	hsZIPK.NVPNPR.+2y5	57	10	16	17
348.693	483.267	18.7	hsZIPK.NVPNPR.+2y4	57	10	16	17
348.693	386.215	18.7	hsZIPK.NVPNPR.+2y3	57	10	16	17
348.693	214.119	18.7	hsZIPK.NVPNPR.+2b2	57	10	16	17
348.693	425.214	18.7	hsZIPK.NVPNPR.+2b4	57	10	16	17
348.693	522.267	18.7	hsZIPK.NVPNPR.+2b5	57	10	16	17
507.293	900.494	33.5	hsZIPK.LIDFGIAHK.+2y8	68	10	24	17
507.293	787.410	33.5	hsZIPK.LIDFGIAHK.+2y7	68	10	24	17
507.293	672.383	33.5	hsZIPK.LIDFGIAHK.+2y6	68	10	24	17
507.293	525.314	33.5	hsZIPK.LIDFGIAHK.+2y5	68	10	24	17
507.293	227.175	33.5	hsZIPK.LIDFGIAHK.+2b2	68	10	24	17
951.510	1356.727	41.8	hsZIPK.LIDFGIAHKIEAGNEFK.+2y12	101	10	46	17
951.510	1243.643	41.8	hsZIPK.LIDFGIAHKIEAGNEFK.+2y11	101	10	46	17
951.510	1172.606	41.8	hsZIPK.LIDFGIAHKIEAGNEFK.+2y10	101	10	46	17
951.510	1035.547	41.8	hsZIPK.LIDFGIAHKIEAGNEFK.+2y9	101	10	46	17

(continued)

Table 1
(continued)

Q1	Q3	RT	Tryptic peptide sequence	DP	EP	CE	CXP
454.230	794.368	24.8	hsZIPK.IEAGNEFK.+2y7	64	10	22	17
454.230	665.325	24.8	hsZIPK.IEAGNEFK.+2y6	64	10	22	17
454.230	594.288	24.8	hsZIPK.IEAGNEFK.+2y5	64	10	22	17
454.230	537.267	24.8	hsZIPK.IEAGNEFK.+2y4	64	10	22	17
454.230	614.278	24.8	hsZIPK.IEAGNEFK.+2b6	64	10	22	17
406.766	699.440	25.5	hsZIPK.LLVKDPK.+2y6	61	10	19	17
406.766	586.356	25.5	hsZIPK.LLVKDPK.+2y5	61	10	19	17
406.766	487.287	25.5	hsZIPK.LLVKDPK.+2y4	61	10	19	17
406.766	359.193	25.5	hsZIPK.LLVKDPK.+2y3	61	10	19	17
406.766	227.175	25.5	hsZIPK.LLVKDPK.+2b2	61	10	19	17
780.398	1127.584	40.1	hsZIPK.M[Oxi]TIAQSLEHSWIK.+2y9	88	10	38	17
780.398	999.526	40.1	hsZIPK.M[Oxi]TIAQSLEHSWIK.+2y8	88	10	38	17
780.398	912.494	40.1	hsZIPK.M[Oxi]TIAQSLEHSWIK.+2y7	88	10	38	17
780.398	799.410	40.1	hsZIPK.M[Oxi]TIAQSLEHSWIK.+2y6	88	10	38	17
780.398	249.090	40.1	hsZIPK.M[Oxi]TIAQSLEHSWIK.+2b2	88	10	38	17
495.242	875.433	17.8	hsZIPK.NVRGEDSGR.+2y8	67	10	24	17
495.242	776.365	17.8	hsZIPK.NVRGEDSGR.+2y7	67	10	24	17
495.242	620.263	17.8	hsZIPK.NVRGEDSGR.+2y6	67	10	24	17
495.242	563.242	17.8	hsZIPK.NVRGEDSGR.+2y5	67	10	24	17
310.635	563.242	15.6	hsZIPK.GEDSGR.+2y5	54	10	14	17
310.635	434.199	15.6	hsZIPK.GEDSGR.+2y4	54	10	14	17
310.635	319.172	15.6	hsZIPK.GEDSGR.+2y3	54	10	14	17
310.635	232.140	15.6	hsZIPK.GEDSGR.+2y2	54	10	14	17
310.635	389.130	15.6	hsZIPK.GEDSGR.+2b4	54	10	14	17
310.635	446.152	15.6	hsZIPK.GEDSGR.+2b5	54	10	14	17
565.781	944.491	14.2	hsZIPK.GEDSGRKPER.+2y8	72	10	27	17
565.781	829.464	14.2	hsZIPK.GEDSGRKPER.+2y7	72	10	27	17
565.781	742.432	14.2	hsZIPK.GEDSGRKPER.+2y6	72	10	27	17
565.781	685.410	14.2	hsZIPK.GEDSGRKPER.+2y5	72	10	27	17
565.781	401.214	14.2	hsZIPK.GEDSGRKPER.+2y3	72	10	27	17
447.768	781.445	25.2	hsZIPK.LKEYTIK.+2y6	64	10	21	17

(continued)

Table 1
(continued)

Q1	Q3	RT	Tryptic peptide sequence	DP	EP	CE	CXP
447.768	653.350	25.2	hsZIPK.LKEYTIK.+2y5	64	10	21	17
447.768	524.308	25.2	hsZIPK.LKEYTIK.+2y4	64	10	21	17
447.768	361.245	25.2	hsZIPK.LKEYTIK.+2y3	64	10	21	17
910.914	1309.581	32.6	hsZIPK.SHSSLPPNNSYADFER.+2y11	98	10	44	17
910.914	1212.528	32.6	hsZIPK.SHSSLPPNNSYADFER.+2y10	98	10	44	17
910.914	1115.475	32.6	hsZIPK.SHSSLPPNNSYADFER.+2y9	98	10	44	17
910.914	1001.432	32.6	hsZIPK.SHSSLPPNNSYADFER.+2y8	98	10	44	17
910.914	887.389	32.6	hsZIPK.SHSSLPPNNSYADFER.+2y7	98	10	44	17
910.914	225.098	32.6	hsZIPK.SHSSLPPNNSYADFER.+2b2	98	10	44	17
679.352	816.421	32.6	hsZIPK.VLEEAAAAEEGLR.+2y8	81	10	33	17
679.352	603.310	32.6	hsZIPK.VLEEAAAAEEGLR.+2y5	81	10	33	17
679.352	474.267	32.6	hsZIPK.VLEEAAAAEEGLR.+2y4	81	10	33	17
679.352	345.224	32.6	hsZIPK.VLEEAAAAEEGLR.+2y3	81	10	33	17
679.352	613.319	32.6	hsZIPK.VLEEAAAAEEGLR.+2b6	81	10	33	17
679.352	755.393	32.6	hsZIPK.VLEEAAAAEEGLR.+2b8	81	10	33	17
394.722	659.395	18.0	hsZIPK.ELQRSR.+2y5	60	10	19	17
394.722	546.311	18.0	hsZIPK.ELQRSR.+2y4	60	10	19	17
394.722	418.252	18.0	hsZIPK.ELQRSR.+2y3	60	10	19	17
394.722	262.151	18.0	hsZIPK.ELQRSR.+2y2	60	10	19	17
394.722	243.134	18.0	hsZIPK.ELQRSR.+2b2	60	10	19	17
1023.502	1479.721	38.8	hsZIPK.RLC[CAM]HEDVEALAAIYEEK.+2y13	106	10	50	17
1023.502	1350.679	38.8	hsZIPK.RLC[CAM]HEDVEALAAIYEEK.+2y12	106	10	50	17
1023.502	1235.652	38.8	hsZIPK.RLC[CAM]HEDVEALAAIYEEK.+2y11	106	10	50	17
1023.502	1136.583	38.8	hsZIPK.RLC[CAM]HEDVEALAAIYEEK.+2y10	106	10	50	17
945.451	1350.679	38.9	hsZIPK.LC[CAM]HEDVEALAAIYEEK.+2y12	100	10	46	17
945.451	1235.652	38.9	hsZIPK.LC[CAM]HEDVEALAAIYEEK.+2y11	100	10	46	17
945.451	1136.583	38.9	hsZIPK.LC[CAM]HEDVEALAAIYEEK.+2y10	100	10	46	17
945.451	1007.541	38.9	hsZIPK.LC[CAM]HEDVEALAAIYEEK.+2y9	100	10	46	17
945.451	274.122	38.9	hsZIPK.LC[CAM]HEDVEALAAIYEEK.+2b2	100	10	46	17
624.789	788.426	27.4	hsZIPK.EESDSLGQDLR.+2y7	77	10	30	17
624.789	701.394	27.4	hsZIPK.EESDSLGQDLR.+2y6	77	10	30	17

(continued)

Table 1
(continued)

Q1	Q3	RT	Tryptic peptide sequence	DP	EP	CE	CXP
624.789	588.310	27.4	hsZIPK.EESDSLGQDLR.+2y5	77	10	30	17
624.789	403.230	27.4	hsZIPK.EESDSLGQDLR.+2y3	77	10	30	17
624.789	288.203	27.4	hsZIPK.EESDSLGQDLR.+2y2	77	10	30	17
450.287	786.483	28.5	hsZIPK.LRQELLK.+2y6	64	10	21	17
450.287	630.382	28.5	hsZIPK.LRQELLK.+2y5	64	10	21	17
450.287	502.324	28.5	hsZIPK.LRQELLK.+2y4	64	10	21	17
450.287	373.281	28.5	hsZIPK.LRQELLK.+2y3	64	10	21	17
586.848	1044.630	31.2	hsZIPK.QELLKTEALK.+2y9	74	10	28	17
586.848	915.587	31.2	hsZIPK.QELLKTEALK.+2y8	74	10	28	17
586.848	802.503	31.2	hsZIPK.QELLKTEALK.+2y7	74	10	28	17
586.848	689.419	31.2	hsZIPK.QELLKTEALK.+2y6	74	10	28	17
480.249	803.389	16.4	hsZIPK.RQAQEEAK.+2y7	66	10	23	17
480.249	675.331	16.4	hsZIPK.RQAQEEAK.+2y6	66	10	23	17
480.249	604.294	16.4	hsZIPK.RQAQEEAK.+2y5	66	10	23	17
480.249	476.235	16.4	hsZIPK.RQAQEEAK.+2y4	66	10	23	17
402.198	675.331	15.7	hsZIPK.QAQEEAK.+2y6	60	10	19	17
402.198	604.294	15.7	hsZIPK.QAQEEAK.+2y5	60	10	19	17
402.198	476.235	15.7	hsZIPK.QAQEEAK.+2y4	60	10	19	17
402.198	347.193	15.7	hsZIPK.QAQEEAK.+2y3	60	10	19	17
402.198	200.103	15.7	hsZIPK.QAQEEAK.+2b2	60	10	19	17
402.198	457.204	15.7	hsZIPK.QAQEEAK.+2b4	60	10	19	17
458.777	675.404	31.9	hsZIPK.GALLGTSGLK.+2y7	65	10	22	17
458.777	562.320	31.9	hsZIPK.GALLGTSGLK.+2y6	65	10	22	17
458.777	505.298	31.9	hsZIPK.GALLGTSGLK.+2y5	65	10	22	17
458.777	404.250	31.9	hsZIPK.GALLGTSGLK.+2y4	65	10	22	17
458.777	242.150	31.9	hsZIPK.GALLGTSGLK.+2b3	65	10	22	17
461.249	774.422	24.3	hsZIPK.FSRLENR.+2y6	65	10	22	17
461.249	687.390	24.3	hsZIPK.FSRLENR.+2y5	65	10	22	17
461.249	531.289	24.3	hsZIPK.FSRLENR.+2y4	65	10	22	17
461.249	418.204	24.3	hsZIPK.FSRLENR.+2y3	65	10	22	17
347.692	531.314	22.4	hsZIPK.YEALAK.+2y5	57	10	16	17

(continued)

Table 1
(continued)

Q1	Q3	RT	Tryptic peptide sequence	DP	EP	CE	CXP
347.692	402.271	22.4	hsZIPK.YEALAK.+2y4	57	10	16	17
347.692	331.234	22.4	hsZIPK.YEALAK.+2y3	57	10	16	17
347.692	293.113	22.4	hsZIPK.YEALAK.+2b2	57	10	16	17
347.692	364.150	22.4	hsZIPK.YEALAK.+2b3	57	10	16	17
347.692	477.234	22.4	hsZIPK.YEALAK.+2b4	57	10	16	17
756.380	1148.609	31.1	hsZIPK.YEALAKQVASEM[Oxi]R.+2y10	86	10	37	17
756.380	1035.525	31.1	hsZIPK.YEALAKQVASEM[Oxi]R.+2y9	86	10	37	17
756.380	964.488	31.1	hsZIPK.YEALAKQVASEM[Oxi]R.+2y8	86	10	37	17
756.380	836.393	31.1	hsZIPK.YEALAKQVASEM[Oxi]R.+2y7	86	10	37	17
418.700	708.335	20.7	hsZIPK.QVASEM[Oxi]R.+2y6	62	10	20	17
418.700	609.266	20.7	hsZIPK.QVASEM[Oxi]R.+2y5	62	10	20	17
418.700	538.229	20.7	hsZIPK.QVASEM[Oxi]R.+2y4	62	10	20	17
418.700	451.197	20.7	hsZIPK.QVASEM[Oxi]R.+2y3	62	10	20	17
418.700	228.134	20.7	hsZIPK.QVASEM[Oxi]R.+2b2	62	10	20	17
438.751	630.357	32.0	hsZIPK.FVQDLVR.+2y5	63	10	21	17
438.751	502.298	32.0	hsZIPK.FVQDLVR.+2y4	63	10	21	17
438.751	387.271	32.0	hsZIPK.FVQDLVR.+2y3	63	10	21	17
438.751	274.187	32.0	hsZIPK.FVQDLVR.+2y2	63	10	21	17
438.751	247.144	32.0	hsZIPK.FVQDLVR.+2b2	63	10	21	17
359.193	646.341	17.8	hsZIPK.ALEQEK.+2y5	57	10	17	17
359.193	533.257	17.8	hsZIPK.ALEQEK.+2y4	57	10	17	17
359.193	404.214	17.8	hsZIPK.ALEQEK.+2y3	57	10	17	17
359.193	276.155	17.8	hsZIPK.ALEQEK.+2y2	57	10	17	17
359.193	185.128	17.8	hsZIPK.ALEQEK.+2b2	57	10	17	17
359.193	442.230	17.8	hsZIPK.ALEQEK.+2b4	57	10	17	17
516.269	918.446	27.8	hsZIPK.LQGVEC[CAM]GLR.+2y8	69	10	25	17
516.269	790.388	27.8	hsZIPK.LQGVEC[CAM]GLR.+2y7	69	10	25	17
516.269	733.366	27.8	hsZIPK.LQGVEC[CAM]GLR.+2y6	69	10	25	17
516.269	634.298	27.8	hsZIPK.LQGVEC[CAM]GLR.+2y5	69	10	25	17
516.269	242.150	27.8	hsZIPK.LQGVEC[CAM]GLR.+2b2	69	10	25	17
687.865	1058.584	26.2	G3P_PICPA.TASGNIIPSSTGAAK.+2y11.	81	10	33	17

(continued)

Table 1 (continued)

Q1	Q3	RT	Tryptic peptide sequence	DP	EP	CE	CXP
687.865	944.541	26.2	G3P_PICPA.TASGNIIPSSTGAAK.+2y10.	81	10	33	17
687.865	831.457	26.2	G3P_PICPA.TASGNIIPSSTGAAK.+2y9	81	10	33	17
687.865	718.373	26.2	G3P_PICPA.TASGNIIPSSTGAAK.+2y8	81	10	33	17
563.343	941.557	39.0	b-TUb-PICPA.LAVNLVPFPR.+2y8	72	10	27	17
563.343	842.488	39.0	b-TUb-PICPA.LAVNLVPFPR.+2y7	72	10	27	17
563.343	728.445	39.0	b-TUb-PICPA.LAVNLVPFPR.+2y6	72	10	27	17
563.343	615.361	39.0	b-TUb-PICPA.LAVNLVPFPR.+2y5	72	10	27	17
563.343	516.293	39.0	b-TUb-PICPA.LAVNLVPFPR.+2y4	72	10	27	17

Peptides and their respective ions as entered into Analyst. The table is generated by Skyline and edited in Microsoft Excel. Abbreviations are *Q1* precursor mass (mass/charge), *Q3* fragment mass (mass/charge), *RT* retention time in min, *DP* declustering potential, *EP* entrance potential, *CE* collision energy, and *CXP* cell exit potential. The tryptic peptide sequences are provided along with associated transition (y-ion or b-ion) and charge state (+2). M(Ox) indicates an oxidized methionine residue. C(CAM) indicates a carbamidomethyl cysteine residue

2. Select an MRM detection window of 5 min (*see* **Note 24**).
3. Create a "Dionex Chromatography MS Link" HPLC program (*see* **Note 25**).
4. The HPLC is set up with a trap column under reverse elution conditions. This ensures that the injected peptides are sequestered onto the trap column, washed, and then reverse-eluted onto the separation column during the ensuing gradient.
5. Set an HPLC program with a flow rate of 50 μL/min and a gradient from 2 to 50% (v/v) ACN with 0.1% (v/v) formic acid and eluting peptides analyzed in-line by the MS.
6. Use 15 μL injection volume in the partial inject mode.
7. Once an Analyst and HPLC program is created, load the samples into the autosampler, and create a batch to submit the samples. Ensure that there are no air bubbles in the sample tubes. As a first sample, inject 5% ACN, 0.5% TFA to equilibrate the system, using the same HPLC method (*see* **Note 26**).

3.6 Data Analysis

1. Copy the Analyst-created data files (.wiff and .wiff.scan) to an analysis computer that has Skyline installed.
2. Open the original Skyline file containing all sequences and peptides and import the results.
3. Start analyzing the yeast loading control peptides from tubulin and G3P. Use the Skyline function "Retention Time," "Comparison between replicates" to check for detected peptides. The

retention time should be very similar (within ~20 s) across all runs (*see* **Note 27**).

4. Use the function "Peak Area Comparison" to quickly compare peak areas across samples. As shown in Fig. 1, the peak area of the yeast control peptides should increase with growth time and parallel yeast OD readings.
5. Analyze the ZIPK expression data by checking for green balls beside the ZIPK peptides in Skyline. This indicates that the peptide was found in the highlighted sample. The peak is

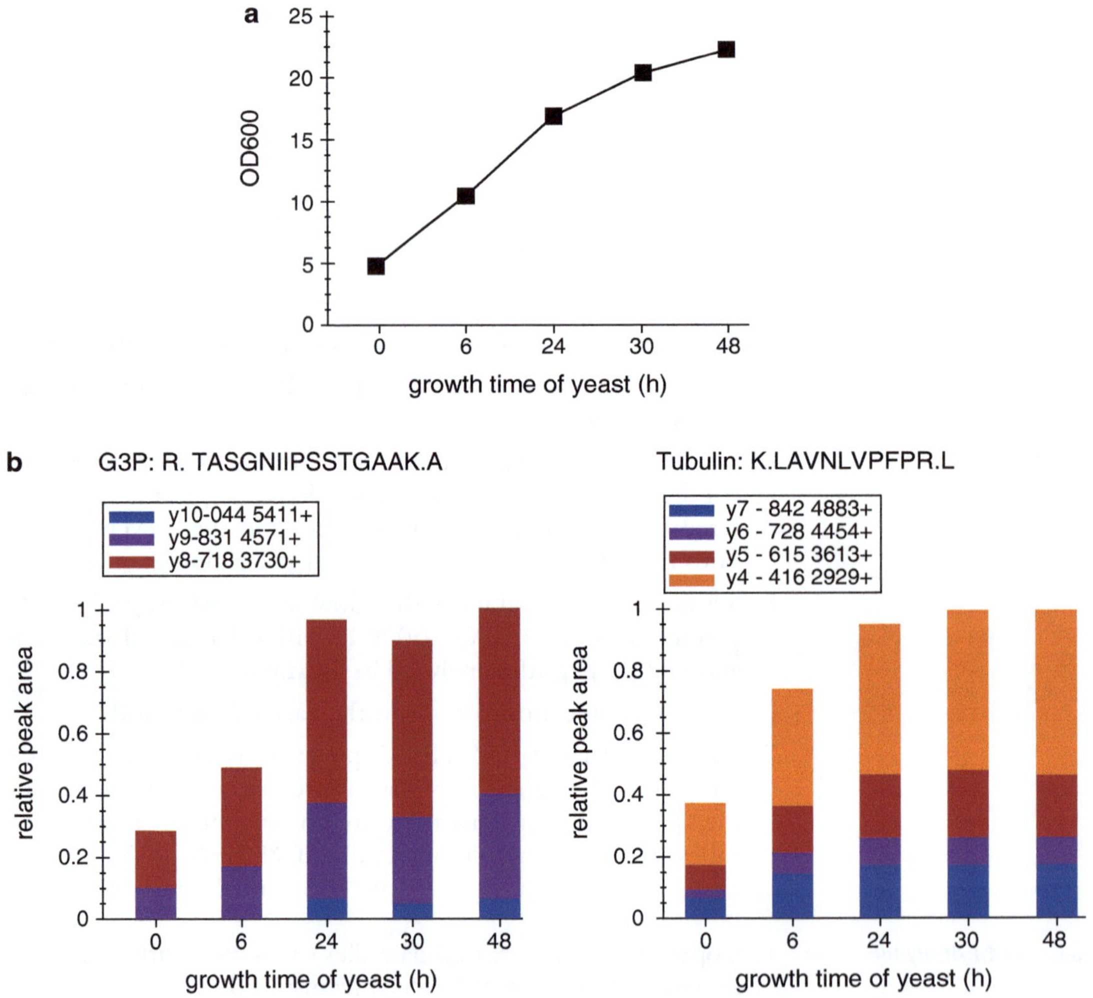

Fig. 1 Tracking typical yeast growth with representative MRM-MS analyses. (**a**) Growth curves of a recombinant ZIPK-producing *P. pastoris* X-33 strain based on visible (OD_{600}) absorptions. (**b**) MRM-MS signals from G3P (*left*) and tubulin (*right*). These peptides from yeast increase with growth time following induction of protein production. The peptides were detected in whole yeast cell extracts sampled at the indicated time. Each color represents a different transition ion of the respective peptide as indicated on the *top*. The peak areas of each transition are summed to represent the relative amount of G3P or tubulin in the sample

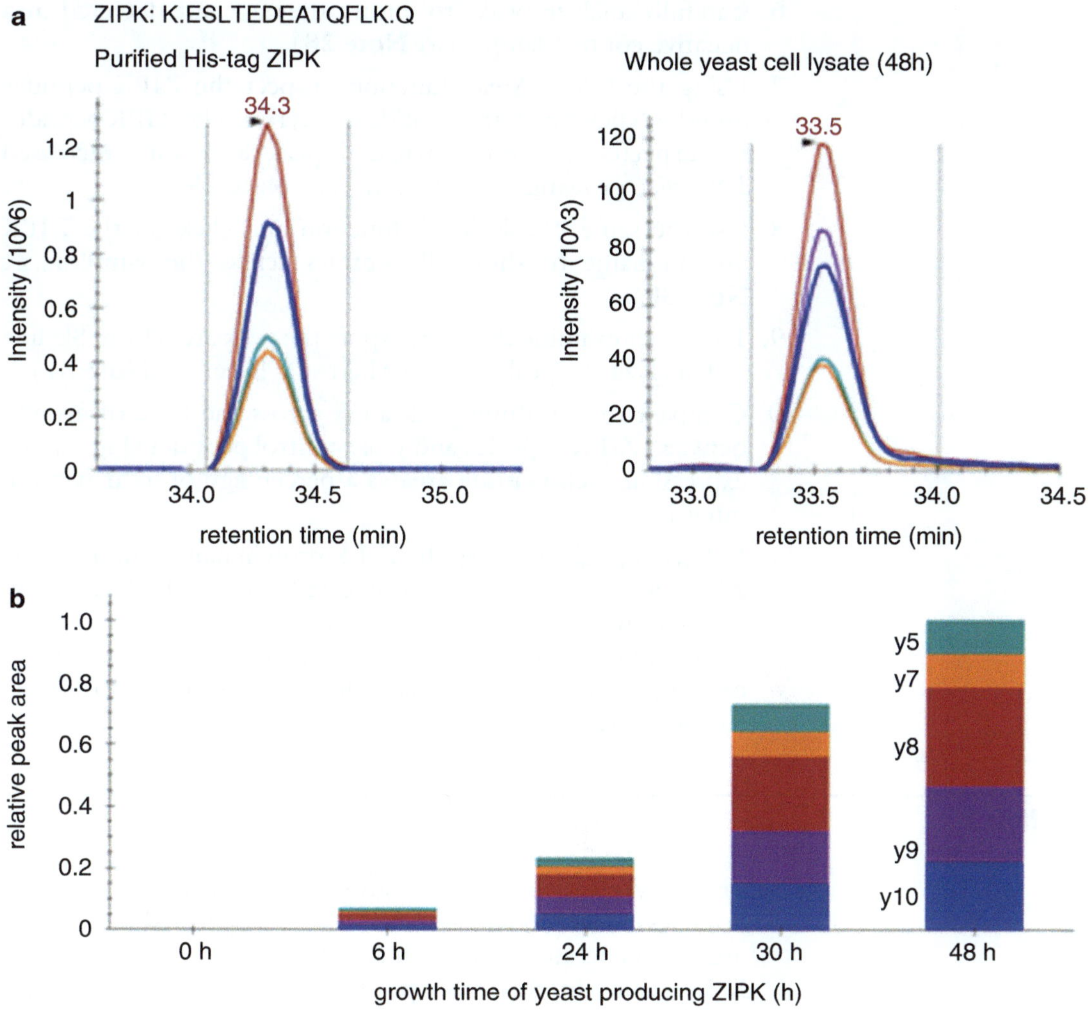

Fig. 2 Profiling recombinant ZIPK production in *P. pastoris*. (**a**) The ZIPK peptide ESLTEDEATQFLK (amino acids 108–120, charge $z = 2$, $m/z = 755.87$) and its y-type transitions were examined. *Left*, signal from purified His-tagged ZIPK, and *right*, signal from ZIPK detected in whole yeast cell extract after 48 h growth. The different transitions are represented in different colors. The peak borders are marked by dashed lines and an arrow head. Note that the retention times differ slightly since the two samples were analyzed at different times (*see* **Note 27**). (**b**) Increase of ZIPK peptide ESLTEDEATQFLK signal intensity with growth time following induction. Shown is the sum of the peak areas from the different y-ion transitions obtained from the parent peptide

bordered by dashed lines and marked with an arrow in the viewer window but needs to be visually inspected. Peaks are only acceptable if more than three transitions are detected at identical retention time across samples. The retention time as well as relative intensity of the different transitions needs to agree with each other and with the positive control. *See* Fig. 2a as an example.

6. Carefully analyze peaks to ensure they are not detected in a negative control sample (*see* **Note 28**).
7. Using the "Peak Area" function, inspect the ZIPK peptides found by clicking on the peptide of interest. The ZIPK peptides are expected to show increased peak area with increased *P. pastoris* growing time (Fig. 2b) (*see* **Note 29**).
8. Use the same "Peak Area" function and click on the ZIPK protein name to show all peptides across the samples (*see* **Note 30**).
9. To further examine the data, export the peak areas from Skyline and analyze the peak areas in Microsoft Excel (*see* **Note 31**).
10. Compare the resulting peak areas across the time of growth between ZIPK peptides and yeast control peptides (Fig. 3a) by expressing each peptide area as a percentage of its maximum intensity.
11. In Skyline, right-clicking the ZIPK protein name will show the ZIPK protein sequence with detected peptides highlighted in blue and not detected peptides (despite being selected) highlighted in black (Fig. 3b). Use this function to verify sequence coverage and to indicate that full-length protein is produced (*see* **Note 32**).

4 Notes

1. The EasySelect *Pichia* Expression Kit (Invitrogen-Life Technologies) as well as all additional reagents specified therein was used to generate a recombinant ZIPK-expressing *P. pastoris* X-33 strain. In this case, positive X-33 transformants were selected by screening cell pellets for ZIPK protein expression by Coomassie-stained SDS-PAGE and western blot. However, it is possible to apply the outlined MRM-MS protocol to screen for high-expressing *P. pastoris* recombinant clones in the absence of any additional functional assays or antibody detection reagents.
2. The dissolved ABC should have a pH of approximately 8. The pH does not need to be further adjusted.
3. Purified protein is not necessary, and the entire step can be omitted. However, using purified protein as a positive control does generally increase the quality of the dataset. Be aware that protein produced in mammalian cells may be subject to different posttranslational modifications which in turn might influence the peptide sizes and likelihood of detection by MRM-MS.

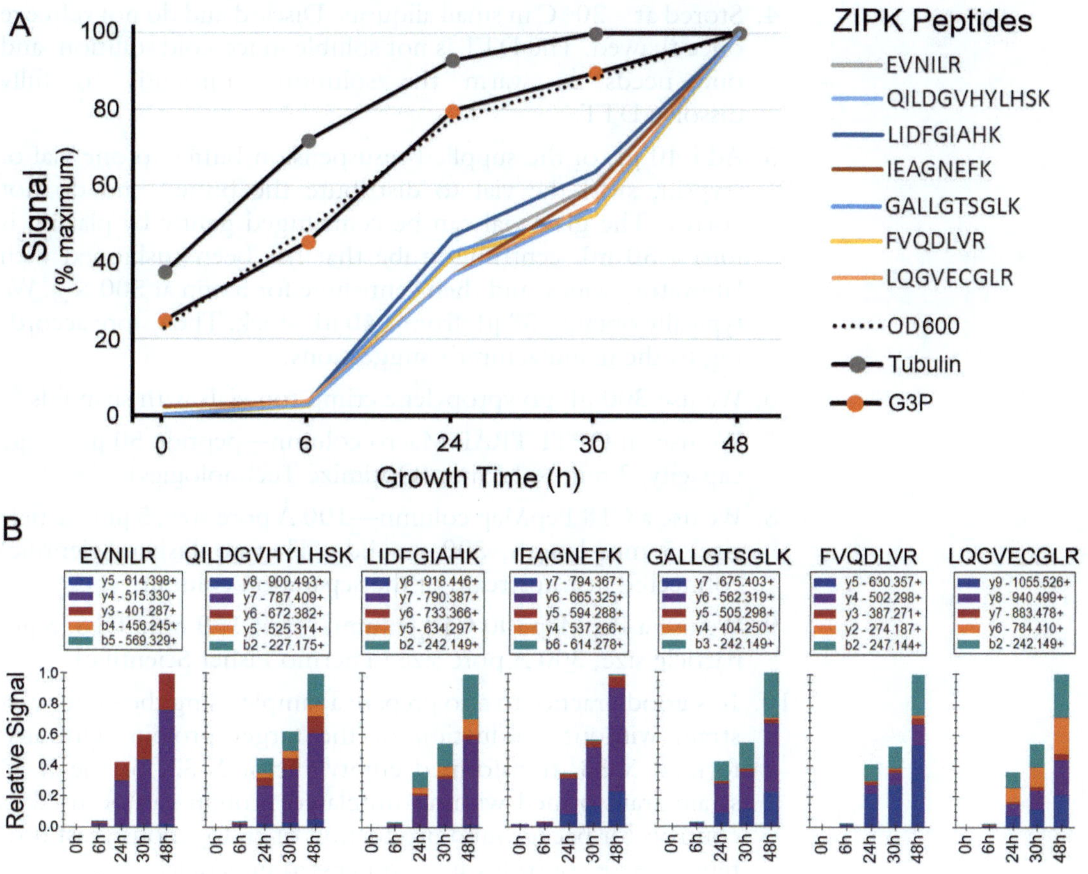

C

ZIPK Sequence

MSTFRQEDVEDHYEMGEELGSGQFAIVRKCRQKGTGKEYAAKFIKKRRLSSSRRGVSREEIEREVNILREIR
HPNIITLHDIFENK**TDVVLILELVSGGELFDFLAEK**ESLTEDEATQFLKQILDGVHYLHSKR**IAHFDLKPEN**
IMLLDKNVPNPRIKLIDFGIAHKIEAGNEFKNIFGTPEFVAPEIVNYEPLGLEADMWSIGVITYILLSGASP
FLGETKQETLTNISAVNYDFDEEYFSNTSELAKDFIRRLLVKDPKRR**MTIAQSLEHSWIK**AIRRRNVR**GEDS**
GRKPERRRLKTTRLKEYTIK**SHSSLPPNNSYADFER**FSK**VLEEAAAAEEGLR**ELQRSRRLCHEDVEALAAIY
EEKEAWYREESDSLGQDLRRLRQELLKTEALKR**QAQEEAK**GALLGTSGLKRRFSRLENRYEALAKQVASEMR
FVQDLVR**ALEQEK**LQGVECGLR

Fig. 3 Recombinant ZIPK protein production in yeast over 48 h growth. (**a**) The peak areas of the strongest y-ion transitions of seven peptides of ZIPK (colored lines) are shown. The intensities of the strongest y-ion transition from G6P and tubulin signature peptides are in black, and the dashed black line shows the optical density (OD_{600}) obtained for the yeast culture. All values are expressed as percentage of their respective maximum. (**b**) Shown are the sum of the peak areas for the different transitions obtained from seven representative parent peptides spanning the ZIPK sequence. (**c**) The human ZIPK protein sequence is provided. Peptides detected by the MRM-MS protocol are colored in blue. The seven representative peptides displayed graphically in panels (**a**) and (**b**) are underlined. Peptides that were included in the MRM-MS protocol but were not detected are in bold

4. Stored at −20 °C in small aliquots. Discard and do not refreeze once thawed. The DTT is not soluble in ice-cold solution, and one needs to warm the solution sufficiently to fully dissolve DTT.
5. Add 40 μL of the supplied resuspension buffer to one vial of trypsin, swirl the vial to distribute the buffer, and do not vortex. The glass vial can be centrifuged gently by placing it into a 50 mL centrifuge tube that has been cushioned with laboratory wipes, and then centrifuge for 5 min at 500 × *g*. We typically obtain ~38 μL from a 40 μL stock. Then store according to the manufacturer's suggestions.
6. We use 300 μL polypropylene crimp top vials with snap lids.
7. We use an OPTI-TRAP Macro column—peptide 50 μL, large capacity, 3 mm × 12 mm (Optimize Technologies).
8. We use a C18 PepMap column—100 Å pore size, 5 μm particle size, 5 mm length, 300 μm i.d. (Thermo Fisher Scientific) connected just upstream of the separation column.
9. We use a PepMap300 C18 column—1 × 150 mm, C18, 5 μm particle size, 300 Å pore size (Thermo Fisher Scientific).
10. It is good practice to also prepare a sample using the same yeast strain without production of the target protein. Untransformed X-33, transformed empty vector X-33, or the X-33 strain transformed with an unrelated protein can be used as controls for background production of endogenous yeast proteins. These controls can be processed in parallel.
11. Sufficiently vigorous shaking is important to oxygenate the yeast cells.
12. The OD_{600} is a good measure of yeast cell growth. However, a significant number of dead cells might be present at very late growth times. Thus, the number of "productive" yeast cells can be overestimated. Dilute the culture as required to ensure the OD_{600} readings are maintained within the linear range of the Beer-Lambert law.
13. We recommend the Skyline tutorial "Targeted Method Editing."
14. Oxidation occurs on methionines, altering its mass. Carbaminomethyl cysteine is created during alkylation/reduction of the proteins. If alkylation steps were not performed, the "Carbamidomethyl Cysteine" can be omitted.
15. A library (e.g., NIST) can be used if the protein and the organism are common. Refer to the Skyline user tutorial on how to employ library data effectively. Skyline also supplies pre-programmed collision energies and declustering potentials for the tandem MS instrument being used. We do not use a proteome background for proteins expressed in *P. pastoris.*

However, it can be applied by the user to assist with the removal of redundant peptides that originate from more than one protein.

16. The range of *m/z* depends on the particular tandem MS instrument employed and its calibrated mass range. The nomenclature in Skyline equates "transition" with "product ion."
17. Control peptides from other housekeeping proteins or yeast strains can be derived using a similar method to the one used for G3P and tubulin.
18. Skyline uses retention time predictor regression (i.e., SSRCarc 100A or others) using previous HPLC run data. For this, the retention times of detected peaks are related to SSRCalc calculated values, which are based on peptide length and amino acid composition. Skyline then creates a linear regression which can be used for new runs with unknown retention times. Skyline provides a good tutorial on this matter.
19. Various protein sources can be used as positive controls: (a) (partially) purified protein from the same species as the target protein, or (b) purchased protein, or (c) protein overexpressed in another system such as *E. coli*, mammalian cell lines such as HEK293, insect cells, or cell-free translation systems. The digestion process should be adapted for these protein sources. If no control protein is available, the entire positive control can be omitted. However, since false positives are possible, we recommend that the results are analyzed very critically by increasing the number of transitions.
20. The reduction and alkylation steps can be omitted if the target protein does not possess disulfide bridges.
21. If this trypsin concentration does not give satisfactory results, then increase the trypsin/protein ratio in the digestion reaction. The optimal trypsin/protein ratio depends on several factors, including the biophysical and structural nature of the protein target. Some proteins digest very easily, some not. We find that ratios between 1:50 and 1:100 routinely work.
22. We routinely omit prior purification of peptides using C18 ZIP-TIP style methods. The HPLC-loading scheme using a trap column achieves the same peptide purification and removal of unwanted contaminants without additional sample processing.
23. Ensure that "MRM Method" and "Scheduling" is selected in Analyst and that the methods table is the same format as the transition list. Refer to the "Analyst Getting Started Guide."
24. The retention time predictor in Skyline estimates retention times with a certain error due to the inherent inaccuracy of the method; thus, we initially select a wide retention time window (~5 min) to ensure all peptides are detected. Because

SSRCalc only provides a rough estimate of retention, a wider time window needs to be used initially. The retention time predictor in Skyline indicates how wide the window should be in order to catch all peptides (at least from the previous run training set). The detection window can then be narrowed for future analyses, if desired. Having wider retention windows is not an issue unless very large numbers (>200) of transitions are programmed for analysis.

25. Refer to the "DCMS Quick Start Guide" and ensure the gradient is identical to the retention time training run.
26. If a positive control is used (i.e., purified protein), this sample should be loaded last to avoid possible bleed-through into subsequent runs.
27. Some retention time drift across many sample injections is possible and is often a sign of issues with the HPLC (e.g., variability in pressure and/or flow). Retention time drift will also occur with replacement of HPLC plumbing and/or columns and/or running buffers. In this case, all peptides in a sample shift. An altered retention time for a single peptide often signifies interferences (false positives). In this case, the peaks need to be critically investigated.
28. It is reasonable to use empty vector transfection controls. Alternatively, samples originating from transfections with some unrelated protein provide a useful negative control sample. Not all peptides are expected to be detected, since the selected MRM peptides and their transitions are only based on predictions rather than previous observations. Furthermore, unknown posttranslational modifications may prevent the detection of the characteristic signatures by MRM-MS. The identified peptides are more reliable if they have been previously detected and reported in an annotated proteomic library. It is possible to install peptide libraries for the human plasma proteome and yeast from the National Institute of Standards and Technology (NIST, http://chemdata.nist.gov), the Global Proteome Machine (GPM, http://gpmdb.thegpm.org), or the Institute for Systems Biology (ISB Peptide Atlas, http://www.peptideatlas.org) with an ion match tolerance of 0.5 *m/z*. Refer to the Skyline tutorials for more specific information regarding these options.
29. Individual peptides that do not follow the increase of the remaining ZIPK peptides are likely interferences and should not be used.
30. It is an inherent property of mass spectrometry that the response of one peptide is different from the next peptide. Thus, peptide peak areas can only be compared for the same peptide across the different samples.

31. This is easiest done by editing the data in Skyline prior to export. For this, save a copy of the Skyline file, then remove all ZIPK peptides that were not detected. Further, remove all but the highest transition for each detected peptide after double-checking the peak borders. Export a "report" with the following information: Peptide Sequence, Protein Name, Replicate Name, Precursor Mz, Product Mz, Area, Begin Pos, and End Pos (amino acid location of peptide).
32. The identified peptides could originate from more than one polypeptide species present in the sample (i.e., full-length protein degraded into two or more fragments where each is detected individually). In light of this, it is possible to use stoichiometry ratios of all the peptides identified in the MRM-MS run to imply whether the protein produced is full-length or truncated.

Acknowledgments

This work was supported by grants from the Canadian Institutes of Health Research (CIHR, MOP#133543 to JAM), the European Commission Research Executive Agency (Marie Sklodowska-Curie International Incoming Fellowship to JAM and RMB), and the Biotechnology and Biological Sciences Research Council (International Partnering Award BB/P025927/1 to JAM and RMB).

References

1. Gillette MA, Carr SA (2013) Quantitative analysis of peptides and proteins in biomedicine by targeted mass spectrometry. Nat Methods 10(1):28–34. https://doi.org/10.1038/nmeth.2309
2. Ebhardt HA, Root A, Sander C, Aebersold R (2015) Applications of targeted proteomics in systems biology and translational medicine. Proteomics 15(18):3193–3208. https://doi.org/10.1002/pmic.201500004
3. Chen YT, Chen HW, Wu CF et al (2017) Development of a multiplexed liquid chromatrography multiple-reaction monitoring mass spectrometry (LC-MRM/MS) method for evaluaiton of salivary proteins as oral cancer biomarkers. Mol Cell Proteomics 16(5):799–811. https://doi.org/10.1074/mcp.M116.064758
4. Bill RM (ed) (2012) Recombinant protein production in yeast—methods and protocols. In: Methods in molecular biology. New York: Human Press
5. MacLean B, Tomazela DM, Shulman N et al (2010) Skyline: an open source document editor for creating and analyzing targeted proteomics experiments. Bioinformatics 26(7):966–968. https://doi.org/10.1093/bioinformatics/btq054
6. Pino LK, Searle BC, Bollinger JG et al (2017) The Skyline ecosystem: informatics for quantitative mass spectrometry proteomics. Mass Spectrom Rev. https://doi.org/10.1002/mas.21540
7. Carlson DA, Franke AS, Weitzel DH et al (2013) Fluorescence linked enzyme chemoproteomic strategy for discovery of a potent and selective DAPK1 and ZIPK inhibitor. ACS Chem Biol 8(12):2715–2723. https://doi.org/10.1021/cb400407c
8. Jung S, Danziger SA, Panchaud A, von Haller P, Aitchison JD, Goodlett DR (2015) Systematic analysis of yeast proteome reveals peptide detectability factors for mass spectrometry. J Proteomics Bioinform 8(10):231–239. https://doi.org/10.4172/jpb.1000374

Chapter 9

CRISPR/Cas9-Mediated Homology-Directed Genome Editing in *Pichia pastoris*

Thomas Gassler, Lina Heistinger, Diethard Mattanovich, Brigitte Gasser, and Roland Prielhofer

Abstract

State-of-the-art strain engineering techniques for the methylotrophic yeast *Pichia pastoris* (syn. *Komagataella spp.*) include overexpression of endogenous and heterologous genes and deletion of host genes. For efficient gene deletion, methods such as the split-marker technique have been established. However, synthetic biology trends move toward building up large and complex reaction networks, which often require endogenous gene knockouts and simultaneous overexpression of individual genes or whole pathways. Realization of such engineering tasks by conventional approaches employing subsequent steps of transformations and marker recycling is very time- and labor-consuming. Other applications require tagging of certain genes/proteins or promoter exchange approaches, which are hard to design and construct with conventional methods. Therefore, efficient systems are required that allow precise manipulations of the *P. pastoris* genome, including simultaneous overexpression of multiple genes. To meet this challenge, we have developed a CRISPR/Cas9-based kit for gene insertions, deletions, and replacements, which paves the way for precise genomic modifications in *P. pastoris*. In this chapter, the versatile method for performing these modifications without the integration of a selection marker is described. A ready-to-use plasmid kit for performing CRISPR/Cas9-mediated genome editing in *P. pastoris* based on the Golden*Pi*CS modular cloning vectors is available at Addgene as CRIS*Pi* kit (#1000000136).

Key words *Pichia pastoris*, Synthetic biology, Genome editing, Golden*Pi*CS, CRISPR/Cas9

1 Introduction

Clustered regularly interspaced short palindromic repeats (CRISPR)/CRISPR-associated (Cas) systems are found in bacteria and archaea where they function as an adaptive immune system for the defense against foreign DNA [1, 2]. The most commonly used CRISPR/Cas system for genome engineering is based on the type II system from *Streptococcus pyogenes*. It requires the CRISPR-associated endonuclease Cas9, a targeting CRISPR RNA (crRNA) complementary to the target sequence and a transactivating crRNA (tracrRNA). For genome editing applications, crRNA and

Brigitte Gasser and Diethard Mattanovich (eds.), *Recombinant Protein Production in Yeast*, Methods in Molecular Biology, vol. 1923, https://doi.org/10.1007/978-1-4939-9024-5_9,

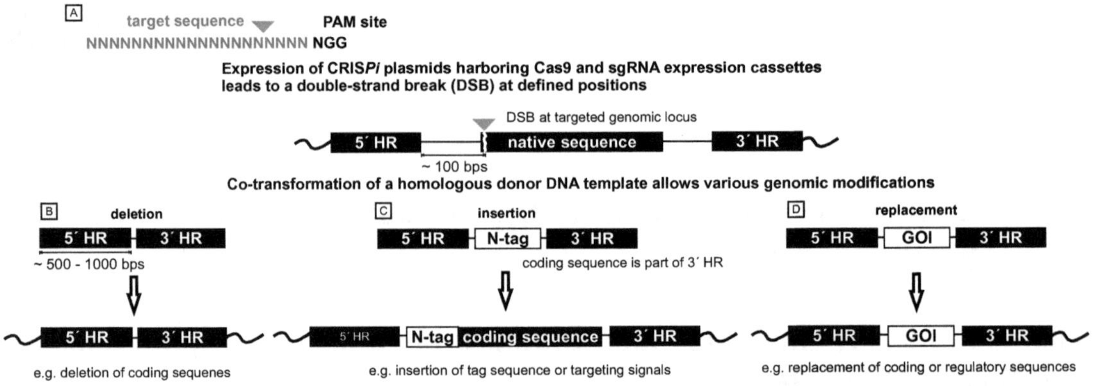

Fig. 1 Homology-directed repair of a double-strand break (DSB) allows precise genomic deletions, integrations, or replacements by co-transformation of a linear double-stranded donor DNA template. (**A**) A DSB (gray triangle) is introduced between the fourth and fifth nucleotides in the targeting sequence (marked in gray) upstream of the PAM site 5′-NGG-3′ in the targeted genomic locus upon expression of a sgRNA/Cas9 plasmid. Homology-directed DNA repair of the DSB is exploited for precise genome engineering, by co-transformation of a double-stranded donor DNA template containing homologous sequences to the targeted locus. (**B**) By providing a donor DNA template consisting of homologous regions (HR) which are fused together, native sequences can be deleted efficiently; (**C**) native sequences can be modified by flanking the modified sequences (e.g., addition of a N-terminal tag or a targeting signal); (**D**) co-transformation of a donor DNA template with heterologous sequences between the 5′ HR and the 3′ HR allows the replacement of native sequences (e.g., replacement of a coding sequence of a gene while maintaining the native promoter activity or gene regulation by exchange of a promoter)

tracrRNA are usually fused to generate a so-called single-guide RNA (sgRNA) [3, 4]. The crRNA/sgRNA recognizes a complementary target sequence of 20 bp, which must be followed by the three nucleotide protospacer adjacent motif (PAM) 5′-NGG-3′. Upon interaction of Cas9 with the sgRNA, it is guided to the target sequence where it introduces a double-strand break. This double-strand break is subsequently repaired by either error-prone nonhomologous end joining (NHEJ), leading to nucleotide insertions and deletions (InDels), or homologous recombination. By providing a homologous repair template, the later mechanism allows precise genome modifications (Fig. 1).

Since the discovery of the CRISPR/Cas9 system, it has been applied for genome editing in a wide range of host cells. For yeasts, the use of CRISPR/Cas9 for genome editing was first described for *Saccharomyces cerevisiae* in 2013 [5] and has since been established for many other relevant yeast species like *Kluyveromyces lactis*, *Kluyveromyces marxianus*, *Yarrowia lipolytica*, *Schizosaccharomyces pombe*, and *Pichia pastoris* (syn. *Komagataella phaffii*) [6–14]. Astonishingly, despite their phylogenetic relationship, extensive fine-tuning was required to establish an efficient CRISPR/Cas9 method for each of the yeast hosts.

Simultaneous expression of sgRNA and Cas9 creates InDel mutations by repair via the NHEJ pathway. By additionally providing a DNA donor template with 5′ and 3′ homology regions (HR), precise sequence editing like the introduction of stop codons or whole gene replacements and deletions are possible. The expression of inactive Cas9 variants (dCas9) allows tagging of genomic regions and has been used to study regulatory components interacting with a specific chromosomal region by a method called CRISPR-ChAP-MS [4, 15]. Fusion proteins of different activation or repression domains with dCas9 have successfully been used for the modulation of gene expression in *S. cerevisiae* [16–19].

CRISPR/Cas9-mediated genome editing offers several advantages over classical methods. Once established for a specific host cell, it can be easily adapted for many different target sequences. Also, the PAM site 5′-NGG-3′ generally occurs with a rather high frequency in the genome, which offers flexibility for the choice of target sequence. But its real potential is determined by its capability to efficiently introduce precise genomic changes without the integration of a selective marker.

Several processes influence CRISPR/Cas9-mediated genome editing efficiency and therefore require optimization. Targeting efficiency depends on the genomic locus and the choice of sgRNA [20]. The choice of promoter for Cas9 expression has little influence on CRISPR/Cas9 efficiency; however a negative effect on cell growth has been described for constructs with strong promoters [13, 21, 22]. Sufficient transcription of the sgRNA is critical for CRISPR/Cas9 efficiency, and RNA polymerase III (RNAPIII) promoters have successfully been used for yeasts like *S. cerevisiae*, *K. lactis*, and *Y. lipolytica* [5, 6, 8]. In *P. pastoris*, commonly used *S. cerevisiae* RNAPIII promoters, like the SNR52 promoter, did not provide functional sgRNA expression (own unpublished data, [11]). As an alternative, RNAPII promoters in combination with self-splicing ribozymes (5′ hammerhead (HH) and 3′ hepatitis delta virus (HDV)) for excision of the sgRNA from the transcript are used [9, 11, 14, 23].

One of the possible drawbacks of using CRISPR/Cas9 for genome editing is the introduction of off-target mutations, which has been reported in some cases [24, 25]. However, studies in yeast have shown that in the relatively small yeast genome off-target effects are rare and unlikely when a specific sgRNA is used [21, 26].

In this protocol we describe all important steps for efficient CRISPR/Cas9-mediated genome editing in *P. pastoris* and provide necessary elements and plasmids as CRIS*Pi* kit through Addgene (#1000000136). The plasmids are based on the recently published Golden*Pi*CS kit (Addgene #1000000133), which is a Golden Gate-derived *P. pastoris* cloning system [13]. Using the described method, we assembled episomal backbone 3 (BB3) plasmids for Cas9 and sgRNA expression (Fig. 2). Each CRIS*Pi* BB3 carries the *S. cerevisiae* CEN6/ARS4 sequence for episomal plasmid

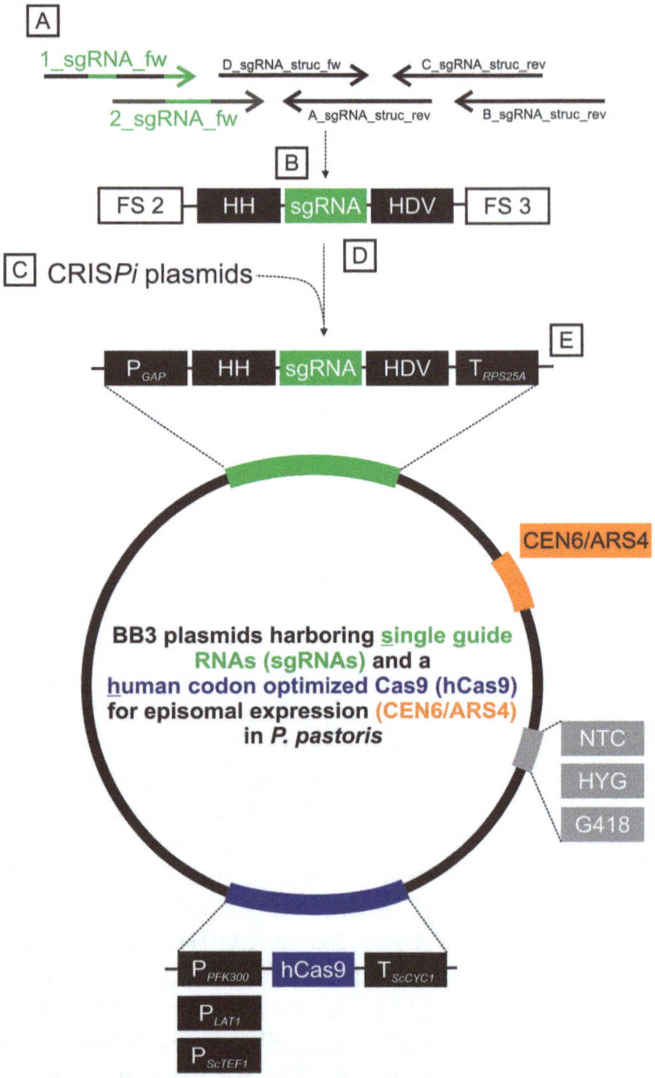

Fig. 2 Single-guide (sg) RNA construction and assembly into CRIS*Pi* plasmids. (**A**) sgRNA constructs are assembled by overlap extension (OE) PCR of six overlapping primers. The variable sequence (*green*) is covered by two primers (1_sgRNA_fw and 2_sgRNA_fw), and the structural part (*black*) containing sgRNA and flanking ribozymes is made up of four primers (D_sgRNA_struc_fw and A, B, and C_sgRNA_struc_rev). (**B**) The final HH–sgRNA–HDV construct is flanked by *Bpi*I restriction sites with open fusion site Fs2 (5′) and Fs3 (3′). (**C**) CRIS*Pi* plasmids harbor all elements required for hCas9 and sgRNA expression in *P. pastoris* and an empty HH–sgRNA–HDV expression cassette with internal *Bpi*I sites between the sgRNA promoter and terminator (Fs 2,3 linker), allowing direct assembly of the final plasmids (**D**). (**E**) Final plasmids feature an CEN6/ARS4 sequence for episomal expression, a resistance marker for positive selection in *E. coli* and *P. pastoris* (NTC...natMX, HYG...hphMX, G418...kanMX), an expression unit for the hCas9 (three different promoters P_{PFK300}, P_{LAT1}, and P_{ScTEF1} available), and the expression unit for the HH–sgRNA–HDV unit under control of the P_{GAP} promoter and the T_{RPS25A} terminator

maintenance, enabling transient gene expression without integration into the genome, a human codon-optimized *hCAS9* gene fused to a nuclear localization signal (NLS) under control of a constitutive promoter (Cas9 sequence from DiCarlo et al. [5], Addgene ID 43802), an insertion site for the sgRNA with flanking ribozyme sequences [23] under control of the *P. pastoris* P_{GAP} promoter, and a marker cassette for yeast/*E. coli* selection (*nat*MX (nourseothricin (NTC) selection), *hph*MX (hygromycin B) or *kan*MX (geneticin (G418)).

2 Materials

2.1 Media and Chemicals

YPD 20 g/L soy peptone, 10 g/L yeast extract, and 20 g/L glucose.

LB (low salt) 10 g/L soy peptone, 5 g/L yeast extract, and 5 g/L NaCl.

For agar plates, 20 g/L agar-agar is added to the media.

NaOH for preparation of a 0.02 M solution for the colony PCR lysis buffer.

Antibiotics: Kanamycin/Geneticin® (kanMX resistance cassette, 500 μg/mL working concentration for *E. coli* and *P. pastoris*), Nourseothricin (natMX resistance cassette, 50 μg/mL for *E. coli* and 200 μg/mL for *P. pastoris*), and Hygromycin (hphMX resistance cassette, 50 μg/mL for *E. coli* and 200 μg/mL for *P. pastoris*).

2.2 Kits and Plasmids

Gel and DNA purification kit.

Plasmid isolation kit.

CRIS*Pi* kit (Addgene #1000000136).

3 Methods

Cloning Workflow

Golden Gate cloning [27–29], a modular cloning strategy, was set up for combining several expression cassettes in one single plasmid allowing the simultaneous overexpression of multiple genes independent of the microorganism (GoldenMOCS) and further developed for application in *P. pastoris* (Golden*Pi*CS) recently [13, 30]. Briefly, the system is comprised of three backbone (BB) levels. BB1 vectors harbor basic modules (promoters, coding sequences, or terminators), BB2 vectors are used to combine these basic BB1 modules to expression cassettes (promoter + CDS + terminator), and BB3 plasmids are used to assemble multiple expression cassettes. Golden Gate cloning employs type IIs restriction enzymes (*Bsa*I and *Bpi*I) which cut outside of their recognition site, and it enables seamless cloning, assembly of multiple DNA fragments, and efficient one-pot cloning reactions.

We applied Golden*Pi*CS to assemble plasmids for CRISPR/Cas9-mediated gene and genome editing in *P. pastoris* (CRIS*Pi* plasmids). CRIS*Pi* plasmids (Addgene IDs: 104906–104912) are BB3 vectors for direct construction of sgRNA/Cas9 plasmids in a one-step cloning procedure, by insertion of the sgRNAs. The sgRNAs are assembled as fusion genes (HH–sgRNA–HDV construct) in an overlap extension (OE) PCR (Fig. 2). Correct genomic insertions, replacements, or deletions are later verified by sequencing of the target site using a set of specific verification primers (*see* Subheading 3.1 for design).

3.1 Plasmid Design and Generation

As outlined in Fig. 2, sgRNAs are designed and generated for each target site individually and inserted into CRIS*Pi* plasmids. The final sgRNA/Cas9 plasmids harbor an expression unit for hCas9 under control of either the P_{ScTEF1}, P_{LAT1}, or P_{PFK300} promoter and a sgRNA expression unit. These plasmids also contain all standard parts for cloning in *E. coli*, an antibiotic resistance cassette and a CEN6/ARS4 sequence, allowing for episomal plasmid maintenance in *P. pastoris.* A suitable 20 bp target site can either be identified automatically by online tools (e.g., ChopChop) (*see* **Note 1**) or manually by seeking a suitable PAM site within the targeted loci. Due to locus-specific variations in targeting efficiencies, it is suggested to design at least two sgRNAs for each target.

The position for InDel creation within the target gene (by transformation of sgRNA/Cas9 plasmids without donor DNA templates) should be close to the start codon ATG in order to prevent partial transcription/translation of the target gene. For homology-directed genome editing, it is necessary to introduce the double-strand break (DSB) within a certain target locus. For optimal efficiency of homology-directed deletions, integrations, or substitutions, it is recommended to keep the distance between the DSB (occurring 4–5 bps upstream of the PAM site) and the beginning of the closest homologous region below 100 bps. Due to the limited sequence space, the choice is usually restricted to just a few possible PAM sites.

For verification of the mutations and alterations created in the genome, several primers are designed and used for sequencing of the targeted site as outlined below.

Procedure

1. Select a Cas9 target sequence either by manual sequence search or using an online tool (*see* **Note 1**).

Table 1
Primer sequences for construction of HH–sgRNA–HDV fusion genes[a]

Primer name	Sequence
A_sgRNA_struc_rev	CGCCATGCCGAAGCATGTTGCCCAGCCGGCGCCAGCGAGGAGGCTGGGACCATGCCGGCC
B_sgRNA_struc_rev	**agaagacg**caagcaGTCCAAAGCTGTCCCATTCGCCATGCCGAAGCATGTTGCCCAGCCG
C_sgRNA_struc_rev	AGGCTGGGACCATGCCGGCC**AAAAGCACCGACTCGGTGCCACTTTTTCAAGTTGATAACG**
D_sgRNA_struc_fw	**GTTTTAGAGCTAGAAATAGCAAGTTAAAATAAGGCTAGTCCGTTATCAACTTGAAAAAGT**
1_sgRNA_fw	**tgaagacg**ccatg**NNNNNN**CTGATGAGTCCGTGAGGACGAAACGAGTAAGCTCGTC**NNNN**
2_sgRNA_fw	AAACGAGTAAGCTCGTC**NNNNNNNNNNNNNNNNNNNNGTTTTAGAGCTAGAAATAGCAAG**

[a]Primer sequences for construction of HH–sgRNA–HDV fusion genes with overlapping sequences (designed according to [22]) with *Bpi*I recognition sites (gaagac) and respective fusion sites (Fs) (Fs2: catg and Fs3: gctt, reverse complement sequence on primer B_sgRNA_struc_rev aagc). The structural sequences (normal uppercase letters) include the hammerhead (HH) ribozyme and the hepatitis delta virus (HDV) ribozyme, the mature sgRNA sequence after self-cleaving (bold uppercase letters), and the target sequence (represented by N)

2. The ribozyme–sgRNA–fusion gene is generated by overlap extension (OE) PCR of six primers (Table 1 and Fig. 2a) which then consists of an invariant structural part and target-specific part. The invariant region is covered by four primers (A_ to C_sgRNA_struc_rev and D_sgRNA_struc_fw), and the target-specific sequence is covered by two primers which are designed individually (1_sgRNA_fw and 2_sgRNA_fw, alterable target gene-specific positions marked as "N" in Table 1). The first 6 "N" nucleotides on primer 1_sgRNA_fw are reverse complementary to the first 6″ N″ on primer 2_sgRNA_fw, which is essential for hammerhead processing. The primer 2_sgRNA_fw contains the 20 bps sequence defining the targeted locus (*see* **Note 2**).
3. OE PCR is done using Q5 polymerase and the following cycle setup: 98 °C/30 s, 35 cycles of 98 °C/10 s, 60 °C/20 s, 72 °C/10 s; and 72 °C/2 min (*see* **Note 3**); PCR products are purified via gel electrophoresis (correctly assembled sgRNAs have a size of 250 bps), and the concentration is adjusted to 40 nM for subsequent cloning.
4. The purified PCR products contain *Bpi*I sites at their 5′ and 3′ ends with fusion sites Fs2 (CATG) and Fs3 (GCTT) enabling direct assembly into tailor-made CRIS*Pi* vectors, which contain *Bpi*I restriction sites with compatible fusion sites for sgRNA insertion between the P_{GAP} promoter and T_{RPS25A} terminator from *P. pastoris.* The final vectors are created by a Golden Gate assembly reaction using *Bpi*I: 1 μL *Bpi*I (10 U), 0.1 μL T4 ligase (40 U), 2 μL CutSmart® Buffer (10×, NEB), 2 μL ATP (10 mM, NEB), and 1 μL of 40 nM dilutions of the PCR fragments and the recipient backbone in 20 μL total volume.
5. The Golden Gate assembly reaction is performed as follows: 45 cycles (depending on insert number—for sgRNA assembly in CRIS*Pi* plasmids, the cycle number can be reduced to 20) of each 2 min at 37 °C and 2.5 min at 16 °C, followed by 10 min at 37 °C, 30 min at 55 °C, and 10 min at 80 °C. Final plasmids should be checked by restriction digestion using appropriate enzymes and by Sanger sequencing.
6. To save time, primers for verification of InDel mutations in the *P. pastoris* genome (enabling the amplification of the targeted locus) should already be designed during the cloning procedure. The forward and reverse primer should bind approximately 300 bp up- and downstream of the introduced DSB.
7. For verification of homology-directed insertions, deletions, or replacements, it is recommended to design several pairs of primers. The primers for verification of integration should cover the 5′ and the 3′ part as well as the entire length of the

engineered locus with at least one oligo binding outside of the respective homology region. Primers covering the 5′ or the 3′ end of the targeted site can be used for screening a large number of clones, while primers covering the entire locus can be used for final verification of the modifications (*see* **Note 4**).

3.2 Generating Gene Disruptions by InDel Mutations

Single-gene disruptions can be obtained by expression of the sgRNA/Cas9 from the BB3 plasmids, leading to a double-strand break (DSB) between the fourth and fifth nucleotides upstream of the PAM site. DSB repair via the error-prone NHEJ pathway introduces insertions or deletions (InDel mutations). Ideally, this leads to a functional knockout of the targeted gene due to a shift in the open reading frame (*see* **Note 5**). For introduction of a single-gene knockout, we alternatively recommend to use conventional split-marker-based techniques followed by marker recycling [31, 32]. This allows selection for the integration on respective antibiotics and reduces the number of clones which have to be screened. Difficult targets may be knocked out by a combined usage of split-marker and CRISPR/Cas9-mediated techniques, where the additional introduction of a DSB by Cas9 is used to increase the frequency of homologous integration of a split-marker cassette at the target locus.

Procedure

1. The sgRNA/Cas9 plasmids are designed and assembled as outlined in Subheading 3.1 (*see* **Note 6**).
2. Between 0.5 and 1 μg of circular plasmid is transformed into electro-competent *P. pastoris* cells (prepared according to, e.g., [32] or [33]), regenerated, and plated (*see* **Note 7**) on selective YPD agar. Afterward, plates are incubated at 30 °C for approximately 48 h (*see* **Note 8**).
3. Cells are streaked out two more times (48 h cultivation each) on selective YPD agar plates (*see* **Note 9**).
4. Screening PCR: By touching a single *P. pastoris* colony with a sterile pipette tip, inoculation loop, or toothpick, a small amount of biomass is resuspended in 10 μL 0.02 M NaOH in a PCR reaction tube. The mixture is then boiled for 10 min at 99 °C and cooled down to room temperature. After brief centrifugation, 2 μL of the supernatant can be used as template for colony PCR (*see* **Note 10**).
5. The targeted locus is amplified by PCR with appropriate primers, and bands are purified by agarose gel electrophoresis followed by verification by Sanger sequencing.
6. Clones with successfully introduced InDel mutations are further restreaked on YPD agar plates without selection pressure

at least twice to promote the loss of the sgRNA/Cas9 plasmid (*see* **Note 11**).

3.3 Generation of Marker-Free Integrations, Deletions, and Replacements Using CRISPR/Cas9-Mediated Homology-Directed Repair

The protocol outlined here should be considered as a brief guideline for applying homology-directed gene and genome editing in *P. pastoris* mediated by CRISPR/Cas9. Functional expression of sgRNA/Cas9 allows the integration or replacement of genome sequences if a suitable donor DNA template for homology-directed DNA repair (Fig. 1) is provided. The following protocol describes the procedure of replacing a native genomic sequence by a heterologous sequence. This procedure can be adapted to perform gene deletions or simple gene insertions by homology-directed repair without integration of selection markers in *P. pastoris*. The method can be performed in any genetic background of P. *pastoris* and does not require additional genetic modifications (e.g., knockout of ku70). The establishment of large synthetic metabolic networks, which demand for integration of various genes at precisely defined positions and very often have to be combined with knockouts of endogenous genes (e.g., such as the pioneering glycoengineering approaches [34]), is facilitated by this workflow.

Procedure (Replacement of a Gene with a Heterologous Sequence)

1. The verification primers and sgRNA/Cas9 plasmids are designed as outlined in Subheading 3.1. The position of the DSB is chosen manually by selecting a PAM site downstream and in close proximity of the start codon of the native gene. The donor DNA template for homology-directed repair consists of 500–1000 bp regions which are homologous to the sequences upstream of the start codon (5′ homology region (HR)) and downstream of the stop codon (3′ homology region (HR)), with the heterologous sequence in between (*see* **Note 12**).
2. The final donor DNA template constructs (*see* Fig. 1) are produced by OE PCR, a suitable cloning strategies or DNA synthesis. For advanced projects, we suggest the use of the Golden*PiCS* kit [13]. *See* **Note 13** for an exemplary description of a gene replacement by two adjacent expression cassettes.
3. Between 80 and 160 nmol of the sgRNA/Cas9 plasmid (corresponding to approximately 0.5–1.0 μg of DNA) together with at least 1500 nmol of donor DNA template (corresponding to approximately 5 μg for a 5000 bps donor DNA template) are transformed into electro-competent *P. pastoris*, plated on selective YPD agar, and incubated at 30 °C for approximately 48 h (*see* **Note 8**).
4. Cells are streaked out two more times on selective YPD agar plates (*see* **Note 9**).
5. Screening PCR is done as described in Subheading 3.2, **step 4**.

6. Transformants are checked individually at 5′ and 3′ ends of the integrated sequences as well as over the entire engineered locus by PCR using the verification primers for homology-directed insertions, deletions, and replacements (*see* Subheading 3.1) (*see* **Note 4**).
7. Clones with correct insertions are further restreaked on YPD agar plates without selection pressure at least twice to promote the loss of the sgRNA/Cas9 plasmid (*see* **Note 11**).

4 Notes

1. Most online tools (for an overview of available tools, *see* Stovicek et al. [35]) are well suited for identification of sgRNAs, which are intended for usage in InDel creations. The tools look at PAM sites within the coding region of a target gene and rank several sgRNAs according to their predicted off-target effects. If the experimental setup allows a gene knockout by InDel introduction (i.e., loss on gene function can serve as a selectable marker), this procedure can facilitate the sgRNA design. If loss in gene function cannot serve as selection marker, the initial screening has to be done by PCR.
2. It is necessary to check the final sgRNA sequence for unwanted *Bpi*I sites.
3. In the OE PCR assemblies, the concentration of the forward (1_sgRNA_fw) and reverse (B_sgRNA_struc_rev) primers is 0.5 μM and for the inner template primers (2_sgRNA_fw, A_sgRNA_struc_rev, C_sgRNA_struc_rev, and D_sgRNA_struc_fw) 0.01 μM.
4. Primers for integration verification must bind outside of the 5′ HR and/or the 3′ HR. By using primers binding only within the donor DNA, false-positive results will be obtained by randomly integrated DNA. It is recommended to use primers covering either the 5′ or 3′ site of the targeted sequence for first screening PCRs (PCR products should be in the size of 1000 bps for a quick workflow). Positive clones are then in the next step analyzed using a screening PCR covering the entire locus, which allows differentiation from the wild-type sequence.
5. In cases where the loss of gene function cannot be exploited for selection, introduced InDel mutations have to be verified by PCR amplification of the respective locus followed by Sanger sequencing.
6. We highly recommend to select at least two sgRNAs for each target. This can save time in case of unsuccessful targeting with a selected sgRNA.

7. According to our experience, electro-competent cells prepared according to the protocol described by Gasser et al. [32] take up episomal plasmids very efficiently. Plating small aliquots (e.g., 10 or 100 μL) of the transformation mixture is usually sufficient; plating all cells most likely causes overgrowth on the selection plates.
8. When using P_{LAT1} or P_{PFK300} promoters for hCas9 expression, colonies appear after 48 h. If the stronger P_{ScTEF1} promoter is used, incubation time has to be prolonged.
9. Isolation of successfully modified strains from non-mutated wild-type cells can be laborious, so we recommend restreaking one clone per plate for proper separation.
10. Integration efficiencies of homology-directed integrations vary to a great extent. It is recommended to screen at least 20 clones per transformation.
11. Selective and nonselective YPD agar plates are used in parallel, to check for plasmid loss. According to our unpublished data, plasmids carrying the *S. cerevisiae* CEN6/ARS4 sequence are lost at a rate of around 14% per generation.
12. The sequences of 5′ HR and 3′ HR must not contain the targeting sequence including the PAM site because Cas9 will not only cut in the genome but also cut the donor DNA template. If necessary, a silent mutation has to be introduced to remove the PAM site on the donor DNA template. Also, shorter homology regions of lengths down to 300 bps can be used but may lead to reduced integration efficiencies.
13. For the replacement of a gene by two other expression cassettes, the promoter and terminator sequence of the native gene (e.g., a protease gene to be deleted, [36]) can be used as 5′ and 3′ HR, respectively. The sequences are amplified from the genome by PCR and cloned into BB1_12 and BB1_34 plasmids, respectively (with flanking fusion sites Fs 1,2 and Fs 3,4). The two genes (X, Y) to be inserted are cloned into BB1_23 plasmids with Fs 2,3 and are further combined with promoter and terminator sequences in BB2 plasmids (Fs AB and BC), with the 5′ homology region in front of the first gene (Fs 1,2) and the 3′ HR as terminator of the second gene (Fs 3,4). The other promoter and terminator sequences can be chosen freely but must be free from sequences homologous to the targeted loci and to the sgRNA/Cas9 plasmid. The expression units contained in BB2 are cloned into a modified BB3 plasmid (BB3nK_ext_AC) (Addgene ID 108686), with fusion sites A and C and external *Bpi*I sites. This modification allows the excision of the final donor DNA constructs (containing two expression units) by a simple *Bpi*I digest. A modified version of this BB3 plasmid with fusion sites A and D

(BB3nK_ext_AD) allowing cloning and excision of three expression units is also available at Addgene (Addgene ID 108687). Be aware that all sequences have to be cured by removing all internal *Bpi*I and *Bsa*I sites, if the donor DNA templates are constructed by Golden*Pi*CS. Sufficient amounts of donor DNA can be obtained by purification of an appropriate amount of plasmid DNA in a Midi prep, followed by restriction digest and gel purification.

Acknowledgments

This work was supported by the Federal Ministry for Digital and Economic Affairs (BMDW), the Federal Ministry of Traffic, Innovation and Technology (BMVIT), the Styrian Business Promotion Agency SFG, the Standortagentur Tirol, the Government of Lower Austria and ZIT-Technology Agency of the City of Vienna through the COMET-Funding Program managed by the Austrian Research Promotion Agency FFG; and the Austrian Federal Ministry for Digital and Economic Affairs (BMDW), the National Foundation for Research, Technology and Development and the Christian Doppler Research Association. TG and LH were supported by the Austrian Science Fund (FWF): Doctoral Program BioToP—Biomolecular Technology of Proteins (FWF W1224). We further want to thank Franz Zehetbauer and Dariusz Yarych for technical support as well as Corinna Rebnegger and Matthias Steiger for initial inspiration and fruitful discussions.

References

1. Wiedenheft B, Sternberg SH, Doudna JA (2012) RNA-guided genetic silencing systems in bacteria and archaea. Nature 482 (7385):331–338. https://doi.org/10.1038/nature10886
2. Hille F, Charpentier E (2016) CRISPR-Cas: biology, mechanisms and relevance. Philos Trans R Soc Lond Ser B Biol Sci 371 (1707):20150496. https://doi.org/10.1098/rstb.2015.0496
3. Gasiunas G, Barrangou R, Horvath P, Siksnys V (2012) Cas9-crRNA ribonucleoprotein complex mediates specific DNA cleavage for adaptive immunity in bacteria. Proc Natl Acad Sci U S A 109(39):E2579–E2586. https://doi.org/10.1073/pnas.1208507109
4. Jinek M, Chylinski K, Fonfara I, Hauer M, Doudna JA, Charpentier E (2012) A programmable dual-RNA-guided DNA endonuclease in adaptive bacterial immunity. Science 337 (6096):816–821. https://doi.org/10.1126/science.1225829
5. DiCarlo JE, Norville JE, Mali P, Rios X, Aach J, Church GM (2013) Genome engineering in *Saccharomyces cerevisiae* using CRISPR-Cas systems. Nucleic Acids Res 41(7):4336–4343. https://doi.org/10.1093/nar/gkt135
6. Horwitz AA, Walter JM, Schubert MG, Kung SH, Hawkins K, Platt DM, Hernday AD, Mahatdejkul-Meadows T, Szeto W, Chandran SS, Newman JD (2015) Efficient multiplexed integration of synergistic alleles and metabolic pathways in yeasts via CRISPR-Cas. Cell Syst 1 (1):88–96. https://doi.org/10.1016/j.cels.2015.02.001
7. Lobs AK, Engel R, Schwartz C, Flores A, Wheeldon I (2017) CRISPR-Cas9-enabled genetic disruptions for understanding ethanol and ethyl acetate biosynthesis in *Kluyveromyces marxianus*. Biotechnol Biofuels 10:164.

https://doi.org/10.1186/s13068-017-0854-5

8. Schwartz CM, Hussain MS, Blenner M, Wheeldon I (2016) Synthetic RNA polymerase III promoters facilitate high-efficiency CRISPR-Cas9-mediated genome editing in *Yarrowia lipolytica*. ACS Synth Biol 5 (4):356–359. https://doi.org/10.1021/acssynbio.5b00162
9. Gao S, Tong Y, Wen Z, Zhu L, Ge M, Chen D, Jiang Y, Yang S (2016) Multiplex gene editing of the *Yarrowia lipolytica* genome using the CRISPR-Cas9 system. J Ind Microbiol Biotechnol 43(8):1085–1093. https://doi.org/10.1007/s10295-016-1789-8
10. Jacobs JZ, Ciccaglione KM, Tournier V, Zaratiegui M (2014) Implementation of the CRISPR-Cas9 system in fission yeast. Nat Commun 5:5344. https://doi.org/10.1038/ncomms6344
11. Weninger A, Hatzl AM, Schmid C, Vogl T, Glieder A (2016) Combinatorial optimization of CRISPR/Cas9 expression enables precision genome engineering in the methylotrophic yeast *Pichia pastoris*. J Biotechnol 235:139–149. https://doi.org/10.1016/j.jbiotec.2016.03.027
12. Weninger A, Fischer J, Raschmanova H, Kniely C, Vogl T, Glieder A (2017) Expanding the CRISPR/Cas9 toolkit for *Pichia pastoris* with efficient donor integration and alternative resistance markers. J Cell Biochem. https://doi.org/10.1002/jcb.26474
13. Prielhofer R, Barrero JJ, Steuer S, Gassler T, Zahrl R, Baumann K, Sauer M, Mattanovich D, Gasser B, Marx H (2017) Golden*Pi*CS: a Golden Gate-derived modular cloning system for applied synthetic biology in the yeast *Pichia pastoris*. BMC Syst Biol 11(1):123. https://doi.org/10.1186/s12918-017-0492-3
14. Heistinger L, Gasser B, Mattanovich D (2018) Creation of stable heterothallic strains of *Komagataella phaffii* enables dissection of mating gene regulation. Mol Cell Biol 38(2). https://doi.org/10.1128/mcb.00398-17
15. Waldrip ZJ, Byrum SD, Storey AJ, Gao J, Byrd AK, Mackintosh SG, Wahls WP, Taverna SD, Raney KD, Tackett AJ (2014) A CRISPR-based approach for proteomic analysis of a single genomic locus. Epigenetics 9 (9):1207–1211. https://doi.org/10.4161/epi.29919
16. Qi LS, Larson MH, Gilbert LA, Doudna JA, Weissman JS, Arkin AP, Lim WA (2013) Repurposing CRISPR as an RNA-guided platform for sequence-specific control of gene expression. Cell 152(5):1173–1183. https://doi.org/10.1016/j.cell.2013.02.022
17. Gilbert LA, Larson MH, Morsut L, Liu Z, Brar GA, Torres SE, Stern-Ginossar N, Brandman O, Whitehead EH, Doudna JA, Lim WA, Weissman JS, Qi LS (2013) CRISPR-mediated modular RNA-guided regulation of transcription in eukaryotes. Cell 154 (2):442–451. https://doi.org/10.1016/j.cell.2013.06.044
18. Farzadfard F, Perli SD, Lu TK (2013) Tunable and multifunctional eukaryotic transcription factors based on CRISPR/Cas. ACS Synth Biol 2(10):604–613. https://doi.org/10.1021/sb400081r
19. Deaner M, Alper HS (2017) Systematic testing of enzyme perturbation sensitivities via graded dCas9 modulation in *Saccharomyces cerevisiae*. Metab Eng 40:14–22. https://doi.org/10.1016/j.ymben.2017.01.012
20. Smith JD, Suresh S, Schlecht U, Wu M, Wagih O, Peltz G, Davis RW, Steinmetz LM, Parts L, St Onge RP (2016) Quantitative CRISPR interference screens in yeast identify chemical-genetic interactions and new rules for guide RNA design. Genome Biol 17:45. https://doi.org/10.1186/s13059-016-0900-9
21. Ryan OW, Skerker JM, Maurer MJ, Li X, Tsai JC, Poddar S, Lee ME, DeLoache W, Dueber JE, Arkin AP, Cate JH (2014) Selection of chromosomal DNA libraries using a multiplex CRISPR system. elife 3. https://doi.org/10.7554/eLife.03703
22. Generoso WC, Gottardi M, Oreb M, Boles E (2016) Simplified CRISPR-Cas genome editing for *Saccharomyces cerevisiae*. J Microbiol Methods 127:203–205. https://doi.org/10.1016/j.mimet.2016.06.020
23. Gao Y, Zhao Y (2014) Self-processing of ribozyme-flanked RNAs into guide RNAs in vitro and in vivo for CRISPR-mediated genome editing. J Integr Plant Biol 56 (4):343–349. https://doi.org/10.1111/jipb.12152
24. Zhang XH, Tee LY, Wang XG, Huang QS, Yang SH (2015) Off-target effects in CRISPR/Cas9-mediated genome engineering. Mol Ther Nucleic Acids 4(11):e264. https://doi.org/10.1038/mtna.2015.37
25. O'Geen H, Yu AS, Segal DJ (2015) How specific is CRISPR/Cas9 really? Curr Opin Chem Biol 29:72–78. https://doi.org/10.1016/j.cbpa.2015.10.001
26. Jakociunas T, Bonde I, Herrgard M, Harrison SJ, Kristensen M, Pedersen LE, Jensen MK, Keasling JD (2015) Multiplex metabolic pathway engineering using CRISPR/Cas9 in *Saccharomyces cerevisiae*. Metab Eng 28:213–222.

https://doi.org/10.1016/j.ymben.2015.01.008

27. Engler C, Kandzia R, Marillonnet S (2008) A one pot, one step, precision cloning method with high throughput capability. PLoS One 3 (11):e3647. https://doi.org/10.1371/journal.pone.0003647
28. Engler C, Gruetzner R, Kandzia R, Marillonnet S (2009) Golden gate shuffling: a one-pot DNA shuffling method based on type IIs restriction enzymes. PLoS One 4(5):e5553. https://doi.org/10.1371/journal.pone.0005553
29. Werner S, Engler C, Weber E, Gruetzner R, Marillonnet S (2012) Fast track assembly of multigene constructs using Golden Gate cloning and the MoClo system. Bioeng Bugs 3 (1):38–43. https://doi.org/10.1371/journal.pone.001676510.4161/bbug.3.1.18223
30. Sarkari P, Marx H, Blumhoff ML, Mattanovich D, Sauer M, Steiger MG (2017) An efficient tool for metabolic pathway construction and gene integration for *Aspergillus niger*. Bioresour Technol 245 (Pt B):1327–1333. https://doi.org/10.1016/j.biortech.2017.05.004
31. Fairhead C, Llorente B, Denis F, Soler M, Dujon B (1996) New vectors for combinatorial deletions in yeast chromosomes and for gap-repair cloning using 'split-marker' recombination. Yeast 12(14):1439–1457. https://doi.org/10.1002/(SICI)1097-0061(199611)12:14<1439::AID-YEA37>3.0.CO;2-O
32. Gasser B, Prielhofer R, Marx H, Maurer M, Nocon J, Steiger M, Puxbaum V, Sauer M, Mattanovich D (2013) *Pichia pastoris*: protein production host and model organism for biomedical research. Future Microbiol 8:191–208. https://doi.org/10.2217/fmb.12.133
33. Wu S, Letchworth GJ (2004) High efficiency transformation by electroporation of *Pichia pastoris* pretreated with lithium acetate and dithiothreitol. BioTechniques 36(1):152–154
34. Hamilton S, Gerngross T (2007) Glycosylation engineering in yeast: the advent of fully humanized yeast. Curr Opin Biotechnol 18 (5):387–392. https://doi.org/10.1016/j.copbio.2007.09.001
35. Stovicek V, Holkenbrink C, Borodina I (2017) CRISPR/Cas system for yeast genome engineering: advances and applications. FEMS Yeast Res 17(5). https://doi.org/10.1093/femsyr/fox030
36. Marsalek L, Gruber C, Altmann F, Aleschko M, Mattanovich D, Gasser B, Puxbaum V (2017) Disruption of genes involved in CORVET complex leads to enhanced secretion of heterologous carboxylesterase only in protease deficient *Pichia pastoris*. Biotechnol J 12(5). https://doi.org/10.1002/biot.201600584

Chapter 10

Tailoring N-Glycan Biosynthesis for Production of Therapeutic Proteins in *Saccharomyces cerevisiae*

Mari A. Piirainen, Heidi Salminen, and Alexander D. Frey

Abstract

The ability to control and adjust the N-glycosylation pathway of *Saccharomyces cerevisiae* is a key step toward production of therapeutic glycoproteins such as antibodies or erythropoietin. The focus of this chapter is to describe the road from yeast-type N-glycosylation to human-type complex N-glycosylation. The chapter describes the cell engineering and provides the detailed analytical procedures required to perform glycan analysis using MALDI-TOF mass spectrometry.

Key words *S. cerevisiae*, Glycoengineering, Glycosyltransferases, Glycosylation efficiency, Glycan analysis, MALDI-TOF MS

1 Introduction

Most of the therapeutic recombinant proteins used in humans are glycoproteins that require N-glycosylation for proper folding, function, and activity. Therefore, for production they need an expression system with proper glycosylation machinery. For this reason, they are usually produced in mammalian cells [1]. However, due to the disadvantages of this system such as possible viral contaminations, high cost, long fermentation times, and heterogeneous products, yeast has recently become an interesting production system for therapeutic glycoproteins. This system could be cheaper and faster, and higher yield of proteins could be obtained. In recent years, several approaches have been undertaken to humanize the N-glycans in *Pichia pastoris* and *Saccharomyces cerevisiae* [2–6]. This article will explicate the road to produce and analyze proteins carrying human-type N-glycans in *S. cerevisiae*.

1.1 N-Glycosylation in Yeast and Mammalian Cells

The N-glycosylation pathway in the endoplasmic reticulum (ER) is highly conserved in eukaryotic cells, while modifications carried out in the Golgi apparatus vary across species. In yeast, the N-glycan

Brigitte Gasser and Diethard Mattanovich (eds.), *Recombinant Protein Production in Yeast*, Methods in Molecular Biology, vol. 1923, https://doi.org/10.1007/978-1-4939-9024-5_10,

structures are typically of hypermannose type, whereas in humans, hybrid- and complex-type structures are prevailing.

In eukaryotes, N-linked glycosylation is initiated at the cytosolic side of the ER membrane, on which the lipid-linked oligosaccharide (LLO) $Man_5GlcNAc_2$ is formed, flipped into the ER lumen, and further glycosylated. The final $Glc_3Man_9GlcNAc_2$ glycan structure is transferred to the asparagine in the glycosylation sequon (N-X-S/T; X is not P) in substrate proteins. This transfer is catalyzed by the oligosaccharyltransferase (OST) complex. The N-linked $Glc_3Man_9GlcNAc_2$ structure is subsequently trimmed to $Man_8GlcNAc_2$, after which the glycoprotein is transported to the Golgi apparatus [7].

Golgi modifications in yeast start with the addition of an alpha-1,6-mannose by Och1p to the alpha-1,3-mannose of the trimannosyl core. To form high-mannose glycan structures, this outer chain is elongated with alpha-1,6-mannoses by sequential action of two protein complexes, mannan polymerases I and II. The outer chain is further glycosylated by addition of alpha-1,2-mannoses by Mnn2p and Mnn5p, mannosylphosphate by Mnn4p and Mnn6p, and finally by capping alpha-1,3-mannoses by Mnn1p. Core-type N-glycans do not have the elongated outer chain, and they contain only an alpha-1,2-mannose, mannosylphosphate, and alpha-1,3-capping mannose [8, 9].

In mammals, the $Man_8GlcNAc_2$ glycan is trimmed to Man_5-$GlcNAc_2$ by alpha-1,2-mannosidase enzymes. Then, the first GlcNAc is added by *N*-acetylglucosaminyltransferase I (GnTI). The $GlcNAcMan_5GlcNAc_2$ glycan is further trimmed by a mannosidase enzyme, and the second GlcNAc is added by *N*-acetylglucosaminyltransferase II (GnTII). In addition, a branching GlcNAc or up to two additional antennae can be introduced by *N*-acetylglucosaminyltransferases III, IV, and V (GnTIII, GnTIV, and GnTV) [10]. Next, galactoses are transferred by galactosyltransferase (GalT). The final step is the terminal capping of the N-glycans with sialic acid by sialyltransferase. In addition, the core structure can be fucosylated.

1.2 Elimination of Hypermannosylation in Yeast

In order to obtain human-type N-glycosylation in yeast, the first objective is to eliminate or reduce the extensive mannosylation specific to yeast glycans and to create a glycan structure that is compatible with further mammalian-type modifications. The $Man_3GlcNAc_2$ core structure is present in all eukaryotic glycans, and this truncated glycan structure can be used directly as an acceptor for mammalian-type glycan modifications. A yeast strain producing this glycan structure has been created by deleting two ER-localized mannosyltransferase genes, *ALG3* and *ALG11* [11]. A MALDI-TOF MS spectrum of the respective strain is shown in Fig. 1a. Additional deletion of the *MNN1* gene further eliminated interfering glycan structures [6].

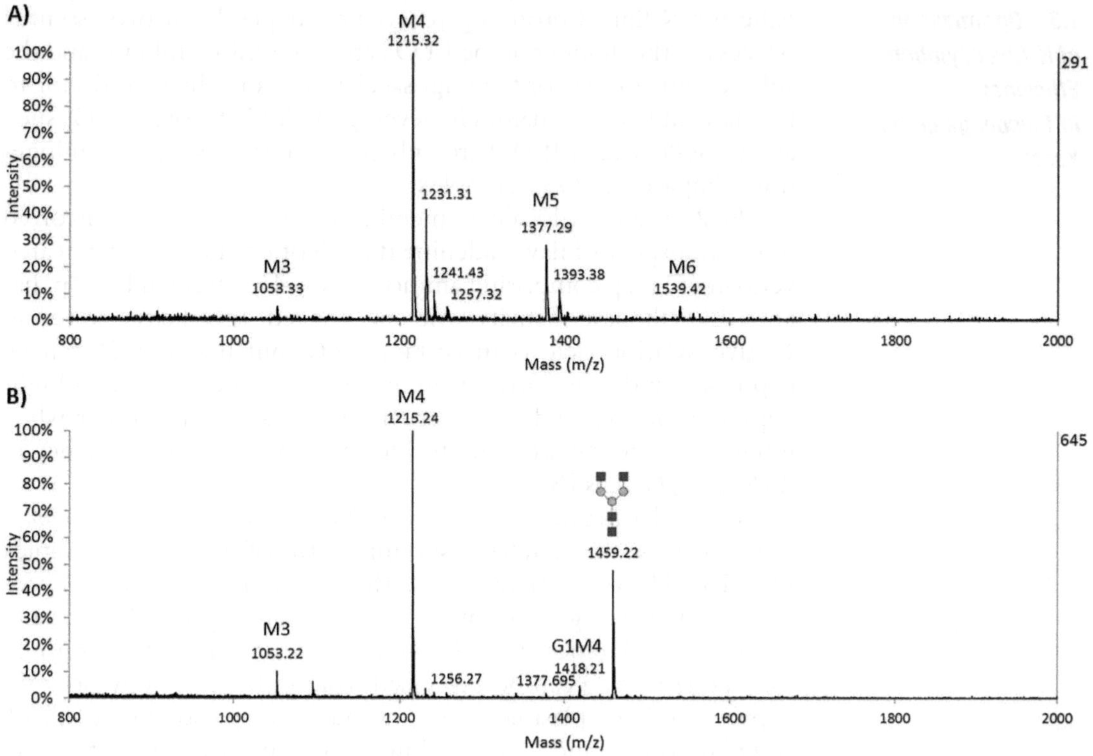

Fig. 1 MALDI-TOF MS spectra of 2-AB-labeled N-glycans isolated from the cell wall glycoproteins of (**a**) YG1429 and (**b**) YMP14 carrying plasmid pMP10 encoding GnTI, GnTII, and *KlYEA4* and an empty *URA3* plasmid. The peaks at m/z 1053, 1215, 1377, and 1599 and at m/z 1418 and 1539 correspond to the sodium adducts of the glycan structures $Man_3GlcNAc_2$ (M3), $Man_4GlcNAc_2$ (M4), $Man_5GlcNAc_2$ (M5), and Man_6-$GlcNAc_2$ (M6) and the hybrid and complex glycan structures $GlcNAcMan_4GlcNAc_2$ (GM4) and $GlcNAc_2Man_3Glc$-NAc_2 (G2M3), respectively. Squares, GlcNAc; circles, mannose

An alternative strategy to eliminate hypermannosylation and to generate substrates for GnTI and GnTII is to delete mannosyltransferase genes involved in the N-glycan modifications in the Golgi apparatus, including *OCH1*, *MNN1*, and *MNN4*, or a combined strategy containing OCH1, MNN1, and ALG3 deletions [12]. However, to obtain a suitable glycan substrate for GnTI and GnTII, all of these alternative strategies require the expression of at least one heterologous mannosidase.

When reducing the extent of hypermannosylation, it should be noted that glycoproteins also have a role in the structure and integrity of the cell wall. Modification of glycan structures can have negative effects on the yeast cell wall, affecting the temperature tolerance and growth rate. In order to overcome these challenges, UV-mutagenesis or other methods can be utilized to generate more robust mutants [5, 12, 13].

1.3 Optimization of N-Glycosylation Efficiency in Glycoengineered Yeast

Efficient N-linked protein glycosylation depends on two essential processes: the flipping of the LLO across the ER membrane and the subsequent transfer of the oligosaccharide from the lipid donor to the nascent peptide chain. However, genetically tailored LLOs such as $Man_3GlcNAc_2$-PP-Dol are only poor substrates for the endogenous flippase and OST complex.

Flc2p is an ER-localized membrane protein, which is involved in the transport of flavin adenine dinucleotide (FAD). A truncated version of Flc2p comprising amino acids 1–452, termed Flc2*p, has an LLO flippase activity with a relaxed specificity, restoring N-glycosylation defects in certain *ALG* mutants. *FLC2** can be expressed under its native promoter when expressed from a high-copy number plasmid or under the strong *PGK1* promoter when integrated into the genome to increase glycosylation efficiency in *Δalg3Δalg11* cells [5].

Particular protozoa such as *Leishmania* sp. and *Trypanosomas* only possess the catalytic subunit Stt3 of the OST complex [14, 15]. However, they possess three or four different Stt3 proteins with varying substrate specificities in regard of the LLO accepted. *Stt3* genes from *L. major*, *L. braziliensis*, *L. infantum, and Trypanosoma* can be expressed from either low-copy or high-copy number plasmids in various yeast strains (*Δalg11, Δalg3Δalg11*), and their ability to improve growth and N-glycosylation can be analyzed. In a *Δalg3Δalg11* strain background, the optimal construct comprised *L. braziliensis* LbSTT3_3 under the control of the *TEF1* promoter [5]. However, it is conceivable that the optimal expression levels should be determined experimentally as they might vary in different genetic backgrounds. Combined together, coexpression of Flc2* and LbSTT3_3 completely restores N-glycosylation defects when integrated into the genome of a *Δalg3Δalg11* strain [5].

1.4 Introduction of Human-Type Modifications in Yeast N-Glycans

Glycosyltransferases present in the Golgi apparatus are type II transmembrane proteins, comprising of a short N-terminal cytosolic tail, a transmembrane domain, a stem region, and a C-terminal catalytic domain. When expressed in yeast, mammalian glycosyltransferases may not localize correctly in the Golgi apparatus. To ensure correct localization, the catalytic domain of the desired glycosyltransferase can be genetically fused with a transmembrane domain of a yeast Golgi protein [2, 16].

Human GnTI (alpha-1,3-mannosyl-glycoprotein 2-beta-*N*-acetylglucosaminyltransferase) is a medial Golgi enzyme which is essential for initiation of the synthesis of hybrid and complex N-glycans. For proper localization of GnTI to the yeast Golgi apparatus, the catalytically active domain of hGnTI (amino acids 103–445) has been genetically fused to the N-terminal cytosolic, transmembrane, and stem domains of the yeast Kre2p (amino acids 1–100) [2, 5, 16]. The yeast Kre2p is a type II transmembrane protein localized to the early Golgi.

Human GnTII (alpha-1,6-mannosyl-glycoprotein 2-beta-*N*-acetylglucosaminyltransferase) is a Golgi enzyme catalyzing the addition of the beta-1,2-linked GlcNAc to the alpha-1,6-mannose. To localize GnTII to the Golgi apparatus in yeast, the catalytically active domain of GnTII (amino acids 88–443) has been genetically fused to the N-terminal cytosolic, transmembrane, and stem domain of the yeast Mnn2p (amino acids 1–36) [2, 5, 16]. The yeast Mnn2p is a type II transmembrane protein localized to the early Golgi. For convenient coexpression, GnTI and GnTII fusion constructs were placed under the bidirectional *GAL1-GAL10* promoter in the episomal pRS426 plasmid [5].

When introducing human-type glycan modifications in yeast, merely expressing mammalian glycosyltransferases may not be sufficient to obtain high efficiency in glycan modification. Expression of enzymes contributing to the transport or biosynthesis of UDP-GlcNAc may be necessary. *Kluyveromyces lactis* Yea4 (*KlYEA4*) is a Golgi-localized transmembrane protein which catalyzes the import of UDP-GlcNAc into the lumen of Golgi apparatus [17]. Coexpression of *KlYEA4* with GnTI and GnTII in *S. cerevisiae* increased the efficiency of GlcNAc addition to glycans [6]. A MALDI-TOF MS spectrum of the respective strain is shown in Fig. 1b.

2 Materials

2.1 Growth Media

1. Synthetic drop-out (SD) medium: 6.7 g/L yeast nitrogen base without amino acids, 2 g/L Hopkins drop-out mix lacking uracil or leucine and uracil, 20 g/L glucose, 0.2 M sorbitol. Adjust pH to 5.8 with KOH. For plates, add 20 g/L agar.
2. Expression medium: 6.7 g/L yeast nitrogen base without amino acids, 2 g/L Hopkins drop-out mix lacking uracil or leucine and uracil, 10 g/L raffinose, 0.2 M sorbitol. Adjust pH to 5.8 with KOH. For plates, add 20 g/L agar.
3. 200 mM sodium phosphate buffer, pH 6.5.
4. 20% (w/v) galactose.
5. 10 mg/mL BSA in water.
6. cOmplete™ Mini EDTA-free Protease Inhibitor Cocktail (Roche).

2.2 Yeast Strains and Plasmids

All strains are derived from the parental strain SS328 (MATα *ade2-*101 *his3Δ200 lys2-*801 *ura3-*52) and are described in Table 1. All plasmids are listed in Table 2. Most plasmids are derived from pRS4XX plasmid series carrying either LEU2 (pRS4X5) or URA3 (pRS4X6) selection marker and CEN/ARS (pRS41X) or 2 μ (pRS42X) origin of replication.

Table 1
Yeast strains

Strain	Genotype/features	References
YG1363	Matα *ade2*-101 *ura3*-52 *his3*D200 *lys2*-801 Δ*alg3*::*HIS3* Δ*alg11*::*HIS3*	[11]
YG1429	Derived from YG1363, UV-mutagenized and selected for better growth. Matα *ade2*-101 *ura3*-52 *his3*D200 *lys2*-801 Δ*alg3*::*HIS3* Δ*alg11*::*HIS3*	[5]
YAF39	Derived from YG1429, Matα *ade2*-101 *ura3*-52 *his3*D200 *lys2*-801 Δ*alg3*::*HIS3* Δ*alg11*::*HIS3* Δ*leu2*::KanMX4::LbSTT3_3::Flc2*	[5]
YMP01	YG1429, Δ*mnn1*::*KanMX4*	[6]
YMP14	YAF39, Δ*mnn1*::*NatNT2*	[6]

Overview of yeast strains with tailored N-glycan biosynthesis

Table 2
Plasmids

Plasmid	Features	References
pRS426-Gal1-Gal10	Episomal 2 μ origin plasmid, bidirectional *GAL1-GAL10* promoter, *URA3* selection	[23]
pRS415-Gal1	Episomal CEN/ARS plasmid, *GAL1* promoter, *LEU2* selection	[23]
pRS305K	Integrative vector targeting *LEU2* locus, G418 selection	[24]
pAX453	Derivative of pRS305K for coexpression of Flc2* and LbSTT3_3	[5]
pAX428	GnTI and GnTII expression under *GAL1-GAL10* promoter, *URA3* selection	[5]
pAX466	AP expression under *GAL1* promoter, *LEU2* selection	[5]
pAX549	IgG expression under *GAL1* promoter, *LEU2* selection	[5]
pMP002	*KlYEA4* expression under *GPD* promoter, *LEU2* selection	[6]
pMP010	GnTI and GnTII expression under *GAL1-GAL10* promoter and *KlYEA4* expression under *GPD* promoter, *LEU2* selection	[6]

Overview of expression plasmids for expression of Flc2* and LbSTT3_3, Golgi apparatus-targeted glycosyltransferases, and UDP-GlcNac transporter and model glycoproteins

2.3 Protein Purification

2.3.1 Purification of Yeast Acid Phosphatase

1. 100 mM phenylmethylsulfonyl fluoride (PMSF) in ethanol (*see* **Note 1**).
2. cOmplete™ Mini EDTA-free Protease Inhibitor Cocktail (Roche).
3. 0.22 μm filter unit.
4. 1 M Tris–HCl, pH 7.5.
5. 5 M NaCl.
6. 200 mM imidazole.

7. Ni-NTA agarose column (Qiagen) with 1 mL bed volume.
8. Wash buffer: 300 mM NaCl, 10 mM imidazole, 20 mM Tris–HCl, pH 7.5.
9. Elution buffer: 300 mM NaCl, 100 mM imidazole, 20 mM Tris–HCl, pH 7.5.

2.3.2 Purification of Human IgG

1. 100 mM phenylmethylsulfonyl fluoride (PMSF) in ethanol (*see* **Note 1**).
2. cOmplete™ Mini EDTA-free Protease Inhibitor Cocktail (Roche).
3. 0.22 μm filter unit.
4. 1 M Tris–HCl, pH 7.5.
5. 5 M NaCl.
6. Protein G sepharose column (GE Healthcare) with 1 mL bed volume.
7. Wash buffer: 300 mM NaCl, 20 mM Tris–HCl, pH 7.5.
8. Elution buffer: 300 mM NaCl, 100 mM glycine, pH 2.7.
9. Neutralization buffer: 2.5 M Tris–HCl, pH 8.5.

2.4 Glycan Sample Preparation

1. 10 mM Tris–HCl pH 7.4.
2. Lysis buffer: 10 mM Tris–HCl pH 7.4, cOmplete™ EDTA-free Protease Inhibitor Cocktail (Roche), 1 mM phenylmethylsulfonyl fluoride (PMSF) (*see* **Note 1**).
3. Acid-washed glass beads, 425–600 μm.
4. Disruptor Genie® cell disruptor (Scientific Industries).
5. Denaturing buffer: 2 M thiourea, 7 M urea, 2% SDS, 50 mM Tris–HCl pH 8 prepared freshly.
6. Reduction buffer: denaturing buffer with 10 mM dithiothreitol (DTT).
7. Iodoacetamide (IAA), 0.5 M in 50 mM Tris–HCl pH 8 (*see* **Note 2**).
8. PNGase F, glycerol-free (New England Biolabs), including 10× glycoprotein denaturing buffer, 10× GlycoBuffer 2, and 10% NP-40.
9. C18 SPE columns: Supelclean™ ENVI™-18 SPE tubes, bed wt. 100 mg (Sigma-Aldrich).
10. Graphitized carbon SPE columns: Supelclean™ ENVI-Carb™ SPE tubes, bed wt. 100 mg (Sigma-Aldrich).
11. Acetonitrile.
12. 2% (v/v) acetonitrile.
13. 95% (v/v) acetonitrile.

14. 70% (v/v) acetonitrile.
15. Vacuum concentrator.
16. 2-Aminobenzamide (2-AB) labeling solution: 0.35 M 2-AB, 1 M sodium cyanoborohydride ($NaCNBH_3$) in 7:3 (v/v) anhydrous dimethylsulfoxide (DMSO)/glacial acetic acid (*see* **Note 3**).
17. Whatman™ 3MM Chr chromatography paper.
18. 2 mL syringes.

2.5 Permethylation

1. Glass tubes with screw cap.
2. Sodium hydroxide (NaOH) pellets.
3. Dimethylsulfoxide (DMSO, anhydrous).
4. Methyl iodide (CH_3I).
5. Chloroform (CH_3Cl).
6. Acetonitrile.
7. 50% (v/v) acetonitrile.
8. Low protein binding microcentrifuge tubes.

2.6 MALDI-TOF

1. DHB matrix: 20 mg/mL 2,5-dihydroxybenzoic acid (DHB), 1 mM NaCl, 0.1% (v/v) trifluoroacetic acid (TFA) in 30% (v/v) acetonitrile.
2. Matrix-assisted laser desorption/ionization (MALDI) time-of-flight (TOF) mass spectrometer (MS) or tandem MS (MS/MS), e.g., 4800 Proteomics Analyzer (Applied Biosystems) or UltrafleXtreme III (Bruker Daltonics).

3 Methods

3.1 Growth of Yeast Strains

Plasmid-containing strains are maintained on solidified SD-medium. The temperature sensitive strain YG1363 is grown at 25 °C. The UV-mutagenized strains YG1429 and strains derived from YG1429 are grown at 28 °C. All cultivations are done in batch mode using shake flasks with 50 mL medium at 180 rpm. All liquid cultures are conducted in expression medium.

3.1.1 Growth of Yeast for Expression of Model Glycoproteins

1. Inoculate a single colony into 5 mL expression medium and grow until an OD of 2 is reached.
2. Prepare expression medium supplemented with 20 mM phosphate buffer, 0.05 mg/mL BSA, and half-strength cOmplete™ Mini EDTA-free Protease Inhibitor Cocktail.
3. Inoculate 50 mL supplemented expression medium to a starting A_{600} of 0.2.

4. Grow until OD of 0.8–1 is reached and induce culture with galactose to final concentration of 2%.
5. Grow for 36 h at 20 °C.
6. Continue with Subheading 3.2.

3.1.2 Growth of Yeast for Isolation of Cell Wall Proteins

1. Inoculate a single colony into 5 mL expression medium and grow until an OD of 2 is reached.
2. Inoculate the culture in 50 mL expression medium to a starting A_{600} of 0.2, and grow until A_{600} of 1.0 is reached.
3. Collect 50 OD units of cells by centrifugation (7 min 3200 × *g*), and resuspend the cells in expression medium containing 2% galactose.
4. Grow for 24 h.
5. Measure OD and collect 50 OD units of cells by centrifugation at 3200 × *g*.
6. Continue with Subheading 3.3 or store pellet at −20 °C.

3.2 Purification of Model Glycoproteins

The protocols given below are suitable for manual small-scale purifications of two model glycoproteins, yeast acid phosphatase (AP) and human IgG. All steps are performed at +4 °C.

3.2.1 Purification of Yeast Acid Phosphatase

Acid phosphatase (AP) is a highly glycosylated protein which is secreted into the culture supernatant. A His-tagged AP can be expressed in the glycoengineered strains under control of the inducible *GAL1* promoter. A His-tagged AP can be purified using immobilized metal affinity chromatography.

1. Clear culture supernatant by centrifugation at 15,000 × *g*, 4 °C for 20 min.
2. Add 5 mM PMSF and protease inhibitor cocktail at the recommended concentration to the cleared supernatant.
3. Adjust cleared culture supernatant to 300 mM NaCl, 10 mM imidazole, 20 mM Tris–HCl, pH 7.5.
4. Pass the supernatant through a 0.22 μm filter unit.
5. Equilibrate a Ni-NTA agarose column with 5–10 column volumes of wash buffer.
6. Load filtrate on the column at a flow rate of 1 mL/min.
7. Wash the column with 10 column volumes of loading buffer.
8. Elute AP with elution buffer, and collect 1 mL fractions.
9. Test fractions for the presence of AP using activity assay [18].

3.2.2 Purification of Human IgG

A second suitable model protein is human IgG. Fused to MAT alpha signal sequence, IgG is secreted into the culture supernatant. IgG can be expressed in the glycoengineered strains under control of the inducible *GAL1* promoter. Secreted IgG is purified using protein G sepharose affinity chromatography.

1. Clear culture supernatant by centrifugation at 15,000 × *g*, 4 °C for 20 min.
2. Add 5 mM PMSF and protease inhibitor cocktail at the recommended concentration to the cleared supernatant.
3. Adjust cleared culture supernatant to 300 mM NaCl, 20 mM Tris–HCl, pH 7.5.
4. Pass the supernatant through a 0.22 μm filter unit.
5. Equilibrate a protein G sepharose column with 5–10 column volumes of wash buffer.
6. Load filtrate on the column at a flow rate of 1 mL/min.
7. Wash the column with 10 column volumes of loading buffer.
8. Elute the bound IgG with elution buffer, and collect 1 mL fractions into tubes containing 50 μL neutralization buffer.
9. Test fractions for the presence of IgG using ELISA assay [19].

3.3 Isolation of Cell Wall Proteins

In this protocol, proteins covalently linked to the yeast cell wall are prepared for glycan release and purification. Yeast cells are lysed using glass beads, and the cell wall proteins are reduced and alkylated. **Steps 1–4** should be performed at 4 °C.

1. Resuspend a cell pellet corresponding to 50 OD units in 500 μL of lysis buffer, and add 500 μL of acid-washed glass beads.
2. Break the cells by vortexing with a cell disruptor for 1 h at 4 °C.
3. Let the glass beads settle and collect the cell suspension in a microcentrifuge tube.
4. Wash the glass beads twice with 500 μL of 10 mM Tris–HCl pH 7.4 and vortex briefly. Let the glass beads settle and collect the suspension (pool with the cell suspension from **step 3**).
5. Centrifuge for 1 min at 16,000 × *g* and discard supernatant.
6. Resuspend the pellet in 1 mL of reduction buffer.
7. Incubate at 37 °C for 30 min to 2 h with shaking at 1400 rpm.
8. Add 50 μL of 0.5 M IAA in 50 mM Tris–HCl pH 8 to the sample.
9. Incubate at 37 °C for 1 h with shaking at 1400 rpm protected from light.
10. Centrifuge for 1 min at 16,000 × *g* and discard supernatant.
11. Wash the pellet five times with denaturing buffer: Resuspend the pellet in 1 mL of denaturing buffer, vortex for approx. 30 s, centrifuge for 1 min at 16,000 × *g*, and discard supernatant.

12. Wash the pellet five times with water: resuspend the pellet in 1 mL of water, vortex briefly, centrifuge for 1 min at 16,000 × *g*, and discard supernatant (*see* **Note 4**).

3.4 Deglycosylation of Purified Glycoproteins with PNGase F

N-glycans from purified proteins from Subheading 3.2 are released using PNGase F.

1. Concentrate fractions containing proteins to 200 μL using spin concentrators with a cutoff of 10 kDa.
2. Add 20 μL of 10× glycoprotein denaturing buffer to 160 μL of sample.
3. Denature proteins for 10 min at 95 °C.
4. Add 20 μL of 10× GlycoBuffer 2 and NP-40 to a final concentration of 1%.
5. Add 1 μL of PNGase F to the digestion reaction.
6. Incubate overnight at 37 °C with shaking at 230 rpm.

3.5 Deglycosylation of Cell Wall Glycoproteins with PNGase F

1. Resuspend the pellet obtained in **step 12** in Subheading 3.3 in 200 μL of PNGase F digestion mix containing 10 μL of 10× glycoprotein denaturing buffer, 20 μL of GlycoBuffer 2, 1% of NP-40, and 1 μL of PNGase F.
2. Incubate overnight at 37 °C with shaking at 230 rpm.

3.6 Purification of Glycans

After PNGase F digestion, the glycans released from cell wall proteins or purified glycoproteins are purified from the reaction mix with C18 and graphitized carbon SPE columns. Before loading samples in the C18 and graphitized carbon SPE columns, samples need to be adjusted to 2% acetonitrile and insoluble material needs to be removed.

3.6.1 Preparation of Cell Wall Protein Samples for Glycan Purification After PNGase F Digest

1. Add 4 μL of acetonitrile to the deglycosylated sample (volume 200 μL) to obtain 2% acetonitrile concentration.
2. Centrifuge for 1 min at 10,000 × *g* and collect supernatant.
3. Wash the pellet twice with 100 μL of 2% acetonitrile, centrifuge for 1 min at 10,000 × *g*, and collect the supernatant.
4. Purify the combined supernatants with C18 and graphitized carbon SPE columns.

3.6.2 C18 Purification

1. Wash a C18 SPE column twice with 500 μL of 95% acetonitrile.
2. Equilibrate the column twice with 500 μL of 2% acetonitrile.
3. Load the sample (adjusted to 2% acetonitrile) to the column and collect the flow-through.
4. Wash the column with 200 μL of 2% acetonitrile and collect the flow-through.

5. Purify the collected flow-through with a graphitized carbon SPE column (*see* Subheading 3.6.3).

3.6.3 Graphitized Carbon Purification

1. Wash a graphitized carbon SPE column twice with 500 μL of 95% acetonitrile.
2. Equilibrate the column twice with 500 μL of 2% acetonitrile.
3. Load the sample (adjusted to 2% acetonitrile) to the column and discard the flow-through.
4. Wash the column twice with 500 μL of 2% acetonitrile.
5. Elute the glycans with two times 200 μL of 70% acetonitrile.
6. Dry the sample in a centrifugal evaporator at 65 °C (*see* **Note 5**). If permethylation is performed after drying, dry the samples in a glass tube with a screw cap.

3.7 2-AB Labeling of Glycans and Post-Labeling Cleanup

Labeling of glycans with 2-aminobenzamide (2-AB) improves sensitivity in MALDI-TOF MS analysis. The reducing end of the glycan is derivatized with 2-AB by reductive amination using sodium cyanoborohydride ($NaCNBH_3$) as the reducing agent [20].

1. Add 20 μL of 2-AB labeling solution to the dried glycan sample and dissolve by vortexing well.
2. Incubate the samples for 2 h at 65 °C (*see* **Note 6**).
3. Prepare a cleanup cartridge by cutting two round disks of chromatography paper matching the size of a 2 mL syringe (diameter approx. 1 cm). Remove the piston from a 2 mL syringe and insert the paper disks to the bottom of the syringe.
4. Wash and equilibrate the cartridge twice with 450 μL of 30% acetic acid, twice with 450 μL of water, twice with 450 μL of 100% acetonitrile, and twice with 450 μl of 95% acetonitrile.
5. Add 380 μL of 100% acetonitrile to the labeled and cooled sample to obtain 95% acetonitrile concentration, and load the sample to the cartridge.
6. Wash the sample tube with 250 μL of 95% acetonitrile and load to the cartridge.
7. Wash the cartridge six times with 450 μL of 95% acetonitrile.
8. Elute the labeled glycans with two times 100 μL and once with 50 μL of water.
9. Dry the samples in a centrifugal evaporator at 65 °C (*see* **Note 5**).
10. Dissolve the samples in 200 μL of 2% acetonitrile, and purify with a graphitized carbon SPE column (*see* Subheading 3.6.3).

3.8 Permethylation of Glycans

Structural analysis of carbohydrates requires the determination of molecular mass, composition and sequence of monosaccharides, and their glycosidic linkages [21]. Permethylation enhances ionization efficiency and aids linkage analysis.

1. Grind a NaOH pellet and add 250 μL of DMSO.
2. Add 200 μL of the NaOH-DMSO slurry to a glass tube containing the dried glycan sample.
3. Add 200 μL of CH_3I.
4. Shake for 20 min at room temperature.
5. Add 200 μL of distilled water to quench the reaction.
6. Add 200 μL of chloroform.
7. Extract with distilled water and discard the aqueous phase. Repeat until the chloroform phase is clear.
8. Evaporate the chlorofom phase to dryness at 95 °C.
9. Add 200 μL of 100% acetonitrile and transfer to a low protein binding microcentrifuge tube.
10. Add 200 μL of 50% acetonitrile, vortex, and transfer to the same microcentrifuge tube.
11. Evaporate the sample to dryness at 95 °C.

3.9 MALDI-TOF MS Analysis of Glycans

MALDI-TOF mass spectrometry in MS or tandem MS (MS/MS) mode is used to analyze permethylated and 2-AB-labeled glycans.

1. Dissolve the dry glycan sample in 10 μL of water.
2. Mix the glycan sample 1:1 with DHB matrix.
3. Spot 1 μL of the sample-matrix mix on a MALDI target plate.
4. Measure the glycan samples with MALDI-TOF MS using a positive reflectron mode.
5. To obtain structural data, analyze permethylated glycans with CID tandem MS using argon as the collision gas.
6. Perform structural assignment for MS spectra using *GlycoMod* tool (http://web.expasy.org/glycomod/) and for MS/MS spectra as described in detail [22].

4 Notes

1. Prepare 100 mM PMSF stock solution in ethanol, store at −20 °C for up to 3 months, and add to lysis buffer briefly before use.
2. Iodoacetamide is sensitive to light. Prepare the solution briefly before use, and protect from light.

3. Prepare 2-AB labeling solution briefly before use by first dissolving 2-AB in 7:3 DMSO/acetic acid and then dissolving $NaCNBH_3$ in the 2-AB solution. The labeling solution can be heated to 65 °C to improve dissolution of $NaCNBH_3$.
4. After washing, continue with PNGase F digest or store pellets at −20 °C.
5. If a centrifugal evaporator is not available, samples can also be dried in a heat block at 65 °C under a stream of air.
6. Avoid contact with moisture during the 2-AB labeling reaction.

References

1. Walsh G, Jefferis R (2006) Post-translational modifications in the context of therapeutic proteins. Nat Biotechnol 24:1241–1252. https://doi.org/10.1038/nbt1252
2. Jacobs PP, Geysens S, Vervecken W et al (2009) Engineering complex-type N-glycosylation in *Pichia pastoris* using GlycoSwitch technology. Nat Protoc 4:58–70. https://doi.org/10.1038/nprot.2008.213
3. Hamilton SRR, Bobrowicz P, Bobrowicz B et al (2003) Production of complex human glycoproteins in yeast. Science 301 (5637):1244–1246. https://doi.org/10.1126/science.1088166
4. Li H, Sethuraman N, Stadheim TA et al (2006) Optimization of humanized IgGs in glycoengineered *Pichia pastoris*. Nat Biotechnol 24:210–215. https://doi.org/10.1038/nbt1178
5. Parsaie Nasab F, Aebi M, Bernhard G, Frey AD (2013) A combined system for engineering glycosylation efficiency and glycan structure in *Saccharomyces cerevisiae*. Appl Environ Microbiol 79:997–1007. https://doi.org/10.1128/AEM.02817-12
6. Piirainen MA, Boer H, De Ruijter JC, Frey AD (2016) A dual approach for improving homogeneity of a human-type N-glycan structure in *Saccharomyces cerevisiae*. Glycoconj J 33:189–199. https://doi.org/10.1007/s10719-016-9656-4
7. Helenius A, Aebi M (2004) Roles of N-linked glycans in the endoplasmic reticulum. Annu Rev Biochem 73:1019–1049. https://doi.org/10.1146/annurev.biochem.73.011303.073752
8. Gemmill TR, Trimble RB (1999) Overview of N- and O-linked oligosaccharide structures found in various yeast species. Biochim Biophys Acta 1426:227–237
9. Munro S (2001) What can yeast tell us about N -linked glycosylation in the Golgi apparatus? FEBS Lett 498:223–227
10. Moremen KW, Tiemeyer M, Nairn AV (2012) Vertebrate protein glycosylation: diversity, synthesis and function. Nat Rev Mol Cell Biol 13:448–462. https://doi.org/10.1038/nrm3383
11. Helenius J, Ng DT, Marolda CL et al (2002) Translocation of lipid-linked oligosaccharides across the ER membrane requires Rft1 protein. Nature 415:447–450. https://doi.org/10.1038/415447a
12. Abe H, Takaoka Y, Chiba Y, et al (2009) Development of a valuable yeast strain using a novel mutagenesis technique for the effective production of therapeutic glycoproteins. Glycobiology 19:428–436. https://doi.org/10.1007/978-1-4939-2760-9_9
13. Xu S, Zhang G-Y, Zhang H et al (2016) Effects of Rho1, a small GTPase on the production of recombinant glycoproteins in *Saccharomyces cerevisiae*. Microb Cell Factories 15:179. https://doi.org/10.1186/s12934-016-0575-7
14. Kelleher DJ, Banerjee S, Cura AJ et al (2007) Dolichol-linked oligosaccharide selection by the oligosaccharyltransferase in protist and fungal organisms. J Cell Biol 177:29–37. https://doi.org/10.1083/jcb.200611079
15. Samuelson J, Banerjee S, Magnelli P et al (2005) The diversity of dolichol-linked precursors to Asn-linked glycans likely results from secondary loss of sets of glycosyltransferases. Proc Natl Acad Sci U S A 102:1548–1553. https://doi.org/10.1073/pnas.0409460102
16. Nett JH, Stadheim TA, Li H et al (2011) A combinatorial genetic library approach to target heterologous glycosylation enzymes to the endoplasmic reticulum or the Golgi apparatus

of *Pichia pastoris*. Yeast 28:237–252. https://doi.org/10.1002/yea.1835

17. Abeijon C, Robbins PW, Hirschberg CB (1996) Molecular cloning of the Golgi apparatus uridine diphosphate-N-acetylglucosamine transporter from *Kluyveromyces lactis*. Proc Natl Acad Sci U S A 93:5963–5968
18. Frey AD, Aebi M (2015) An enzyme-based screening system for the rapid assessment of protein N-glycosylation efficiency in yeast. Glycobiology 25:252–257. https://doi.org/10.1093/glycob/cwu134
19. de Ruijter JC, Frey AD (2015) Analysis of antibody production in *Saccharomyces cerevisiae*: effects of ER protein quality control disruption. Appl Microbiol Biotechnol 99:9061–9071. https://doi.org/10.1007/s00253-015-6807-7
20. Bigge JC, Patel TP, Bruce JA et al (1995) Nonselective and efficient fluorescent labeling of glycans using 2-amino benzamide and anthranilic acid. Anal Biochem 230:229–238. https://doi.org/10.1006/abio.1995.1468
21. Ciucanu I (2006) Per-O-methylation reaction for structural analysis of carbohydrates by mass spectrometry. Anal Chim Acta 576:147–155. https://doi.org/10.1016/j.aca.2006.06.009
22. Yu SY, Wu SW, Khoo KH (2006) Distinctive characteristics of MALDI-Q/TOF and TOF/-TOF tandem mass spectrometry for sequencing of permethylated complex type N-glycans. Glycoconj J 23:355–369. https://doi.org/10.1007/s10719-006-8492-3
23. Mumberg D, Muller R, Funk M (1994) Regulatable promoters of *Saccharomyces cerevisiae*: comparison of transcriptional activity and their use for heterologous expression. Nucleic Acids Res 22:5767–5768
24. Taxis C, Knop M (2006) System of centromeric, episomal, and integrative vectors based on drug resistance markers for *Saccharomyces cerevisiae*. BioTechniques 40:73–78

Chapter 11

Improving the Production of Cofactor-Containing Proteins: Production of Human Hemoglobin in Yeast

Olena P. Ishchuk, José L. Martínez, and Dina Petranovic

Abstract

Human hemoglobin is an essential protein, whose main function as an oxygen carrier is indispensable for life. Hemoglobin is a cofactor-containing protein with heme as prosthetic group. Same as in humans, heme is synthesized in many organisms in a complex pathway involving two cellular compartments (mitochondria and cytosol), which is tightly regulated. Red blood cells (erythrocytes) are specialized and adapted for production and transport of the hemoglobin molecules. In addition to oxygen binding, hemoglobin can participate in a variety of chemical reactions by its iron and heme and may become toxic when released from erythrocytes. Hemoglobin is a major target for the development of blood substitutes/oxygen carriers, and therefore its microbial production is attractive, as it may provide a cheap and reliable source of human hemoglobin. Significant efforts have been dedicated to this task for the last three decades. Moreover since the first generation of cell-free blood substitutes based on unmodified hemoglobin failed human trials, mutant forms became of great interest.

In this chapter we summarize the existing knowledge about human hemoglobin, challenges of its microbial production, and its improvement, with a particular focus upon yeast as production host.

Key words Hemoglobin, Recombinant protein production, Metabolic engineering, Heme, Yeast cell factories, *Saccharomyces cerevisiae*

1 Introduction to Hemoglobin

Cofactors and prosthetic groups are important components of many enzymes and proteins and play crucial roles in their activity. The loss of these cofactors or prosthetic groups is usually associated with loss of function and/or catalytic activity. These molecules can be covalently or non-covalently bound to the peptide and have different nature: can be inorganic (metals, non-covalent interaction, coordination bonds) or organic [e.g., heme in hemoglobin (non-covalently bond), vitamin derivatives such as thiamine pyrophosphate (in pyruvate dehydrogenase and decarboxylase, non-covalently bound), pyridoxal phosphate (in 5-aminolevulinate synthase, covalently bound)]. Due to the complexity of some groups and their biosynthetic pathways, cofactor-containing

Brigitte Gasser and Diethard Mattanovich (eds.), *Recombinant Protein Production in Yeast*, Methods in Molecular Biology, vol. 1923, https://doi.org/10.1007/978-1-4939-9024-5_11,

proteins can be challenging to produce. The most abundant class of cofactor-containing proteins of industrial importance are oxidoreductases: flavoproteins containing flavin mononucleotide (FMN) and flavin adenine dinucleotide (FAD) such as hydrogenases, oxidases, and flavocytochromes; metalloproteins, which contain metal ions (Fe, Fe-S, Mo, Cu, Zn, Mn, Ni, Va) such as nitrogenases, nitrate reductases, sulphite oxidases, superoxide dismutases, hydrogenase, and urease; quinoproteins containing quinone such as quinoprotein glucose dehydrogenases (sGDH, mGDH), quinoprotein glycerol dehydrogenase, and methanol dehydrogenase; and heme proteins such as hemoglobins, myoglobins, peroxidases, cytochromes, and catalases [1]. Catalases, which perform hydrogen peroxide dismutation, are used in the textile industry for bleaching [2]. Glucose oxidase (flavoprotein oxidase), which catalyzes the glucose oxidation to gluconic acid, has been recombinantly produced and used in different industries (food, pharmaceuticals, clinical, etc.) [3]. Hydrogenases, which catalyze hydrogen production, are of interest to biofuel industry [4]. Several other cofactor proteins are used for the development of biosensors.

Hemoglobin (Hb) is an essential protein of vital importance to many animals, including humans. It is the main component of red blood cells (hemoglobin represents more than 90% of the protein content in erythrocytes and up to 35% of their total weight [5, 6]). Its main physiological role is the transport of oxygen from lungs throughout the body into tissues. Mammals and birds possess tetrameric hemoglobins, which are composed of two molecules of two similar globin chains (Fig. 1c,d), each globin chain bearing one non-covalently bound heme group (Fig. 1a,b).

The heme molecule consists of one protoporphyrin IX with one central iron atom coordinated by four nitrogen atoms that are part of the porphyrin ring. There are three major biologically important types of heme (*a*, *b*, *c*), which can be found in different hemeproteins of eukaryotes, and they differ by the modifications of porphyrin ring and attachment to its apoprotein (covalent or non-covalent). While heme *b* [found in hemoglobin, myoglobin, mitochondrial *bc*1 complex (Fig. 1a)] and heme *a* (cytochrome *c* oxidase) prosthetic groups are bound non-covalently, heme *c* (cytochromes *c* and *c*1) is covalently bound to the protein part [10]. The iron atom of heme can act either as an electron acceptor or donor and thus alternates between the ferrous (Fe^{2+}) and ferric (Fe^{3+}) states. Under normal physiological conditions, the iron of heme is present in its ferrous form (Fe^{2+}). This hemoglobin form is called oxyhemoglobin and is responsible for oxygen binding during physiological respiration (Fig. 2). Oxyhemoglobin binds oxygen with high affinity in the lungs [where the partial pressure of oxygen (pO_2) is high (~100 torr)] and releases in the tissues [where pO_2 is lower (20 torr)] [11]. Hemoglobin has an allosteric regulator 2,3-bisphospho-glycerate (2,3-BPG), which is normally present in

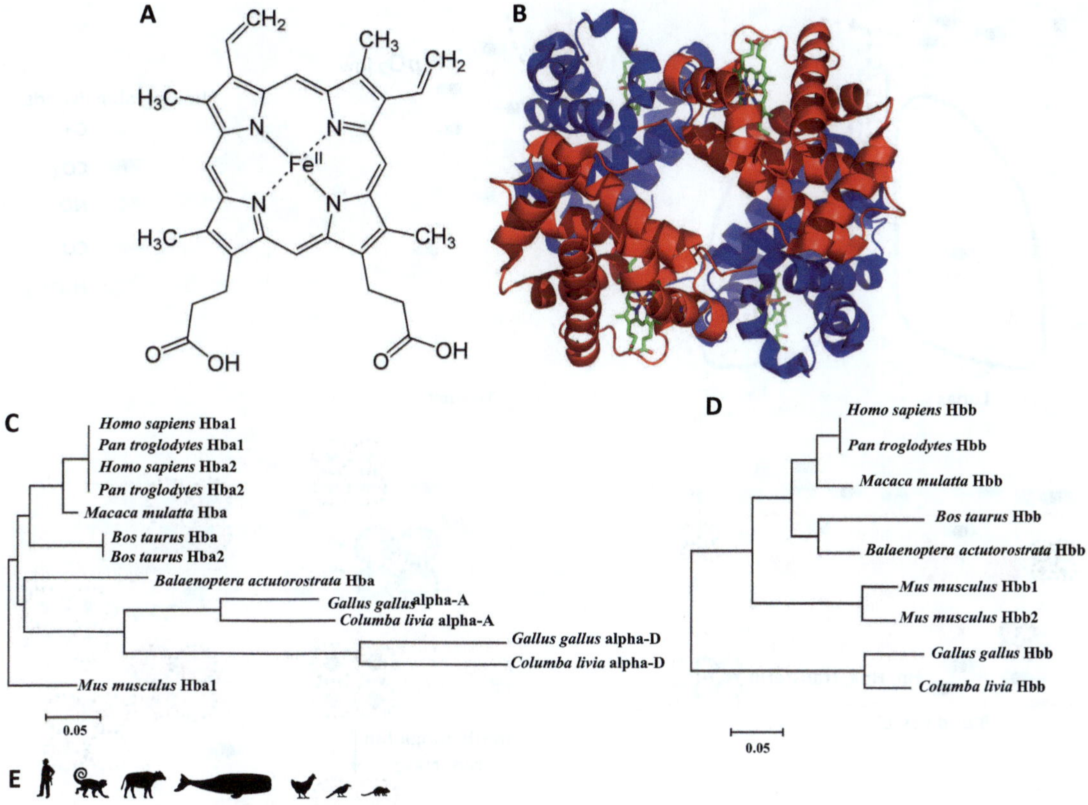

Fig. 1 (**a**) The structure of heme *b* (image by Yikrazuul 2010), which is protoporphyrin IX with ferrous iron in the center. (**b**) The tertiary structure of human hemoglobin HbA (tetramer). The heme prosthetic groups are shown in green, α-globin chains are shown in red, and β-globin chains are shown in blue [image by Richard Wheeler (Zephyris) 2007]. The evolutionary relationships of α-globin protein sequences (**c**) and β-globin protein sequences (**d**). The evolutionary analysis of sequences of globin proteins of different organisms was performed in MEGA7 software [7]. The evolutionary history was inferred using neighbor-joining method [8]; the evolutionary distances were computed using Poisson correction method [9] and are in units of the number of amino acid substitutions per site. Sequences of α-globins [*Homo sapiens HBA1* (accession number AAC97373), *H. sapiens HBA2* (AAB59407), *Mus musculus HBA1* (NP_032244), *Bos taurus HBA* (NP_001070890), *B. taurus HBA2* (AAI33478), *Pan troglodytes HBA1* (NP_001036092), *P. troglodytes HBA2* (NP_001036091), *Gallus gallus* alpha-A (AAL35404), *G. gallus* alpha-D (AAL35403), *Macaca mulatta HBA* (NP_001038189), *Columba livia* alpha-A globin (BAA19669), *C. livia* alpha-D globin (BAA19668), *Balaenoptera acutorostrata HBA* (P18971)] and β-globins [*H. sapiens HBB* (AAA16334), *M. musculus HBB1* (P02088), *M. musculus HBB2* (P02089), *B. taurus HBB* (NP_776342), *P. troglodytes HBB* (P68873), *G. gallus HBB* (P02112), *M. mulatta HBB* (NP_001157900), *C. livia HBB* (P11342), and *B. acutorostrata HBB* (P18984)] were used to build the phylogenetic tree. (**e**) Graphic representation of the organisms, which sequences were used for analysis

erythrocytes. It binds preferentially to oxyhemoglobin and decreases its oxygen affinity. Thus it enables the efficient oxygen release by hemoglobin in tissues and provides the efficient oxygen carrier properties of the red blood cells [12]. The oxygen affinity of the human hemoglobin (Hb) is inversely related to the pH and the carbon dioxide (CO_2) concentration, which is known as the Bohr

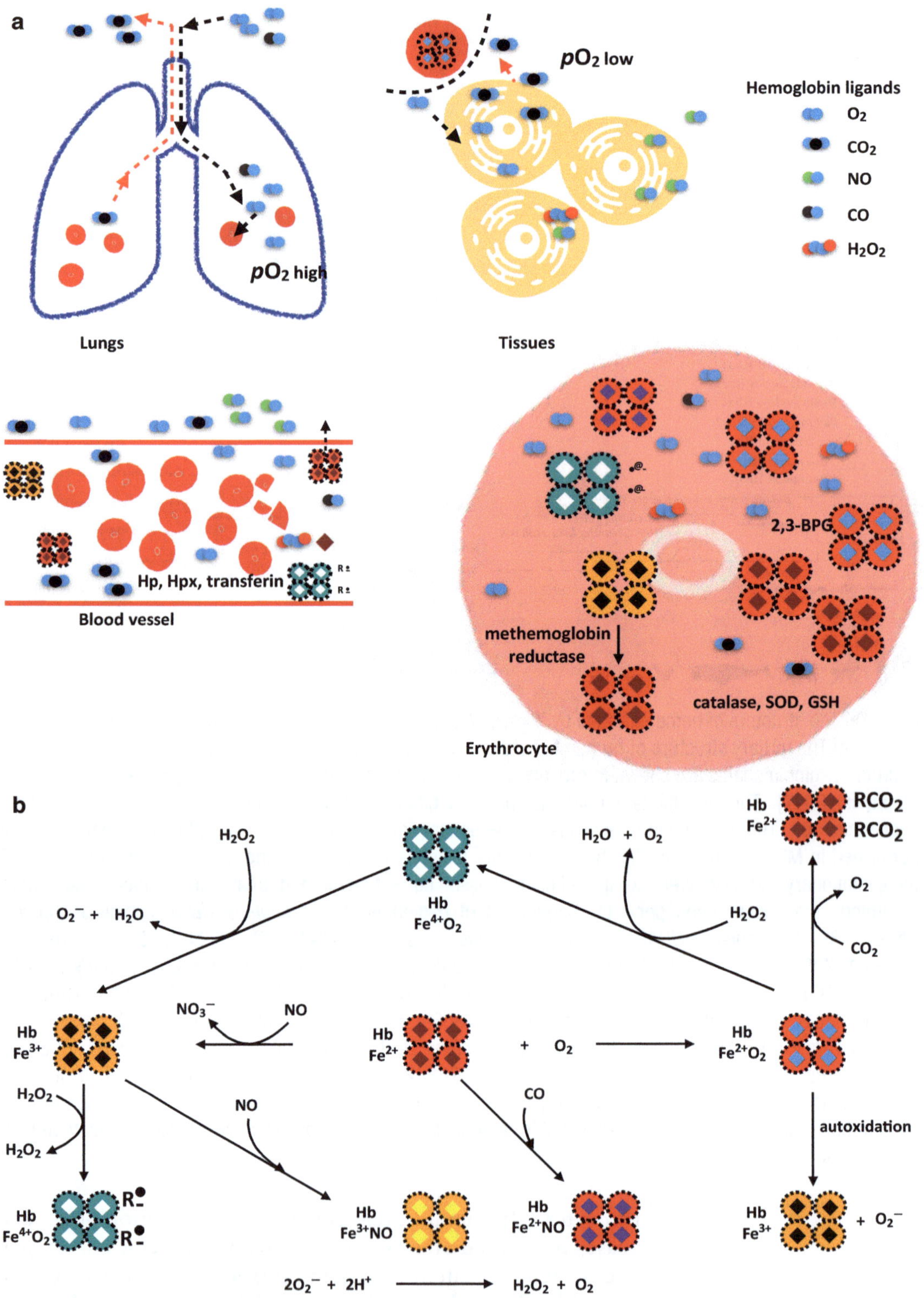

Fig. 2 Heme and hemoglobin reactions. (**a**) Human hemoglobin functions. It is the main protein of erythrocytes, and its main function is oxygen carrier [it binds oxygen in the lungs (high pO_2) and releases oxygen in the tissues (low pO_2)]. Binding of carbon dioxide to the protein part of hemoglobin enhances the oxygen release in

effect [13]. The binding of oxygen changes the quaternary structure of hemoglobin and thus enables the "communication" between sub-units in hemoglobin tetramer and provides the cooperative oxygen binding [11]. Hemoglobin also functions as CO_2 carrier; however only 10% of the CO_2 is carried this way, as the plasma transports the majority of CO_2. Hemoglobin binds CO_2 at N-terminus of the α-chain, which destabilizes the oxyhemoglobin conformation, and this leads to oxygen release by Hb (Fig. 2).

The autoxidation of oxyhemoglobin leads to the formation of ferric hemoglobin (methemoglobin), which loses the ability to bind oxygen and superoxide anions. The superoxide anions are the source of hydrogen peroxide and cause further oxidation of iron in hemoglobin (Fig. 2). Under physiological conditions, the amount of methemoglobin in blood is not higher than 1–2%, and its level is regulated by the enzyme methemoglobin reductase [5]. In addition to oxygen, hemoglobin can react with other ligands, like nitric oxide (NO), hydrogen peroxide (H_2O_2), and carbon monoxide (CO) (Fig. 2). NO, which is produced in endothelial cells, is an important agent in the regulation of blood pressure [14, 15]. When Hb is released from erythrocytes (under pathological conditions, e.g., hemolysis), it rapidly binds NO and forms methemoglobin ($HbFe^{3+}$) and nitrate or iron-nitrosyl hemoglobin. Methemoglobin further reacts with NO resulting in HbNO formation [15]. The scavenging of NO by Hb results in hypertension [16, 17]. The iron atom of the methemoglobin can react with H_2O_2-generating ferryl hemoglobin (Fe^{4+}) (Fig. 2), which is a strong oxidant shown to be able to oxidize various substrates in vitro and can lead to oxidative damage in the cell [18]. Although CO is normally produced in cells in low amounts during heme catabolism, the exposure to this gas is toxic, since ferrous Hb binds CO with high affinity and this ligand alters its properties. The dissociation of bound CO from hemoglobin is low, thus CO binding decreases the oxygen saturation and oxygen content of blood [19].

In erythrocytes, the oxidative damage (methemoglobin and ferryl hemoglobin formation) is decreased mainly by the action of NADPH-dependent methemoglobin reductase, which reduces the methemoglobin (Fig. 2) [20]. Among the other factors, hemoglobin contributes to generation of ROS. The ROS level is reduced by

Fig. 2 (continued) the tissues. In addition to oxygen and carbon dioxide, hemoglobin can react with other ligands encountered inside or outside the erythrocytes (NO, CO, H_2O_2) causing toxicity by generating ROS and hemoglobin molecules, which cause further oxidative damage ($HbFe^{3+}$, methemoglobin and $HbFe^{4+}$, ferryl hemoglobin) or unable to bind oxygen ($HbFe^{3+}$). When released from erythrocytes (e.g., hemolysis) or used cell-free (HBOCs), hemoglobin can react with ligands in blood plasma or/and migrate to the surrounding tissue causing lipid oxidation and NO scavenging (resulting in hypertension). Natural scavengers of toxic effects of hemoglobin and heme are present in erythrocytes (methemoglobin reductase, catalase, SOD, GSH) and blood plasma (Hp, haptoglobin and Hpx, hemopexin). (**b**) Reactions mediated by hemoglobin and heme antioxidant systems of catalase, superoxide dismutase, glutathione peroxidase, peroxiredoxin-2, and vitamin C [20, 21]. Natural scavengers diminish toxic effects of cell-free Hb in blood: haptoglobin (Hp, binds hemoglobin), hemin-binding plasma glycoprotein hemopexin (Hpx, binds heme), and transferrin (binds iron) (Fig. 2) [22].

There are different hemoglobin types in human, which dominate during different ontogeny periods. In addition to embryonic hemoglobin (HbE), three types of other hemoglobins are found in healthy adults and are tetramers of different globin chains: HbA (~97%), HbA2 (~2–3%), and HbF (~1%) [23–25]. Embryonic hemoglobin has four forms depending on the combination of five different globin chains ζ-, ϵ-, α-, β-, and γ [HbE Gower 1 (ζ2ϵ2), HbE Gower 2 HbE Portland I (ζ2γ2), and HbE Portland II (ζ2β2)]. HbEs are present in developing erythroblasts during first several weeks after fertilization [26–29]. After the 6th–7th week of gestation, the HbE is switched to HbF (fetal hemoglobin, α2γ2), which is further dominated by HbA (α2β2) in adults (Fig. 1b) due to the sequential changes in their gene expression during different ontogeny periods [26, 27, 29]. Whereas at the time of birth the HbF represents the 70% of the total hemoglobin, by the age of 1 year, the HbA dominates and comprises 97% [24]. The expression of both α- and β-globin genes is tightly coordinated and is required for efficient HbA assembly; otherwise the changes in the expression lead to blood disorders, such as thalassemia [30].

There is a difference in 39 amino acid residues in the sequence of β and γ chains, and this influences the physical and biochemical properties of HbF and HbA. Both forms differ in their affinity to oxygen, stability, and redox activity. HbF shows a higher affinity to oxygen due to a more stable T conformation (deoxy state) of the quaternary structure of hemoglobin. The T conformation has an increased number of favorable electrostatic and hydrophobic protein-solvent interactions, and a larger hydrophobic area is exposed to the solvent. The amino acid substitution at H21(143) His β → Ser γ stabilizes HbF T conformation and increases the binding of organic phosphates (e. g., 2,3-BPG) [31]. The HbF is also considerably more stable as tetramer compared to HbA. Its dissociation constant is 68-fold lower. The sequence of N-termini is likely to contribute to the interactions between the molecules and tetramer stability [32]. HbF causes less DNA damage when cell-free [33]. These features may have evolutionary advantages for the developing fetus and for the newborn [33, 34]. Therefore, HbF is a more attractive target for the development of anemia therapies (e.g., for sickle-cell disease and β-thalassemia caused by mutations in β-globin-encoding gene) [35–38]. In addition, HbF is an attractive target for the development of cell-free oxygen carriers due to these superior properties.

1.1 Need for Development of Recombinant Hemoglobins

Patients who experience massive blood losses due to injury or major surgery, or suffer from anemia, cancer, severe infections, or bone marrow failure, require blood restoration. Currently, blood transfusion is a common medical procedure to treat these pathologies and implies the use of red blood cells, platelets, cryoprecipitate, or plasma isolated from donor blood [39]. Donated blood may be stored in a refrigerated container for a maximum of 6 weeks. After this period, it has to be discarded and is no longer efficient for medical purposes. Moreover, approximately 15–20% of the red blood cell population decreases within the first 24 h of storage. Donated blood is in demand. According to the World Health Organization, from 112.5 million blood donations collected globally, half of the donations are collected in high-income countries, which correspond to 19% of the world population. Red blood cells and plasma carry several antigens, which define the so-called blood groups. Although the most widely recognized and therefore taken into account for blood transfusions are the A, B, and the Rh antigens, there are currently over 40 different types of antigens recognized that are not catalogued as a proper blood group. Thus patients that are in need of a blood transfusion are often limited to compatible blood group types. In some cases blood transfusion procedures are rejected based on religious views of the patients. Currently, one unit of the transfused blood in US hospitals costs $155 in average [40], and one unit is the typical amount of blood collected from one blood donor. For car accident victims with severe blood loss, approximately 50 units of blood are needed. Blood bank supplies are shrinking in developed countries, and prices for donated blood are rising. This market is difficultly regulated by conventional means, taking into account the availability of donors. Thus, the usage of blood transfusions has several limitations and imposes serious problems in acute situations and in the event of disasters.

There are two major types of blood substitutes, which are currently under development: perfluorocarbon-based (PFCOCs) and hemoglobin-based oxygen carriers (HBOCs). Unfortunately, neither of them meets all blood-related functions but exclusively oxygen transport. In PFCOCs, perfluorocarbons are inert, nontoxic liquids with straight or cyclic hydrocarbon chains (C_nF_{2n+2}). They carry oxygen dissolved in liquid, and their capacity to carry oxygen has direct linear relationship to oxygen pressure [41]. PFCOCs have only been shown to be effective in acute transfusions, as they can only provide marginal oxygen supply to tissues.

HBOC substitutes are prepared either from dated human blood (not allowed for medical utilization anymore) or bovine blood, or produced by microbial hosts, and carry cell-free hemoglobin molecules [42]. The blood substitutes based on cell-free recombinant hemoglobin (rHb) could have some advantages over the erythrocyte-based blood products, as their supply would be stable and continuous. It would not have the limitation of compatibility, as it would be antigen free. rHb would not impose any risks of known and unknown blood-transmitted diseases (hepatitis B, HIV, etc.) and would have longer storage time. Additionally, its functional properties (ligand binding properties, stability, turnover in vivo, etc.) could be tailor-made to target different patient groups and their specific needs. The main disadvantages of first-generation cell-free HBOCs are their very high oxygen affinity (do not release bound oxygen effectively), oxidative toxicity, NO scavenging, and short blood circulation time. All of this is not optimal and toxic, in animal models and human trials, for example, causing rise in blood pressure and renal and neuronal toxicity [42, 43]. Unprotected by erythrocytes and its natural scavengers, cell-free Hb is toxic. Further development of HBOCs to diminish their toxic effects by directed mutagenesis and the improvement of their production in microbial hosts are needed.

1.2 Microbial Production of Hemoglobins

Both prokaryotic and eukaryotic microbial platforms are used in modern bio-based economy, for example, to produce numerous fuels, chemicals, nutraceuticals, enzymes, pharmaceuticals, and biologicals [44, 45]. Advances in metabolic and genetic engineering, and recently in synthetic biology, are creating possibilities to produce large amounts of recombinant proteins, ideally using cheap and renewable substrates [44–46]. Modern biotechnology uses bacterial, fungal, mammalian, and insect cell cultures as platforms for different protein production [44, 46]. Researchers have also developed different hosts to produce human hemoglobin starting decades ago, e.g., bacteria (*Escherichia coli* [47]), yeast (*S. cerevisiae* [48]), animals (transgenic mice [49] and pigs [50]), and plants (tobacco [51]). The production of recombinant hemoglobin is challenging, as it requires the incorporation of heme and correct folding of the hemoglobin peptides in the host cells. Although there is a need for heterologous expression of human globin-encoding genes, most organisms, including microorganisms, have their own heme biosynthesis pathway. The inclusion of heme into the globin chains is co-translational [52]. The availability of both types of globin chains and heme molecule at the same time was shown to facilitate correct folding of hemoglobin [53–55]. Hoffman and co-authors [54] co-expressed α- and β-globin-encoding genes in *E. coli* from a single operon to obtain stoichiometric amounts of both chains. To enhance the expression of both globin genes, their sequence was optimized for *E. coli* codon usage. The rHb expressed in *E. coli* had biochemical properties similar to native human Hb totaling >5% of total cell protein of the host. Early expression systems of globins in *E. coli* were associated with production of polypeptide chains with an unprocessed methionine on N-termini [54, 56], which altered some of the biochemical characteristics of rHb like reduced Bohr effect [57, 58]. The expression of heterologous proteins in *E. coli* often results in production of products with extra N-methionine since they are less efficiently removed, particularly from recombinant proteins [59]. Further, the co-expression of methionine amino-peptidase (Met-AP) and α- and β-globin-encoding genes in *E. coli* improved the processing of N-terminal methionine of both globins [58]. Since the purified rHb lacks 2,3-BPG, it displays very high oxygen affinity (inefficient release of bound oxygen); site-directed mutagenesis was used to reduce the oxygen affinity normally regulated by 2,3-BPG in erythrocytes [56, 60] and to improve the soluble protein production in *E. coli* [61]. The cell-free Hb tetramer easily dissociates, and resulting dimers are rapidly cleared from the circulatory system by kidneys, which could result in kidney damage [60]. To prevent the dissociation of hemoglobin tetramer into αβ dimers, di-α-globin fusion protein and β-globin polypeptides were constructed and expressed in *E. coli* [60–62]. Free α-globin often precipitates, which inhibits the di- and tetramer assembly [63]. The α-hemoglobin-stabilizing protein (encoded by the *AHSP* gene), which is present in human erythrocytes, decreases the oxidation and precipitation of α-globin by interaction with the N-terminus of its H helix and can be easily displaced by β-globin chain to form HbA [63–65]. The C-terminus of α-globin interacts with heme, which is also important for its stability [65]. The precipitation of α-globin chain was a problem during the expression of HbA in microbial hosts. The co-expression of human *AHSP* gene was the solution for the precipitation of globins in *E. coli* resulting in high HbA yields [66]. The expression of Hb in *E. coli* is also limited by the availability of free heme and requires the addition of external heme in combination with expression of the heme transport system [43] or supplementation with heme precursor, 5-aminolevulinic acid [33, 67].

The expression of mutant forms of hemoglobin by targeted mutagenesis of globin-encoding genes, which affect its oxygen affinity [56, 60] and NO scavenging [17], addresses the needs for hemoglobin with reduced toxicity and needs further development for the improved recombinant products.

1.3 Yeast as a Cell Factory for Recombinant Hemoglobin

Besides being a valuable and extensively used model organism for eukaryal systems, yeast is the most preferred industrial production platform with high yields and low costs. *S. cerevisiae* is generally regarded as safe (GRAS) and is tolerant to low pH and many harsh industrial conditions. Several studies reported successful expression of human hemoglobin in this yeast [6, 44, 48, 68–73]. In addition to wild-type form, the low O_2-affinity mutant and sickle hemoglobins were produced [48, 70]. The recombinant hemoglobin in yeast was produced at the range of 3–5% of the total protein and had identical, structural, and biochemical properties of native human hemoglobin [48, 69]. Some portion of the recombinant hemoglobin was folded improperly and had decreased stability and cooperativity of oxygen binding [68]. Yeast synthesizes heme endogenously, and metabolic engineering can be used to further improve the heme production. Heme biosynthesis starts with glycine and succinyl coenzyme A (CoA) condensation catalyzed by 5-aminolevulinate synthase (encoded by *HEM1* gene) in the mitochondrial matrix. The entire pathway consists of eight reactions and is strictly regulated by the carbon source, oxygen, and heme availability [74, 75]. The product of the first reaction, 5-aminolevulinate, is exported from mitochondria to the cytosol, where it is converted into coproporphyrinogen III by four subsequent reactions. Further, the pathway proceeds in mitochondria, where coproporphyrinogen III undergoes the oxidative decarboxylation and oxidation to yield protoporphyrin IX. The final reaction of the insertion of ferrous iron generates heme [76]. The *HEM2*, *HEM3*, and *HEM12* coding for aminolevulinate dehydratase, porphobilinogen deaminase, and uroporphyrinogen decarboxylase, respectively, were determined as rate limiting in the pathway, as they display lower specific activity and higher K_m [75]. These genes can be further used to create strains with elevated heme and hemoglobin production. In our lab we work on improving *S. cerevisiae* strains for the production of human hemoglobin using different strategies. To enhance the heme biosynthesis, we overexpressed each of the limiting genes (*HEM2*, *HEM3*, and *HEM12*) in different combinations on two micron-based episomal plasmids and found that the overexpression of *HEM3* gene alone resulted in the highest porphyrin level [72]. By balancing the expression of α- and β-globin genes with the enhanced biosynthesis of endogenous heme (by *HEM3* gene overexpression), the recombinant active hemoglobin constituted 4% of the total soluble protein extracted [72]. The additional metabolic engineering of oxygen sensing mechanisms proved to be beneficial for the recombinant hemoglobin production [73]. In response to high environmental oxygen levels, there is an inherent increase in intracellular heme availability, and in response to that, the transcription factor Hap1 (heme activator protein) activates the expression of both set of genes responsible for the aerobic metabolism and genes involved in oxidative stress response in the yeast [74, 77–79]. Hap1 regulates the heme biosynthesis by also activating the expression of the repressor protein Rox1, which inhibits the *HEM13* gene transcription [80]. The activity of Hap1 is increasing with increasing concentrations of heme and when heme concentration decreases the Hap1 activity drops [74, 81, 82]. Thus heme plays a main role in oxygen sensing by affecting the Hap1 activity [79]. It has been also shown that the deletion of the *HAP1* gene increases the heterologous inulinase production in *Kluyveromyces lactis* [83]. Recent studies performed in out lab showed that deletion of *HAP1* in *S. cerevisiae* improves the recombinant hemoglobin production up to 7% of the total soluble protein [73].

1.4 Outlook

Using different approaches significant advances have been made in the heterologous production of human hemoglobin. The use of yeast as cell factory provides several advantages over bacteria. Using *S. cerevisiae* as a factory for hemoglobin production is cheaper as simple minimal media may be used, and furthermore it does not require the addition of expensive precursors to the media (e.g., 5-aminolevulinic acid), which would significantly increase the overall production costs. We have been able to engineer strains, which have heme and hemoglobin production significantly improved by overexpressing one of the limiting genes of heme biosynthesis (*HEM3*), balancing globin gene expression, and engineering heme and oxygen sensing mechanisms (*Δhap1* strain). Further engineering efforts involving heme biosynthesis pathway, heme and hemoglobin degradation, and its stability in the microbial host could result on even higher yields of the production. Reducing toxic effects of heme/hemoglobin by the overexpression of its natural scavengers could have additional benefits, resulting in better cell performance of the microbial host during the fermentation process and thus improved production.

2 Materials

2.1 Multi-copy Gene Expression Vectors and Host Strains for Heme and Globin Genes

1. pIYC04 plasmid (2 μ vector containing bidirectional gene expression cassettes *TEF1–ADH1* and *PGK1–CYC1* and *HIS3* as selectable marker).
2. pSP-GM1 plasmid (2 μ vector containing bidirectional gene expression cassettes *TEF1–ADH1* and *PGK1–CYC1* and *URA3* as selectable marker).
3. *Saccharomyces cerevisiae* CEN.PK 113-11C (*MATα his3*Δ1 *ura3*-52 *MAL2*-8c *SUC2*).
4. *Escherichia coli* DH5α.
5. *HEM3* gene sequence from CEN.PK 113-11C strain flanked by restriction sites *NotI* (5′end) and *SacI* (3′end).
6. Human α-globin gene sequence, codon optimized for yeast expression systems (GenScript Inc.), flanked by restrictions sites *NotI* (5′end) and *SacI* (3′end).
7. Human β-globin gene sequence, codon optimized for yeast expression systems (GenScript Inc.), flanked by restriction sites *SalI* (5′end) and *KpnI* (3′end).
8. FastDigest® restriction enzymes, Thermo Fisher Scientific (as listed above).
9. T4 DNA ligase, Thermo Fisher Scientific, #EL0012.
10. FastAP® alkaline phosphatase, Thermo Fisher Scientific, #EF0651.

11. QIAquick PCR purification kit, Qiagen, #28104.
12. YPD medium.
13. Synthetic dextrose (SD) his^{-} and/or ura^{-} medium.
14. LB (Luria-Bertani) + ampicillin (100 mg/mL) medium.
15. PEG 4000.
16. Lithium acetate.
17. Tris–HCl, pH 7.5.
18. 0.5 M EDTA, pH 8.0.
19. ssDNA (single stranded).
20. Water bath.
21. Incubator.

2.2 Cell-Free Extracts: Inhibitors of Hemoglobin Degradation

1. Glass beads, 425–600 μm (30–40 US sieve), Sigma-Aldrich, #G8772.
2. PBS buffer, pH 7.0.
3. FastPrep® 24.
4. High-speed centrifuge with cooling.
5. Halt Protease Inhibitor Single-Use Cocktail (×100), Thermo Fisher Scientific, #78430.
6. 0.5 M EDTA, pH 8.0.

2.3 Human Hemoglobin Detection by Immunoblotting

1. Hemoglobin Assay Kit, Sigma-Aldrich, #MAK115.
2. Microtiter plates, Nunc™ MicroWell™ 96-Well, Thermo Fisher Scientific, #260860.
3. Mini-PROTEAN® TGX™ 4–20% gels, BIO-RAD, #456-1093 or NuPAGE™ 12% Bis–Tris gels, Invitrogen, Thermo Fischer Scientific, #NP0343BOX.
4. Trans-Blot® Turbo Mini PVDF Transfer Packs, BIO-RAD, #170-4156.
5. BCIP/TNBT Single Reagent, Blue/brown, Alkaline Phosphatase Substrate (insoluble), MILLIPORE, #ES007-500ML.
6. PBS buffer, pH 7.0.
7. Loading dye, ×4 (0.252 M Tris–HCl, pH 6.8; 40% glycerol; 8% SDS; 0.1% bromophenol blue).
8. Western Blocker™ Solution, Sigma-Aldrich, #ES007-500ML.

2.4 Determination of Heme/Porphyrin Content

1. Solutions (2 M and 20 mM) of oxalic acid Sigma-Aldrich, #75688.
2. Black microtiter plates, Nunc™ MicroWell™ 96-Well, Thermo Fisher Scientific, #265301.

3. Heat block.
4. Microtiter plate spectrophotometer equipped for fluorescent measurements (FLUOstar Omega plate reader spectrophotometer).

3 Methods

3.1 Construction of Recombinant Strains Carrying Both Heme and Human Globin Genes

Recombinant vectors were constructed separately for overexpression of *HEM3* gene from *S. cerevisiae* and the human globins, respectively (Fig. 3a), by the procedures described in the following sections.

3.1.1 Construction of the H3 Plasmid

1. Digest the DNA sequence corresponding to *HEM3* and flanked by the restriction sequences (as described in Subheading 2.1), in a final reaction volume of 30 μL as recommended by the manufacturer [2 μL of ×10 reaction buffer, 1 μL of DNA (0.1–1 μg), 1 μL *NotI*, 1 μL *SacI*, 25 μL water]. Place the tube containing the reaction mix in a water bath at 37 °C during 30 min followed by 5 min at 65 °C for enzyme inactivation.
2. Digest the plasmid pIYC04 in a final reaction volume of 20 μL as recommended by the manufacturer [2 μL of 10× reaction buffer, 4 μL of plasmid DNA (~1 μg), 1 μL *NotI*, 1 μL *SacI*, 1 μL FastAP, 11 μL water]. Place the tube containing the reaction mix in a water bath at 37 °C during 30 min followed by 5 min at 65 °C for enzyme inactivation.
3. Purify both reactions using a QIAquick PCR purification kit, to remove enzymes and concentrate the DNA amount. At the last step of the manufacturer's instructions, elute the DNA from the column in 10 μL to increase the DNA concentration prior to the ligation step.
4. Ligate the purified *HEM3* fragment and the pIYC04 vector using T4 DNA ligase, adding 100 ng of vector and calculating a vector/insert ratio of 3:1 reaction mix (20 μL) as follows: 2 μL of ×10 reaction buffer, 1 μL (100 ng) of vector DNA, 1.9 μL (69 ng) of insert DNA, 1 μL (1 U) of T4 ligase, and 14.1 μL of water. Incubate the reaction overnight at room temperature.
5. The following day, use the ligation mix to transform *E. coli* DH5α competent cells, and plate them in LB with ampicillin for further selection of the positive transformants, after 1-day incubation at 37 °C.

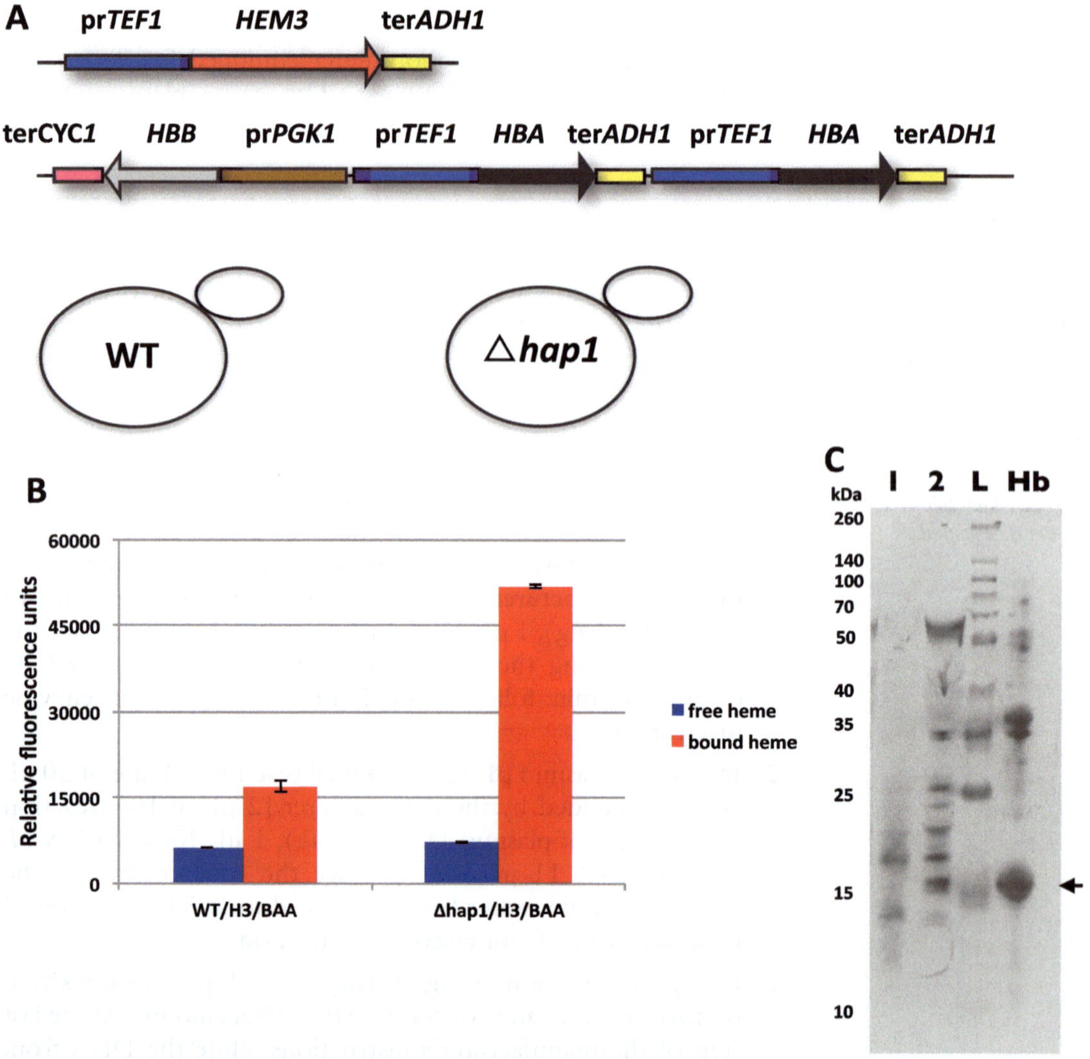

Fig. 3 Construction of improved hemoglobin producers in *S. cerevisiae*. (**a**) Hemoglobin expression cassettes [promoter (pr) *TEF1* and *PGK1*; terminator (ter) *ADH1* and *CYC1*] of *S. cerevisiae HEM3* gene and human globin genes *HBA* (encoding α-chains, two copies) and *HBB* (encoding β-chain) were used for yeast transformation as part of multi-copy plasmids (H3 and B/A/A) (Adapted from [72]). Strains of *S. cerevisiae* CEN.PK 113-11C used: WT and *Δhap1*. (**b**) Porphyrin synthesis (free heme and bound heme) at 24 h of glucose fermentation by transformants WT and *Δhap1* carrying hemoglobin expression plasmids. (**c**) Western blotting of crude extracts of transformants carrying plasmids H3 and B/A/A using antibody for α-globin chain. 1-WT, 2-*Δhap1*, L-protein molecular weight marker (Thermo Fisher Scientific, 26634), Hb control, and purified human hemoglobin (Sigma-Aldrich, H7379). The α-chain monomer is arrowed

3.1.2 Construction of the B/A/A Plasmid

As described by Liu et al. [72], a double ratio of alpha globin overexpression over beta globin yields the higher recombinant hemoglobin production. Thus, for construction of such vector (B/A/A plasmid), the procedure is as follows:

1. Digest the human β-globin sequence using restriction enzymes *SalI* and *KpnI*, following the procedure exactly as described in **step 1** in Subheading 3.1.1.
2. Digest the plasmid pSP-GM1 using restriction enzymes *SalI* and *KpnI*, following the procedure exactly as described in **step 2** in Subheading 3.1.1.
3. Follow **steps 3–5** in Subheading 3.1.1 for ligation of β-globin gene under *PGK1* promoter and further obtaining the B plasmid.
4. Digest the human α-globin sequence using restriction enzymes *NotI* and *SacI*, following instructions in **step 1** at the previous Subheading 3.1.1.
5. Digest the B plasmid (obtained after **step 3** of Subheading 3.1.2) using restriction enzymes *NotI* and *SacI* as previously described at **step 2** of Subheading 3.1.1.
6. Follow **steps 3–5** in Subheading 3.1.1 for ligation of the α-globin gene under *TEF1* promoter and further obtaining the B/A plasmid.
7. Digest the human α-globin sequence using restriction enzymes *PstI* and *MfeI*, following instructions in **step 1** in Subheading 3.1.1.
8. Digest the B/A plasmid (obtained after **step 6** of Subheading 3.1.2) using restriction enzymes *PstI* and *MfeI*, following instructions in **step 1** in Subheading 3.1.1.
9. Follow **steps 3–5** in Subheading 3.1.1 for ligation of the α-globin gene under *TEF1* promoter and further obtaining the B/A/A plasmid.

3.1.3 Construction of S. cerevisiae *Strains Containing H3 and B/A/A Plasmids*

1. Prepare seed cultures of *S. cerevisiae* CEN.PK 113-11C in YPD media (5 mL, use a falcon tube) from fresh agar plates, and incubate overnight, 200 rpm and 30 °C.
2. The following day, pour the 5 mL of culture into 50 mL of pre-warmed (30 °C) shake flask containing YPD, and incubate for 3–4 h until exponential growth phase is reached ($OD_{600} \geq 2$).
3. Pellet yeast cells by centrifugation (~5000 × *g*), and resuspend in 25 mL of sterile water.
4. Pellet cells again by centrifugation, and resuspend in 1 mL of sterile water. Transfer the cell suspension to a 1.5 mL tube.
5. Pellet cells one more time by centrifuging and resuspend in 100 mM lithium acetate in a final volume of 1 mL. Incubate during 10 min at 30 °C in a water bath (alternatively use a heating block, but the heat transfer occurs more efficiently in water).

6. Take 100 μL of the cell mix per transformation reaction, and transfer to a new 1.5 mL tube.
7. Centrifuge for 1 min each tube (max speed) and discard the supernatant.
8. To each tube with pelleted cells, add 500 μL of the following mix: 40% PEG 4000, lithium acetate 0.1 M, Tris–HCl 10 mM (pH 7.5), EDTA 1 mM, 50 μg ssDNA, 1 μg of H3 plasmid, and 1 μg of B/A/A plasmid. Tap gently to mix (do not use vortex).
9. Incubate the tube with the cell suspension during 30 min at 30 °C (preferably in a water bath).
10. Transfer the tubes immediately to 42 °C (preferably in water bath), and incubate for additional 30 min.
11. Collect the cells by centrifugation (5000 × *g*, 5 min).
12. Resuspend the cells in 100 μL of water and plate in SD $ura^{-}his^{-}$ medium for further selection of the transformant yeast strains.

3.2 Fermentations in Bioreactors

For an optimal production of recombinant human hemoglobin by yeast cells, tightly controlled batch fermentations in stirred tank bioreactors (STR) need to be performed.

The ideal working volume for these strains as they are currently designed is 1–3 L. At larger volumes, mass transfer becomes a limiting factor, and therefore further optimization steps of the fermentation conditions are required. Nevertheless, the fermentation setup described in the current section, corresponding to 0.5 L of working volume, is sufficient to yield an amount of recombinant hemoglobin corresponding to 4% of the total cell protein being fully active human hemoglobin molecules. The procedure for bioreactor cultivation is simple and requires only for a few steps as follows:

1. Prepare seed cultures of transformant *S. cerevisiae* cells (obtained from Subheading 3.1.3) in shake flasks containing selective SD $ura^{-}his^{-}$ medium (50–100 mL), and incubate overnight (200 rpm, 30 °C).
2. The following day, use the seed culture to inoculate the bioreactors at an initial OD_{600} of 0.01, for a total working volume of 500 mL.
3. The setup for the bioreactors should meet the following criteria: 1 v.v.m. (volume of vessel per minute) of airflow (21% oxygen, standard aerobic conditions), 600 rpm agitation, 30 °C, pH 5 controlled with potassium hydroxide (do not use sodium hydroxide), and dissolved oxygen ≥30%. A cascade control may be implemented in the setup of the bioreactor in order to keep dissolved oxygen levels over the minimum

parameters, by changes in the agitation speed and/or the airflow; however it is recommended not to change the percentage of oxygen in the air blend, as it may affect the heme biosynthesis and therefore have a negative effect on the final yields.

3.3 Determination of Heme/Porphyrin Content

The intracellular levels of heme in yeast transformants were determined by cell fluorescence after the treatment of the cells with oxalic acid (Fig. 3b). It is known that the iron of porphyrin ring quenches the fluorescence; however, under acidic conditions, the heme moiety of hemoglobin losses iron and becomes fluorescent (more direct and high-throughput assays using perhaps molecular sensors are needed). This is a modified protocol of the method described before [84, 85].

1. Collect 8 OD of cells (usually 2–5 mL of cell suspensions from the fermentation experiments are used) by the centrifugation, and resuspend cells in 500 μL of 20 mM oxalic acid. Incubate samples overnight at +4 °C in the dark.
2. On the next day, add the 500 μL of preheated 2 M oxalic acid (2 M oxalic acid does not dissolve in water at room temperature; for this reason, the heating is recommended before the use of this concentrated solution). Incubate one half of the mixture at room temperature (corresponds to free porphyrins). Incubate the other half of the mixture at 95–98 °C for 30 min (corresponds to bound heme and free porphyrins).
3. After the treatment, pellet the cells by centrifugation. Transfer 100 μL of supernatant to black microtiter plates.
4. Determine free and bound porphyrin content by their fluorescence with excitation at $\lambda = 400$ nm and emission at $\lambda = 600$ nm on FLUOstar Omega plate reader spectrophotometer. The lowest detection limit of heme concentration by this method is 0.5×10^{-12} mol [84].

3.4 Preparation of Cell-Free Extracts: Inhibition of Hemoglobin Degradation

It is not known which proteolytic enzymes cause the recombinant hemoglobin degradation upon its purification from yeast. To reduce the degradation of the heterologous hemoglobin during the preparation of cell-free extracts, we use the protease inhibitor cocktail [Halt Protease Inhibitor Cocktail (×100), Thermo Fisher Scientific], which is active against a wide range of proteases (serine proteases, cysteine proteases, aspartic acid proteases, and aminopeptidases). Carry out all procedures on ice.

1. Resuspend yeast cells in equal volume of ice-cold PBS buffer pH 7.0 with Halt Protease Inhibitor Cocktail (×1).
2. Add glass beads (425–600 μm, Sigma-Aldrich) in an equal volume of yeast suspension.

3. Perform the cell disruption in 5 vortexing pulses of 60 s (6 m/s speed) on FastPrep 24 instrument (MP Biomedicals). Keep cells on ice between vortexing cycles.
4. Following the disruption, remove the cell debris by centrifugation (20,817 × *g*, +4 °C). Transfer the supernatant to a new tube, and perform protein quantification (DC™ protein assay, BIO-RAD) and the hemoglobin assay. The hemoglobin assay is performed directly after preparation of cell-free extracts.

3.5 Hemoglobin Assay and Immunoblotting

The hemoglobin concentrations are assessed by Hemoglobin Assay Kit and by Western blotting.

Hemoglobin Assay Kit (Sigma-Aldrich) is based on the improved Triton/NaOH method in which hemoglobin is converted into colorimetric product detected at 400 nm (the calibrator solution absorbance data of the kit is used to convert the absorbance into hemoglobin concentration values of mg/dL). The kit is used to determine the hemoglobin in cell-free extracts.

1. Add 100 μL of the reagent to the microtiter plate wells.
2. Add 25 μL of freshly prepared cell-free extract of yeast transformants.
3. Incubate 5 min at room temperature. Measure the absorbance at 400 nm.

For Western blotting the proteins of cell-free extracts are separated on either 12% or gradient 4–20% SDS-PAGE with further electrotransfer to PVDF membrane and incubation with antibodies raised against alpha chain of human hemoglobin (Fig. 3c).

1. Denaturate 50–100 μg of the protein of cell-free extracts in ×1 loading dye at 96 °C for 6 min, and further cool the mixture on ice.
2. Load samples and separate denatured proteins on SDS-PAGE gel.
3. Proceed to protein staining and/or transfer to PVDF membrane (Trans-Blot® Turbo Mini PVDF Transfer Packs) using Trans-Blot® Turbo Transfer System, BIO-RAD (program MIXED MW, 7 min, 1.3 A constant up to 25 A).
4. Block the membrane with Western Blocker™ Solution.
5. Add 0.1 μg/mL anti-hemoglobin antibodies [Hemoglobin α Antibody (D-16): sc-31110, goat polyclonal, Santa Cruz Biotechnology] in blocking buffer; incubate 1 h at room temperature on the rotary shaker.
6. Remove primary antibodies by washing the membrane three times for 15 min with PBS buffer containing 0.05% Tween-20.

7. Add secondary anti-goat antibodies (anti-goat IgG, Sigma A4187) diluted ×4000 in blocking buffer; incubate 30 min at room temperature on the rotary shaker.
8. Remove secondary antibodies by washing the membrane three times for 15 min with PBS buffer containing 0.05% Tween-20.
9. Add the BCIP/TNBT to the membrane; incubate 5 min at room temperature.
10. After the development of the signal, rinse the membrane with deionized water three times. Dry the membrane.

References

1. Prabhulkar S, Tian H, Wang X et al (2012) Engineered proteins: redox properties and their applications. Antioxid Redox Signal 17 (12):1796–1822
2. Lončar N, Fraaije MW (2015) Catalases as biocatalysts in technical applications: current state and perspectives. Appl Microbiol Biotechnol 99(8):3351–3357
3. Bankar SB, Bule MV, Singhal RS et al (2009) Glucose oxidase – an overview. Biotechnol Adv 27(4):489–501
4. Kim DH, Kim MS (2011) Hydrogenases for biological hydrogen production. Bioresour Technol 102(18):8423–8431
5. Antonini E, Brunori M (1971) Hemoglobin and myoglobin in their reactions with ligands. In: Neuberger A, Tatum EL (eds) Frontiers of biology, vol 21. North-Holland Publishing Company, Amsterdam, p 436
6. Messerschmidt A (2001) Handbook of metalloproteins. Wiley, Chichester
7. Kumar S, Stecher G, Tamura K (2016) MEGA7: molecular evolutionary genetics analysis version 7.0 for bigger datasets. Mol Biol Evol 33:1870–1874
8. Saitou N, Nei M (1987) The neighbor-joining method: a new method for reconstructing phylogenetic trees. Mol Biol Evol 4:406–425
9. Zuckerkandl E, Pauling L (1965) Evolutionary divergence and convergence in proteins. In: Bryson V, Vogel HJ (eds) Evolving genes and proteins. Academic Press, New York, pp 97–166
10. Moraes CT, Diaz F, Barrientos A (2004) Defects in the biosynthesis of mitochondrial heme c and heme a in yeast and mammals. Biochim Biophys Acta 1659(2–3):153–159
11. Berg JM, Tymoczko JL, Stryer L (eds) (2002) Biochemistry. W. H. Freeman, New York
12. Benesch R, Benesch RE (1967) The effect of organic phosphates from the human erythrocyte on the allosteric properties of hemoglobin. Biochem Biophys Res Commun 26 (2):162–167
13. Bohr C, Hasselbalch K, Krogh A (1904) Concerning a biologically important relationship – the influence of the carbon dioxide content of blood on its oxygen binding. Skand Arch Physiol 16:402–412 [In German]
14. Farber HW, Loscalzo J (2004) Pulmonary arterial hypertension. N Engl J Med 351 (16):1655–1665
15. Schaer DJ, Buehler PW (2013) Cell-free hemoglobin and its scavenger proteins: new disease models leading the way to targeted therapies. Cold Spring Harb Perspect Med 3(6):a013433
16. Doherty DH, Doyle MP, Curry SR (1998) Rate of reaction with nitric oxide determines the hypertensive effect of cell-free hemoglobin. Nat Biotechnol 16(7):672–676
17. Olson JS, Foley EW, Rogge C et al (2004) No scavenging and the hypertensive effect of hemoglobin-based blood substitutes. Free Radic Biol Med 36(6):685–697
18. Alayash AI, Patel RP, Cashon RE (2001) Redox reactions of hemoglobin and myoglobin: biological and toxicological implications. Antioxid Redox Signal 3(2):313–327
19. Piantadosi CA (2002) Biological chemistry of carbon monoxide. Antioxid Redox Signal 4 (2):259–270
20. Tsantes AE, Bonovas S, Travlou A et al (2006) Redox imbalance, macrocytosis, and RBC homeostasis. Antioxid Redox Signal 8 (7–8):1205–1216
21. Johnson RM, Ho YS, Yu DY et al (2010) The effects of disruption of genes for peroxiredoxin-2, glutathione peroxidase-1, and catalase on erythrocyte oxidative metabolism. Free Radic Biol Med 48(4):519–525
22. Smith A, McCulloh RJ (2015) Hemopexin and haptoglobin: allies against heme toxicity from hemoglobin not contenders. Front Physiol

6:187. https://doi.org/10.3389/fphys.2015.00187

23. Rochette J, Craig JE, Thein SL (1994) Fetal hemoglobin levels in adults. Blood Rev 8 (4):213–224
24. Manca L, Masala B (2008) Disorders of the synthesis of human fetal hemoglobin. IUBMB Life 60(2):94–111
25. Steinberg MH, Rodgers GP (2015) HbA2: biology, clinical relevance and a possible target for ameliorating sickle cell disease. Br J Haematol 170(6):781–787
26. Peschle C, Mavilio F, Carè A et al (1985) Haemoglobin switching in human embryos: asynchrony of zeta-alpha and epsilon-gamma-globin switches in primitive and definite erythropoietic lineage. Nature 313(5999):235–238
27. Al-Mufti R, Hambley H, Farzaneh F et al (2000) Fetal and embryonic hemoglobins in erythroblasts of chromosomally normal and abnormal fetuses at 10–40 weeks of gestation. Haematologica 85(7):690–693
28. He Z, Russell JE (2001) Expression, purification, and characterization of human hemoglobins Gower-1 (zeta(2)epsilon(2)), Gower-2 (alpha(2)epsilon(2)), and Portland-2 (zeta (2)beta(2)) assembled in complex transgenic-knockout mice. Blood 97(4):1099–1105
29. Wilber A, Nienhuis AW, Persons DA (2011) Transcriptional regulation of fetal to adult hemoglobin switching: new therapeutic opportunities. Blood 117:3945–3953
30. Hardison RC (2012) Evolution of hemoglobin and its genes. Cold Spring Harb Perspect Med 2(12):a011627
31. Militello V, Vitrano E, Cupane A (1991) The effect of organic cosolvents on the oxygen affinity of fetal hemoglobin. Relevance of protein-solvent interactions to the functional properties. Biophys Chem 39(2):161–169
32. Dumoulin A, Manning LR, Jenkins WT et al (1997) Exchange of subunit interfaces between recombinant adult and fetal hemoglobins. Evidence for a functional inter-relationship among regions of the tetramer. J Biol Chem 272 (50):31326–31332
33. Chakane S, Matos T, Kettisen K et al (2017) Fetal hemoglobin is much less prone to DNA cleavage compared to the adult protein. Redox Biol 12:114–120
34. Ratanasopa K, Strader MB, Alayash AI et al (2015) Dissection of the radical reactions linked to fetal hemoglobin reveals enhanced pseudoperoxidase activity. Front Physiol 6:39
35. Lettre G, Bauer DE (2016) Fetal haemoglobin in sickle-cell disease: from genetic epidemiology to new therapeutic strategies. Lancet 387(10037):2554–2564
36. Saki N, Abroun S, Soleimani M et al (2016) MicroRNA expression in β-thalassemia and sickle cell disease: a role in the induction of fetal hemoglobin. Cell J 17(4):583–592
37. Sripichai O, Fucharoen S (2016) Fetal hemoglobin regulation in β-thalassemia: heterogeneity, modifiers and therapeutic approaches. Expert Rev Hematol 9(12):1129–1137
38. Habara AH, Shaikho EM, Steinberg MH (2017) Fetal hemoglobin in sickle cell anemia: the Arab-Indian haplotype and new therapeutic agents. Am J Hematol 92(11):1233–1242. https://doi.org/10.1002/ajh.24872
39. Tobian AA, Heddle NM, Wiegmann TL et al (2016) Red blood cell transfusion: 2016 clinical practice guidelines from AABB. Transfusion 56(10):2627–2630
40. Forbes JM, Anderson MD, Anderson GF et al (1991) Blood transfusion costs: a multicenter study. Transfusion 31(4):318–323
41. Moradi S, Jahanian-Najafabadi A, Roudkenar MH (2016) Artificial blood substitutes: first steps on the long route to clinical utility. Clin Med Insights Blood Disord 9:33–41
42. Alayash AI (2014) Blood substitutes: why haven't we been more successful? Trends Biotechnol 32(4):177–185
43. Graves PE, Henderson DP, Horstman MJ et al (2008) Enhancing stability and expression of recombinant human hemoglobin in *E. coli*: progress in the development of a recombinant HBOC source. Biochim Biophys Acta 1784 (10):1471–1479
44. Martínez JL, Liu L, Petranovic D et al (2012) Pharmaceutical protein production by yeast: towards production of human blood proteins by microbial fermentation. Curr Opin Biotechnol 23(6):965–971
45. Jullesson D, David F, Pfleger B et al (2015) Impact of synthetic biology and metabolic engineering on industrial production of fine chemicals. Biotechnol Adv 33(7):1395–1402
46. Gutierrez JM, Lewis NE (2015) Optimizing eukaryotic cell hosts for protein production through systems biotechnology and genome-scale modeling. Biotechnol J 10(7):939–949
47. Nagai K, Perutz MF, Poyart C (1985) Oxygen binding properties of human mutant hemoglobins synthesized in *Escherichia coli*. Proc Natl Acad Sci U S A 82(21):7252–7255
48. Wagenbach M, O'Rourke K, Vitez L et al (1991) Synthesis of wild type and mutant human hemoglobins in *Saccharomyces cerevisiae*. Biotechnology (N Y) 9(1):57–61

49. Behringer RR, Ryan TM, Reilly MP et al (1989) Synthesis of functional human hemoglobin in transgenic mice. Science 245 (4921):971–973
50. Swanson ME, Martin MJ, O'Donnell JK et al (1992) Production of functional human hemoglobin in transgenic swine. Biotechnology (N Y) 10(5):557–559
51. Dieryck W, Pagnier J, Poyart C et al (1997) Human haemoglobin from transgenic tobacco. Nature 386(6620):29–30
52. Komar AA, Kommer A, Krasheninnikov IA et al (1993) Cotranslational heme binding to nascent globin chains. FEBS Lett 326 (1–3):261–263
53. Yip YK, Waks M, Beychok S (1977) Reconstitution of native human hemoglobin from separated globin chains and alloplex intermediates. Proc Natl Acad Sci U S A 74(1):64–68
54. Hoffman SJ, Looker DL, Roehrich JM et al (1990) Expression of fully functional tetrameric human hemoglobin in *Escherichia coli*. Proc Natl Acad Sci U S A 87(21):8521–8525
55. Komar AA, Kommer A, Krasheninnikov IA et al (1997) Cotranslational folding of globin. J Biol Chem 272(16):10646–10651
56. Hernan RA, Hui HL, Andracki ME et al (1992) Human hemoglobin expression in *Escherichia coli*: importance of optimal codon usage. Biochemistry 31(36):8619–8628
57. Kavanaugh JS, Rogers PH, Arnone A (1992) High-resolution X-ray study of deoxy recombinant human hemoglobins synthesized from beta-globins having mutated amino termini. Biochemistry 31(36):8640–8647
58. Shen TJ, Ho NT, Simplaceanu V et al (1993) Production of unmodified human adult hemoglobin in *Escherichia coli*. Proc Natl Acad Sci U S A 90(17):8108–8112
59. Walsh G (2002) Proteins: biochemistry and biotechnology. Wiley, Chichester
60. Looker D, Abbott-Brown D, Cozart P et al (1992) A human recombinant haemoglobin designed for use as a blood substitute. Nature 356(6366):258–260
61. Weickert MJ, Pagratis M, Glascock CB et al (1999) A mutation that improves soluble recombinant hemoglobin accumulation in *Escherichia coli* in heme excess. Appl Environ Microbiol 65(2):640–647
62. Fronticelli C, Arosio D, Bobofchak KM et al (2001) Molecular engineering of a polymer of tetrameric hemoglobins. Proteins 44 (3):212–222
63. Feng L, Gell DA, Zhou S et al (2004) Molecular mechanism of AHSP-mediated stabilization of alpha-hemoglobin. Cell 119(5):629–640
64. Gell D, Kong Y, Eaton SA et al (2002) Biophysical characterization of the alpha-globin binding protein alpha-hemoglobin stabilizing protein. J Biol Chem 277(43):40602–40609
65. Domingues-Hamdi E, Vasseur C, Fournier JB et al (2014) Role of α-globin H helix in the building of tetrameric human hemoglobin: interaction with α-hemoglobin stabilizing protein (AHSP) and heme molecule. PLoS One 9 (11):e111395
66. Vasseur-Godbillon C, Hamdane D, Marden MC et al (2006) High-yield expression in *Escherichia coli* of soluble human alpha-hemoglobin complexed with its molecular chaperone. Protein Eng Des Sel 19(3):91–97
67. Ratanasopa K, Cedervall T, Bülow L (2016) Possibilities of using fetal hemoglobin as a platform for producing hemoglobin-based oxygen carriers (HBOCs). Adv Exp Med Biol 876:445–453
68. Adachi K, Konitzer P, Lai CH et al (1992) Oxygen binding and other physical properties of human hemoglobin made in yeast. Protein Eng 5(8):807–810
69. Coghlan D, Jones G, Denton KA et al (1992) Structural and functional characterisation of recombinant human haemoglobin A expressed in *Saccharomyces cerevisiae*. Eur J Biochem 207 (3):931–936
70. Martin de Llano JJ, Schneewind O, Stetler G et al (1993) Recombinant human sickle hemoglobin expressed in yeast. Proc Natl Acad Sci U S A 90(3):918–922
71. Ogden JE, Harris R, Wilson MT (1994) Production of recombinant human hemoglobin A in *Saccharomyces cerevisiae*. Methods Enzymol 231:374–390
72. Liu L, Martínez JL, Liu Z et al (2014) Balanced globin protein expression and heme biosynthesis improve production of human hemoglobin in *Saccharomyces cerevisiae*. Metab Eng 21:9–16
73. Martínez JL, Liu L, Petranovic D et al (2015) Engineering the oxygen sensing regulation results in an enhanced recombinant human hemoglobin production by *Saccharomyces cerevisiae*. Biotechnol Bioeng 112(1):181–188
74. Zhang L, Hach A (1999) Molecular mechanism of heme signaling in yeast: the transcriptional activator Hap1 serves as the key mediator. Cell Mol Life Sci 56(5–6):415–426
75. Hoffman M, Góra M, Rytka J (2003) Identification of rate-limiting steps in yeast heme biosynthesis. Biochem Biophys Res Commun 310 (4):1247–1253
76. Dailey HA (2002) Terminal steps of haem biosynthesis. Biochem Soc Trans 30(4):590–595

77. Pfeifer K, Kim KS, Kogan S et al (1989) Functional dissection and sequence of yeast *HAP1* activator. Cell 56(2):291–301
78. Kwast KE, Burke PV, Poyton RO (1998) Oxygen sensing and the transcriptional regulation of oxygen-responsive genes in yeast. J Exp Biol 201(Pt 8):1177–1195
79. Ter Linde JJ, Steensma HY (2002) A microarray-assisted screen for potential Hap1 and Rox1 target genes in *Saccharomyces cerevisiae*. Yeast 19(10):825–840
80. Keng T (1992) *HAP1* and *ROX1* form a regulatory pathway in the repression of *HEM13* transcription in *Saccharomyces cerevisiae*. Mol Cell Biol 12(6):2616–2623
81. Guarente L, Mason T (1983) Heme regulates transcription of the *CYC1* gene of *S. cerevisiae* via an upstream activation site. Cell 32 (4):1279–1286
82. Zhang L, Guarente L (1994) Evidence that *TUP1/SSN6* has a positive effect on the activity of the yeast activator *HAP1*. Genetics 136 (3):813–817
83. Yu J, Jiang J, Fang Z et al (2010) Enhanced expression of heterologous inulinase in *Kluyveromyces lactis* by disruption of hap1 gene. Biotechnol Lett 32(4):507–512
84. Sassa S (1996) Sequential induction of heme pathway enzymes during erythroid differentiation of mouse Friend leukemia virus-infected cells. J Exp Med 143(2):305–315
85. Michener JK, Nielsen J, Smolke CD (2012) Identification and treatment of heme depletion attributed to overexpression of a lineage of evolved P450 monooxygenases. Proc Natl Acad Sci U S A 109(47):19504–19509

Chapter 12

Membrane Protein Production in Yeast: Modification of Yeast Membranes for Human Membrane Protein Production

Anita Emmerstorfer-Augustin, Tamara Wriessnegger, Melanie Hirz, Guenther Zellnig, and Harald Pichler

Abstract

Approximately 30% of the genes in the human genome code for membrane proteins, and yet we know relatively little about these complex molecules. Therefore, the biochemical and structural characterization of this challenging class of proteins represents an important frontier in both fundamental research and advances in drug discovery. However, due to their unique physical properties and requirement for association with cellular membranes, expression in heterologous systems is often daunting. In this chapter we describe how to engineer the yeast *Pichia pastoris* to obtain humanized sterol compositions. By implementing some simple genetic engineering approaches, *P. pastoris* can be reprogrammed to mainly produce cholesterol instead of ergosterol. We show how to apply mass spectrometry to confirm the production of cholesterol instead of ergosterol and how we have further analyzed the strain by electron microscopy. Finally, we delineate how to apply and test the cholesterol-forming *P. pastoris* strain for functional expression of mammalian Na,K-ATPase α3β1 isoform. Na,K-ATPases have been shown to specifically interact with cholesterol and phospholipids, and, obviously, the presence of cholesterol instead of ergosterol was the key to stabilizing correct localization and activity of this ion transporter.

Key words Yeast, *Pichia pastoris*, Sterols, Human membrane protein expression, Membrane engineering

1 Introduction

1.1 Membrane Structure: Lipids and Membrane Proteins

Embedded in the lipid bilayers of the membranes of all organelles and cells, integral membrane proteins (IMPs) are the transport vehicles for ions, nutrients, hormones, drugs, and even large molecules such as proteins and DNA. To fulfil these highly diverse functionalities, cells had to evolve a significant variety of these proteins; important examples of this protein class include, for example, ion channels, certain signal transduction receptors, cell adhesion molecules, or, as focused on in this method description, transmembrane ATPases. Mostly due to the hydrophobic nature of

Brigitte Gasser and Diethard Mattanovich (eds.), *Recombinant Protein Production in Yeast*, Methods in Molecular Biology, vol. 1923, https://doi.org/10.1007/978-1-4939-9024-5_12,

their transmembrane segments, IMPs are notoriously difficult to handle and study and, therefore, are one of the most important remaining frontiers of structural biology and drug research. In addition, their natural abundance is commonly rather low [1, 2], which makes purification and functional studies challenging. Therefore, the application of yeasts for recombinant protein expression is highly attractive as they comprise low-cost cultivation and high-quantity production platforms meeting the demand for criteria of safety and authentically processed proteins [3, 4]. Yeasts also display many tools desired for correct folding, cellular translocation, and posttranslational modifications of higher eukaryotic proteins. However, yeasts and higher eukaryotes exhibit some cellular differences that can restrict high-level expression of recombinant IMPs (extensively reviewed by [5, 6]). For example, correct functionality of IMPs was shown to be directly influenced by the membranous lipids they are surrounded with, which is barely surprising considering their close spatial interactions (reviewed by [7]). In general, plasma membranes are constructed from a highly diverse mixture of lipids, such as phosphatidylcholine, phosphatidylethanolamine, phosphatidylinositol, and phosphatidylserine but also phosphatidic acid, sphingolipids, and sterols [8]. However, the general composition and molecular properties of the lipids involved significantly differ from lower to higher eukaryotes [9]. The most notable difference between mammalian cells, plant cells, and fungi—including yeast—is the sterols they produce and incorporate into their plasma membranes. Yeast cells produce ergosterol, whereas mammalian cells produce cholesterol and plant cells sitosterol, phytosterol and campesterol as their main sterols [10]. The first steps in sterol biosynthesis are the same in animals, plants, and fungi [10]. Several highly conserved enzymatic reactions lead to the formation of the uniform precursor cholesta-5,7,24(25)-trienol. In fungi, a sterol C-24 methyl transferase (Erg6p) adds an additional methyl group at C-24 in order to synthesize ergosterol. Furthermore, a double bond is introduced by sterol C-22 desaturase (Erg5p) (Fig. 1a). In mammals, by contrast, sterols are saturated at positions C-7 and C-24 by dehydrocholesterol reductases 7 (DHCR7) and 24 (DHCR24), respectively (Fig. 1b). Plant cells have been reported to produce more than 250 different sterols and related compounds [11]. Sitosterol is the principal sterol found in plant materials, followed by stigmasterol, campesterol, and brassicasterol [12]. Chemical structures of these sterols are similar to cholesterol, differing only in the side chain. Sitosterol and stigmasterol have an ethyl group at C-24, and campesterol and brassicasterol have a methyl group at the same position, which is catalyzed by C-24 ethyl- and methyltransferases SMT2 and SMT1, respectively. Stigmasterol and brassicasterol are additionally dehydrated at C-22 by CYP710A [13] (Fig. 1c). Despite the close structural similarities between the different sterol species found in fungi and

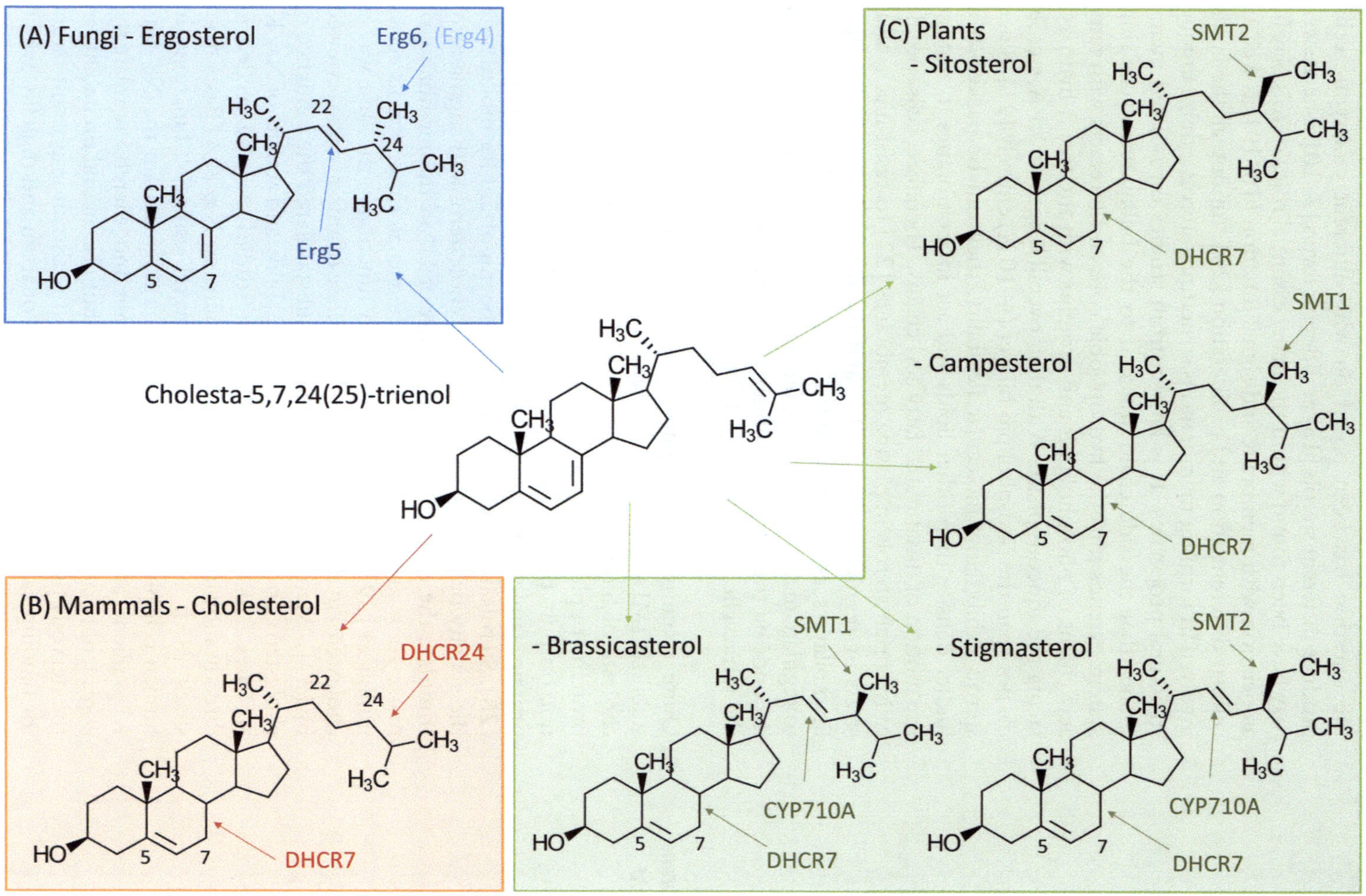

Fig. 1 Stereochemical structures of most abundant sterols in fungi, mammals, and plants. In all eukaryotes, sterols are synthesized from the conserved precursor cholesta-5,7,24(25)-trienol. Genes coding for sterol modification activities are indicated in capitals and italics; arrows indicate regioselectivity of sterol modification. *ERG4* sterol C-24(28) reductase, *ERG5* sterol C-22 desaturase (yeast), *ERG6* sterol C-24 methyl transferase (yeast), *SMT1/SMT2* sterol methyl and ethyl transferase, respectively (plant), *DHCR7* and *DHCR24* dehydrocholesterol reductases 7 and 24, respectively, *CYP710A* sterol C-22 desaturase (plant)

mammalian and plant cells, they display distinct functions in biological systems as well as in artificial membranes [14]. Early on the question has been posed on why different eukaryotic cells produce different sterols [15] at different levels [7, 10]. The answer may have been found within the many essential roles sterols play in association with membrane and signal function [16–19], as well as cellular differentiation and proliferation [20]. Among all the lipids involved in forming membranous structures, sterols exhibit unique structural properties. It is known from studies on isolated membranes that the side chain attached to the bulky, tetracyclic ring system extends into the hydrophobic core and interacts with fatty acyl chains of phospholipids and proteins with the free 3-hydroxyl facing the aqueous phase. With their overall length of 2.1 Å, sterols do not span the phospholipid bilayer (4–10 Å) entirely [21]. Since sterols exhibit low molecular flexibility, they can be packed at a higher density than other lipids and act in membranes to restrict the motion of unsaturated fatty acyl chains. Therefore, cells specifically enrich sterols in plasma membranes [22] to resist mechanical stress by regulating the fluidity and the permeability of phospholipid bilayers, especially at extreme temperatures. All sterols are able to regulate membrane fluidity, but with different efficiencies, which depends on the size and configuration of the side chain attached to the tetracyclic ring system [13, 21, 23].

1.2 Membrane Protein: Lipid Interaction

Quite recently, new insights have been derived on how sterols stabilize IMPs such as G protein-coupled receptors [24–26]. A cholesterol-binding consensus motif had been proposed for G protein-coupled receptors [27]. Moreover, it was described that not only in the β2-adrenergic receptor but also in the ATPase protein family cholesterol-binding sites are strongly conserved [28]. Sterols have also been shown to be important for modulating the activity of ATPases. For example, cholesterol and stigmasterol stimulate the ATPase catalyzed export of H^+ in the membrane of maize roots at low concentrations, whereas all other sterols act as inhibitors [29]. In this regard they may function in the same way as cholesterol in activating the Na/K-ATPase of animal cells (reviewed by [29]). Many independent studies substantiated the influence of cholesterol and phospholipids on the stability and activity of Na/K-ATPases [30–34], and crystallization studies even succeeded in directly showing that cholesterol is bound to Na/K-ATPases from shark [35]. Na/K-ATPases fulfil several essential functions in human cell physiology. Most notably, they maintain the Na^+ and K^+ gradients across the plasma membrane and, thereby, regulate the contractility of heart and muscle cells but also neural excitability in the nervous tissue [36]. The ion pump is also an important target for binding cardiac glycosides such as ouabain and digitalis, which are used in the treatment of heart failure [36]. For structural and functional studies, diverse alternative expression hosts have been applied to express different isoforms of the protein (summarized in [35]). Na/K-ATPases consist of two subunits; the catalytic α subunit and the plasma membrane-bound β subunit [37–39]. However, high-level co-expression of the two subunits was found to be difficult [34, 40].

As published by our group, the use of cholesterol-producing *P. pastoris* significantly improved functional expression, stability, and translocation of human Na^+/K^+-ATPases α3β1 isoform [35]. Here, we describe how we have engineered *P. pastoris* to produce cholesterol instead of ergosterol. For efficient conversion of cholesta-5,7,24(25)-trienol to cholesterol, *ERG6* and *ERG5* were replaced by constitutive DHCR7 and DHCR24 overexpression cassettes [35] (Figs. 1b and 2a). We delineate the method used to isolate sterols from yeast extracts and how to confirm the formation of different sterols by GC–MS. Quantification of sterols extracted from wild-type *P. pastoris* revealed a composition of 85% ergosterol, 10% ergosta-5,7,22-24(28)-tetraenol, and 5% zymosterol (Fig. 3a), whereas the latter two simply represent prominent intermediate products of the yeast ergosterol biosynthesis pathway. Analysis of the most advanced cholesterol strain revealed that 83% of total sterols were cholesterol; 10% 7-dehydrocholesterol, an intermediate resulting from low activity of DHCR7; 5% cholesta-5,7,24(25)-trienol; and 2% zymosterol (Fig. 3b). Furthermore, our cholesterol-forming strain has been analyzed by electron microscopy, revealing that the cholesterol yeast built a markedly thicker cell wall than the ergosterol-producing control strain (Fig. 4). The sterol-modified strain has successfully been applied to express human Na/K-ATPase α3β1 isoform at high levels, and protein production has been monitored by immunoblot analysis using microsomal preparations (Fig. 2d). We propose that our cholesterol-producing *P. pastoris* strain can be used for improved functional expression of many mammalian or plant membrane proteins.

2 Materials

2.1 Strains, Vectors, and Reagents

1. *Escherichia coli* TOP 10 F′ cells (F′[lacI^q Tn10(tet^R)] mcrA Δ(mrr-hsdRMS-mcrBC) φ80lacZΔM15 ΔlacX74 deoR nupG recA1 araD139 Δ(ara-leu)7697 galU galK rpsL(Str^R) endA1 λ) (Life Technologies, Vienna, Austria).
2. Host strains of *Pichia pastoris* CBS7435 *Δhis4* and CBS7435 *Δhis4 Δku70* (*see* **Note 1**).
3. Vectors: pGAPZ A containing a Zeocin™ selection marker; pPpKan_S (GenBank Accession: JQ519694.1) with a kanamycin/geneticin selection marker (*see* **Note 2**).

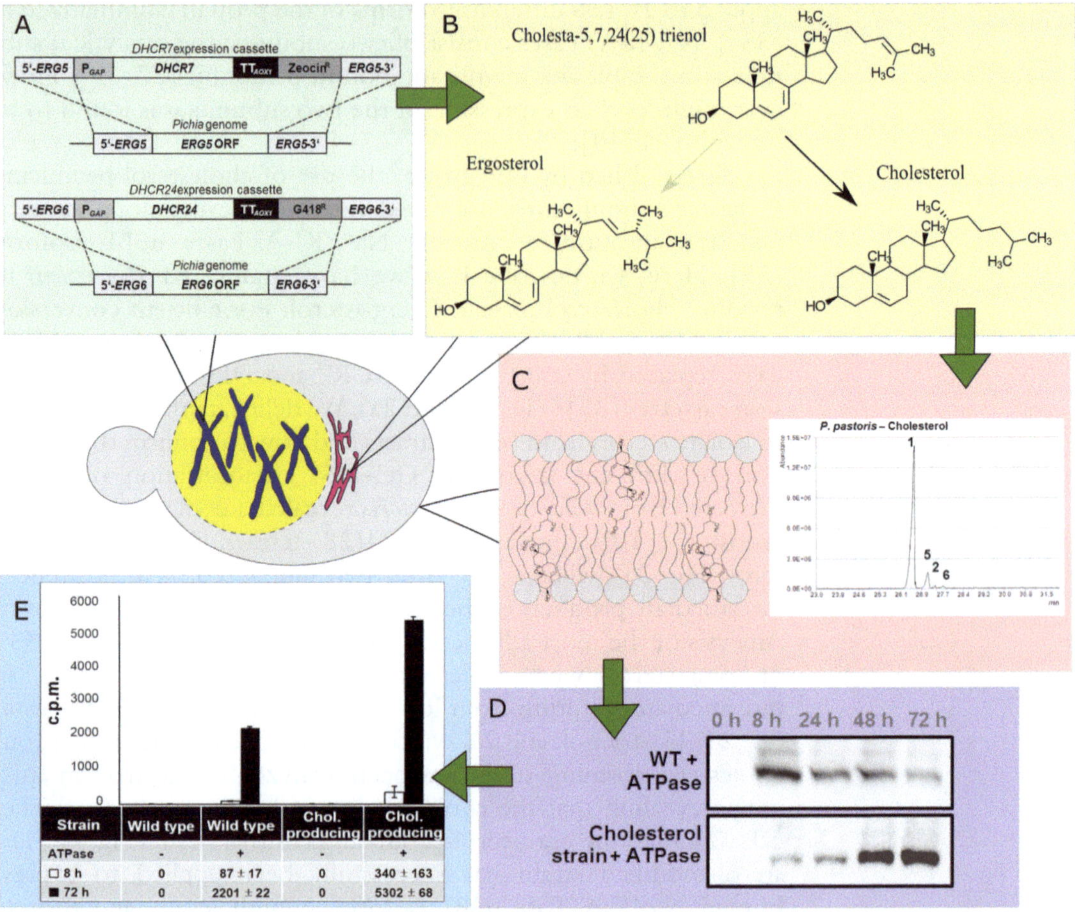

Fig. 2 Schematic overview of how *P. pastoris* has been engineered to produce cholesterol and further applied and tested for functional expression of human Na/K-ATPase. After inserting DHCR7 and DHCR24 expression cassettes into the *ERG5* und *ERG6* loci, respectively (**a**), *P. pastoris* produced cholesterol instead of ergosterol (**b**), which has been confirmed by GC–MS analysis (**c**). Na,K-ATPase α3β1 isoform has been expressed in wild-type (WT) and cholesterol-producing strain, and expression levels have been monitored after indicated hours of methanol induction by immunoblot analysis (**d**). The cholesterol strain expressing Na,K-ATPases has been confirmed to show higher [^{3}H]-ouabain binding capacities than the wild-type control (**e**). Figure has been adapted from Hirz et al. [35]

4. Synthetic genes for dehydrocholesterol reductases from *Danio rerio* (zebra fish) specific for positions C-7 (DHCR7) and C-24 (DHCR24) in the sterol molecule (*see* **Note 3**).
5. Genes for Na,K-ATPase subunits α3 and β1 were used as an example and were kindly provided by Laura Popolo (*see* **Note 4**).
6. *Gel* and PCR cleanup system.
7. MicroPulser™ electroporator.
8. Phusion® High-Fidelity DNA polymerase.
9. Restriction endonucleases: *Eco*RI, *Not*I, *Xho*I, and *Bgl*II.

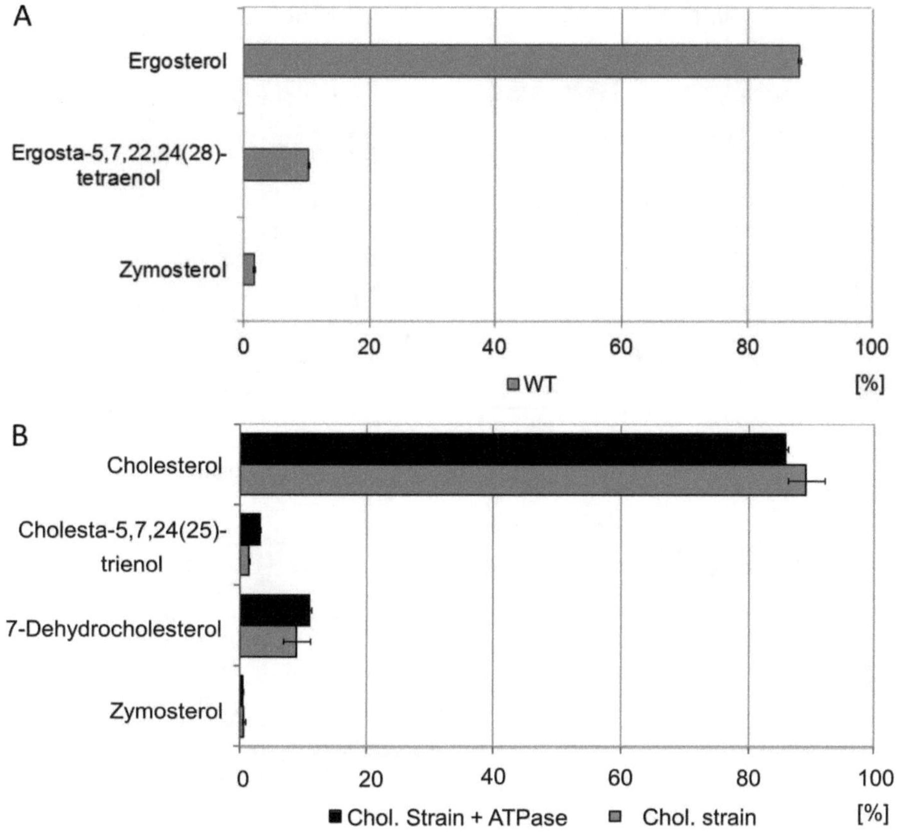

Fig. 3 Sterol composition of *P. pastoris* wild-type cells (**a**) and *P. pastoris* cholesterol strain, with and without overexpression of the human Na,K-ATPase (**b**) by GC–MS analysis. Cells were analyzed after 72 h of methanol induction (adopted from Hirz et al. [35]). Mean standard deviations of three biological replicates are given

2.2 Culture Media, Chemicals, and Cultivation Equipment

1. Luria Bertani (LB) Lennox medium.
2. Buffered glycerol complex medium (BMGY) and buffered methanol complex medium (BMMY): 10 g/L yeast extract, 20 g/L peptone, 100 mM potassium phosphate, pH 6.0, 13.4 g/L Difco™ Yeast nitrogen base (YNB) without amino acids, 0.04 mg/L biotin, and 10 g/L glycerol or methanol, respectively. 1 M potassium phosphate buffer, pH 6.0: combine 132 mL of 1 M K_2HPO_4 and 868 mL of 1 M KH_2PO_4. Confirm the pH and use phosphoric acid or KOH to adjust the pH, if necessary. Autoclave and store at room temperature.
3. Agar plates: Media for plates were solidified by addition of agar to 15 g/L. YPD-plates: 10 g/L yeast extract, 20 g/L peptone and 20 g/L D-glucose; minimal dextrose (MD) plates: 13.4 g/L YNB (without amino acids), 0.04 mg/L biotin, 20 g/L D-glucose. Antibiotics were used at the following concentrations: Zeocin™ (100 mg/L); geneticin sulfate (300 mg/L).
4. 300 mL shake flasks with baffles.

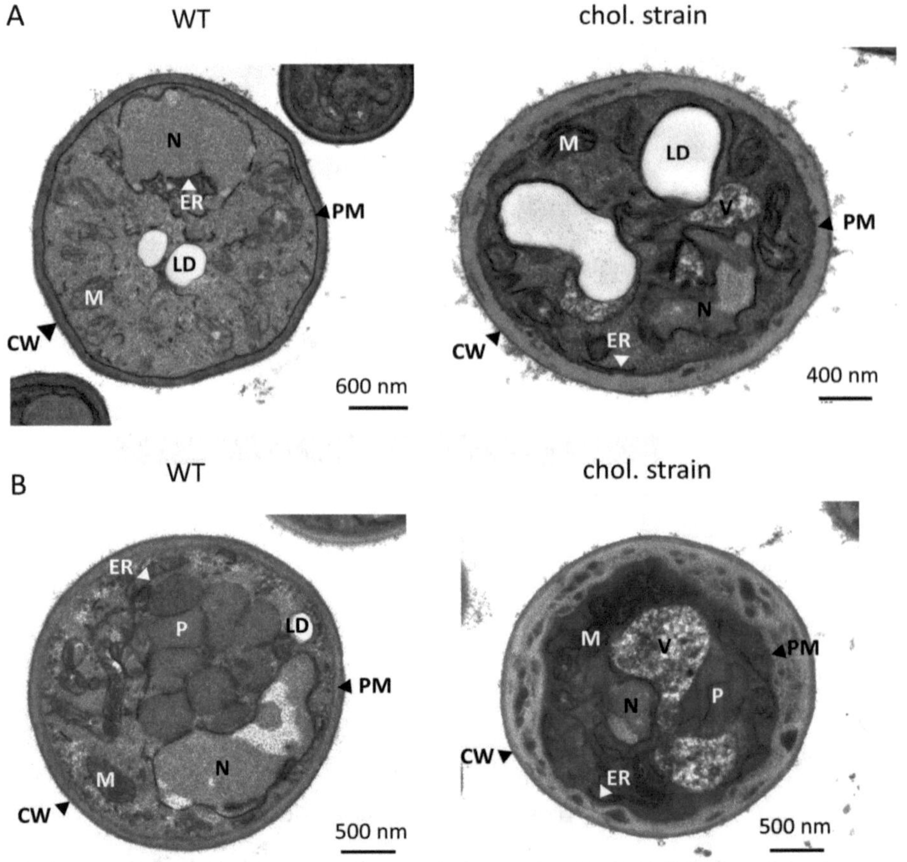

Fig. 4 Transmission electron microscopy of *P. pastoris* wild-type (WT) and sterol-modified (chol. strain) strains. Cells were either grown in YPD medium for 48 h (**a**) or pre-grown on glycerol medium (BMGY) for 48 h followed by induction with methanol for 48 h (BMMY medium) (**b**). *N* nucleus, *ER* endoplasmic reticulum, *LD* lipid droplet, *V* vacuole, *M* mitochondria, *P* peroxisome, *PM* plasma membrane, *CW* cell wall

5. Nunc™ OmniTray™ with lid (rectangular petri dish compatible with microtiter plates).
6. 96-deep well plates.
7. Shakers at 25 mm orbit, 120 rpm, 28 °C, 80% humidity.

2.3 Sterol Analyses of *P. pastoris* Cells

2.3.1 Sterol Extraction

1. Methanol (99.9%).
2. 0.5% (w/v) pyrogallol (≥98%) dissolved in methanol.
3. 60% (w/v) aqueous potassium hydroxide (KOH).
4. Cholesterol (≥99%) (2 mg/mL, dissolved in ethanol) (*see* **Note 5**).
5. Pyrex® tubes (Glass tube with Teflon fitting and cap; total volume approximately 15 mL).
6. Heating plate with sand bath and thermometer.
7. *n*-Heptane.

8. Nitrogen.
9. Pyridine (99.8%).
10. *N*,*O*-bis(trimethylsilyl)-trifluoroacetamide (BSTFA) ($\geq$99%).
11. Ethyl acetate (>99.9%).
12. Eppendorf centrifuge 5810 R.
13. IKA® VXR Vibrax orbital shaker.

2.3.2 Gas Chromatography–Mass Spectrometry (GC–MS) Analysis

1. GC crimp neck vials with respective micro inserts and aluminum crimp seals with septa.
2. Handheld crimper.
3. Gas chromatograph equipped with a mass selective detector.

2.4 Electron Microscopy of *P. pastoris* Cells

1. 50 mL plastic tubes.
2. 15 mL conical plastic centrifuge tube.
3. 0.5 mL plastic tubes.
4. Eppendorf centrifuge 5810 R.
5. Heraeus Biofuge Stratos.
6. Potassium permanganate ($KMnO_4$).
7. Uranyl acetate.
8. Ethanol.
9. Propylene oxide.
10. Agar 100 resin kit.
11. Lead citrate.
12. Sodium hydroxide.
13. Pioloform (polyvinylbutyral).
14. Drying cabinet (Memmert basic, Schwabach, Germany).
15. Diamond knife, type ultra, (DiATOME, Nidau, Switzerland).
16. Reichert ultracut S microtome (Leica Microsystems, Vienna, Austria).
17. TEM grids: hexagonal 200 mesh copper (Agar Scientific, Stansted, United Kingdom).
18. Philips CM10 transmission electron microscope.

2.5 *P. pastoris* Cell Fractionation

2.5.1 Total Cell Lysates for Determination of Protein Expression Level

1. 1.5 mL plastic tubes.
2. Table top centrifuge.
3. Vortex shaker.
4. 50 mM sodium phosphate, pH 7.4.
5. EDTA (ethylenediaminetetraacetic acid).
6. Glycerol.

7. 1 M phenylmethylsulfonyl fluoride (PMSF) in dimethyl sulfoxide (DMSO) (*see* **Note 6**).
8. Breaking buffer: 50 mM sodium phosphate, pH 7.4, 1 mM EDTA, 5% (w/v) glycerol, 1 mM PMSF (*see* **Note 6**).
9. Glass beads (0.25–0.5 mm diameter, Carl Roth GmbH, Karlsruhe, Germany).
10. TE buffer: 10 mM Tris–HCl, 1 mM EDTA, pH 7.4.
11. Braun MSK cell homogenizer (Sartorius, Goettingen, Germany) with CO_2 cooling system.
12. Braun MSK cell homogenizer vessel (*see* **Note 7**).
13. Glass beads (0.25–0.5 mm diameter, Carl Roth GmbH, Karlsruhe, Germany).
14. Avanti centrifuge J-26XP series with rotors JA 10 and JA 25.50 (Beckman Coulter™, USA).
15. Centrifugation vessels for JA 10 and JA 25.50 rotors.
16. Ultracentrifuge.
17. Polycarbonate tubes with aluminum cap assembly for use in ultracentrifuge rotor.

2.6 Example: Na,K-ATPase Expression in P. pastoris Cholesterol Strain

1. Bio-Rad protein assay.
2. Bovine serum albumin (BSA).
3. Table top centrifuge.
4. SDS-PAGE loading dye.
5. SDS-PAGE gel system (e.g., Novex® NuPAGE® 4–12% Bis–Tris precast gel from Life Technologies, Vienna, Austria).
6. Hybond ECL nitrocellullose membrane.
7. TBS (tris-buffered saline): 0.25 M Tris–HCl, pH 7.5, 1.5 M NaCl.
8. TBS-T: TBS buffer with 0.003% (w/v) Tween-20.
9. Primary antibodies: Rabbit anti-KETYY and anti-GERK antisera recognizing Na,K-ATPase α subunit and β subunit, respectively (kindly provided by Steven J. D. Karlish (Weizmann Institute of Sciences, Rehovot, Israel); 1:2000 dilution in 20 g/L BSA.
10. HRP-conjugated secondary antibodies (anti-rabbit); 1:10,000 dilution in TBS with 25 g/L milk powder.
11. SuperSignal® West Pico Chemiluminescent Substrate (Thermo Fisher Scientific, Vienna, Austria).
12. G:BOX gel documentation and analysis system (Syngene, Cambridge, UK).

3 Methods

The construction of the cholesterol-producing *P. pastoris* strain has been described in detail by Hirz et al. [35]. Hence, we only give a brief overview of the modifications to generate sterol-modified *P. pastoris* strains but lay a special focus on methods used to characterize the sterol-modified strains.

3.1 Strategy for Creating P. pastoris Strains with Modified Sterol Profiles

The first steps in sterol biosynthesis are the same in animals, plants, and fungi [10]. Cholesta-5,7,24(25)-trienol is the common precursor of all sterols. Then, the pathways diverge into the distinct sterol-producing branches (Fig. 1) (*see* **Note 8**). For a successful replacement of ergosterol by cholesterol, four genetic modifications have to be performed in *P. pastoris*: The inactivation of the *ERG5* and *ERG6* genes for the elimination of the double bond at position C-22 and the methyl group at position C-24, respectively. Additionally, two double bonds are reduced at positions C-7 and C-24 by the actions of DHCR7 and DHCR24, respectively. Using a similar approach, *P. pastoris* can also be engineered toward the production of phytosterols (Fig. 1c). The plant sterol campesterol (ergosta-5en-3β-ol) is created upon replacing the *ERG5* gene by DHCR7. Furthermore, the expression of DHCR7 in wild-type *P. pastoris* may yield brassicasterol (ergosta-5,22-dienol) as a major sterol compound. The generation of strains with different sterol profiles might have a beneficial effect on membrane-associated protein functions. As published by our group, overexpression of the membrane protein Na,K-ATPase, composed of two subunits α3 and β1, was enhanced in the *P. pastoris* strain with a mammalian-like sterol composition [35]. In the following sections, methods for characterizing the sterol-modified *P. pastoris* strains are described in detail, and the expression of the mammalian Na,K-ATPase exemplifies the applicability of membrane alterations for improving expression levels and activity of recombinant membrane proteins.

3.2 Sterol Analysis

Sterol analysis of *P. pastoris* membranes was adapted from previously described methods [41–43]. Sterol amounts of wild-type and engineered strains are indicated in Fig. 3.

1. For methanol induction conditions: Cultivate cells in 25 mL of BMGY in 300 mL baffled flasks for 48 h at 28 °C and 120 rpm. Add 25 mL of BMMY (containing 1% methanol), and continue the cultivation for 48 h. Add 2.5 mL of 10% aqueous methanol in 8/16 h cycles (*see* **Note 9**).
2. Determine the OD_{600} value of the culture, and transfer 15 OD_{600} units to a 15 mL Pyrex tube. Pellet the cells at 2916 × *g* for 5 min in an Eppendorf 5810R centrifuge. Remove the supernatant.

3. Add to the pellet a mixture of 0.6 mL methanol, 0.4 mL of 0.5% (w/v) pyrogallol (dissolved in methanol), and 0.4 mL of 60% (w/v) aqueous KOH solution for alkaline hydrolysis of ester and amide bonds. Mix by vortexing.
4. Add 5 μL of cholesterol solution (or another sterol of choice) as internal standard, and mix (*see* **Note 6**).
5. Close the Pyrex tubes tightly, and heat them in a sand bath for 2 h at 90 °C to complete the alkaline hydrolysis reactions.
6. Let the tube cool down to room temperature, and add 1 mL *n*-heptane for extraction of sterols.
7. Put the tubes on an IKA® VXR Vibrax orbital shaker, and extract by mixing for 10 min.
8. Centrifuge tubes for 5 min at 2916 × *g*, and transfer the upper phases into new Pyrex tubes.
9. Re-extract the lower aqueous phases with 1 mL *n*-heptane two more times (i.e., repeat **steps 6–8**).
10. Dry the combined organic phases under a stream of nitrogen to avoid oxidation of sterols, and dissolve the resulting lipid film in 10 μL pyridine.
11. Add 10 μL of BSTFA for derivatization (*see* **Note 10**), briefly mix the solution, and incubate at room temperature for 10 min.
12. Dilute the samples with 50 μL ethyl acetate and analyze by GC–MS.

3.2.1 Gas Chromatography–Mass Spectrometry (GC–MS) Analysis

1. GC–MS analysis is performed, for example, on a Hewlett–-Packard 5690 gas chromatograph equipped with a mass selective detector (HP 5972) using an HP5-MS capillary column (30 m × 0.25 mm i.d. × 0.25 μM film). Temperature program: 100 °C for 1 min, ramp of 10 °C/min to 250 °C and ramp of 3–310 °C. Inject 1 μL aliquots of samples onto the column.
2. Identify the sterols according to their mass fragmentation pattern, and use the internal standard for quantitative analysis (*see* **Note 6**).

3.3 Electron Microscopy of Sterol-Modified Strains

Phenotypes of sterol-engineered strains, grown either under methanol—induction (BMMY medium) or standard YPD conditions were investigated (Fig. 4). Preparation of *P. pastoris* cells for electron microscopy was adopted from [44] and is described in detail in the following section.

1. For methanol induction phenotype, cultivate cells as described in **step 1** in Subheading 3.2.
2. For standard conditions, cultivate cells in 50 mL of YPD medium in 300 mL baffled flasks at 28 °C and 120 rpm for 48 h (*see* **Note 11**).

3. Pellet cells at 3220 × *g* for 5 min.
4. Wash the cell pellets with 25 mL of distilled water for 10 min.
5. Resuspend the cell pellet in 10 mL of distilled water by vortexing, and transfer the suspension into a 15 mL conical plastic centrifuge tube (*see* **Note 12**).
6. Pellet cells at 2420 × *g* for 5 min.
7. For fixation resuspend cells in 5 mL of 1% aqueous $KMnO_4$, and incubate at room temperature for 5 min (*see* **Notes 12** and **13**).
8. Wash the cells with 10 mL of distilled water for 5 min.
9. Add 5 mL of 1% aqueous $KMnO_4$, and incubate at room temperature for 20 min.
10. Wash the fixed cells four times for 10 min each with 10 mL of distilled water.
11. Resuspend pellet in 5 mL of 0.5% aqueous uranyl acetate for staining purposes, and agitate overnight at 4 °C (*see* **Note 12**).
12. Dehydrate the sample by adding 10 mL of a graded series of ethanol (50%, 70%, 90%, and 100%), and incubate each concentration for 20 min at room temperature.
13. Pure ethanol is replaced by 10 mL of a 1:1 mixture of pure ethanol and propylene oxide for 10 min.
14. Pure ethanol/propylene oxide mixture is replaced by 10 mL of pure propylene oxide for 10 min.
15. Gradually infiltrate the specimen with 5 mL of increasing concentrations of Agar 100 resin (30%, 50%, 70%, and 100%) mixed with propylene oxide. Incubate at room temperature for at least 3 h per step (*see* **Note 14**).
16. Infiltrate the samples in 3–5 mL (depending on the volume of the cell pellet) fresh pure Agar 100 resin, resuspend the cells, and aliquot them into 0.5 mL plastic tubes.
17. Centrifuge the samples at 3000 × *g* for 3 min, and solidify the samples uncapped at 60 °C for 48 h in a drying cabinet.
18. Cut ultrathin sections of 80 nm using an ultramicrotome, collect them on pioloform-coated 200 mesh copper grids, and counterstain the grids for 5 min with 0.02 M lead citrate dissolved in freshly prepared 0.16 M NaOH. Grids are air-dried before investigation in the transmission electron microscope.
19. Investigate the stained sections with a transmission electron microscope (Philips CM10).

3.4 Cell Disruption and Fractionation

Different subcellular fractions of (yeast) cells can be enriched by differential centrifugation techniques. Upon lysing the cells, centrifugations are carried out at increasing speed to separate total cellular membrane fractions into enriched populations of subcellular membranes (i.e., organelles). Marker enzyme measurements, or immunoblotting techniques, targeting proteins representative of defined intracellular membranes (markers) allow for identification and rough quantification of the enrichment of certain organellar membranes in the fractions obtained by differential centrifugation. Highly purified organelles may be obtained by combining differential and density gradient centrifugation techniques.

3.4.1 Preparation of Total Cell Lysates

1. To prepare total cell lysates for protein expression analyses, cultivate *P. pastoris* as described under Subheading 3.2 (*see* **Note 15**).
2. Take 1 mL aliquots of the yeast culture and spin for 5 min at 3000 × *g* at 4 °C.
3. Resuspend the cell pellet in 200 μL of ice-cold breaking buffer, and maintain proper cooling throughout the whole procedure.
4. Add 0.2 μL of 1 M PMSF.
5. Add 200 μL of glass beads, and disrupt the cells by vortexing for 30 s followed by cooling for 30 s on ice. Repeat the disruption and cooling cycles eight times.
6. Centrifuge the samples at 3000 × *g* and 4 °C for 5 min.
7. Transfer the supernatant containing the total cell lysate into a fresh 1.5 mL tube, and store at −20 °C until use.

3.4.2 Preparation of Membrane Fractions

1. Harvest 200 mL of *P. pastoris* culture for 5 min at 3000 × *g* and 4 °C (*see* **Note 16**).
2. Discard the culture supernatant, and wash the cell pellet with 25 mL of ice-cold water.
3. Discard the supernatant and determine cell wet weight (CWW).
4. Resuspend the pellet in 1 mL TE buffer per g of CWW.
5. Add 2 μL of DMSO per g of CWW.
6. Transfer the suspension into 60 mL Merckenschlager homogenizer (Braun MSK) vessel (filled 1/3 with glass beads).
7. Disrupt the cells with the Merckenschlager homogenizer with CO_2 cooling for 3 min with 30 s cooling intervals.
8. Transfer the solution into a JA 25.50 centrifugation tube, and spin the samples for 10 min at 3000 × *g* and 4 °C to remove cell debris and glass beads.
9. Take a 1 mL sample of the total cell lysate (soluble fraction).
10. Transfer the total cell lysate into a fresh centrifugation tube, and spin for 15 min at 12,000 × *g* and 4 °C to obtain supernatant S12 (soluble part) and pellet P12 fractions.

11. Take a 1 mL sample of S12 fraction, and transfer the rest of the S12 fraction to a fresh centrifuge tube. Resuspend the pelleted P12 fraction in 2 mL of TE buffer.
12. Centrifuge the transferred S12 fraction for 15 min at 20,000 × *g* and 4 °C to obtain S20 (soluble part) and pellet P20 fractions.
13. Take a 1 mL sample of S20 fraction, transfer the rest of the S20 fraction to a fresh centrifuge tube, and resuspend the pelleted P20 fraction in 2 mL of TE buffer.
14. Transfer the supernatant S20 into a clean ultracentrifuge tube, and spin for 45 min at 100,000 × *g* and 4 °C to obtain S100 (soluble part) and pellet P100 fractions.
15. Take a 1 mL sample of S100 fraction, discard the rest of the S100 fraction, and resuspend the P100 fraction in 2 mL of TE buffer.
16. Store all samples at −80 °C until further analyses, e.g., immunoblotting (*see* Subheading 3.5.2).

3.5 Example: Na,K-ATPase Expression in *P. pastoris* Cholesterol Strain

The Na,K-ATPase was chosen as ideal enzyme for testing the cholesterol-producing *P. pastoris* strain as expression host, since the cholesterol dependency of this membrane protein has been described previously [28]. Expression studies of the α3 and β1 subunits of Na,K-ATPase were carried out in the cholesterol and the wild-type strains. The α3 subunit shows a size of 110 kDa on immunoblots using an antibody specifically recognizing the KETYY amino acid sequence. The expression level significantly increased with induction time in the cholesterol-forming strain compared to the wild-type strain, indicating a positive influence of the altered sterol composition on the stability of recombinant Na, K-ATPase α3β1 (Fig. 1d). The expression level for the β1 subunit was usually weaker than for α3 in all strains, but the co-localization of both subunits in the plasma membrane of *P. pastoris* cells was confirmed [35].

3.5.1 *P. pastoris* Transformation

1. Linearize 5 μg of the expression vector pAO815-α3β1 with *Bgl*II, which cuts in the *AOX1* promoter region.
2. Purify linear plasmid DNA by agarose gel electrophorese followed by gel cleanup (*Wizard*® SV *gel* and PCR cleanup system). Elute the linear fragment in 30 μL of sterile water.
3. Add ~2 μg of linearized plasmid to 80 μL of freshly prepared electrocompetent *P. pastoris* cells (wild-type and cholesterol strain), and transfer to a 2 mm gap electroporation cuvette (keep on ice) (*see* **Note 17**).
4. Pulse the cells in a MicroPulser™ electroporator using the program for *Pichia* (2 kV, 1 pulse). Immediately add 500 μL

of cold 1 M sorbitol and 500 µL of YPD to the pulsed cells. Regenerate prospective transformants for at least 2 h.

5. Plate cells on YPD-zeocin agar plates to select for pAO815-α 3β1 integration (*see* **Note 18**).
6. Incubate plates for 3–5 days at 28 °C. Longer time span is required for slowly growing cholesterol yeast colonies.
7. Fill 96-deep well plates (96-DWPs) with 500 µL BMGY medium per well. Pick single colonies and transfer them to 96-DWP wells.
8. Incubate at 28 °C, 320 rpm, and 80% humidity for 24 h. Humidification counteracts detrimental evaporation from small 96-DWP media volumes.
9. Use a 96 pin-stamp to pin transformants from 96-DWP onto respective agar plates for selection of genetic integrations (MD-his, YPD-zeocin, or YPD-G418 in Nunc™ OmniTray™ plates) (*see* **Note 19**).
10. Pick transformants growing well after pinning onto OmniTray plates for determination of protein expression level (*see* Subheading 3.5.2).

3.5.2 Determination of Protein Expression Level by Immunoblotting

1. Determine the protein concentration in the cell lysates by using the Bio-Rad (Bradford) protein assay using bovine serum albumin as standard for quantification. For membrane preparation samples, protein quantification by Lowry procedure yielded more accurate results.
2. Load 20 µg of total protein onto a NuPAGE SDS gel. SDS-PAGE and immunoblotting is carried out according to the instructions of the supplier (e.g., Novex® NuPAGE® SDS-PAGE gel manual).
3. Block the nitrocellulose membrane with TBS containing 5% milk powder for 1 h at room temperature.
4. Incubate membrane with primary antibodies (anti-KETYY and anti-GERK) for 1 h at room temperature.
5. Wash five times with TBS-T.
6. Incubate membrane with secondary antibodies for 1 h at room temperature.
7. Wash five times with TBS-T.
8. Detect proteins by applying the chemiluminescence solution for 3 min, and visualize protein bands, e.g., in the G:BOX.

4 Notes

1. *Komagataella phaffii* strain CBS7435 Δhis4, commonly known as *Pichia pastoris* CBS7435 Δhis4, is derived from the *P. pastoris* wild-type strain NRRL-Y11430 (ATCC 76273). In the publication of Hirz et al. [35], the *P. pastoris* strain CBS7435 *Δhis4 Δku70* is used as background strain for sterol modification. The deletion of the *KU70* homologue has been shown to improve gene targeting efficiency in *P. pastoris* [45]. Recently we confirmed that the effects of the sterol-modified strains are similar when using the *P. pastoris* CBS7435 *Δhis4* strain not carrying the *KU70* deletion (unpublished data). The data presented in this manuscript are predominately based on *P. pastoris* strain CBS7435 *Δhis4*.
2. For constitutive expression of DHCR24, the *AOX1* promoter of the vector pPpKan_S was replaced by the glyceraldehyde-3-phosphate dehydrogenase (*GAP*) promoter obtained from plasmid pGAPZ A using *Eco*RI and *Bgl*II restriction sites for cloning.
3. To construct cholesterol-producing *P. pastoris*, we followed a strategy that has initially been used in *S. cerevisiae* [46]. Genes for DHCR7 and DHCR24, codon optimized for expression in *S. cerevisiae*, were kindly provided by Howard Riezman (University of Geneva, Switzerland).
4. The strain *P. pastoris* S-α3β1 containing the expression plasmid pAO815-α3β1 served as control for our experiments [40]. Both strain and isolated expression plasmid were a kind gift of Laura Popolo (Università degli Studi di Milano, Italy).
5. Cholesterol was added to the extracts as internal standard for GC–MS analysis, but not to the extracts of the cholesterol-forming strain. Internal standard amounts to 10 μg per extraction experiment. All sterols available as highly purified compounds can be employed as internal standards.
6. Prepare the PMSF stock solution (1 M PMSF in DMSO) always freshly, and add the solution to the cells resuspended in the breaking buffer.
7. Fill the Merckenschlager vessel 1/3 with glass beads and pre-cool on ice. Ideally add cold cell suspension to a total volume of 2/3 to the Merckenschlager vessels for most efficient cell disruptions.
8. Cholesta-5,7,24(25)-trienol is hardly detectable in wild-type *P. pastoris* cells due to the high activity of Erg6p (C-24-methyltransferase).
9. In 8/16 h cycles—taking into account regular working hours—10% aqueous methanol is fed instead of pure methanol

to increase accuracy and reproducibility of methanol addition. For practical reasons, pre-culture duration and induction times are kept constant despite divergent proliferation rate of wild-type and cholesterol-producing yeast strains. In shake flasks, wild-type cultures will typically reach OD_{600} values of 50–70, whereas cholesterol-producing strain cultures will reach OD_{600} values of 30–45.

10. Make sure that BSTFA is fresh, since BSTFA used for longer than 1–2 months loses its ability to efficiently derivatize sterols. Incompletely silylated sterols are not properly separated and quantified using our GC–MS method.
11. Cholesterol-producing yeast will grow about 30% slower than the ergosterol-producing wild-type in YPD. Therefore, we have cultivated all *P. pastoris* strains for at least 48 h to let cells grow to stationary phase and thereby minimize differences in final OD_{600}.
12. During the change of the media (**steps 7–17**), the samples should not run dry. The samples should be placed on a lab shaker during the incubation with the different media in order to ensure homogeneous mixing of cells and medium.
13. For changing the different media, cells are centrifuged at 800 × *g* (**steps 8–13**) and 1500 × *g* (**step 14**), respectively, for 1 min, the supernatant is removed, and the cell pellet is resuspended by vortexing in the new medium. The volume of the cell pellet should not exceed 1 mL.
14. Long incubation steps may be required to spread these steps over 2 days.
15. For the determination of protein expression levels at different time points, *P. pastoris* cells have been induced with methanol for 72 h. One mL samples have been taken at time points 0, 8, 24, 48, and 72 h of methanol induction in this example. The duration of the induction phase may be different for other membrane proteins, e.g., 24–96 h.
16. For the determination of cell wet weight (CWW), do not forget to weigh the empty centrifugation bottles before harvesting the cells.
17. Electrocompetent *P. pastoris* CBS7435 cells are prepared according to the protocol described by Lin-Cereghino et al. [47].
18. Plate 50 μL, 100 μL and the resuspended cell pellet after short centrifugation on the respective selection plates to ensure single colonies growing.
19. The *P. pastoris* cholesterol strain contains the zeocin and geneticin (G418) resistances due to knock-ins of DHCR7 and DHCR24 at the *ERG5* and *ERG6* loci, respectively.

Acknowledgments

This work has been supported by the Federal Ministry for Digital and Economic Affairs (bmwd); the Federal Ministry for Transport, Innovation, and Technology (bmvit); the Styrian Business Promotion Agency SFG, the Standortagentur Tirol, Government of Lower Austria, and ZIT Technology Agency of the City of Vienna through the COMET-funding program managed by the Austrian Research Promotion Agency FFG and by the Austrian Science Fund (FWF) project J3787 (A. E.-A.). The funding agencies had no influence on the conduct of this research.

References

1. Blonder J, Conrads TP, Veenstra TD (2004) Characterization and quantitation of membrane proteomes using multidimensional MS-based proteomic technologies. Expert Rev Proteomics 1:153–163. https://doi.org/10.1586/14789450.1.2.153
2. Zhang L-J, Wang X-E, Peng X, Wei Y-J, Cao R, Liu Z, Xiong J-X, Yin X-f, Ping C, Liang S (2006) Proteomic analysis of low-abundant integral plasma membrane proteins based on gels. Cell Mol Life Sci 63:1790–1804. https://doi.org/10.1007/s00018-006-6126-3
3. Gellisson G (2004) Key and criteria to the selection of an expression platform. https://doi.org/10.1002/3527603670
4. Gellissen G (2005) Production of recombinant proteins: novel microbial and eukaryotic expression systems. Wiley, Weinheim
5. Freigassner M, Pichler H, Glieder A (2009) Tuning microbial hosts for membrane protein production. Microb Cell Fact 8:69
6. Emmerstorfer A, Wriessnegger T, Hirz M, Pichler H (2014) Overexpression of membrane proteins from higher eukaryotes in yeasts. Appl Microbiol Biotechnol 98:7671–7698. https://doi.org/10.1007/s00253-014-5948-4
7. van Meer G, Voelker DR, Feigenson GW (2008) Membrane lipids: where they are and how they behave. Nat Rev Mol Cell Biol 9:112–124. https://doi.org/10.1038/nrm2330
8. van der Rest ME, Kamminga AH, Nakano A, Anraku Y, Poolman B, Konings WN (1995) The plasma membrane of *Saccharomyces cerevisiae*: structure, function, and biogenesis. Microbiol Rev 59:304–322
9. Schneiter R, Brügger B, Sandhoff R, Zellnig G, Leber A, Lampl M, Athenstaedt K, Hrastnik C, Eder S, Daum G, Paltauf F, Wieland FT, Kohlwein SD (1999) Electrospray ionization tandem mass spectrometry (ESI-MS/MS) analysis of the lipid molecular species composition of yeast subcellular membranes reveals acyl chain-based sorting/remodeling of distinct molecular species en route to the plasma membrane. J Cell Biol 146:741–754
10. Nes WD (2011) Biosynthesis of cholesterol and other sterols. Chem Rev 111:6423–6451. https://doi.org/10.1021/cr200021m
11. Tamura T, Akihisa T, Kokke W (1992) Naturally occurring sterols and related compounds from plants. In: Patterson GW, Nes WD (eds) Physiology and biochemistry of sterols. AOCS, Champaign, IL, pp 172–228
12. Piironen V, Lindsay DG, Miettinen TA, Toivo J, Lampi A (2000) Plant sterols: biosynthesis, biological function and their importance to human nutrition. J Sci Food Agric 80:939–966. https://doi.org/10.1002/(SICI)1097-0010(20000515)80:7<939::AID-JSFA644>3.0.CO;2-C
13. Marsan M-P, Muller I, Milon A (1996) Ability of clionasterol and poriferasterol (24-epimers of sitosterol and stigmasterol) to regulate membrane lipid dynamics. Chem Phys Lipids 84(2):117–121. https://doi.org/10.1016/S0009-3084(96)02628-X
14. Xu X, Bittman R, Duportail G, Heissler D, Vilcheze C, London E (2001) Effect of the structure of natural sterols and sphingolipids on the formation of ordered sphingolipid/sterol domains (rafts). Comparison of cholesterol to plant, fungal, and disease-associated sterols and comparison of sphingomyelin, cerebrosides, and ceramide. J Biol Chem 276:33540–33546. https://doi.org/10.1074/jbc.M104776200
15. Parks LW, Casey WM (1995) Physiological implications of sterol biosynthesis in yeast.

Annu Rev Microbiol 49:95–116. https://doi.org/10.1146/annurev.mi.49.100195.000523

16. Rodriguez RJ, Low C, Bottema CDK, Parks LW (1985) Multiple functions for sterols in *Saccharomyces cerevisiae*. Biochim Biophys Acta 837:336–343. https://doi.org/10.1016/0005-2760(85)90057-8
17. Bloch KE (1983) Sterol structure and membrane function. CRC Crit Rev Biochem 14:47–92
18. Xu F, Rychnovsky SD, Belani JD, Hobbs HH, Cohen JC, Rawson RB (2005) Dual roles for cholesterol in mammalian cells. Proc Natl Acad Sci 102:14551–14556. https://doi.org/10.1073/pnas.0503590102
19. Schaller H (2003) The role of sterols in plant growth and development. Prog Lipid Res 42:163–175
20. Guo DA, Venkatramesh M, Nes WD (1995) Developmental regulation of sterol biosynthesis in *Zea mays*. Lipids 30:203–219
21. Nes WR (1987) Multiple roles for plant sterols. In: The metabolism, structure, and function of plant lipids. Springer, New York, pp 3–9
22. Klose C, Surma MA, Simons K (2013) Organellar lipidomics—background and perspectives. Curr Opin Cell Biol 25:406–413. https://doi.org/10.1016/j.ceb.2013.03.005
23. Armstrong MJ, Carey MC (1987) Thermodynamic and molecular determinants of sterol solubilities in bile salt micelles. J Lipid Res 28:1144–1155
24. Jafurulla M, Rao BD, Sreedevi S, Ruysschaert J-M, Covey DF, Chattopadhyay A (2014) Stereospecific requirement of cholesterol in the function of the serotonin1A receptor. Biochim Biophys Acta 1838:158–163. https://doi.org/10.1016/j.bbamem.2013.08.015
25. Goddard AD, Watts A (2012) Regulation of G protein-coupled receptors by palmitoylation and cholesterol. BMC Biol 10:27. https://doi.org/10.1186/1741-7007-10-27
26. Oates J, Watts A (2011) Uncovering the intimate relationship between lipids, cholesterol and GPCR activation. Curr Opin Struct Biol 21:802–807. https://doi.org/10.1016/j.sbi.2011.09.007
27. Hanson MA, Cherezov V, Griffith MT, Roth CB, Jaakola V-P, Chien EYT, Velasquez J, Kuhn P, Stevens RC (2008) A specific cholesterol binding site is established by the 2.8 Å structure of the human β2-adrenergic receptor. Structure 16:897–905. https://doi.org/10.1016/j.str.2008.05.001
28. Adamian L, Naveed H, Liang J (2011) Lipid-binding surfaces of membrane proteins: evidence from evolutionary and structural analysis. Biochim Biophys Acta 1808:1092–1102. https://doi.org/10.1016/j.bbamem.2010.12.008
29. Grandmougin-Ferjani A, Schuler-Muller I, Hartmann MA (1997) Sterol modulation of the plasma membrane H+-ATPase activity from corn roots reconstituted into soybean lipids. Plant Physiol 113:163–174
30. Cornelius F, Turner N, Christensen HRZ (2003) Modulation of Na,K-ATPase by phospholipids and cholesterol. II. Steady-state and presteady-state kinetics. Biochemistry 42:8541–8549. https://doi.org/10.1021/bi034532e
31. Cornelius F (2001) Modulation of Na,K-ATPase and Na-ATPase activity by phospholipids and cholesterol. I. Steady-state kinetics. Biochemistry 40:8842–8851
32. Lifshitz Y, Petrovich E, Haviv H, Goldshleger R, Tal DM, Garty H, Karlish SJD (2007) Purification of the human α2 isoform of Na,K-ATPase expressed in *Pichia pastoris*. Stabilization by lipids and FXYD1. Biochemistry 46:14937–14950. https://doi.org/10.1021/bi701812c
33. Haviv H, Cohen E, Lifshitz Y, Tal DM, Goldshleger R, Karlish SJD (2007) Stabilization of Na^+,K^+-ATPase purified from *Pichia pastoris* membranes by specific interactions with lipids. Biochemistry 46:12855–12867. https://doi.org/10.1021/bi701248y
34. Cohen E, Goldshleger R, Shainskaya A, Tal DM, Ebel C, le Maire M, Karlish SJD (2005) Purification of Na+,K+-ATPase expressed in *Pichia pastoris* reveals an essential role of phospholipid-protein interactions. J Biol Chem 280:16610–16618. https://doi.org/10.1074/jbc.M414290200
35. Hirz M, Richter G, Leitner E, Wriessnegger T, Pichler H (2013) A novel cholesterol-producing *Pichia pastoris* strain is an ideal host for functional expression of human Na,K-ATPase α3β1 isoform. Appl Microbiol Biotechnol 97(21):9465–9478. https://doi.org/10.1007/s00253-013-5156-7
36. Aperia A (2007) New roles for an old enzyme: Na,K-ATPase emerges as an interesting drug target. J Intern Med 261:44–52. https://doi.org/10.1111/j.1365-2796.2006.01745.x
37. Hasler U, Wang X, Crambert G, Béguin P, Jaisser F, Horisberger JD, Geering K (1998) Role of beta-subunit domains in the assembly, stable expression, intracellular routing, and functional properties of Na,K-ATPase. J Biol Chem 273:30826–30835

38. Geering K (2001) The functional role of beta subunits in oligomeric P-type ATPases. J Bioenerg Biomembr 33:425–438
39. Béguin P, Hasler U, Beggah A, Horisberger JD, Geering K (1998) Membrane integration of Na,K-ATPase alpha-subunits and beta-subunit assembly. J Biol Chem 273:24921–24931
40. Reina C, Padoani G, Carotti C, Merico A, Tripodi G, Ferrari P, Popolo L (2007) Expression of the alpha3/beta1 isoform of human Na,K-ATPase in the methylotrophic yeast *Pichia pastoris*. FEMS Yeast Res 7:585–594. https://doi.org/10.1111/j.1567-1364.2007.00227.x
41. Quail MA, Kelly SL (1996) The extraction and analysis of sterols from yeast. Methods Mol Biol 53:123–131. https://doi.org/10.1385/0-89603-319-8:123
42. Tuller G, Nemec T, Hrastnik C, Daum G (1999) Lipid composition of subcellular membranes of an FY1679-derived haploid yeast wild-type strain grown on different carbon sources. Yeast 15:1555–1564. https://doi.org/10.1002/(SICI)1097-0061(199910)15:14<1555::AID-YEA479>3.0.CO;2-Z
43. Wriessnegger T, Gübitz G, Leitner E, Ingolic E, Cregg J, de la Cruz BJ, Daum G (2007) Lipid composition of peroxisomes from the yeast *Pichia pastoris* grown on different carbon sources. Biochim Biophys Acta 1771:455–461
44. Wriessnegger T, Augustin P, Engleder M, Leitner E, Müller M, Kaluzna I, Schürmann M, Mink D, Zellnig G, Schwab H, Pichler H (2014) Production of the sesquiterpenoid (+)-nootkatone by metabolic engineering of *Pichia pastoris*. Metab Eng 24:18–29. https://doi.org/10.1016/j.ymben.2014.04.001
45. Näätsaari L, Mistlberger B, Ruth C, Hajek T, Hartner FS, Glieder A (2012) Deletion of the *Pichia pastoris* KU70 homologue facilitates platform strain generation for gene expression and synthetic biology. PLoS One 7:e39720. https://doi.org/10.1371/journal.pone.0039720
46. Souza CM, Schwabe TME, Pichler H, Ploier B, Leitner E, Guan XL, Wenk MR, Riezman I, Riezman H (2011) A stable yeast strain efficiently producing cholesterol instead of ergosterol is functional for tryptophan uptake, but not weak organic acid resistance. Metab Eng 13:555–569. https://doi.org/10.1016/j.ymben.2011.06.006
47. Lin-Cereghino J, Wong WW, Xiong S, Giang W, Luong LT, Vu J, Johnson SD, Lin-Cereghino GP (2005) Condensed protocol for competent cell preparation and transformation of the methylotrophic yeast *Pichia pastoris*. Biotechniques 38:44, 46, 48

Chapter 13

Yeast Surface Display and Cell Sorting of Antigen-Binding Fc Fragments

Flávio Sádio, Gerhard Stadlmayr, Katharina Stadlbauer, Florian Rüker, and Gordana Wozniak-Knopp

Abstract

Since the introduction of the yeast display platform, this method has increasingly gained popularity for the discovery and affinity maturation of antibodies and other protein scaffolds intended for antigen recognition. Yeast display is particularly well suited for the selection of antigen-binding Fc fragments (Fcabs) as it allows rapid combinatorial library construction via gap repair-driven homologous recombination and an efficient display of a glycosylated Fc able to interact with Fcγ receptors. Apart from expression-related normalization, isolation of properly folded Fcabs can be guided efficiently by simultaneous staining with ligands such as protein A, FcγRI, or the conformation-sensitive anti-FigC$_H$2 antibody, whose binding is critically dependent on the integrity of the Fc structure. The particular properties of the Fcab scaffold, such as its homodimeric state which can promote binding to multiple antigen molecules, require modifications of traditional affinity maturation strategies. Preferred to equilibrium selections are kinetically driven antigen selections, designed to specifically influence the binding off-rate, which in many cases augments the desired biological effect. A simple design of a yeast-displayed heterodimeric Fc fragment is described and can be used as a general guideline for affinity selection of Fcabs with an asymmetric binding site. Overall, this chapter underlines the importance of the versatile yeast display technique for the optimization of the novel Fcab scaffold for antigen recognition.

Key words Fcab, Yeast display, Affinity maturation, Directed evolution, Heterodimer

1 Introduction

1.1 Yeast Display of Fcab Libraries

Since the validation of the Fc fragment with antigen-binding activity (Fcab) as a scaffold [1], the rapid development of these valuable biologicals has enabled their entry into clinical trials [2]. Fcabs display several properties of a complete antibody: binding to Fc receptors eliciting cellular effector functions; pH-dependent interaction with the neonatal Fc receptor, which provides them with a long in vivo half-life; and antigen binding, which endows the fragments with the ability of specific targeting (Fig. 1a, b). These properties underline the conceptual value of the Fcab as a

Brigitte Gasser and Diethard Mattanovich (eds.), *Recombinant Protein Production in Yeast*, Methods in Molecular Biology, vol. 1923, https://doi.org/10.1007/978-1-4939-9024-5_13, © Springer Science+Business Media, LLC, part of Springer Nature 2019

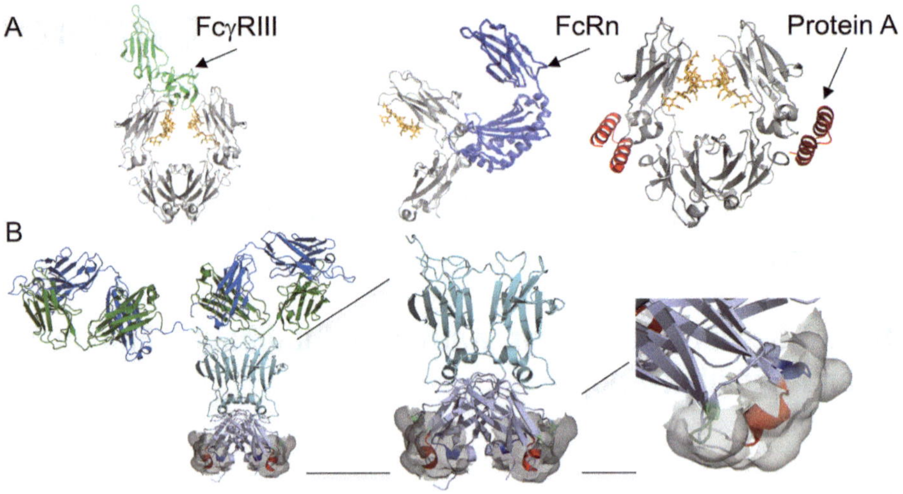

Fig. 1 (**a**) Fc fragment with its biologically important ligands: FcγRIII (left), FcRn (center), and protein A (right) (source: 1T89.pdb, 4N0F.pdb, 1OQO.pdb). (**b**) Cartoon diagram of a mAb^2 (left) and an antigen-binding Fc (Fcab) (center) with surfaced novel antigen-binding sites; closer view of C-terminal loops of the C_H3 domain typically targeted for randomization for construction of Fcab binder libraries and their potential antigen interaction surface (right). Green, antibody light chains; blue, variable domains of the heavy chain and C_H1 domain; cyan, C_H2 domains; light blue, C_H3 domains. The three C-terminal loops of the C_H3 domains are highlighted in dark blue (AB loop), green (CD loop), and red (EF loop)

therapeutically valuable antigen-targeting scaffold [2]; however their application is extended by the possibility of their integration in a complete monoclonal antibody, which yields a multivalent targeting molecule (mAb^2) by a simple domain exchange (Fig. 1b). The two binding paratopes of a mAb^2 can either recognize different epitopes, which leads to a bispecific agent, or aim at an identical target and thereby mediate an avidity-enhanced binding effect.

Few mutated amino acid residues, typically concentrated at the C-terminal loops of the C_H3 domain to form a binding surface, can mediate novel antigen recognition by an IgG1 Fc fragment with high affinity (Fig. 1b). These mutations can be introduced without a large negative impact on the stability of the Fc fragment. Several studies have been used to delineate the precise sites permissive to mutagenesis and insertion, some of them involving whole-library-scale screening [3].

Due to the posttranslational modifications typical of an Fc fragment, involving N-linked glycosylation at the residue Asn297 of the heavy chain of human IgG1 (EU numbering), yeast display [4] was the first method of choice for display and selection of antigen-binding Fcabs. An efficient Fc display on the *S. cerevisiae* EBY100 surface was achieved over the linkage of A-agglutinin cell wall attachment subunits Aga1p and Aga2p with the C-terminally fused Fc fragment (Fig. 2a). This orientation was chosen because

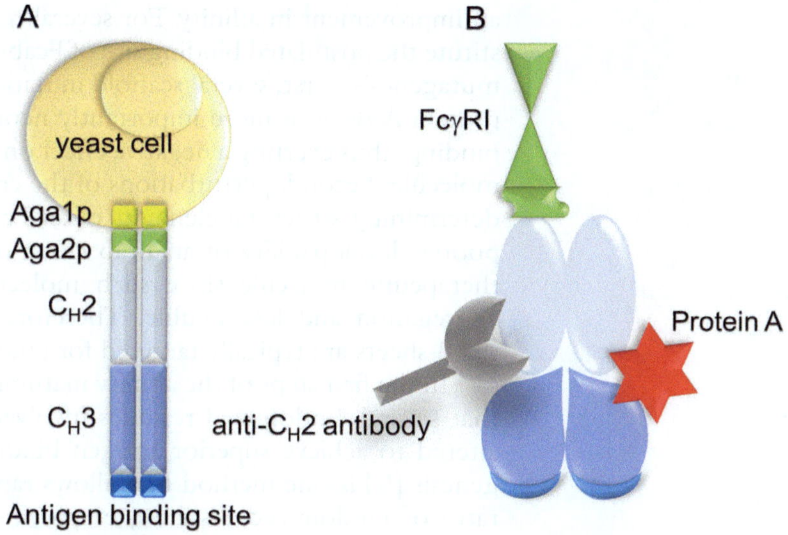

Fig. 2 Fcab display on yeast. (**a**) Organization of yeast-displayed Fcab construct. (**b**) Schematic of Fcab with bound structural reporter ligands: FcγRI, anti-C_H2 antibody, and protein A

the mutated surface designed for the interaction with the novel antigen is positioned at the C-terminus of the molecule. Multiparameter analysis supported by fluorescence-activated cell sorting (FACS)-based methods has allowed screening where several properties of the displayed molecule can be analyzed simultaneously. This readout was invaluable in screening for specific antigen binders that should initiate the response of effector cells, where the ability of simultaneous binding to the Fc receptor and the target antigen is a prerequisite for the selection of a biologically active molecule. Further studies have shown that yeast display is optimally suited for optimization of Fcab binder properties such as stability [5] or alteration in antigen-recognition mode [6].

1.2 Affinity Maturation of a Binding Fcab Clone

1.2.1 Preferred Methods for Construction of Fcab Affinity Maturation Libraries

Once a binder clone with the desired biological activity has been identified, its binding affinity can be increased to augment its activity. The binders isolated in a primary library selection usually populate several different epitope bins, but their binding affinity can be inferior to a typical binding affinity required to exert an effect in an in vivo setting. The optimal affinity is dependent on the particular biological situation studied and is typically in the range between 10^{-8} and 10^{-9} M for naturally occurring antibodies [7], about 10^{-9} for clinical antibodies targeting receptors, and even higher for antibodies targeting cytokines [8]. There are several possibilities to modify the sequence of a binding clone to achieve an improvement in affinity. For several reasons, residues that constitute the postulated binding site of Fcabs are the primary target of mutagenesis. First, several scaffold mutations can be detrimental to protein A or even more importantly neonatal Fc receptor (FcRn) binding, thus exerting a negative effect on the in vivo half-life of the molecule. Second, perturbations of the crucial conserved stability-determining structural elements of an Fc fragment can result in a poorer developability of an Fcab clone or its derivate mAb^2 as a therapeutic molecule since such molecules are more prone to aggregation and less soluble. Therefore the regions connecting the β-sheets are typically targeted for randomization.

In the first steps of the affinity maturation procedure, it is likely that several randomized residues involved in binding need to be altered to achieve superior antigen binding. Parsimonious mutagenesis [9] is one method that allows rapid exchange of a desired ratio of randomized residues, employing the proportion of the original nucleotides as a tool to stochastically govern the number of the exchanged amino acids in the randomized sequence. For the construction of an affinity maturation library, typically one third of amino acids of the modified Fcab loops can be exchanged by setting the percentage of the first two original nucleotides in each codon to 85% and the third to 50%, which results approximately in $0.85 \times 0.85 = 71\%$ probability of reversion of the original amino acid (Table 1). The library size required to achieve complete coverage of such library is difficult to achieve, and sorting such library in FACS selection, where the minimum number of participating yeast cells is 20-fold the library size to achieve 95% representation, would involve several 10 or 100 mL of yeast suspension. Prior processing with magnetic-activated cell sorting (MACS) reduces the input of yeast population, but does not allow normalization to correct the antigen-binding signal for the number of displayed binding sites on each yeast cell. For practical reasons, yeast Fcab libraries are therefore constructed with 10^8–10^9 independent members when parsimonious mutagenesis is performed.

Another useful method for a rapid improvement of the affinity of Fcabs is the isolation of binders with superior antigen-recognition properties from a pool-expanded library created by loop shuffling (Table 1). Here the randomized C-terminal loop sequences of the selected affinity-matured pools of Fcab clones are combined to make use of the additive effect of the mutations which enhance the contact to the antigen. Improved mutants are then identified by additional FACS sorting rounds and subsequent screening. This approach is preferred to testing single binders constructed by shuffling the randomized loops of single improved clones, first because the additivity does not have to apply [10] and second to avoid the selection of Fcab clones with unfavorable biophysical properties resulting from sterically interfering amino acid residues in close proximity.

Table 1
Examples of oligonucleotides that can be used for yeast Fcab library construction using parsimonious mutagenesis and loop shuffling

	Sequence with homology to vector	Randomized sequence	Priming (annealing) sequence
Forward oligonucleotide	GAACCACAGGTGTACACCC TGCCCCCATCCCGGGATGAG	245125115115215	GTCAGCCTGACCTGCCTGGTCAAAG
wild-type C_H3 sequence	GAACCACAGGTGTACACCC TGCCCCCATCCCGGGATGAG	CTGACCAAGAACCAG	GTCAGCCTGACCTGCCTGGTCAAAG
	Priming (annealing) sequence	Randomized sequence	Sequence with homology to vector
Reverse oligonucleotide	CACGGTGAGCTTGCTGTAGA	512544522543543521524524544542	GTGGTTGTGCAGAGCCTCATGCA TCACGGAGCATGAGAA
wild-type C_H3 sequence	CACGGTGAGCTTGCTGTAGA	GACGTTCCCCTGCTGCCACCTGCTCTTGTC	GTGGTTGTGCAGAGCCTCATGCA TCACGGAGCATGAGAA
Nucleotide code: 1: 85%A, 5% each C,G,T 2: 85%C, 5% each A,G,T 3: 85%G, 5% each A,C,T 4: 85%T, 5% each A,G,C 5: 50%C, 50%G			
Forward oligonucleotide		Reverse oligonucleotide	
ABnest	CCAAGGGCCAGCC TCGAGAACCACAGGTGTACAC	3AB	GATAGAAGCCTTTGACCAGGCAGG TCAGGCTGAC
5CD	GTCAGCCTGACCTGCCTGG TCAAAGGCTTCTATC	3CD	GGAAGAAGGAGCCGTCGGAG TCCAGCAC
5EF	GTGCTGGACTCCGACGGCTCCTTC TTCC	UPnest	GTGTGTAGTGGTTGTGCAGAGCCTCA TGCATCACGGAG

An attractive possibility of mutagenesis of the targeted segments of C-terminal loops is error-prone PCR that operates directionally on these regions. Here the PCR fragments are produced with long oligonucleotides overlapping with the remaining segments of the C_H3 domain, and the loop regions are amplified with error-prone Taq polymerase or a mixture of nucleotide analogs. PCR fragments are then allowed to assemble in yeast via homologous recombination. This method is used less extensively as the libraries constructed in this way tend to have a lower degree of correctness due to a higher error rate resulting from the synthesis of long oligonucleotides.

1.2.2 Equilibrium Antigen Selections

Provided the number of binding sites on a yeast cell does not exceed the number of molecules of the antigen present at labeling, yeast clones displaying Fcabs with a higher affinity will be able to bind more antigen molecules and thereby display stronger fluorescence intensity than a binder with a lower affinity. When an affinity maturation library of 2×10^7 members is to be sorted, the recommended antigen concentration for labeling is half-logarithm of the threshold concentration required for the parental clone to show a positive signal. In the setup of the sorting procedure, several measures can be introduced to minimize artifactual hits (Subheading 3.7 and Fig. 3), such as stringent gating of single yeast cells and gating out of yeast cells with high autofluorescence. When the enrichment of an antigen-positive population is observed, further steps can be taken to isolate the best candidate clones. For example, several sorting gates of various stringencies can be set to collect different subsets of antigen-labeled yeast population normalized for Fcab expression. Alternatively, sorting outputs can individually be titrated and antigen concentration further decreased for subsequent sorting rounds.

1.2.3 Kinetically Guided Selections

Kinetically guided selections of yeast display libraries have been used in the past to specifically influence the selection of antibody fragments with an improved on- and off-rate of antigen binding and resulted in the selection of extremely high-affinity antibodies [11]. The selection of binders with a faster on-rate can be performed by staining at a low temperature for a short period of time, whereas library members with an improved off-rate of binding are mostly selected for by limiting the time of competition of labeled antigen with an excess of the unlabeled antigen. Another important reason to resort to kinetic selection protocol is the fact that an efficient multivalent yeast display results in several tens of thousands copies of displayed binding fragments, and the antigen concentration has to exceed the number of binding sites on yeast cells in an equilibrium selection. As the number of binding Fcabs displayed on a yeast cell amounts to 50,000 or more, staining with an antigen concentration below 10 nM implies that the yeast cells

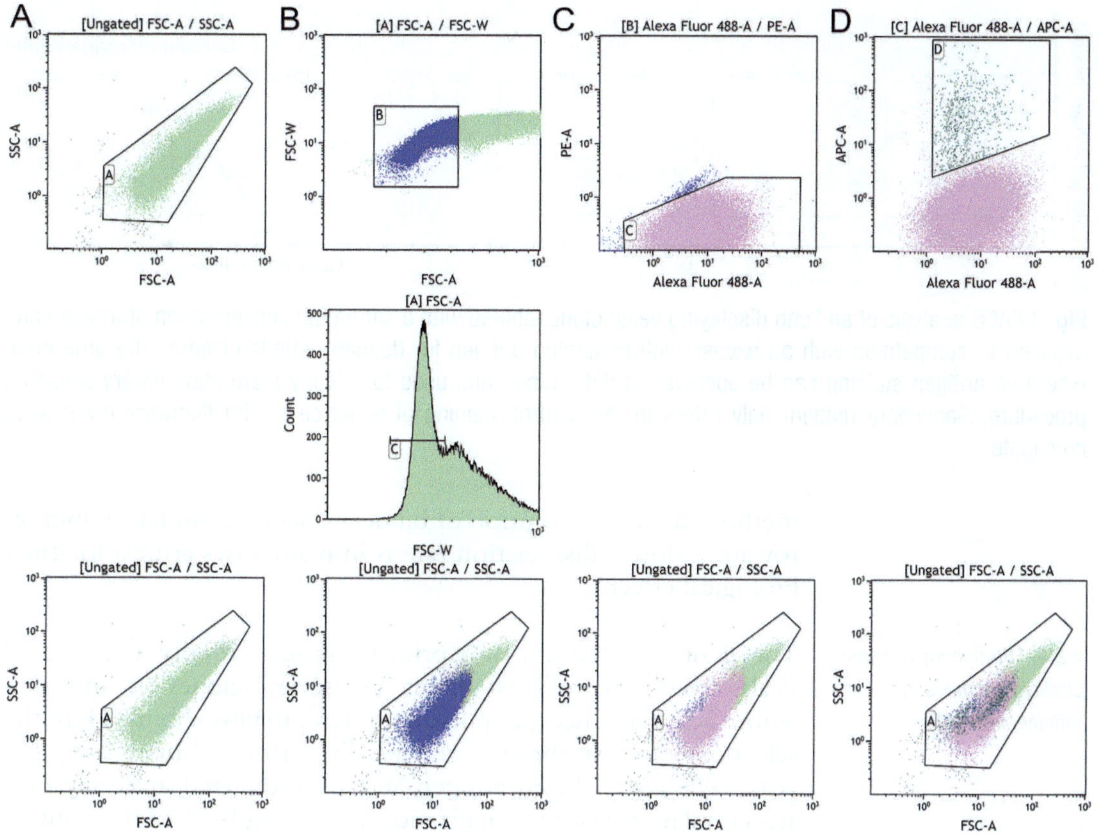

Fig. 3 Setup of sorting procedure with serial gating. Presented are FACS plots used for gating Fcab-displaying yeast cells in antigen selections. (**a**) FSC-A/SSC-A plot with gated yeast cell population; (**b**) FSC-A/FSC-W plot and FSC-W histogram with gated single yeast cells to avoid selection of aggregates; (**c**) FL1/FL2 plot to remove the yeast cells with high autofluorescence; (**d**) FL1/FL6 plot to collect double (antigen binding and structural marker binding) positive yeast cells. In the downmost row, the gated cells are projected back onto the total yeast population

should be diluted in several liters of antigen solution, which makes the selection difficult to perform. At the same time, the avidity effect resulting from multivalent binding that can be mediated by homodimeric Fcab can lead to a phenomenon where the antigen titration curve determined for yeast binders largely deviates from the sigmoid curve describing 1:1 antigen interaction typical for yeast-displayed single-chain Fv or Fab fragments [12]. An attractive solution therefore is the off-rate selection, where yeast cells of an Fcab library are first labeled with a saturating concentration of labeled antigen and then incubated with an excess of unlabeled antigen for a defined period of time (Fig. 4). The competition is stopped by adding cold buffer to the labeled yeast cells. After the staining step with a detection reagent that reveals bound labeled antigen, yeast cells are analyzed using FACS to determine the time of competition where yeast clones with an improved off-rate still would show antigen binding. Next to the practical value of this

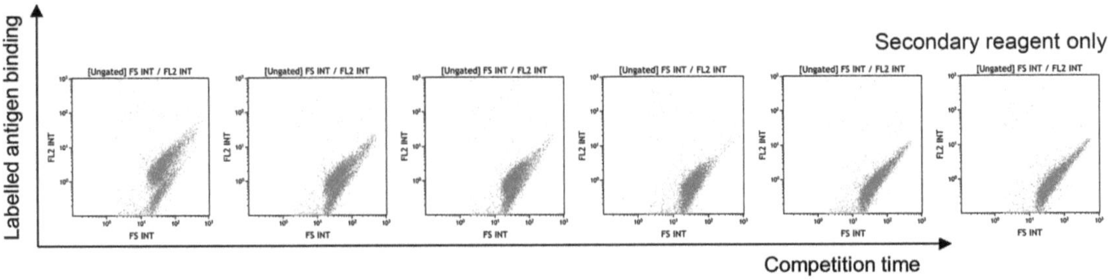

Fig. 4 FACS analysis of an Fcab displaying yeast clone labeled with a saturating concentration of antigen and exposed to competition with an excess with unlabeled antigen for defined periods of time. The time point where no antigen staining can be observed is determined and used for affinity maturation library selection procedure. Secondary reagent only refers to the control staining of yeast cells with fluorescently labeled conjugate

method, directed evolution of binders isolated from naïve libraries toward a slower dissociation rate is in many cases critical for their biological effect.

1.2.4 Stringency of Yeast Library Selections in Affinity Maturation

The desired outcomes of a primary antigen affinity selection of Fcabs from a naïve yeast library are several clones of unrelated amino acid sequences, populating different epitope bins. Here the selections of yeast libraries are typically performed by staining the yeast cells with labeled antigen and a tag-directed antibody or a structural marker for normalization and gating 1–5% of the double-positive population for further propagation in every sorting round. In contrast, the successful selection of affinity-matured clones can be achieved in three to four consecutive sorting rounds by more stringent gating at about 0.1% false-positive cells, where false-positives are determined on the basis of a sample of yeast cells stained with the same procedure as the sample intended for sorting except for the labeled antigen. When the enrichment of a double-positive yeast population is observed, single yeast cells can be sorted onto an agar plate to grow to colonies that can directly enter the screening process.

1.3 Strategies for Labeling of Yeast-Display Fcab Libraries

1.3.1 Detection Agents

The most common method for labeling of the antigen is biotinylation, and the preferred labeling reagent used a biotin moiety attached to a long linker, which minimizes steric hindrance of the epitopes. Detection of antigen binding can be performed using neutravidin or streptavidin as detection reagents labeled with a strong fluorescent conjugate, such as PE, APC, or tandem conjugates. As an alternative, a fluorescently labeled anti-biotin antibody can be used for selection. When displaying human Fc-derived proteins, it is advisable to avoid secondary antibody-based reagents, which are usually applied to detect tag-specific antibodies used for normalization, or Fc-tagged antigens. Sera depleted of antibodies reactive with human Fc variants can be used instead.

1.3.2 Normalization in the Sorting Procedure and Simultaneous Staining with Antigen and Structural Markers

The successful outcome of a yeast library antigen selection is reportedly dependent on normalization, a protocol intended to compensate for the effect of different protein display levels of library members. Yeast cells are stained simultaneously for antigen binding and the reactivity with an antibody directed against a tag, positioned N-terminally or C-terminally to the displayed mutant protein. The selection proceeds with a diagonal sorting window (Fig. 3). The detection of a C-terminal tag harbors an additional informative value as it corroborates the integrity of the reading frame of the displayed protein. As early as at the point of quality control of the library, the percentage of C-terminal tag-positive yeast cells reports on the correctness of the library. In the design of Fcab libraries, the C-terminal tags were obliterated with an introduction of a stop codon following the sequence of the mutated Fc fragment to avoid the interference with the antigen-binding site postulated at the C-terminal loops of the C_H3 domain. Normalization can still be performed with the anti-Xpress antibody, which detects the N-terminally positioned Xpress tag.

Importantly, the selections of correctly folded Fcabs can also be guided by simultaneous staining with an Fc-specific ligand molecule (Table 2, Fig. 2). These protocols have contributed to the selection of Fcab clones that exhibit favorable biophysical properties, especially when the libraries have been induced under stress induction conditions at 37 °C as recommended also for other proteinaceous scaffolds [13]. One of the most attractive possibilities is the staining with high-affinity Fc-receptor FcγRI (CD64), the binding site of which maps to the N-terminal loops of the C_H2 domain and overlaps with the binding site for other molecules eliciting effector functions, most importantly FcγRIII (CD16). Not only is the binding obliterated when the C_H3 domain lacks integrity or is poorly folded, but also the ability of simultaneous binding to the effector molecule and to the target antigen is a

Table 2
Reagents used for normalization in affinity selection of Fcabs

Binding site on yeast-displayed Fcab	Reagent (first step)	Dilution	Reagent (second step)	Dilution
N-terminal Xpress tag	Anti-Xpress ™ antibody	1:2000	Goat anti-mouse $(Fab)_2$-FITC	1:200
C_H2 domain			Anti-C_H2 antibody-FITC	1:200
C_H2 and C_H3 domain			Protein A-FITC	1:200
Fc fragment			Anti-human Fcγ chain-PE	1:800
N-terminal loops of C_H2	FcγRI (his)	1 μg/mL	Anti-penta-His Alexa Fluor 488	1:400

prerequisite for the selected Fcab binder to mediate an antibody-dependent cytotoxicity response to kill the target cell.

Another possibility is the use of the C_H2 domain-specific antibody (clone MCA647, Bio-Rad), whose binding site maps to a properly folded C_H2 domain, which is however dependent on the integrity of the C_H3 domain [14]. Should the mutagenesis-induced perturbation cause the destabilization and hence misfolding or proteolytic degradation of the C_H3 domain, the reactivity of the anti-C_H2 antibody will be diminished or obliterated.

A third agent useful for monitoring the structural properties of the members of Fcab libraries is protein A, an Fc ligand used in manufacturing for large-scale purification of monoclonal antibodies. In an Fc, its binding site maps between C_H2 and C_H3 domains and requires the presence of both properly folded domains. While the staining with fluorescently labeled protein A is a prerequisite for a properly folded Fc molecule, simultaneous staining with the antigen does not necessarily have to test positive as bound protein A can sterically alter the Fcab fragment and influence its antigen-binding properties. Nevertheless, the availability of either unlabeled or fluorescently labeled protein A in comparison with more costly FcγRI or the C_H2 antibody makes it a valuable ligand for preselection of yeast libraries for correctly folded members using MACS or FACS protocols, where large numbers of cells have to be processed and a large amount of structure-reporter molecule is required.

1.3.3 Choice of Blocking Agent

BSA is used as a blocking agent in most yeast surface display selections and is especially popular in FACS-based selections due to its high solubility. If an Fcab yeast library is constructed with oligonucleotides designed to contain the NNK triplet encoding mutated amino acids, a high proportion of charged amino acids, especially arginine, can be found in the randomized sequences of selected Fcab clones. This can be due to the fact that arginine is encoded by six different nucleotide triplets, but high display levels of such proteins can also be due to their specific mode of interaction of arginine strands with cell membranes [15]. Engineered binding proteins that contain such motives usually suffer from structural perturbations, which result in a lower solubility and a higher aggregation propensity [16]. The selection of such yeast clones can be reduced from the antigen-positive pools by using dry milk as a blocking agent (*see* **Note 1**).

1.3.4 Epitope Masking as a Method to Isolate Fcabs Targeting Novel Binding Epitopes

The topology of the postulated binding site in Fcabs is that of a flat paratope as opposed to most antibodies [17]. This fact alone leads to the presumption that Fcab binders can be directionally selected to recognize epitopes different from those recognized by a validated classical antibody. In an antigen selection, labeled antigen is incubated with an antibody masking the undesired epitope and then applied for labeling of an Fcab library. Alternatively, the unbound epitopes can be discovered using a competition method where Fcab libraries are stained with labeled antigen and then incubated with an antibody that can strip the yeast cell of antigen if it is bound to the displayed Fcab by an identical or partially overlapping epitope.

1.4 Screening

The success of directed evolution crucially depends on the number of screened clones, and the throughput can improve drastically with the automatization of the screening process. Fcab clones from naïve selections are typically first screened for antigen binding in yeast display format and at the same time examined for binding to structure-dependent ligands, such as protein A or FcγRI. Candidate binding clones can be selected according to additional screening criteria, e.g., the ability to compete with an antibody recognizing the targeted epitope. The subthreshold antigen concentration used for selections is also applied for screening of affinity-matured Fcab clones. Rare clones can be discovered with high fidelity using high-throughput sequencing of sorting outputs [18]. Supernatants of Fcab yeast display cultures contain roughly 1 mg/L soluble Fcab, resulting either from overexpression or shedding from yeast cell surface, and if Fcab clones are of about 10 nM affinity, the supernatants of induced yeast cultures can be used directly in an ELISA screening procedure that detects soluble Fcabs which can bind target antigen, preferably using a polyclonal anti-human Fc antibody as a detection reagent. Fcab clones can be ranked according to their antigen reactivity in soluble form, accelerating the discovery of Fcab clones with improved binding properties.

1.5 Affinity Maturation Strategies for Heterodimeric Fcab Fragments

Homodimeric Fcab fragments are initially designed with symmetrically positioned mutations in the AB and EF loops in the C_H3 domains (Fig. 1b); and hence a binding stoichiometry of one Fcab binding to one or two antigen molecules is anticipated, as was observed in the crystal structure of the first Fcab binder H10-03-6 directed against HER2/neu [19]. In case of one Fcab dimer binding to one antigen, the symmetry of the C_H3 does not have to apply: using mutagenesis on a single amino acid chain that assembles into a dimer, the same mutations on each chain will not necessarily be beneficial for antigen binding. Hence we have probed the display of a heterodimeric Fcab on the surface of yeast cells, where we have introduced "knobs-into-holes" (T366Y/Y407T) [20] as a heterodimerization motive. To avoid undesired recombination events, we have dehomologized the sequence of the heterodimeric Fc fragment by choosing a different codon usage in each of the two chains. These were cloned into pYD1-based vectors, the "knob" chain into the unmodified pYD1 vector allowing expression as an Aga2p-anchored protein, and the Fc chain cloned in frame with a leader peptide guiding its secretion as a soluble protein. For the feasibility of simultaneous transformation and selection as well as the maintenance in a double auxotrophic yeast cell, the vectors used were with Trp or with Leu selectable markers. Each was then transformed to *S. cerevisiae* EBY100 and selected on minimal medium lacking Trp or Leu, respectively. Further, we have transformed EBY100 harboring the "knob"-Fc chain vector with the vector encoding "hole"-Fc chain and selected for mutants surviving deficiency of both nutritional markers in the culture medium. The constructs were cultured in the minimal medium (2 × 0.67% YNB with ammonium sulfate, 100 mM KH_2PO_4/K_2HPO_4 buffer, pH 6.0, 2% glucose) and induced with the same medium containing galactose and raffinose. Staining with an anti-Fc antibody and FACS analysis revealed that yeast could display the "knob"-Fc fragment, but the Fc chains could not dimerize, since the staining with FcγRI, which can only bind to a dimeric Fc [21], was negative (Fig. 5). As expected, yeast clones transformed with

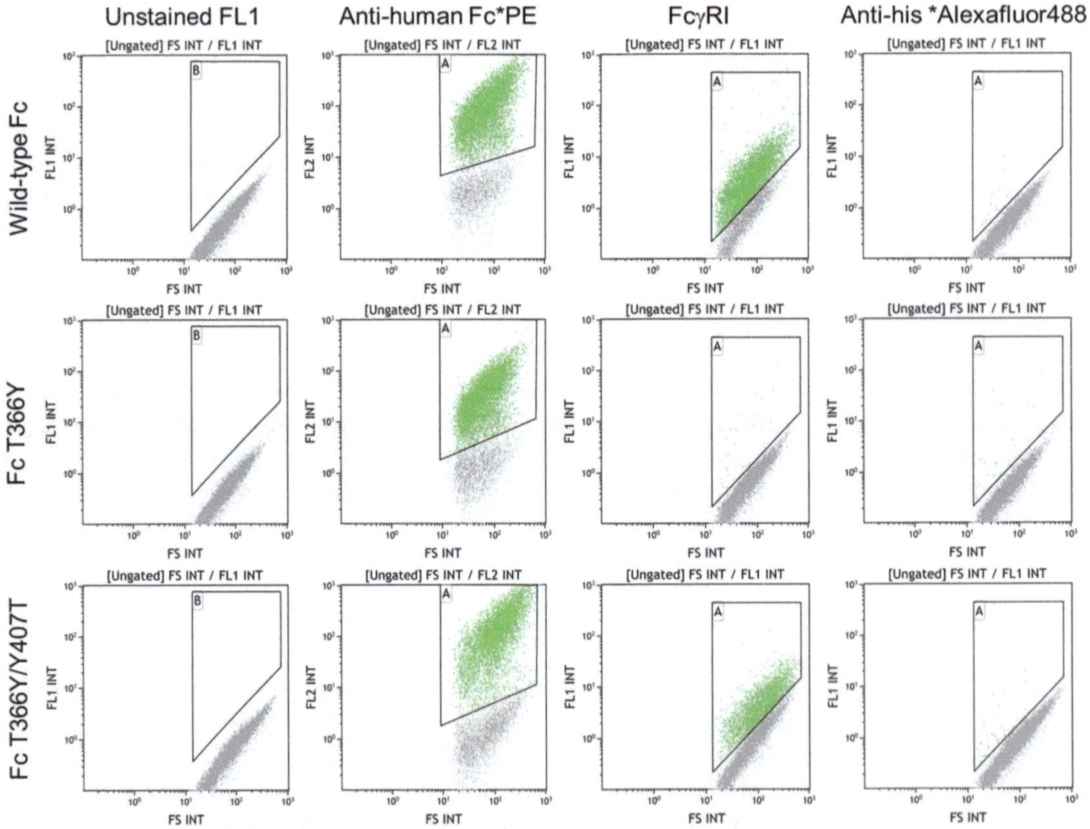

Fig. 5 Display of heterodimeric Fcab on yeast. Yeast cells were transformed with wild-type Fc display construct, Fc "knob" (T366Y) or with both Fc "knob" and "hole" (T366Y/Y407T) chains, and stained with PE-conjugated anti-human Fc serum or FcγRI, the binding of which was detected with an anti-His antibody conjugated with Alexa Fluor 488. Display of heterodimer was confirmed with positive reactivity with FcγRI

both vectors encoding heterologous Fc chains exhibited a positive signal with FcγRI. A noticeable difference to FcγRI staining of a homodimeric Fc was that the percentage of positive cells amounted to 40% for heterodimer display in comparison with 70% usually observed for a homodimer, a phenomenon which can probably be ascribed to recombination events in the vector sequences that were left homologous. This approach offers the attractive possibility to produce Fcab libraries where clones that do not recognize a symmetrical binding site can be augmented with directed evolution targeting different residues on each chain of the heterodimer. An increase in the library size in such cases can be achieved by using a mating procedure of two different yeast-mating types, each harboring a pool of heterodimer chains, analogously to the construction of Fab libraries by mating yeast transformed with pools of light and heavy antibody chains [22].

2 Materials

2.1 Reagents

EZ-Link™ Sulfo-NHS-LC-LC-Biotin (21338, Thermo Fisher Scientific).

Anti-Xpress antibody (46-05-28, Thermo Fisher Scientific).

Anti-C_H2 antibody-FITC conjugate (MCA647F, Bio-Rad).

Anti-mouse Fc $(Fab')_2$ fragment-FITC conjugate (F-2653, Sigma-Aldrich).

Anti-human γ chain-PE conjugate (P-8047, Sigma-Aldrich).

Anti-biotin antibody-APC conjugate (130-090-856, MACS Miltenyi).

Mouse anti-human IgG Fc (ab99762, Abcam).

Goat anti-mouse Fc $F(ab')_2$ FITC (F-2772, Sigma-Aldrich).

FcRγI (Fc-1257, R&D Systems).

Anti-Penta-His antibody Alexa Fluor 488 conjugate (35310, QIAGEN).

Protein A-FITC conjugate (P-8651, Sigma-Aldrich).

Neutravidin-PE (A2660, Thermo Fisher Scientific).

Streptavidin-Alexa Fluor-647 conjugate (S21374, Thermo Fisher Scientific).

Restriction enzymes.

2.2 Media Components

1. Casaminoacids (CAA).
2. Yeast nitrogen base (YNB).
3. Glucose.
4. Galactose.

5. Raffinose.
6. L-leucine.
7. Peptone.
8. Yeast extract.
9. Penicillin/streptomycin.

2.3 Solutions and Buffers

50% PEG3350.

1 M Li-acetate.

2 mg/mL salmon sperm DNA.

Phosphate-buffered saline (PBS): 137 mM NaCl, 2.7 mM KCl, 10 mM Na_2HPO_4, 2 mM KH_2PO_4.

Blocking buffer: 5% bovine serum albumin (BSA) in PBS.

FACS staining solution: 2% BSA in PBS.

Freezing buffer: 30% glycerol in PBS.

2.4 Media

YPD: 20 g/L peptone, 10 g/L yeast extract, and 2% glucose.

SD-CAA: 1% CAA solution with 0.67% YNB with ammonium sulfate, 100 mM KH_2PO_4/ K_2HPO_4 buffer, pH 6.0, and 2% glucose.

SG/R-CAA: 1% CAA solution with 0.67% YNB with ammonium sulfate, 100 mM KH_2PO_4/ K_2HPO_4 buffer, pH 6.0, 2% galactose, and 1% raffinose.

MDL solid medium: 1.5% agar with 0.67% YNB with ammonium sulfate, 2% glucose, and 0.1 mg/mL L-leucine.

YPD solid medium: 1.5% agar with 20 g/L peptone, 10 g/L yeast extract, and 2% glucose.

100× penicillin-streptomycin solution: 100,000 U/mL penicillin and 10 mg/mL streptomycin.

2.5 Kits

Gel and DNA purification kit.

Plasmid isolation kit (for midi-preparation).

Zymoprep II kit.

QFIKIT (Dako K0078).

2.6 Equipment

96-well microtiter plates.

96-well compatible stacked micronic tubes (ScreenMate or similar).

1.5 mL microcentrifuge tubes

1.8 mL cryotubes

15 and 50 mL conical tubes.

90 mm Petri dishes.

Pipettes from P10 to P1000 range with respective tips and multichannel pipettes.

Rotating wheel.

−80 °C refrigerator.

Incubator with shaking platform that can be temperated to 20, 30, and 37 °C.

Heat blocks or water baths.

High-speed sorter such as ARIA I (Becton Dickinson).

FACS analysis apparatus.

3 Methods

3.1 Yeast Fcab Library Construction Using Parsimonious Mutagenesis

The regions of the molecule that are going to be randomized are first determined, and the vector is restricted as close as possible to the randomized region. Oligonucleotides designed for the synthesis of the PCR fragment will contain the region intended for annealing to produce the library amplicon, the randomized region, and 25-base pairs overlaps to allow gap repair-driven homologous recombination with the vector sequence. Minimizing the distance to the homologous sequences is of advantage as it reduces the required length of the oligonucleotide and thereby reduces the probability of synthesis errors and allows a more efficient recombination. Enzymes that cleave outside their binding site (e.g., *Bsa*I, *Bsm*BI, *Fok*I) allow the design of a precise overlap with the ends of the PCR library fragment; a dummy vector designed in this way can be linearized and used without gel purification. As the dummy vector does not encode an integral C_H3 domain, transformants that may result from an incomplete restriction will be excluded from the sorted population due to their non-reactivity with structural marker molecules.

1. Design oligonucleotides to randomize the selected amino acid residues. Table 1 gives an example of an oligonucleotide that randomizes a wild-type C_H3 domain sequence in the residues 358–362 of the AB loop and a reverse oligonucleotide that is used for randomization of the residues 413–415 and 418–419 in the EF loop.
2. Run a PCR using a high-fidelity Taq polymerase and a plasmid-encoding C_H3 as a template.
3. Gel purify the PCR fragment. 70 μg DNA is required for a library of about10^8 independent clones.
4. Digest the recipient vector and purify the DNA. 100 μg DNA is required.

5. Start an overnight culture of *S. cerevisiae* EBY100 in 10 mL YPD, on an orbital shaker at 30 °C.
6. On the next day, determine the OD_{600} of the culture and dilute to 500 mL of OD_{600} of 0.4 in YPD medium.
7. Incubate the yeast culture on an orbital shaker at 30 °C until it reaches an OD_{600} between 2 and 3.
8. Distribute the culture to 50 mL conical tubes.
9. Pellet the yeast by centrifugation at 1000 × *g*, 5 min, at room temperature.
10. Wash pellets with 25 mL distilled water.
11. Collect the cells by centrifugation at 1000 × *g*, 5 min, at room temperature.
12. Resuspend each pellet with 3 mL 100 mM Li acetate solution and let it shake at 30 °C for 15 min.
13. Collect the cells by centrifugation at 1000 × *g*, 5 min, at room temperature. Remove the Li acetate solution.
14. Resuspend the cells in the remains of the Li acetate as the pellets are difficult to resuspend in PEG3350 solution.
15. Add 2.4 mL 50% PEG3350, 360 μL 1 M Li acetate, and 500 μL heat-shocked (5 min at 95 °C) salmon sperm DNA, and insert and vector DNA in a volume that should not exceed 340 μL. Let shake at 30 °C for 30 min.
16. Incubate at 42 °C in a water bath for 45 min. Invert the tubes every 10 minutes.
17. Pellet the yeast and remove the transformation solution.
18. Add 5 mL YPD medium and incubate with shaking at 30 °C for 30 min.
19. Collect the cells by centrifugation at 1000 × *g*, 5 min, at room temperature. Inoculate into 500 mL SD-CAA medium.
20. At this point, plate the aliquots to MDL plates to determine the number of independent transformants in the library. Incubate at 30 °C for at least 3 days.
21. Incubate the yeast library on an orbital shaker at 30 °C for 24 h.
22. Passage 10 mL to 490 mL fresh SD-CAA medium, and incubate the yeast library on an orbital shaker at 30 °C for 24 h.
23. Collect yeast cells with centrifugation, discard medium, and mix with an equal volume of freezing buffer.
24. Aliquot to 1 mL cryotubes. About 20 mL yeast suspension will be produced. To determine the number of viable cells, remove an aliquot of the culture and dilute in SD-CAA before plating to MDL plates. Between 10^9 and 10^{10} cells/mL is expected.
25. Store the library at −80 °C.

3.2 Yeast Fcab Library Pool Expansion Using Loop Shuffling

When several Fcab clones with improved antigen binding result from a selection of an affinity maturation library based on a single parental clone, their randomized regions can be shuffled to increase the diversity of the pool and to select for even more affine variants. Here the gap repair-driven homologous recombination is performed between several fragments with homologous ends and the linearized recipient vector. Libraries of Fcabs that are usually randomized in regions mapping to three C-terminal loops of the C_H3 domain can be transformed with the vector and three PCR fragments with the same efficiency as with a single PCR fragment.

1. Sequences of oligonucleotides that encompass the ends of AB, CD, and EF loop of the C_H3 domain are given in Table 1. These allow an efficient PCR amplification and subsequent homologous recombination of the three parts of the C_H3 domain, each containing one randomized region.
2. Run PCRs using a high-fidelity Taq polymerase and a C_H3 as a template.
3. Gel purify PCR fragments. 25 μg DNA of each fragment is required for a library of about 10^8 independent clones.
4. Proceed with **steps 4–25** as listed in the Subheading 3.1.

3.3 Cultivation and Induction of Yeast Cultures

1. Inoculate the yeast clones from a single colony or yeast libraries (at least 20-fold excess of the number of independent library members) into SD-CAA with penicillin/streptomycin (*see* **Note 2**) at an OD_{600} of 0.2.
2. Let grow at 30 °C overnight, with shaking.
3. Collect the cells by centrifugation at 1000 × *g*, 5 min, at room temperature.
4. Remove the supernatant.
5. Resuspend the cells in SG/R-CAA medium.
6. If libraries are to be induced, use an aliquot of the cells that is at least 20-fold the number of the output cells of the previous selection round for induction. The remaining cells can be frozen by resuspending them in an equal volume of 30% glycerol and placed to −80 °C to serve as a backup.
7. Incubate the yeast cultures with shaking at 20 °C for 48 h or at 37 °C overnight, when stress induction conditions should be used.

3.4 Staining Fcab Yeast Display Cultures for Equilibrium Selection

1. Determine the OD_{600} of induced yeast cultures.
2. Collect the cells by centrifugation at 1000 × *g*, 5 min, at room temperature.
3. Resuspend the cells at an OD_{600} of 5 in a solution of 5% BSA-PBS.

4. Incubate on a rotating wheel at room temperature for 30 min.
5. Collect the cells and remove the blocking solution.
6. Resuspend the cells in blocking solution with antigen of the desired concentration. For naïve selections, this is usually 1 μM and for affinity maturation half-logarithm lower than the threshold concentration where antigen binding of a parental clone can be detected.
7. Incubate on a rotating wheel at room temperature for 1 h.
8. Collect the cells by centrifugating the cells at 1000 × *g*, 5 min, at 4 °C. From now on, washing steps and incubations should be performed on ice to reduce dissociation of the antigen from yeast cell surface.
9. Wash cell pellet briefly in ice-cold PBS and centrifuge at 1000 × *g*, 5 min, at 4 °C.
10. Resuspend the cells in ice-cold solution of detection reagent and fluorescently labeled tag-specific antibody or structural ligand for normalization.
11. Incubate on ice for 30 min, protected from light.
12. Collect the cells and remove the blocking solution.
13. Wash cell pellet briefly in ice-cold PBS and centrifuge at 1000 × *g*, 5 min, at 4 °C.
14. Resuspend the cells at an OD_{600} of 20–50 in ice-cold PBS and keep on ice until sorting (*see* **Note 3**).
15. Before sorting, filter the cell suspension through nylon mesh-capped FACS tubes to remove clumps.

3.5 Staining Fcab Yeast Display Cultures for Kinetic Selection

Steps 1–5 are performed as described in Subheading 3.4.

6. Resuspend the cells in blocking solution with labeled antigen at a saturation concentration determined for the parental clone.
7. Incubate on a rotating wheel at room temperature for 1 h.
8. Add unlabeled antigen at 100-fold excess concentration of the labeled antigen.
9. Incubate at room temperature for the period of time required for dissociation of antigen from the parental clone.
10. Stop the dissociation by diluting yeast cells in tenfold volume of ice-cold blocking buffer.
11. Proceed with the **steps 8–15** of the staining procedure as described in the Subheading 3.4.

3.6 Determining the Number of Displayed Fcabs on the Surface of Yeast Cells

This parameter is important for determining the minimal reaction volume for labeling yeast cells with antigen when a titration is performed to determine the antigen concentration that will be used for sorting. The number of antigen molecules should be at least tenfold of the available binding sites. At the beginning of the antigen selection process, few yeast clones bind the antigen, but this strategy becomes of importance through subsequent enrichment as it allows discrimination between lower- and high-affinity clones. Surface expression levels are determined using the QFIKIT (Dako K0078) employing flow cytometry. Five populations of calibration beads presenting different numbers of mouse mAb molecules are used as a calibration standard.

1. For the blocking of yeast cells, follow **steps 1–5** of the Subheading 3.4.
2. Label 10^5 yeast cells with a saturating concentration of 5 μg/mL of a mouse anti-human Fc antibody (Abcam) in 2% BSA-PBS in 100 μL volume with shaking for 30 min at RT.
3. Collect the cells by centrifugation at 1000 × *g*, 5 min, at room temperature.
4. Wash cell pellet briefly in 100 μL ice-cold PBS and centrifuge at 1000 × *g*, 5 min, at 4 °C.
5. Stain beads and cells with secondary goat anti-mouse Fc F(ab′)$_2$-FITC conjugate at 1:200 dilution in 2% BSA-PBS in 100 μL volume for 30 min, on ice, protected from light.
6. Collect the beads and cells by centrifugation at 1000 × *g*, 5 min, at room temperature.
7. Resuspend the cells and beads in 200 μL PBS.
8. Determine the mean fluorescence intensity (MFI) values by flow cytometry.
9. Plot the MFI values for the standard beads against the number of antibody molecules per bead (on a log/log scale). Fit the values with linear regression using the line as a standard curve.
10. Using this standard curve, convert the MFI values of yeast cells samples to the number of antibody binding sites per cell.

3.7 How to Gate Antigen-Binding Yeast Cells from the Sorting Sample

For clarity, a protocol is described where we assume that the normalization antibody or Fc ligand is labeled with Alexa Fluor 488, and the fluorescence can be detected in FL1, and the antigen binding is detected with APC and can be detected in FL6. An irrelevant fluorescent signal that can only arise from yeast cells with high autofluorescence will be assumed to be observed in FL2 channel.

1. Set the following bidimensional plots: FSC/SSC, FSC-A/FSC-W, FL1/FL2, and FL1/FL6 (example shown in Fig. 3).
2. Draw a gate around the yeast population on the FSC/SSC plot.
3. Observe the gated population on an FSC-A/FSC-W plot. The kink in the dot-plot allows discrimination of the aggregates and budding cells from single yeast cells. Alternatively, display a

histogram with FSC-W as a primary parameter against cell count and separate the single cell peak with a marker. Typically, single cells in yeast cultures induced at 37 °C represent 60–70% of the population.

4. Visualize the gated population on the FL1/FL2 plot. Cells that appear positive in the FL2 channel can be excluded as this irrelevant fluorescent signal originates from their high autofluorescence.
5. Visualize the gated population in an FL1/FL6 plot.

3.8 Setting the Gate for Selection of Antigen-Binding Yeast Cells

1. Prepare the following controls: unlabeled yeast cells, yeast cells labeled with normalization marker only, and yeast cells that were stained with the same procedure as the sample for sorting except for antigen.
2. Compare the sample that was not stained with labeled antigen with the positive sample. If the binding cells are still rare, collect 100,000–500,000 cells to be able to discern the events that are unique and appear double positive in the antigen-labeled sample.
3. Gate and sort the double-positive population.
4. Alternatively, if discrete antigen-binding yeast cells cannot yet be discriminated from the background, position the sorting gate over top 0.1–1% of the "false-positive" population stained with a normalization marker or a structural marker.
5. The level of unspecific staining that can sometimes be observed due to secondary reagents can be determined by comparison with yeast cells stained with normalization marker only. Typically, the concentration of the secondary reagent is then reduced to minimize this effect.
6. Sort at least 20-fold excess number of independent library clones or output cells from the previous sorting round to assure 95% representation into a tube with SD-CAA medium.
7. If a double-positive population can already be observed, those yeast cells can be sorted onto a gridded MDL-plate to form colonies that can directly enter the screening procedure (*see* **Note 4**).
8. Incubate the yeast cells sorted into liquid medium at 30 °C with shaking. When at least 20,000 cells have been collected, an overnight incubation is sufficient for yeast to reach an OD_{600} of 5 and can be induced.
9. Incubate the yeast cells sorted onto solid medium at 30 °C for 2–3 days. The colonies are then inoculated into liquid SD-CAA for screening or preservation. At this point, yeast plasmid DNA can be isolated using Zymoprep II kit, exactly according to manufacturer's instructions.

4 Notes

1. If dry milk is to be used in a FACS-based selection, the reagent should be stirred in PBS at 4 °C overnight, centrifuged at high speed, and filtered through a 0.45 μM filter to avoid the particles that obscure the cell count during the sorting procedure.
2. Sorting is not sterile but an aseptic procedure. It is recommended to supplement culture media with penicillin-streptomycin solution to avoid bacterial contamination.
3. Sorting of labeled cells for longer than 3 h can prove detrimental to the viability of yeast cells. Preferably, fresh aliquots of yeast culture should be stained if incubation on ice and sorting exceeds this period.
4. When yeast cells are sorted directly onto a culture plate, a gridded plate with solid medium is preferred to 96-well plate with liquid medium. A greater proportion of single yeast cells is able to divide and form colonies when spotted onto a solid-medium plate.

Acknowledgments

The financial support by the Austrian Federal Ministry for Digital and Economic Affairs and the National Foundation for Research, Technology and Development is gratefully acknowledged. The company F-star has supported this work as well.

References

1. Wozniak-Knopp G, Bartl S, Bauer A et al (2010) Introducing antigen-binding sites in structural loops of immunoglobulin constant domains: Fc fragments with engineered HER2/neu-binding sites and antibody properties. Protein Eng Des Sel 23(4):289–297. https://doi.org/10.1093/protein/gzq005
2. Leung KM, Batey S, Rowlands R et al (2015) A HER2-specific modified Fc fragment (Fcab) induces antitumor effects through degradation of HER2 and apoptosis. Mol Ther 23 (11):1722–1733. https://doi.org/10.1038/mt.2015.127
3. Hasenhindl C, Traxlmayr MW, Wozniak-Knopp G et al (2013) Stability assessment on a library scale: a rapid method for the evaluation of the commutability and insertion of residues in C-terminal loops of the CH3 domains of IgG1-Fc. Protein Eng Des Sel 26 (10):675–682. https://doi.org/10.1093/protein/gzt041
4. Boder ET, Wittrup KD (1997) Yeast surface display for screening combinatorial polypeptide libraries. Nat Biotechnol 15(6):553–557. https://doi.org/10.1038/nbt0697-553
5. Traxlmayr MW, Lobner E, Antes B et al (2013) Directed evolution of Her2/neu-binding IgG1-Fc for improved stability and resistance to aggregation by using yeast surface display. Protein Eng Des Sel 26(4):255–265. https://doi.org/10.1093/protein/gzs102
6. Traxlmayr MW, Lobner E, Hasenhindl C et al (2014) Construction of pH-sensitive Her2-binding IgG1-Fc by directed evolution. Biotechnol J 9(8):1013–1022. https://doi.org/10.1002/biot.201300483
7. Poulsen TR, Jensen A, Haurum JS et al (2011) Limits for antibody affinity maturation and

repertoire diversification in hypervaccinated humans. J Immunol 187(8):4229–4235. https://doi.org/10.4049/jimmunol.1000928

8. Chames P, Van Regenmortel M, Weiss E et al (2009) Therapeutic antibodies: successes, limitations and hopes for the future. Br J Pharmacol 157(2):220–233. https://doi.org/10.1111/j.1476-5381.2009.00190.x
9. Schier R, Balint RF, McCall A et al (1996) Identification of functional and structural amino-acid residues by parsimonious mutagenesis. Gene 169(2):147–155
10. Chen Y, Wiesmann C, Fuh G et al (1999) Selection and analysis of an optimized anti-VEGF antibody: crystal structure of an affinity-matured Fab in complex with antigen. J Mol Biol 293(4):865–881. https://doi.org/10.1006/jmbi.1999.3192
11. Razai A, Garcia-Rodriguez C, Lou J et al (2005) Molecular evolution of antibody affinity for sensitive detection of botulinum neurotoxin type A. J Mol Biol 351(1):158–169. https://doi.org/10.1016/j.jmb.2005.06.003
12. Chao G, Lau WL, Hackel BJ et al (2006) Isolating and engineering human antibodies using yeast surface display. Nat Protoc 1 (2):755–768. https://doi.org/10.1038/nprot.2006.94
13. Shusta EV, Holler PD, Kieke MC et al (2000) Directed evolution of a stable scaffold for T-cell receptor engineering. Nat Biotechnol 18 (7):754–759. https://doi.org/10.1038/77325
14. Traxlmayr MW, Faissner M, Stadlmayr G et al (2012) Directed evolution of stabilized IgG1-Fc scaffolds by application of strong heat shock to libraries displayed on yeast. Biochim Biophys Acta 1824(4):542–549. https://doi.org/10.1016/j.bbapap.2012.01.006
15. Takechi Y, Yoshii H, Tanaka M et al (2011) Physicochemical mechanism for the enhanced ability of lipid membrane penetration of polyarginine. Langmuir 27(11):7099–7107. https://doi.org/10.1021/la200917y
16. Austerberry JI, Dajani R, Panova S et al (2017) The effect of charge mutations on the stability and aggregation of a human single chain Fv fragment. Eur J Pharm Biopharm 115:18–30. https://doi.org/10.1016/j.ejpb.2017.01.019
17. Novotny J, Bruccoleri R, Newell J et al (1983) Molecular anatomy of the antibody binding site. J Biol Chem 258(23):14433–14437
18. Glanville J, D'Angelo S, Khan TA et al (2015) Deep sequencing in library selection projects: what insight does it bring? Curr Opin Struct Biol 33:146–160. https://doi.org/10.1016/j.sbi.2015.09.001
19. Lobner E, Humm AS, Goritzer K et al (2017) Fcab-HER2 Interaction: a Menage a Trois. Lessons from X-Ray and Solution Studies. Structure 25(6):878–889 e875. https://doi.org/10.1016/j.str.2017.04.014
20. Ridgway JB, Presta LG, Carter P (1996) 'Knobs-into-holes' engineering of antibody CH3 domains for heavy chain heterodimerization. Protein Eng 9(7):617–621
21. Shields RL, Namenuk AK, Hong K et al (2001) High resolution mapping of the binding site on human IgG1 for Fc gamma RI, Fc gamma RII, Fc gamma RIII, and FcRn and design of IgG1 variants with improved binding to the Fc gamma R. J Biol Chem 276(9):6591–6604. https://doi.org/10.1074/jbc.M009483200
22. Weaver-Feldhaus JM, Lou J, Coleman JR et al (2004) Yeast mating for combinatorial Fab library generation and surface display. FEBS Lett 564(1–2):24–34. https://doi.org/10.1016/S0014-5793(04)00309-6

Chapter 14

Guidelines for Small-Scale Production and Purification of Hepatitis B Surface Antigen Virus-Like Particles from Recombinant *Pichia pastoris*

Maria Zahid and Ursula Rinas

Abstract

Virus-like particle (VLP)-based vaccines have been in the market since decades for preventing viral infection and have proven their usefulness also in other areas of biotechnology. Here, we describe in detail simple small-scale production and purification procedures for the generation of hepatitis B surface antigen (HBsAg) VLPs using *Pichia pastoris* as expression host. This protocol may also be applicable with variations to other HBsAg-based VLPs additionally carrying antigens of other pathogens.

Key words HBsAg, *Pichia pastoris*, Vaccine, Virus, Virus-like particle

1 Introduction

Ever since the development of medical sciences, the approach of prevention being better than cure has been a driving motive to develop vaccines. Vaccines against viral diseases have been quite challenging because of the daunting nature of the viruses.

The first virus-like particle (VLP) product to be launched in the market was against hepatitis B virus, e.g., HBsAg VLPs [1]. Since then, VLP technology has become more popular, and currently, there are two commercially available VLP-based vaccines already in the market, vaccines against hepatitis B and human papilloma virus infections [2]. Moreover, vaccine candidates against many emerging infections such as caused by Zika virus are based on VLP technology [3].

A universal vaccination program has led to a remarkably decreased rate of HBV infections in many countries in Asia, Africa, and Europe [4]. There have been approved vaccines against HBV since the 1980s, and nowadays all of them are produced using recombinant DNA technology [5, 6]. Chimeric HBsAg-based VLPs are also used as scaffold to display epitopes of proteins from

Brigitte Gasser and Diethard Mattanovich (eds.), *Recombinant Protein Production in Yeast*, Methods in Molecular Biology, vol. 1923, https://doi.org/10.1007/978-1-4939-9024-5_14,

other infective agents such as dengue virus [7] and the malaria parasite [8].

VLPs offer a compact and stable protein structure which is one of the major requirements for a vaccine product. Many viral surface proteins have an inbuilt tendency to form VLPs under appropriate conditions [9] and then display immunogenic properties of the actual virions and hence trigger a protective immune response.

An efficient approach to produce HBsAg VLPs is based on subunit expression using yeast-based expression systems such as *Saccharomyces cerevisiae* or *Pichia pastoris* [10]. The HBsAg is a very hydrophobic protein and not efficiently secreted [11]. Thus, yeast-based production leads to an intracellular maintained product [12–14]. For a long time, it was thought that the HBsAg VLPs are formed inside the expressing host cells. Recently, it has been shown that HBsAg VLPs do not assemble inside the cell but are formed during downstream processing [15, 16].

Production and purification of HBsAg VLPs using *P. pastoris* have been reported many times, e.g. [15–19]. However, most published processes are only realizable in specialized laboratories. Here, we would like to offer a protocol which is practicable in nonspecialized laboratories and also applicable for those not yet familiar with VLP production and purification procedures.

2 Materials

2.1 Production

1. *P. pastoris* harboring gene encoding protein of interest under control the the *AOX1* promoter, e.g., HBsAg [17].
2. Yeast media components (yeast extract, peptone, yeast nitrogen base, biotin). Prepare all solutions with double-deionized water, ddH_2O (*see* **Note 1**).
3. Shake flasks with baffles (250 mL, 1 L, if present also larger flasks).
4. Glycerol and methanol (biological grade) as non-inducing and inducing carbon sources, respectively.
5. 1 M potassium phosphate buffer, pH 6.0 (stock solution).
6. Buffered medium glycerol (BMG): 100 mM potassium phosphate, pH 6.0, 1.34% (w/v) yeast nitrogen base (YNB), 4×10^{-5}% (w/v) biotin, and 1% (v/v) glycerol.
7. Buffered medium methanol (BMM): 100 mM potassium phosphate, pH 6.0, 1.34% (w/v) yeast nitrogen base (YNB), 4×10^{-5}% (w/v) biotin, and 0.5% methanol (add 5 ml to 95 ml methanol-free medium).
8. Shaking incubator preset to 30 °C.
9. Pipette set and tips and tubes for centrifugation and storage.

Table 1
Guidelines for small-scale production of HBsAg VLPs: List of buffers used during downstream processing of HBsAg VLP

Step	Buffer	pH
Lysis	25 mM PB + 5 mM EDTA + 0.6% Tween 20	8.0
PEG precipitation (overnight)	Lysis buffer +500 mM NaCl + 5% PEG 6000	8.0
Aerosil binding	25 mM PB + 500 mM NaCl	7.2
Aerosil washing	25 mM PB	7.2
Aerosil elution	50 mM CBB + 1.2 M Urea	10.8
Aging	50 mM CBB	10.8
IEX binding	50 mM CBB	8.0
IEX washing	50 mM Tris–HCl	8.0
IEX elution	Tris–HCl + 500 mM NaCl	8.0
Concentration (e.g., via Vivaspin column)	50 mM Tris–HCl + 500 mM NaCl	8.0
Gel filtration	PBS	7.2
KSCN	PBS + 1.2 M KSCN	7.2
Dialysis	PBS	7.2

IEX ion exchange chromatography, *PB* (sodium) phosphate buffer, *CBB* sodium carbonate-bicarbonate buffer, *PBS* phosphate-buffered saline

10. Centrifuges with temperature control for harvesting the culture and separating soluble and insoluble fractions.

2.2 Downstream Processing

1. Acid-washed glass beads (0.45 mm diameter).
2. Tween 20 (biological grade).
3. Polyethylene glycol (PEG) 6000.
4. Aerosil 380 (Evonik, Hanau, Germany).
5. FPLC system with ion exchange (DEAE Sepharose FF, 20–200 mL, GE healthcare) and gel filtration (Sephacryl S300, 26/60, 320 mL, GE healthcare) columns.
6. FPLC buffers (for detailed recipes, *see* Table 1).
7. Phosphate buffered saline (PBS): 137 mM NaCl, 2.7 mM KCl, 10 mM Na_2HPO_4, 1.76 mM KH_2PO_4, pH 7.2.
8. Concentrators, 10 kDa molecular weight cutoff (MWCO).
9. Cellulose acetate dialysis membrane (14 kDa MWCO).

2.3 Analytical Methods

1. UV-VIS spectrophotometer for measuring the optical density at 600 nm (OD_{600}) and cuvettes (1 mm path length).
2. Protein quantification reagent (Bradford, BCA, etc.)

3. SDS-PAGE sample buffer: 10 mM Tris–HCl, pH 6.8, 1 mM EDTA, 2.5% (w/v) SDS, 0.2% Bromophenol blue, 45% (v/v) β-mercaptoethanol, and 5% (v/v) glycerol.
4. SDS-PAGE gels (12%), molecular weight marker, and electrophoresis apparatus.
5. Fixing and Colloidal Coomassie Blue staining solution [20].
6. Transfer apparatus for Western blotting.
7. Polyvinylidene difluoride (PVDF) membrane.
8. Buffers for immunoblotting.
   (a) Tris-buffered saline (TBS): 25 mM Tris–HCl, pH 7.5, and 150 mM NaCl.
   (b) TBS-Tween 20 (TBS-T): TBS, pH 7.5, and 0.5% Tween 20.
   (c) Blocking buffer: TBS-T, pH 7.5, and 2% bovine serum albumin (or 2% skim milk powder).
9. Primary and horseradish peroxidase (HRP)-labeled secondary antibodies for HBsAg.
10. Tetramethylbenzidine (TMB) detection substrate ("insoluble" for immunoblotting, generating insoluble product and "soluble" for ELISA, generating soluble product), plate reader ELISA plates, and 96-well ELISA plates.
11. Formvar-coated Cu grids (300 mesh).
12. 2% (w/v) uranyl acetate, pH 4.5.

3 Methods

3.1 Shake Flask Cultivation for HBsAg Production

To test for HBsAg production, shake flask cultures of *P. pastoris* harboring the HBsAg gene are grown at 30 °C and 250 rpm, according to manufacturer's instructions (*Pichia* Expression Kit, Catalog no. K1710-01, Invitrogen USA). As inoculum, 100 μL of glycerol stock culture is introduced to 50 mL BMG medium and incubated for 12–16 h or until the OD_{600} reaches 15. Subsequently, 10 mL of this inoculum is introduced into 1 L BMG medium to start the main culture(s) (*see* **Note 2**), which is further incubated until $OD_{600} = 8$. After approx. 14 h, the cells are harvested and resuspended into BMM to start production. Induction is continued with regular addition of methanol, i.e., 0.5% volume after every 24 h, and HBsAg accumulation is analyzed by SDS-PAGE analysis (Fig. 1, *see* **Note 3**).

3.2 High Cell Density Cultivation for HBsAg Production Using *P. pastoris* GS115

If the laboratory is equipped to run controlled bioreactor cultivations (e.g., equipped for controlled continuous methanol feeding to keep the methanol concentration constant), detailed information on fed-batch procedures to produce HBsAg can be found elsewhere [13] (*see* **Note 4**).

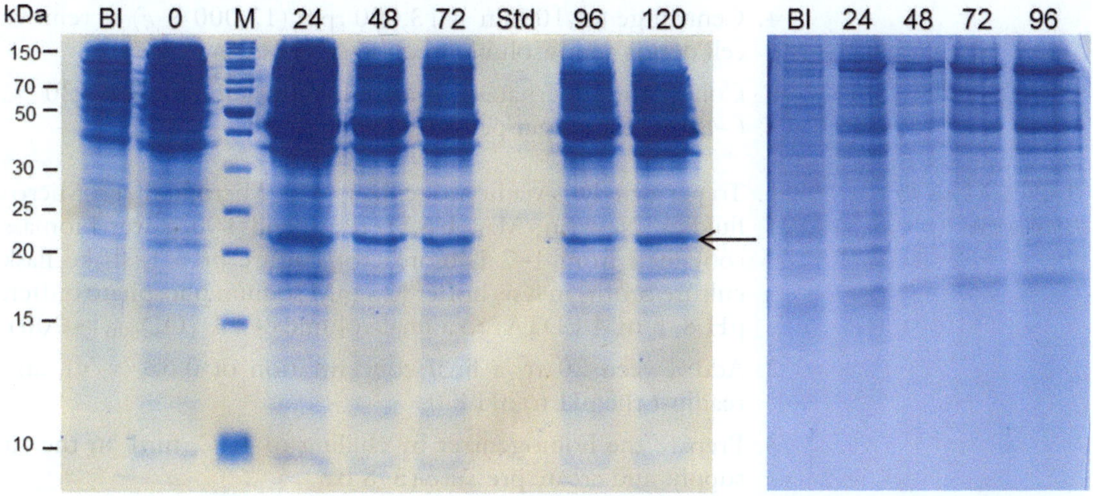

Fig. 1 Guidelines for small-scale production of HBsAg VLPs: Production of HBsAg in shake flask cultures. Cells were grown first in BMG, and induction was started by resuspending cells in BMM containing 0.5% methanol. Cell samples were collected 0, 24, 48, 72, 96, and 120 h post-induction, concentrated/diluted to $OD_{600} = 100$, and lysed via glass beads and the soluble cell fraction analyzed by SDS-PAGE. The arrow on the left gel points toward HBsAg. The gel on the right side shows samples from the nonproducing parental strain GS115 grown under the same conditions as a control culture. Std, purified HBsAg VLPs at a concentration of 0.2 mg/mL used as standard; BI, before induction; M, molecular weight marker

3.3 Culture Harvesting

In shake flask cultures, production continues up to 120 h after induction. The best timing needs to be tested empirically. For harvesting the culture, use continuous-flow centrifuges (for controlled bioreactor fed-batch cultivations) or buckets of 250 mL to pellet the biomass at 4 °C and 4000 rpm (3345 × *g*) for 20 min. Time course samples and biomass for HBsAg purification can be stored at −20 °C for about one week and for longer periods at −80 °C.

3.4 Cell Lysis

To analyze for intracellular HBsAg production, glass beads (0.45 mm diameter) are an efficient and cheap option for lysis and useful for analytical and small-scale purification procedures. Additionally, a homogenizer can be used for cell lysis of suspension volumes equal or larger than 10 mL (*see* **Note 5**).

3.4.1 Cell Lysis Using Glass Beads

1. Wash cell pellet equivalent to a cell pellet from 1 mL culture with an $OD_{600} = 100$ with 25 mM phosphate buffer (PB), pH 8.0, and resuspend in lysis buffer.
2. Add glass beads (0.45 mm diameter) of roughly equal weight of wet cell pellet.
3. Vortex for 40 s at maximum speed, and keep on ice for the next 40 s. Repeat the cycle of vortexing and chilling 8–10 times.

4. Centrifuge for 10 min at 13,000 rpm (17,000 × *g*) to remove cell debris and insoluble fraction of HBsAg.
5. Collect the supernatant in a fresh tube, and store at −20 °C (−80 °C for longer periods).

3.4.2 Cell Lysis via a Homogenizer

1. To perform lysis via homogenizer (e.g., Microfluidizer, Microfluidics, Newton, MA, USA), resuspend 100 g wet biomass (obtained from 1–2 L bioreactor culture or 5 L shake flask culture broth) in lysis buffer (25 mM sodium phosphate buffer, pH 8, 5 mM EDTA) to a final volume of 1 L ($OD_{600} \approx 200$).
2. Add Tween 20 to a final concentration of 0.6% (v/v), and readjust the pH to pH 8.0.
3. Prepare the homogenizer by chilling to 4 °C, turn on the air supply, and create pressure (5–8 bar).
4. Wash the homogenizer chamber with ddH_2O and lysis buffer.
5. Pass cells at 12,000 psi (38 psi at inlet) through the homogenizer 12–14 times.
6. Proceed for the further steps of downstream processing (as shown in Fig. 2).

3.5 Purification of HBsAg

Lysed cells are subjected to further downstream processing mainly based on different chromatographic methods (summarized in Fig. 2).

3.5.1 Clarification of Cell Lysate

1. Perform lysis of 100 g wet biomass using a homogenizer as described in the previous section.
2. Slowly add solid polyethylene glycol 6000 (PEG 6000) to the cell lysate to a final concentration of 5% (w/v) followed by the addition of 5 M NaCl to obtain a final concentration of 500 mM NaCl in about half an hour keeping everything at 4 °C.
3. Stir the mixture for 2 h, and then let it incubate for further 12–16 h at 4 °C to facilitate precipitation.
4. This suspension is clarified by centrifugation at 4 °C and 4000 rpm (3345 × *g*) for 25 min.

3.5.2 Aerosil 380 Adsorption

1. For preparation of pre-equilibrated Aerosil, gently mix the Aerosil in 500 mL binding buffer (0.13 g Aerosil per 1 g wet biomass), centrifuge at 4000 rpm (3345 × *g*), and discard the supernatant (for recipe, *see* Table 1).
2. Add the PEG supernatant to the pre-equilibrated slurry of Aerosil 380 in the Aerosil binding buffer.
3. Stir the suspension obtained by mixing the PEG supernatant and the Aerosil pellet at 4 °C and 300 rpm.

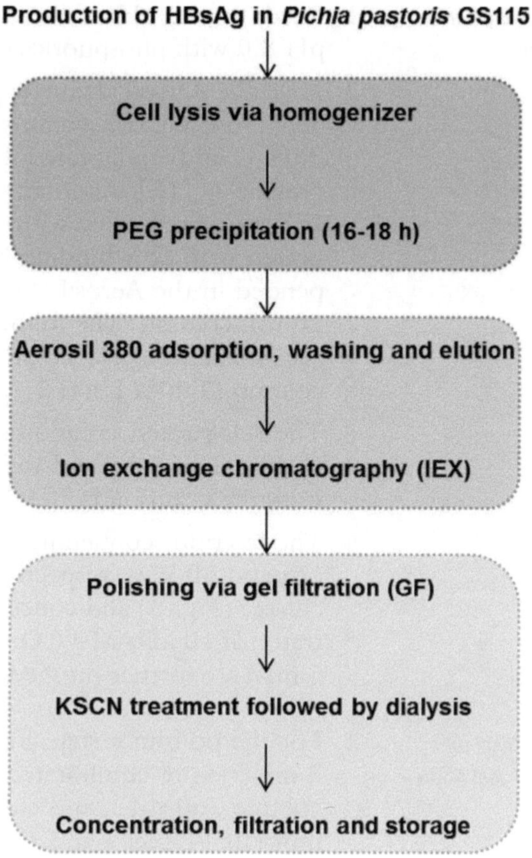

Fig. 2 Guidelines for small-scale production of HBsAg VLPs: Flow sheet illustration of HBsAg purification. The three blocks represent three stages of purification, i.e., lysis and lysate treatment, capturing of HBsAg and semi-purification, and, ultimately, polishing. After salt-mediated PEG precipitation, the soluble part of the cell lysate is subjected to HBsAg capturing (adsorption of HBsAg to fumed silica, Aerosil 380, through hydrophobic binding). The Aerosil eluate is processed via ion exchange (IEX) chromatography to remove further contaminants. The semi-purified product is further purified via gel filtration (GF) chromatography and treated with potassium thiocyanate (KSCN). Finally, the VLP-containing solution is filtered and stored at 4°C until further use

4. After 4 h, centrifuge the Aerosil suspension at 4 °C and 4000 rpm (3345 × *g*) for 25 min.
5. Wash the pellet twice using the 100 mL Aerosil washing buffer (Table 1), resuspend in 800 mL elution buffer (Table 1), and incubate for 12 h at 37 °C and 120 rpm.
6. Separate the Aerosil eluate (supernatant fraction) from the pellet after centrifugation for 150 min at 25 °C and 10,000 rpm (8664.5 × *g*), and further clarify by vacuum filtration (0.2 μm pore size).
7. Store the filtered eluate at 4 °C for 24 h to improve stability of HBsAg, a process called aging [18].

3.5.3 Ion Exchange (IEX) Chromatography

1. Readjust the pH of the Aerosil eluate (approx. 800 mL) to pH 8.0 with phosphoric acid before loading on IEX column.
2. Load the Aerosil eluate (approx. 800 mL) on a 200 mL DEAE Sepharose FF IEX column, prewashed with 1 M NaOH and ddH_2O and equilibrated with IEX binding buffer as described previously [15]. Alternatively, 20 mL of DEAE Sepharose FF IEX resin prewashed with 1 M NaOH and ddH_2O and equilibrated with IEX binding buffer (Tables 1 and 2) can be suspended in the Aerosil eluate and subsequently used to pack a 20 mL column. The following purification steps are based on the utilization of a DEAE Sepharose FF self-packed 20 mL column (Tables 1 and 2).
3. The self-packed column is washed with IEX washing buffer (Tables 1 and 2), and the HBsAg is eluted via isocratic flow using IEX elution buffer (Tables 1 and 2).
4. The fractions containing mono- and dimeric forms of HBsAg (25 and 50 kDa, respectively) are pooled (approx. 30–40 mL eluate) (Fig. 3) and concentrated using ultrafiltration concentrator of 10 kDa MWCO. The concentrated fractions (approx. 5 mL) are further purified via gel filtration chromatography.

3.5.4 Gel Filtration (GF) Chromatography and KSCN Treatment

1. For the polishing step, load the HBsAg concentrate (approx. 5 mL) on pre-equilibrated gel filtration (GF) column (column volume 360 mL), and elute with PBS, pH 7.2 at 1 mL/min flow rate (Tables 1 and 3). For a smaller volume of the concentrate (e.g., 1–2 mL), a GF column with a bed volume of 120 mL is recommended.
2. Pool the fractions containing HBsAg (mono- and dimeric versions) (Fig. 4), and treat with KSCN, to a final molarity of 1.2 M, for partial "surface unfolding."
3. Incubate this mixture at 100 rpm and 37 °C for 5 h in an orbital shaker.
4. Dialyze the KSCN-treated HBsAg against PBS, pH 7.2, using a cellulose acetate dialysis membrane of 14 kDa MWCO. Filter sterilize the purified HBsAg VLPs (0.2 μm pore size), and store at 4 °C.

Table 2
Guidelines for small-scale production of HBsAg VLPs: Protocol for the ion exchange chromatography (IEX) of HBsAg

Method step	Volume (CV)	Flow rate (mL/min)
Washing	5	1
Elution	5	0.5
Washing	5	1

CV (column volume) of self-packed column is 20 mL. The eluate is collected in 5 mL fractions

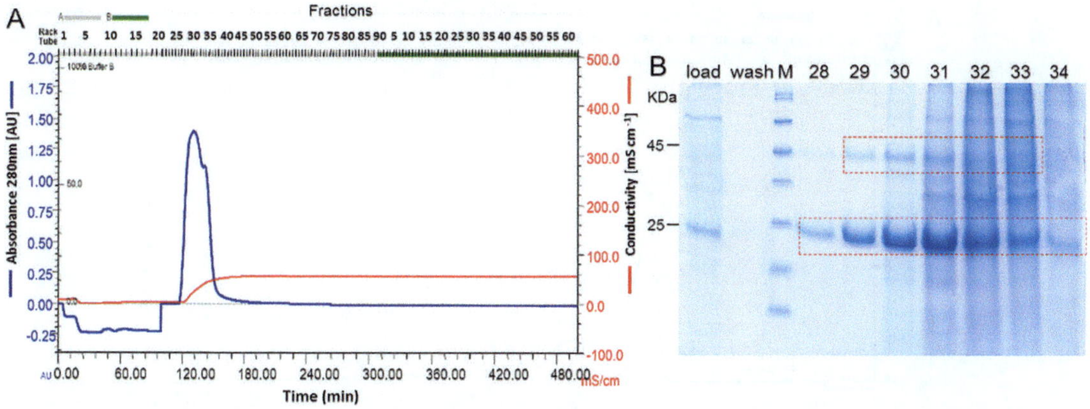

Fig. 3 Guidelines for small-scale production of HBsAg VLPs: Ion exchange chromatography of HBsAg. (**a**) Chromatogram, elution step with 500 mM NaCl. (**b**) SDS-PAGE analysis of peak fractions (numbers given on top of chromatogram and gel). The mono- and dimeric HBsAg-containing fractions (marked with red rectangle) are processed for GF chromatography

Table 3
Guidelines for small-scale production of HBsAg VLPs: Protocol for the gel filtration chromatography of HBsAg VLPs using Sephacryl S300, 26/60 (GE Healthcare)

Step	Volume	Flow rate (mL/min)
Equilibrium	0.1 CV	1
Sample application	5 mL	1
Elution	1.5 CV	1

The total column volume (CV) is 320 mL. The eluate is collected in 5 mL fractions

3.6 General Analytical Methods for Analysis of Cell Growth and HBsAg Concentration

3.6.1 Monitoring Cell Growth

Cell growth is monitored by measuring the optical density at 600 nm wavelength via UV-vis spectrophotometer. For each sample, take an average of three readings (*see* **Note 6**).

3.6.2 SDS-PAGE Analysis for (Soluble) Intracellular HBsAg Production

Cell samples collected during cultivation are lysed by glass beads, and the soluble fraction of cell lysates corresponding to $OD_{600} = 100$ are mixed with an equal volume of the SDS-PAGE sample buffer, vortexed for up to 1 min, and incubated for 10 min at 95 °C. 15 μL of each sample are loaded per well on a 12% SDS-polyacrylamide gel. The gels are stained overnight with the Colloidal Coomassie Blue staining solution.

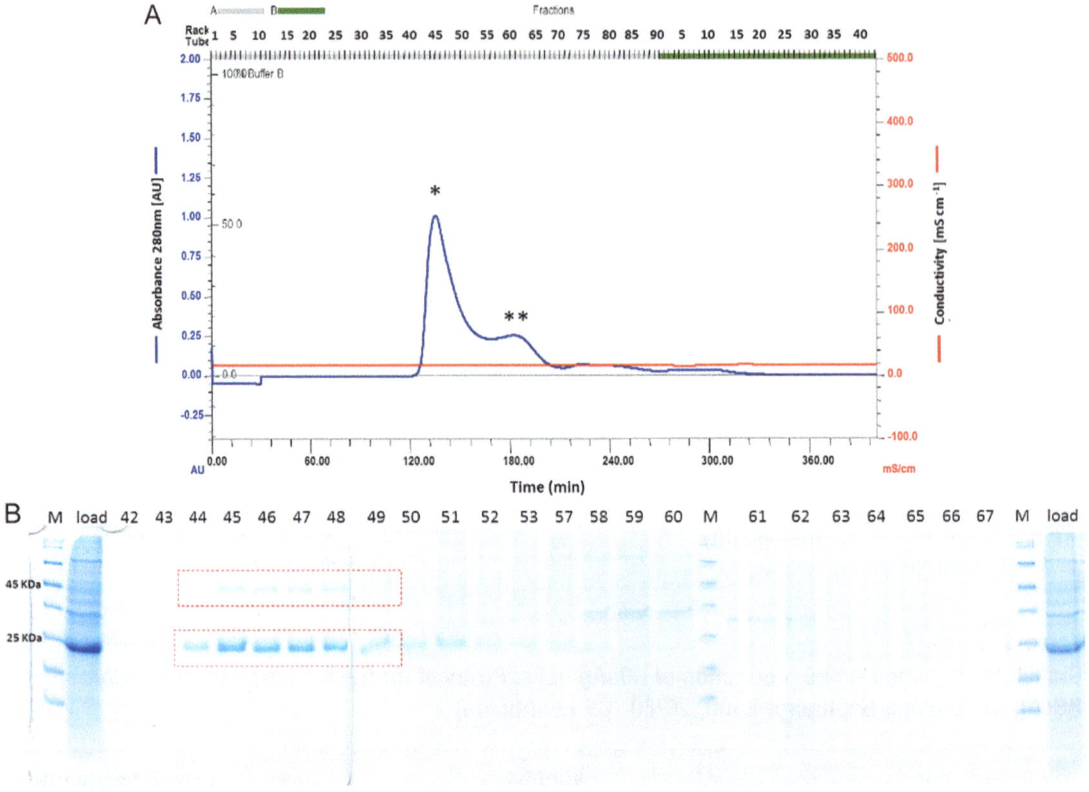

Fig. 4 Guidelines for small-scale production of HBsAg VLPs: Polishing of HBsAg VLPs is done via gel chromatography. (**a**) GF chromatogram of IEX-purified, IEX-pooled, and IEX-concentrated HBsAg. The two peaks of the chromatogram correspond to HBsAg VLPs (*) and HBsAg not assembled into VLPs (**). (**b**) SDS-PAGE analysis of GF load and peak fractions (numbers given on top of chromatogram and gel). HBsAg VLP peak fractions showing both mono- and dimeric HBsAg on reducing SDS-PAGE gels (fractions marked by red rectangles are pooled and used for further processing). *M* protein molecular weight marker

3.6.3 Protein Concentration

Quantify the total amount of protein by bicinchoninic acid (BCA) or Bradford assay. For pure HBsAg samples, the concentration is quantified by measuring the absorbance at 280 nm by using $\varepsilon = 3.2/\text{M}/\text{cm}$ [17] (http://web.expasy.org/protparam/). All protein samples are analyzed in triplicates.

3.6.4 Western Blot and Immunostaining

The cell lysates are also analyzed via Western blot and immunostaining to verify the presence of HBsAg. The samples are electrotransferred from SDS-PAGE gels to an equilibrated PVDF membrane for 60 min. The transfer is verified by co-transfer of a prestained molecular weight marker. After blocking with blocking buffer, the membrane is incubated in a solution with the primary antibody against HBsAg (e.g., linear epitope-specific mouse monoclonal anti-HBsAg antibody) for 1 h at room temperature. After three washing steps with TBS-T, the membrane is incubated for 1 h with horseradish peroxidase (HRP)-labeled goat anti-mouse

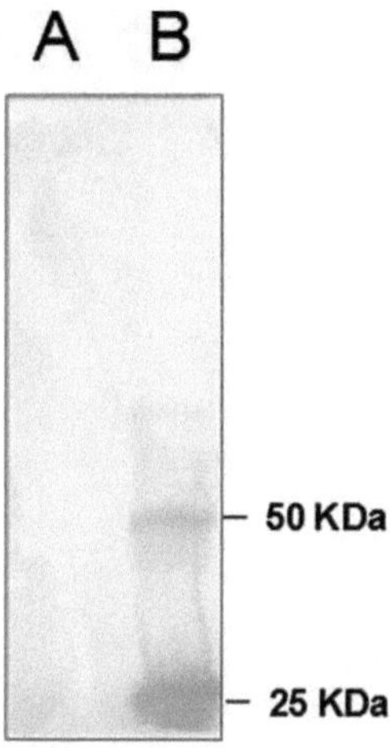

Fig. 5 Guidelines for small-scale production of HBsAg VLPs: Immunoblot of (**a**) the soluble fraction of the GS115 lysate and (**b**) the soluble fraction of the cell lysate of HBsAg-producing cells. Note the bands corresponding to mono- and dimeric HBsAg

secondary antibody and later washed twice with TBS. The immune blot is developed using an "insoluble" TMB substrate. Once the bands are visible, the reaction is stopped by washing the membrane two to four times with ddH_2O. Bands corresponding to mono- and dimeric HBsAg are visible in the soluble fraction of the cell lysate (Fig. 5).

3.6.5 ELISA Analysis

For quantification of soluble HBsAg, the samples collected during the cultivation can be analyzed by ELISA (*see* **Note 7**). Soluble fractions of cell lysate samples corresponding to $OD_{600} = 100$ are diluted 100 times with PBS, pH 7.2, and 100 μL of each sample and control, positive and negative, are loaded on the ELISA plate. A serial dilution of pure HBsAg VLP standard prepared in PBS is also included for the quantification. The plate is incubated at 37 °C for 1 h. The wells are washed with PBS, and 50 μL of the HRP-labeled anti-HBs conjugate is added to each well and further incubated at 37 °C for 1 h. The wells are washed six times with PBS, 100 μL of substrate (e.g., urea peroxide + H_2O_2 or only "soluble" TMB) is added, and the plate is incubated at room temperature for 30 min. 100 μL of 1 M H_2SO_4 is added to stop the reaction, the plate is scanned at 450 nm.

3.6.6 Transmission Electron Microscopy

The best choice to confirm and characterize the presence of HBsAg VLPs is transmission electron microscopy (TEM) (Fig. 6). Briefly, dilute the purified HBsAg VLPs with PBS, pH 7.2 to an appropriate protein concentration (0.2–0.5 mg/mL), adsorb for 1 minute to a glow-discharged Formvar-coated Cu grids (300 mesh), and negatively stain with 2% (w/v) uranyl acetate, pH 4.5. Further

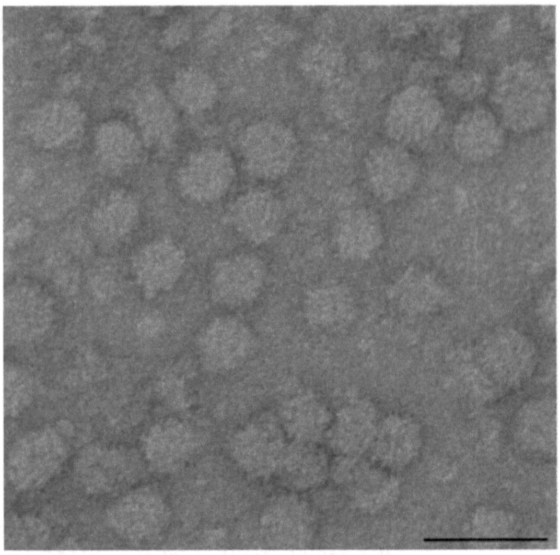

Fig. 6 Guidelines for small-scale production of HBsAg VLPs: TEM image of purified HBsAg VLPs (HBsAg VLPs of 22–25 nm size, scale bar = 50 nm, energy-filtered TEM Libra 120, Zeiss, Oberkochen, Germany) (Photo courtesy: H. Lünsdorf)

details for TEM analysis of HBsAg VLPs are given elsewhere [15, 16]. In addition to TEM, other useful analytical tools for VLP analysis are atomic force microscopy (AFM) and multiangle light scattering (MALS) techniques [21, 22].

4 Notes

1. Sterilize all media components by autoclaving at 121 °C for 20 min. Sensitive components, e.g., biotin, should be filter sterilized.
2. For Pichia cultures, the culture volume should not exceed 20% of the total flask volume to ensure proper aeration.
3. The parental nonproducing strain *P. pastoris* GS115 should be grown as a negative control to analyze for HBsAg production. Purified HBsAg VLPs can show different migration properties on reducing SDS-PAGE gels compared to non-purified intracellular HBsAg presumably because of the presence of different non-reduced disulfide bonds. Please see also the amount of reducing agent needed for sample preparation for SDS-PAGE analysis (*see* also Fig. 1).
4. In case there is no access to bioreactor facilities, multiple shake flask cultures can be used to produce adequate biomass for the purification of VLPs.

5. Complete lysis should be confirmed by probing the cell lysate under a light microscope with 40× magnification.
6. For OD measurement, collect the samples from the cultures regularly, and dilute to a suitable extent (between OD_{600} 0.1 and 0.8) with 0.9% (w/v) NaCl.
7. For ELISA, a commercially available HBsAg ELISA kit (Hepanostika micro ELISA, bioMérieux, France) can be used. However, it is also feasible to use "in-house" ELISA strips with the appropriate specific antibodies.

Acknowledgments

We are grateful to C. Gurramkonda for critical comments.

References

1. Cregg JM, Tschopp JF, Stillman C, Siegel R, Akong M, Craig WS et al (1987) High-level expression and efficient assembly of hepatitis B surface antigen in the methylotrophic yeast, *Pichia pastoris*. Nat Biotechnol 5:479–485
2. Kushnir N, Streatfield SJ, Yusibov V (2012) Virus-like particles as a highly efficient vaccine platform: diversity of targets and production systems and advances in clinical development. Vaccine 31:58–83
3. Boigard H, Alimova A, Martin GR, Katz A, Gottlieb P, Galarza JM (2017) Zika virus-like particle (VLP) based vaccine. PLoS Negl Trop Dis 11:e0005608
4. Meireles LC, Marinho RT, Van Damme P (2015) Three decades of hepatitis B control with vaccination. World J Hepatol 7:2127–2132
5. Roldao A, Mellado MC, Castilho LR, Carrondo MJ, Alves PM (2010) Virus-like particles in vaccine development. Expert Rev Vaccines 9:1149–1176
6. Zhao Q, Li S, Yu H, Xia N, Modis Y (2013) Virus-like particle-based human vaccines: quality assessment based on structural and functional properties. Trends Biotechnol 31:654–663
7. Khetarpal N, Poddar A, Nemani SK, Dhar N, Patil A, Negi P et al (2013) Dengue-specific subviral nanoparticles: design, creation and characterization. J Nanobiotechnology 11:15
8. Nardin EH, Oliveira GA, Calvo-Calle JM, Wetzel K, Maier C, Birkett AJ et al (2004) Phase I testing of a malaria vaccine composed of hepatitis B virus core particles expressing *Plasmodium falciparum* circumsporozoite epitopes. Infect Immun 72:6519–6527
9. Zhang L, Lua LH, Middelberg AP, Sun Y, Connors NK (2015) Biomolecular engineering of virus-like particles aided by computational chemistry methods. Chem Soc Rev 44:8608–8618
10. Kim HJ, Kim HJ (2017) Yeast as an expression system for producing virus-like particles: what factors do we need to consider? Lett Appl Microbiol 64:111–123
11. Wampler DE, Lehman ED, Boger J, McAleer WJ, Scolnick EM (1985) Multiple chemical forms of hepatitis B surface antigen produced in yeast. Proc Natl Acad Sci U S A 82:6830–6834
12. Biemans R, Thines D, Petre-Parent B, De Wilde M, Rutgers T, Cabezon T (1992) Immunoelectron microscopic detection of the hepatitis B virus major surface protein in dilated perinuclear membranes of yeast cells. DNA Cell Biol 11:621–626
13. Gurramkonda C, Adnan A, Gabel T, Lünsdorf H, Ross A, Nemani SK et al (2009) Simple high-cell density fed-batch technique for high-level recombinant protein production with *Pichia pastoris*: application to intracellular production of hepatitis B surface antigen. Microb Cell Factories 8:13
14. Vanz AL, Lünsdorf H, Adnan A, Nimtz M, Gurramkonda C, Khanna N et al (2012) Physiological response of *Pichia pastoris* GS115 to methanol-induced high level production of the hepatitis B surface antigen: catabolic

adaptation, stress responses, and autophagic processes. Microb Cell Factories 11:103

15. Gurramkonda C, Zahid M, Nemani SK, Adnan A, Gudi SK, Khanna N et al (2013) Purification of hepatitis B surface antigen virus-like particles from recombinant *Pichia pastoris* and in vivo analysis of their immunogenic properties. J Chromatogr B 940:104–111
16. Zahid M, Lünsdorf H, Rinas U (2015) Assessing stability and assembly of the hepatitis B surface antigen into virus-like particles during down-stream processing. Vaccine 33:3739–3745
17. Vassileva A, Chugh DA, Swaminathan S, Khanna N (2001) Effect of copy number on the expression levels of hepatitis B surface antigen in the methylotrophic yeast *Pichia pastoris*. Protein Expr Purif 21:71–80
18. Bardiya N (2006) Expression in and purification of Hepatitis B surface antigen (S-protein) from methylotrophic yeast *Pichia pastoris*. Anaerobe 12:194–203
19. Ottone S, Nguyen X, Bazin J, Berard C, Jimenez S, Letourneur O (2007) Expression of hepatitis B surface antigen major subtypes in *Pichia pastoris* and purification for in vitro diagnosis. Protein Expr Purif 56:177–188
20. Candiano G, Bruschi M, Musante L, Santucci L, Ghiggeri GM, Carnemolla B et al (2004) Blue silver: a very sensitive colloidal Coomassie G-250 staining for proteome analysis. Electrophoresis 25:1327–1333
21. Kanno T, Yamada T, Iwabuki H, Tanaka H, Kuroda S, Tanizawa K et al (2002) Size distribution measurement of vesicles by atomic force microscopy. Anal Biochem 309:196–199
22. McEvoy M, Razinkov V, Wei Z, Casas-Finet JR, Tous GI, Schenerman MA (2011) Improved particle counting and size distribution determination of aggregated virus populations by asymmetric flow field-flow fractionation and multiangle light scattering techniques. Biotechnol Prog 27:547–554

Chapter 15

Efficient Development of a Mixed Feed Process for *Pichia pastoris*

David Johannes Wurm and Oliver Spadiut

Abstract

Pichia pastoris is one of the most important host organisms for the recombinant production of proteins in industrial biotechnology. A prominent promoter system for recombinant protein production in *P. pastoris* is the promoter of alcohol oxidase (P_{AOX1}) which is induced by methanol, but repressed by several other carbon sources, like glucose and glycerol. Thus, typical cultivation strategies for such *P. pastoris* strains describe two different phases: growth on a carbon source, like glycerol, to get a high biomass concentration, followed by the induction of recombinant protein production by methanol. However, cells barely grow on methanol resulting in only moderate productivity in such bioprocesses. To enhance productivity, it is common to employ mixed substrate feeding strategies. The knowledge of certain strain-specific parameters is required to be able to set up such mixed feed fed-batch cultivations to avoid methanol accumulation and guarantee highest productivity. Here, we present an efficient strategy comprising only one experiment to determine the settings of such a mixed feed system based on the physiology of the respective yeast strain.

Key words *Pichia pastoris*, Methanol pulse, Specific substrate uptake rate, Dynamic fed-batch strategy, Strain characterization

1 Introduction

The methylotrophic yeast *Pichia pastoris*, also known as *Komagataella phaffii*, is extensively used as host organism for recombinant protein production (e.g., [1–7]). The main advantages of *P. pastoris* are its fast growth, its ability to use the cheap substrate methanol as sole carbon source, its ability to perform typical eukaryotic post-translational modifications, and the possibility of secreting the recombinant product [5, 6, 8–10]. Usually, recombinant protein production in *P. pastoris* is either regulated by a constitutive promoter, like the promoter of glyceraldehyde-3-phosphate dehydrogenase (P_{GAP}), or an inducible promoter, like the promoter of alcohol oxidase (P_{AOX1}) (e.g., [7, 8, 11, 12]).

Brigitte Gasser and Diethard Mattanovich (eds.), *Recombinant Protein Production in Yeast*, Methods in Molecular Biology, vol. 1923, https://doi.org/10.1007/978-1-4939-9024-5_15,

The well-known P_{AOX1} is induced by methanol but repressed by several other carbon sources, like glucose and glycerol, which causes accumulation of methanol [13]. Thus, commonly used cultivation strategies for *P. pastoris* describe two different phases: growth on a carbon source, like glycerol or glucose, to get a high biomass concentration, followed by the induction of recombinant protein production by methanol. However, cells barely grow on methanol resulting in moderate productivity in such bioprocesses. To enhance the productivity of such recombinant *P. pastoris* processes, it is common to employ mixed substrate feeding strategies [14, 15]. A mixed feed strategy gives different benefits, like lower oxygen consumption and lower heat production [16], it facilitates biomass growth due to higher biomass yields on the second substrate [16] and leads to an increased cell density resulting in increased volumetric productivity. A prominent C-source for these approaches is glycerol which represses P_{AOX1} if a critical specific uptake rate is exceeded [17, 18]. Thus, it is crucial to find the physiological limits of the respective strain and thus the optimal substrate feeding ratio of methanol and glycerol [13, 18, 19]. Usually, several fed-batch cultivations and time-consuming continuous cultivations have to be performed for that purpose. Here, we present an efficient strategy comprising only one experiment to determine the settings of a mixed feed system based on the physiology of the respective yeast strain. This dynamic strategy is schematically depicted in Fig. 1.

2 Materials

Prepare all media and solutions with analytical grade reagents and deionized water.

2.1 Medium for Preculture

Yeast nitrogen base (YNB) medium per L: 1.0 M potassium phosphate buffer (dissolve 118.1 g KH_2PO_4 and 23.0 g K_2HPO_4 in 1000 mL distilled water, pH 6.0), 3.4 g YNB without amino acids and ammonium sulfate, 10 g $(NH_4)_2SO_4$, 400 mg biotin, and 20 g glucose. Weigh YNB without amino acids and ammonium sulfate $(NH_4)_2SO_4$, biotin, and glucose in a beaker. Add 100 mL of 1.0 M potassium phosphate buffer (pH 6.0), dissolve by stirring, and set with water to 1 L. Filter-sterilize through a 0.2 μm cutoff filter into a sterile flask (*see* **Note 1**). Store at 4 °C.

2.2 Stock Solutions for Batch and Fed-Batch Cultivation

1. Basal salt medium (BSM) for 1 L of Cultivation Medium: 26.7 mL of 85% (v/v) phosphoric acid, 1.17 g $CaSO_4 \cdot 2H_2O$, 18.2 g K_2SO_4, 14.9 g $MgSO_4 \cdot 7H_2O$, 4.13 g KOH, 44 g $C_6H_{12}O_6 \cdot H_2O$, and 0.3 mL antifoam. Weigh the chemicals in a beaker, dissolve in around 600 mL of water, and then fill up to 725 mL in a measuring cylinder. Fill the bioreactor with this medium and autoclave.

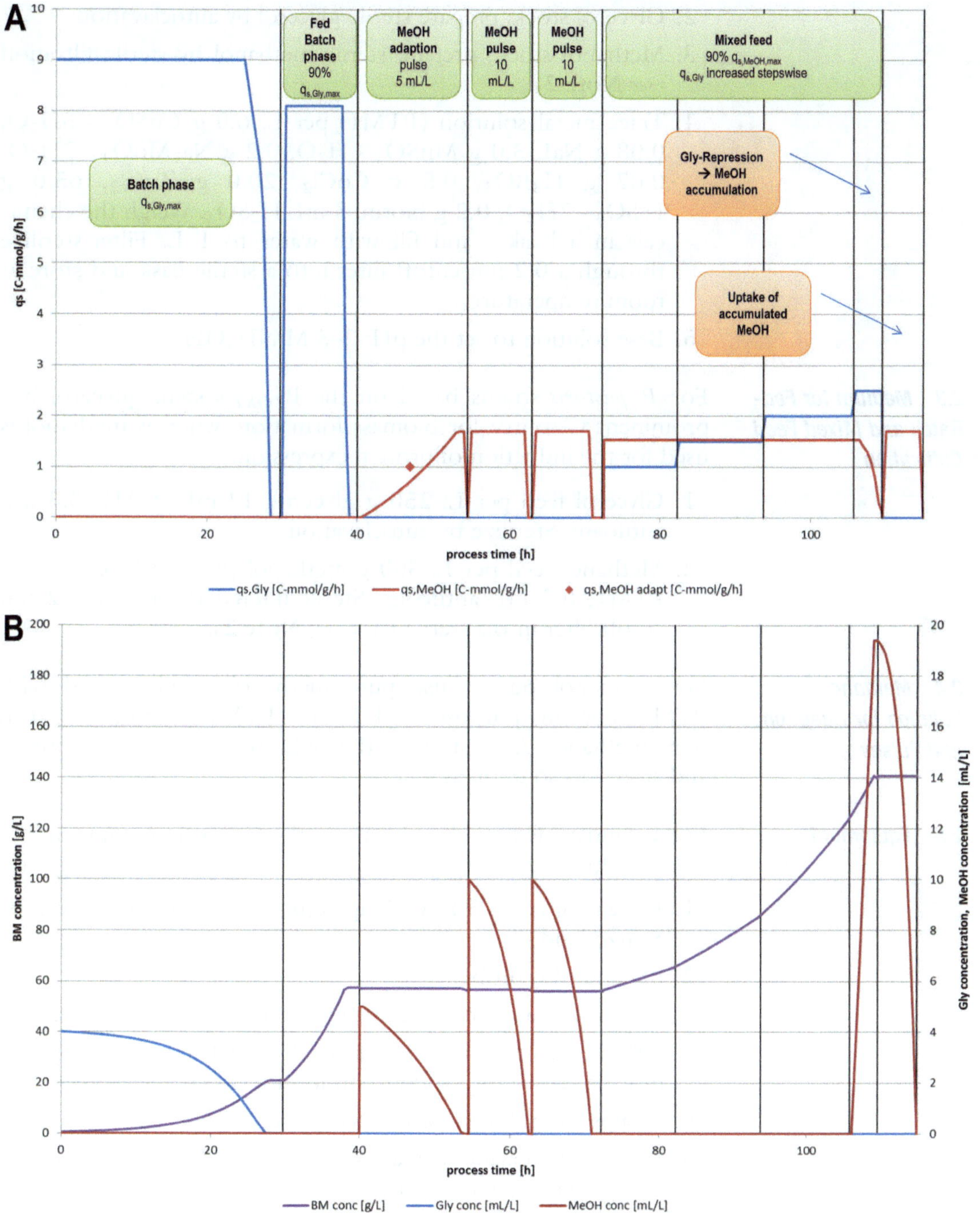

Fig. 1 Dynamic strategy to determine the settings for a mixed feed strategy with glycerol and methanol for a *Pichia pastoris* strain. (**a**) Course of specific uptake rates q_s during cultivation; (**b**) course of concentrations during cultivation

2. Glycerol stock: prepare sterile glycerol by autoclavation.
3. Methanol stock: prepare sterile methanol by sterile filtration (*see* **Note 2**).
4. Trace metal solution (PTM1) per L: 6.0 g $CuSO_4 \cdot 5H_2O$, 0.08 g NaI, 3.0 g $MnSO_4 \cdot H_2O$, 0.2 g $Na_2MoO_4 \cdot 2H_2O$, 0.02 g H_3BO_3, 0.5 g $CoCl_2$, 20.0 g $ZnCl_2$, 65.0 g $FeSO_4 \cdot 7H_2O$, 0.2 g biotin, 5 mL H_2SO_4. Weigh the chemicals in a beaker and fill with water to 1 L. Filter-sterilize through a 0.2 μm cutoff filter into a sterile flask and store at room temperature.
5. Base solution to set the pH: 2–3 M NH_4OH.

2.3 Medium for Fed-Batch and Mixed Feed Cultivation

For *P. pastoris* strains based on the P_{AOX1} system, glycerol is a prominent C-source for biomass formation, whereas methanol is used for the induction of protein expression.

1. Glycerol feed per L: 250 g glycerol, 12 mL PTM1, 0.3 mL antifoam. Sterilize by autoclavation.
2. Methanol feed per L: 300 g methanol (use a balance), 4 mL PTM1, 0.3 mL antifoam. Sterile-filtered through a 0.2 μm cutoff filter into a sterile flask (*see* **Note 2**).

2.4 Methanol Solution for Adaption and Pulses

For methanol pulses, use pure methanol supplemented with PTM1. For this purpose, add 12 mL PTM1 solution into 1 L pure methanol, filter-sterilize this solution into a sterile bottle, and store this solution at 4 °C.

2.5 Equipment

For a standard fed-batch cultivation, the following equipment are at least required:

1. Bioreactor (e.g., 5 L working volume glass bioreactor; Infors, Switzerland).
2. pH and dO_2 probe.
3. Pressurized air and oxygen supply lines.
4. Off-gas analyzer (e.g., infrared cell for CO_2 and a zirconium dioxide sensor for O_2 concentration; DasGip, Germany).
5. Pumps and tubings for base and feed.
6. If available: balances (reactor balance, feed balance, base balance) - connected to the process information management system.
7. Process information management system (PIMS; e.g., Lucullus, Securecell, Switzerland).
8. Spectrophotometer, centrifuge, and dry oven for sample preparation.

9. HPLC for exact determination of methanol concentrations (e.g., Agilent Technologies, USA) equipped with a SUPELCOGEL C-610H ion-exchange column (Sigma-Aldrich, USA), and a refractive index detector (Agilent Technologies, USA).

3 Methods

3.1 Preculture of Pichia pastoris

Start a preculture of the *P. pastoris* strain of interest in 100 mL of YNB medium in 1 L baffled shaking flasks at 220 rpm and 28 °C for maximum 24 h (to guarantee good aeration, only 1/10 of the total volume of the flask is filled with medium). The preculture is inoculated with 1 mL of frozen glycerol stock (*see* **Notes 1**, **3**, and **4**) and should yield an OD_{600} of about 15–20 (ca. 8–10 g/L dry cell weight) after 20–24 h.

3.2 Batch Phase

After autoclaving the BSM in the bioreactor vessel, aseptically add the sterile glycerol to a final concentration of 40 g/L. Then adjust the temperature and the stirring speed to the desired values, before the pH in the bioreactor is adjusted to pH 5.0 using 25% (v/v) ammonia solution (NH_4OH) by manually pumping the solution into the bioreactor. Aseptically transfer sterile PTM1 solution into the bioreactor (4.5 mL/L BSM). Aseptically transfer the preculture into a sterile inoculation flask (a vessel providing a connection to the bioreactor, *see* **Note 5**). The inoculum should be 10% of the final volume in the bioreactor thus leading to a starting OD_{600} of 1.5–2 (about 1 g/L dry cell weight). Dissolved oxygen (dO_2) is measured with a sterilizable polarographic dissolved oxygen electrode. The pH is measured online with a sterilizable electrode and maintained constant with a step controller using 2–3 M NH_4OH which also represents the N-source during cultivation. The exact concentration of NH_4OH in the base bottle is determined by titration with 0.25 M potassium hydrogen phthalate (KHP; *see* **Note 6**). Base consumption is determined gravimetrically by putting the base bottle on a balance and recording the loss in weight over time. Set the cultivation temperature (e.g. 30 °C), and aerate the culture with 2.0 vvm dried air (volume per volume per minute; in 1 L cultivation volume 2.0 vvm correspond to 2 L of in-gas per minute). Always keep the dO_2 above 30% by increases the stirrer speed and, if necessary, by increasing the ratio of oxygen in the in-gas by addition of pure oxygen. Measure the off-gas of the culture by using an infrared cell for CO_2 and a paramagnetic cell for O_2 concentration. Temperature, pH, dO_2, agitation in the vessel, as well as CO_2 and O_2 in the off-gas are measured online and logged in a process information management system.

3.3 Fed-Batch Phase for Biomass Generation

After the complete consumption of the C-source glycerol in the batch (indicated by an increase of dO_2 and a drop in off-gas activity and verifiable by atline HPLC) an exponential fed-batch phase on glycerol, e.g., at 90% of $q_{s,Gly,max}$, is performed (Fig. 1). Assuming a biomass yield $Y_{X/S}$ of about 0.5, this corresponds to a certain specific growths rate μ (*see* **Note 7**). Therefore, the biomass concentration after the batch phase must be known (*see* **Note 7**). If the biomass yield of the strain on glycerol is known (this can be analyzed from the data in the batch; *see* **Note 7**), the biomass concentration can be calculated based on the amount of glycerol used in the batch (*see* **Notes 8** and **9**). The feed rate can then be determined by Eqs. 1 and 2 and controlled using a gravimetrically based PID flow controller.

$$F_0 = \frac{X_0 \times V_0 \times q_s \times \delta_{\text{Feed}}}{c_{\text{Feed}}} \tag{1}$$

$$F = F_0 \times e^{\left(q_s \times \Upsilon_{X/S} \times t\right)} = F_0 \times e^{(\mu \times t)} \tag{2}$$

F_0 = initial feed rate [g/h]; X_0 = calculated biomass concentration at the start of the fed-batch [g/L]; V_0 = volume in the bioreactor at the start of the fed-batch [L]; q_s = specific substrate uptake rate [g/g/h]; μ = specific growth rate [1/h]; δ_{Feed} = density glycerol feed [g/L]; $\Upsilon_{X/S}$ = biomass yield on glycerol [g/g]; c_{Feed} = concentration glycerol feed [g/L]; F = feed rate [g/h]; e = Euler constant; t = time [h].

During the glycerol fed-batch, dO_2 levels should be kept above 30% at all time points, which is why pure oxygen is added in case airflow is not sufficient (*see* **Note 10**). The fed-batch on glycerol is stopped when the biomass concentration reaches about 50–60 g/L (Fig. 1).

3.4 Methanol Adaption

After the complete consumption of the C-source glycerol, which is indicated by an increase of dO_2 and a drop in off-gas activity, a methanol adaptation pulse of a final concentration of 0.5% (v/v) is added using the sterile methanol solution supplemented with PTM1 (Fig. 1).

3.5 Determination of $q_{s,MeOH,max}$ by Methanol Pulses

After the complete uptake of MeOH in the adaptation pulse, the cells are regarded to be adapted to MeOH and to take up MeOH at their maximum specific methanol uptake rate ($q_{s,MeOH,max}$). The complete uptake of the MeOH can be determined in real time by a drop in the CO_2 off-gas signal, an increase in the O_2 off-gas signal, as well as an increase in the dO_2 signal. Furthermore, it can be verified by a MeOH sensor in the off-gas in online mode or by HPLC analytics at-line. To determine the $q_{s,MeOH,max}$, at least two MeOH pulses of a final concentration of 1–2.0% (v/v) are added (Fig. 1). To obtain the specific rates for MeOH uptake during each pulse, a minimum of two samples has to be taken: one directly after methanol addition and the other close to complete methanol depletion. The samples are used to determine the concentration of residual substrate, OD600, and the dry cell weight. Determined values at the beginning and the end of the respective MeOH pulse are used to calculate an average value of the specific substrate uptake rate (q_s) according to Eq. 3 (in short: pulse methanol; take a sample; when the off-gas signal starts to drop, indicating depletion of methanol, take another sample; measure the exact methanol concentration in these two samples by HPLC; calculate the volumetric methanol uptake rate; and relate it to the total biomass content at the latter sample point).

$$q_{s,\mathrm{MeOH},\max} = \frac{\frac{\Delta \mathrm{MeOH}}{\mathrm{BM} \times \Delta t}}{M_{(\mathrm{MeOH})}} \times 1000 \qquad (3)$$

$q_{s,\mathrm{MeOH},\max}$ = maximum specific methanol uptake rate [mmol/g/h]; ΔMeOH = amount of methanol which was taken up [g]; Δt = duration of pulse [h]; BM = average biomass during the pulse [g], $M_{(\mathrm{MeOH})}$ = molecular weight of methanol (32.04 [g/mol]).

The strain-specific physiological parameters which can be determined by this strategy are (1) the adaptation time of *P. pastoris* to methanol, (2) the q_s during the adaptation pulse, and (3) a maximum q_s for methanol ($q_{s,\mathrm{MeOH},\max}$) (*see* **Notes 11** and **12**).

3.6 Mixed Feed

To find the physiological limits of the respective *P. pastoris* strain in a mixed feed environment as well as optimum feeding conditions (highest productivity or q_p), a MeOH feed at 90% of $q_{s,\mathrm{MeOH},\max}$ is fed (Fig. 1). During this MeOH feed another, separately controlled glycerol feed is started. Initially, the glycerol feed is adjusted to a rather low $q_{s,\mathrm{Gly}}$ (e.g., 1.0 mmol/g/h). In regular time intervals (e.g., every 10 h), the feed is adjusted to correspond to a higher $q_{s,\mathrm{Gly}}$; thus, different $q_{s,\mathrm{Gly}}$ steps are performed (Fig. 1). At a certain $q_{s,\mathrm{Gly}}$, P_{AOX} is repressed and MeOH accumulates (*see* **Note 13**). In case MeOH accumulates, the glycerol feed can be stopped, and accumulated MeOH is metabolized. Regular sampling during these different steps allows the determination of physiology, productivity, and MeOH accumulation and thus reveals the optimal mixed feed settings for the respective *P. pastoris* strain (*see* **Note 14**).

3.7 Sampling

During the cultivation, samples have to be taken to characterize the strain regarding biomass growth, substrate uptake, possible metabolite formation, and product formation (*see* **Note 15**).

For dry cell weight determination, harvest 5 mL of culture broth, centrifuge in 10 mL glass tubes (4500 × *g*, 4 °C, 10 min), wash the pellet twice with 5 mL physiological salt solution (0.9% NaCl in deionized water), and determine the dry cell weight (DCW) after drying at 105 °C to a constant weight in an oven (approximately 2–3 days). Optical density of the culture broth throughout the process is measured using a spectrophotometer at a wavelength of 600 nm (OD_{600}). Dry cell weight measurement and OD_{600} have to be correlated to be able to use the measured OD_{600} values for q_s adaptation in subsequent fed-batch cultivations (*see* **Notes 8** and **9**).

For methanol determination, samples are centrifuged (20,000 × *g*, 15 min), and concentration of methanol is determined in cell-free samples by HPLC on a SUPELCOGEL C-610H ion-exchange column (Sigma-Aldrich, USA) with a refractive index detector. The mobile phase is 0.1% H_3PO_4 with a constant flow rate of 0.5 mL/min, and the system is run isocratic. Calibration is done by measuring standard points in the range of 0.1–10 g/L methanol.

4 Notes

1. If necessary, add antibiotics specific for the selection markers harbored by the strain (e.g., Zeocin, Kanamycin) to the preculture in appropriate concentrations (e.g., 100 μg/mL medium) to further reduce the risk of contamination.
2. Do not autoclave methanol because of evaporation.
3. When combining the sterile solutions for the preculture in a baffled shaking flask, work in the laminar flow hood and be careful to work sterile and avoid contaminations.
4. The glycerol cryo-cultures are prepared by mixing 1 mL of a fresh overnight culture of the respective *P. pastoris* strain with 0.5 mL sterile 75% glycerol (v/v), and snap-freezing it in liquid N2. The frozen glycerol stocks are then stored at −80 °C.
5. Before inoculating the bioreactor with the appropriate amount of preculture, the following actions should be taken:
   (a) Aseptically add the C-source to the sterile BSM in the bioreactor.
   (b) Set the desired temperature (typically 30 °C) and stirring speed (e.g., 1500 rpm).
   (c) Set the pH value of the BSM to pH 5.0 with NH_4OH and note the amount of base which is required to determine the overall content in the bioreactor vessel.
   (d) Add PTM1 aseptically to the cultivation broth.
   (e) Calibrate the dO_2 electrode according to manufacturer's instructions.
   (f) Adjust the weight of the bioreactor balance to the weight of the bioreactor content—the bioreactor weight is logged in the process information management system, and by adjusting it correctly at this stage of the bioprocess, the final data analysis will be facilitated.

6. For base titration, the following materials are required: base (NH_4OH), 0.25 M KHP, bromothymol blue (indicator), burette and beaker with magnetic stirrer. Add 2 mL of base to the beaker (dilute NH_4OH 1:10) and use the burette to add 0.25 KHP. At the point of equivalence, the color of the indicator will turn from blue to gray and then to green.
7. The biomass yield on glycerol can be calculated from the data in the batch. It is known how much glycerol was used in the batch, e.g., 60 g/L. Based on the off-gas signals, the time, which the cells needed to consume the substrate glycerol, can be determined, e.g., 60 h. From these values, the volumetric uptake rate of glycerol ($r_{s,glycerol}$ [g/L/h]) can be calculated:

$$r_s = \frac{60\,\frac{g}{L} - 0\,\frac{g}{L}}{60\ h} = 1\,\frac{g}{L \times h}$$

In analogy, the volumetric biomass formation rate (r_x) can be calculated from the dry cell weight values before (e.g., 1.5 g/L) and after the batch (e.g., 31.5 g/L) and the respective time period:

$$r_x = \frac{31.5\,\frac{g}{L} - 1.5\,\frac{g}{L}}{60\ h} = 0.5\,\frac{g}{L \times h}$$

To calculate the biomass yield on glycerol ($Y_{X/S}$ [g/g]), these two volumetric rates are put in relation:

$$Y_{X/S} = \frac{r_X}{r_S} = 0.5\,\frac{g}{g}$$

The specific growth rate during the fed-batch can thus be calculated by the set specific substrate uptake rate and the yield:

$$\mu = q_s \times Y_{X/S}$$

8. It is important to know the biomass concentration before the fed-batch phase to be able to implement a correct feed strategy. There are several ways to measure, calculate or estimate the biomass concentration in a bioreactor. Some of the possibilities are listed here:
   (a) Measure online by near-infrared spectroscopy [20].
   (b) Measure online by capacitance probe [21].
   (c) Calculate by using a soft sensor tool based on the feed and the off-gas signal [22, 23].
   (d) Calculate based on the amount of glycerol used in the batch and the biomass yield on the respective substrate.
   (e) Estimate by correlating the measured OD_{600} values to the biomass dry cell weight (*see* **Note 9**).

9. To be able to use the OD_{600} values to set the feeding rate to the desired q_s set point, it is crucial to have a good and reliable calibration curve of the OD_{600} and the biomass content (dry cell weight) in (g/L). Before starting the fed-batch bioreactor cultivation, generate such a calibration curve by using the biomass from batch cultivations in different dilutions. During cultivations, use the same photometer for OD_{600} measurements as for the calibration curve. Do not switch photometers during the experiment.
10. During batch and fed-batch on glycerol, the dO_2 level should be >30% to avoid anaerobic metabolism and thus the production of undesired metabolites. This can be done by increasing the oxygen ratio in the in-gas by supplementing the airflow with pure oxygen.
11. The strain characteristic parameters, which can be analyzed by the batch experiment with methanol pulses, are:
    (a) $\Delta time_{adapt}$: time period from induction until the off-gas (CO_2) has reached its maximum.
    (b) $q_{s,adapt}$: specific uptake rate for methanol during the adaptation pulse.
    (c) $q_{s,MeOH,max}$: the maximum specific uptake rate for methanol during consecutive pulses.
12. To get even more precise data for q_s, the methanol stripping from the bioreactor can be considered according to the Antoine's equation [24, 25].
13. Methanol accumulation can be easily be detected without great time delay by online HPLC or online GC.
14. This strategy allows high flexibility. During the MeOH pulses as well as during the mixed feed phase, different process parameters (e.g., temperature, dO_2, pH, etc.) can be adjusted and thus tested.
15. To quantify productivity, product-specific assays must be used. In case such assays are not available, the total extracellular protein content can be used to get an idea about productivity, since *P. pastoris* mainly secretes the target product.

References

1. Macauley-Patrick S, Fazenda ML, McNeil B, Harvey LM (2005) Heterologous protein production using the *Pichia pastoris* expression system. Yeast 22:249–270
2. Weinacker D, Rabert C, Zepeda AB, Figueroa CA, Pessoa A, Farias JG (2013) Applications of recombinant *Pichia pastoris* in the healthcare industry. Braz J Microbiol 44:1043–1048
3. Spohner SC, Muller H, Quitmann H, Czermak P (2015) Expression of enzymes for the usage in food and feed industry with *Pichia pastoris*. J Biotechnol 202:118–134
4. Cregg JM (2007) Pichia protocols, Methods Mol Biol, 2nd edn. Springer, New York, p 268
5. Cregg JM, Cereghino JL, Shi JY, Higgins DR (2000) Recombinant protein expression in *Pichia pastoris*. Mol Biotechnol 16:23–52

6. Cereghino JL, Cregg JM (2000) Heterologous protein expression in the methylotrophic yeast *Pichia pastoris*. FEMS Microbiol Rev 24:45–66
7. Cereghino GPL, Cereghino JL, Ilgen C, Cregg JM (2002) Production of recombinant proteins in fermenter cultures of the yeast *Pichia pastoris*. Curr Opin Biotechnol 13:329–332
8. Daly R, Hearn MT (2005) Expression of heterologous proteins in *Pichia pastoris*: a useful experimental tool in protein engineering and production. J Mol Recognit 18:119–138
9. Daly R, Hearn MT (2006) Expression of the human activin type I and II receptor extracellular domains in *Pichia pastoris*. Protein Expr Purif 46:456–467
10. Spadiut O, Herwig C (2014) Dynamics in bioprocess development for *Pichia pastoris*. Bioengineering 5:401–404
11. Zhang WH, Potter KJH, Plantz BA, Schlegel VL, Smith LA, Meagher MM (2003) *Pichia pastoris* fermentation with mixed-feeds of glycerol and methanol: growth kinetics and production improvement. J Ind Microbiol Biotechnol 30:210–215
12. Zhang AL, Luo JX, Zhang TY, Pan YW, Tan YH, Fu CY, Tu FZ (2009) Recent advances on the GAP promoter derived expression system of *Pichia pastoris*. Mol Biol Rep 36:1611–1619
13. Zalai D, Dietzsch C, Herwig C, Spadiut O (2012) A dynamic fed batch strategy for a *Pichia pastoris* mixed feed system to increase process understanding. Biotechnol Prog 28:878–886
14. Arnau C, Casas C, Valero F (2011) The effect of glycerol mixed substrate on the heterologous production of a *Rhizopus oryzae* lipase in *Pichia pastoris* system. Biochem Eng J 57:30–37
15. d'Anjou MC, Daugulis AJ (2001) A rational approach to improving productivity in recombinant *Pichia pastoris* fermentation. Biotechnol Bioeng 72:1–11
16. Jungo C, Marison I, von Stockar U (2007) Mixed feeds of glycerol and methanol can improve the performance of *Pichia pastoris* cultures: a quantitative study based on concentration gradients in transient continuous cultures. J Biotechnol 128:824–837
17. Hellwig S, Emde F, Raven NPG, Henke M, van der Logt P, Fischer R (2001) Analysis of single-chain antibody production in *Pichia pastoris* using on-line methanol control in fed-batch and mixed-feed fermentations. Biotechnol Bioeng 74:344–352
18. Capone S, Horvat J, Herwig C, Spadiut O (2015) Development of a mixed feed strategy for a recombinant *Pichia pastoris* strain producing with a de-repression promoter. Microb Cell Factories 14:101
19. Spadiut O, Zalai D, Dietzsch C, Herwig C (2013) Quantitative comparison of dynamic physiological feeding profiles for recombinant protein production with *Pichia pastoris*. Bioprocess Biosyst Eng 37:1163–1172
20. Finn B, Harvey LM, McNeil B (2006) Near-infrared spectroscopic monitoring of biomass, glucose, ethanol and protein content in a high cell density baker's yeast fed-batch bioprocess. Yeast 23:507–517
21. Fehrenbach R, Comberbach M, Petre JO (1992) On-line biomass monitoring by capacitance measurement. J Biotechnol 23:303–314
22. Sagmeister P, Langemann T, Wechselberger P, Meitz A, Herwig C (2013) A dynamic method for the investigation of induced state metabolic capacities as a function of temperature. Microb Cell Factories 12:94
23. Wechselberger P, Sagmeister P, Herwig C (2013) Real-time estimation of biomass and specific growth rate in physiologically variable recombinant fed-batch processes. Bioprocess Biosyst Eng 36:1205–1218
24. Dietzsch C, Spadiut O, Herwig C (2011) A fast approach to determine a fed batch feeding profile for recombinant *Pichia pastoris* strains. Microb Cell Factories 10:85
25. Dietzsch C, Spadiut O, Herwig C (2011) A dynamic method based on the specific substrate uptake rate to set up a feeding strategy for *Pichia pastoris*. Microb Cell Factories 10:14

Chapter 16

Primary Recovery of Yeast Culture Supernatant for Recombinant Protein Purification

Michael M. Maurer and Harald Schillinger

Abstract

The harvest of a bench scale high cell density fermentation of *Pichia pastoris*, using standard laboratory equipment, to process products from supernatant is described. The process consists of a centrifugation step and a depth filtration, followed by a (sterile) membrane filtration. The procedure yields supernatant ready for the next purification step such as chromatography.

Key words Centrifugation, Depth filtration

1 Introduction

The choice of yeast as the expression host for recombinant protein production is often motivated by its ability to secrete the target protein into the supernatant. Depending on the strain and culture conditions, the overexpressed protein of interest is often the major protein in the supernatant, e.g., *Pichia pastoris* secretes only 20 host proteins [1]. However, to develop efficient production processes and yield high product titers, high cell density fermentations have been established reaching up to 10^9 cells/mL and dry cell weight of as much as 100 g/L (which corresponds to approximately 420 g/L wet cell weight) [2, 3]. Such high biomass concentrations are challenging for the primary recovery, especially for the scale-up to production scale.

In principle, there are three techniques to remove biomass from supernatant (a) centrifugation, (b) depth filtration, and (c) tangential flow filtration (micro filtration). Figure 1 illustrates the performance range of each technique to remove particles. Due to the different capacities of the separation techniques, the combination of centrifugation and filtration is state of the art for the clarification at lab and production scale [3].

Brigitte Gasser and Diethard Mattanovich (eds.), *Recombinant Protein Production in Yeast*, Methods in Molecular Biology, vol. 1923, https://doi.org/10.1007/978-1-4939-9024-5_16, © Springer Science+Business Media, LLC, part of Springer Nature 2019

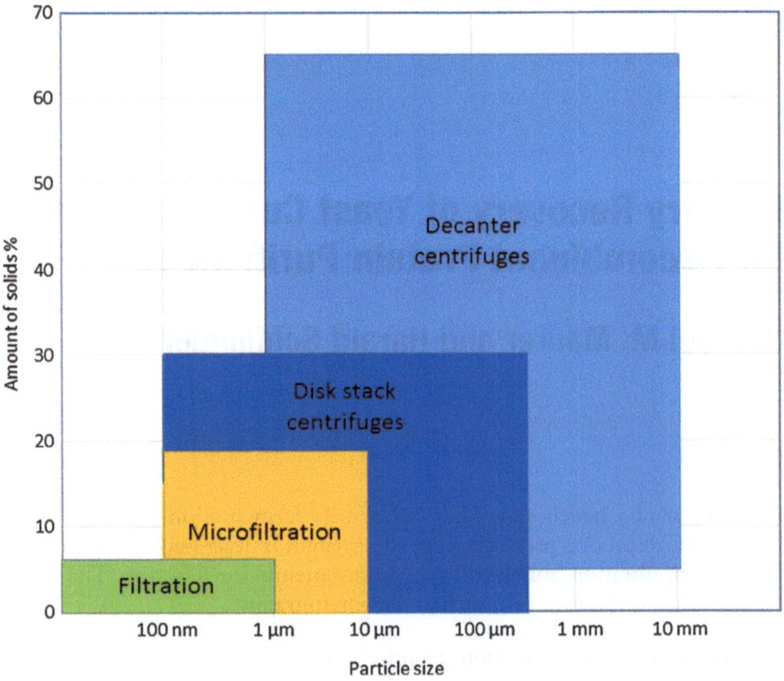

Fig. 1 Working area of separation techniques (data source filtration [8], microfiltration, disk stack, decanter centrifuges [11])

The separation of solid-liquid phases in the gravity field of a centrifuge can be described by the adoption of Stoke's law (Eq. 1):

$$v_S = \frac{\rho_P - \rho_l}{18\,\mu}\, d_P^2\, \omega^2 r \qquad (1)$$

where v_S is the settling velocity of a particle under gravity, ρ_P is the density of the particle, ρ_l is the density of the liquid, μ is the viscosity of the liquid, d_P is the particle diameter, ω is the angular velocity of the bowl (rad/s), and r is the radius of the centrifuge drum. Consequently, the performance of a centrifuge is based on the geometry of the bowl and can be increased by increasing the centrifuge speed ω. Furthermore, the separation efficiency depends on the running time t of laboratory centrifuges in batch mode or the residence time (influenced by the volumetric feed rate Q) for continuous devices, such as a disk stack centrifuge [4].

To compare and scale-up different centrifuges, the sigma factor has been defined, which represents the equivalent settling area in a simple sedimentation chamber. For a tubular centrifuge, where a particle has to settle the length L to the wall, the sigma factor can be calculated by Eq. 2 [5]:

$$\sum_{\text{Bottle}} = \frac{\pi \omega^2 L \left(r_o^2 - r_i^2\right)}{g \ln\left(r_o / r_i\right)} \qquad (2)$$

Table 1
Characteristics of yeast stains for the primary recovery

	Density (kg/m^3)	Mean particle size (μm)	Ref.
Pichia pastoris	1100 (kg/m^3)	2.0–5.0	[Not published]
Saccharomyces cerevisiae	1110 (kg/m^3)	4.0–8.0	[10]

where L is the length of the bottle or settle distance in a fixed angle rotor, r_o and r_i are the outer and inner radius, and g is the gravity acceleration [6]. The combination of the above equations enables the calculation of the smallest particle that can be separated, $d_{\min}$ [5]:

$$d_{\min} = \sqrt{\frac{18\ V\ \mu}{t \sum g(\rho_P - \rho_l)}} \tag{3}$$

Applying typical parameters for yeast to Eq. 3 from Table 1, a culture density of 1005 kg/m^3 [7], and a viscosity of 1.1 mPa s [7], using a standard setup of a laboratory centrifuge (Subheading 2), i.e., a separation for 30 min at 5000 × g, a theoretical minimal particle of 0.12–0.33 μm can be calculated. However, in practice the turbidity can be reduced from 60 NTU +/-30 NTU, but the supernatant still contains cells and a fraction of particles smaller than <0.5 μm.

For this reason, a second clarification step, such as depth filtration, is necessary to yield liquids ready for chromatography. The separation mechanisms of depth filtration are not only based on size exclusion but also van der Waals forces and physical adsorption are proposed to act [8]. That means depth filtration has the potential to remove a fraction of particles but has no specific cutoff of particle size. However, filtration manufacturer improved the performance by developing functionalized material, which acts as an anion exchanger and thereby improves the capacity of the filter material to bind charged particles, bioburden, genomic DNA, host cell protein, and viruses. That implies that one has to carefully choose the buffer system, as is done for anion exchange chromatography, in general one pH unit below the isoelectric point of the product.

Typically yeast supernatant can be clarified below 25 NTU (±5). However, it is recommended to subsequently perform a membrane filtration (cutoff 0.2 μm) to protect the chromatography resin.

2 Materials

2.1 Centrifugation

1. Laboratory floor centrifuge, e.g., SORVALL RC-6 PLUS (Thermo Fisher Scientific).
2. A fixed angle rotor, e.g., Fiberlite™ F10S-6 × 500y (r_i 41 mm, r_o 158 mm, angle 23°, Thermo Fisher Scientific).
3. 8× plastic centrifuge bottles, e.g., 3120-9500 CENT bottle PPCO 500 mL (Thermo Fisher Scientific, O.D. × L: 69.5 × 170.2 mm).
4. Pipettes, 10 mL.
5. Balance 1–6000 g, +/− 0.1 g, e.g., CP6201 (Sartorius, Germany).
6. Crushed ice and a box.
7. Two 5 L glass bottles.
8. Magnetic stirrer.

2.2 Filtration

1. Filter units (capsules) from the 3 M™ Zeta Plus™ MAXIMIZER EXT Series ZB depth filter [9], we recommend E0170FSA (Fig. 2c) with media grades 90ZB08A.
2. Peristaltic pump, e.g., Watson Marlow Pumps Model 323 (15–400 rpm).
3. Tubing dimensions:

   #14 (1/16″ × 3/16″ × 1/16″) for BC0025
   #16 (1/8″ × ¼″ × 1/16″) for E0170.

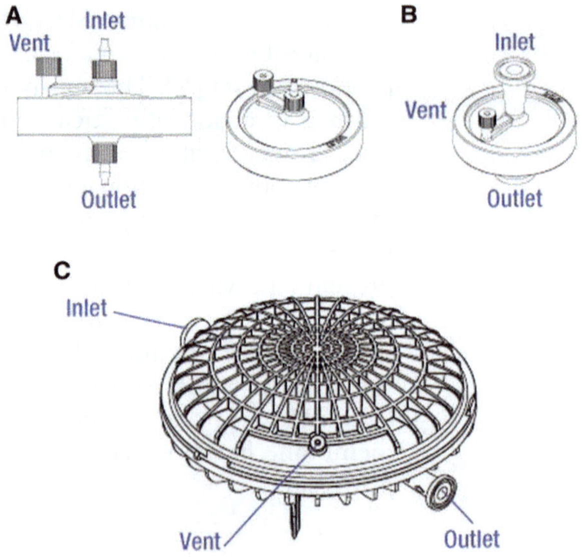

Fig. 2 Filter cartridges form the 3 M depth filter series: 3 M BioCap™ 25 capsule (BC0025), ideal for laboratory and small volume filtration. (**a**) with Luer-Lock, (**b**) with 1/2″ TC fitting, and (**c**) Scale-Up Capsule (E0170FSA) with 1/2″ TC fitting

4. Pressure transducers (if applicable), e.g., PressureMAT (PendoTECH, USA).
5. Balance 1–6000 g, +/− 0.1 g, e.g., CP6201 (Sartorius, Germany).
6. Turbidimeter (0–1000 NTU), Turbiquant 1100 IR (Merck, Germany).
7. Tubing and fittings.

3 Methods

3.1 Centrifugation of the High Cell Density Fermentation Broth

In this section the harvest of 3.0 L *P. pastoris* high cell density culture broth with 100 g dry cell weight per liter by centrifugation will be described. The equipment used is a common lab centrifuge with a fixed angle rotor for large volumes.

1. Insert the correct rotor (F10S-6 × 500y) into the centrifuge (Sorvall RC-6 Plus, Thermo Fisher Scientific) according to the manual, and allow the centrifuge to cool down to +4 °C by starting a precool program (e.g., 30 min, 5000 × *g*).
2. Harvest the culture broth from the bioreactor and collect broth in a 5 L bottle.
3. Immediately chill the culture in an ice bath by continuous stirring, to avoid potential protein degradation.
4. Fill the culture into sterile centrifuge beakers, thereby preparing always a pair of beakers with equal weight (±0.1 g), which will be placed in opposite sites in the rotor. That is done to avoid unbalance during centrifugation. The maximum loading volume depends on the bottle and the rotor specification. In this setup in total eight bottles are filled with 375–400 g culture.
5. Set the centrifugation parameter, such as rotor type (F10S-6 × 500y), the acceleration (9 full speed) and deceleration (9), temperature (4 °C) and duration (30 min), as well as speed (5320 rpm equal to 5000 × *g*).
6. Put the centrifuge beakers in correct order into the rotor (*see* **step 4**), close the lid of the centrifuge, and start the program.
7. After the centrifugation carefully open the bottles and decant the centrate into a clean 5 L bottle and measure the turbidity. The yield should be around 1.8–2.0 L with a turbidity of approx. 70 NTU.

3.2 Filtration of the Culture Broth

The goal of depth filtration is clarifying the centrate from cell debris and small particles to prevent clogging of a 0.2 μm membrane filter or chromatography resins. The optional utilization of Emphaze™ AEX AEX020 Hybrid Purifier following depth filtration in a matched component setup will optimize downstream processes with enhanced clarification of significant insoluble and soluble impurities, like cell debris, aggregates, bacteria, DNA, host cell protein (HCP), and virus.

1. Prepare depth filter units like the 3 M™ Zeta Plus™ MAXIMIZER EXT Series ZB (alternative membrane absorber such as Musanq q (Pall) or Sartobind Q (Sartorius) can be used as well) for filtration in constant flow mode. For 2.5 L yeast broth clarification (70 NTU, ±5), the *90ZB08A filter* unit with 170 cm^2 (E0170FSA, Fig. 2c) is the recommended choice. For smaller aliquots BioCap™ 25 cm^2 (BC0025, Fig. 2a, b), filter units are available (*see* **Note 1**).

   For difficult clarification problems, such as high turbidity coursed by pure viability or pure centrifugation, efficiently *see* **Note 2** for more advanced filtration setup.
2. Prior to each filtration, the filter units require a flush with 54 L/m^2 with purified water or buffer, corresponding to 918 mL for E0170FSA (135 mL/BC0025) (*see* also **Note 3**). Adjust the flow rate to 102 mL/min (6 L/m^2 min) for the E0170FSA module, and purge the device air -free using the Luer-Lock vent on top of each device. The flushing cycle could be used to adjust the proper flow rate, and measure ΔP_0 if a pressure transducer is available for monitoring.
3. After pre-flush, empty the system from remaining buffer, and connect the transfer tubing accordingly to the feeding reservoir, respectively, receiving flask.
4. For more in-depth studies, parameters such as time [min], cumulative weight [g], converted to volume by density, pressure p (psi, Pa, bar), and effluent turbidity (NTU) can be monitored and tabulated for monitoring a potential turbidity breakthrough (equates to >50% of centrate NTU) and further analyses.
5. Allow the system to fill while driving with half pump speed for 10–15 min, purge air-free, and increase to operational flow rate (34–102 mL/min per E0170FSA).
6. As a rule of the thumb, NTU of filtrate pool should be between 5 and 20 NTU, depending on consecutive DSP (0.2 μm sterilizing-grade membrane filtration, any chromatography resin, etc.).
7. Avoid any back pressure as depth filters are quite sensitive for flow in reversed direction (*see* **Note 4**).
8. The depth filtrate can be directly filtered through a membrane filer (0.2 μm), e.g., LifeASSURE™ PDA Series 50 mm Capsule Filter (15 cm^2, PES), applying a flow rate of 100–200 mL/min.
9. The resulting filtrate is ready for the next downstream processing step.

4 Notes

1. If smaller aliquots are processed, throughputs could be expected in a range of 150–300 L/m^2 corresponding to 375–750 mL for BC0025 and 2.6–5.0 L for E0170FSA, respectively.
2. For advanced filtration problems, the depth filters can be set in series:
   (a) 60ZB05A followed by 90ZB08A, for centrates containing a large fraction of reaming cells.
   (b) 90ZB08A followed by 120ZB08A, for centrates containing small cell debris.
   (c) 90ZB08A followed by 120ZB10A for centrates containing a large fraction of small cell debris.
   (d) Emphaze™ AEX020A as the second stage after Zeta Plus will deliver effluents with significantly lower turbidity and enhanced reduction for HCP, DNA, and endotoxin.
3. Emphaze™ AEX020A requires fluids having a minimum conductivity with 3–16 mS/cm corresponding to 25–150 mM NaCl. Do not flush AEX020A with pure water as the media will swell and block. This is reversible with the correct conductivity but is time consuming.
4. Generally, the final differential pressures ΔP of each filter stage should be $\leq$2.4 bar (maximum allowable specified ΔP) and in case of dual path the line pressure $P \leq 2.8$ bar g for BC0025, respectively, 3.1 bar for E0170FSA, both the maximum specified working pressure level for the capsules.

Acknowledgments

The authors would like to acknowledge Robin Ott, Karl Metzger, and Frederik Hoppe for the execution of some of the experiments outlined in this chapter and Nico Lingg for critical discussion. This work has been supported by the Austrian Federal Ministry of Science, Research and Economy (BMWFW), the Federal Ministry of Traffic, Innovation and Technology (bmvit), the Styrian Business Promotion Agency (SFG), the Standortagentur Tirol, the Government of Lower Austria, and ZIT—Technology Agency of the City of Vienna through the Austrian FFG-COMET-Funding Program.

References

1. Mattanovich D, Graf A, Stadlmann J, Dragosits M, Redl A, Maurer M, Kleinheinz M, Sauer M, Altmann F, Gasser B (2009) Genome, secretome and glucose transport highlight unique features of the protein production host *Pichia pastoris*. Microb Cell Factories 8(1):29
2. Stephenne J (1990) Production in yeast versus mammalian cells of the first recombinant DNA human vaccine and its proved safety, efficacy, and economy: hepatitis B vaccine. Adv Biotechnol Process 14:279–299
3. Meyer HP, Schmidhalter DR (2014) Industrial scale suspension culture of living cells. Wiley-Blackwell, Weinheim ISBN:978-3-527-33547-3
4. Doran PM (2013) Bioprocess engineering principles. Academic Press, Cambridge, MA ISBN:978-0-08-091770-2
5. Richardson JF, Harker JF, Backhurst JR, Coulson JM (2002) (first published 2002) Coulson and Richardson's chemical engineering. In: Particle technology and separation processes, vol 2. Butterworth-Heinemann, Oxford
6. Richardson JF, Harker JF, Backhurst JR, Coulson JM (2002) Chapter 9: Centrifugal separations. In: Chemical engineering (first published). Butterworth-Heinemann, Oxford, pp 475–501
7. Varga EG, Titchener-Hooker NJ, Dunnill P (2001) Prediction of the pilot-scale recovery of a recombinant yeast enzyme using integrated models. Biotechnol Bioeng 74 (2):96–107
8. Shukla AA, Etzel MR, Gadam S (2007) Process scale bioseparations for the biopharmaceutical industry. CRC/Taylor & Francis, Boca Raton, FL
9. 3M™ Zeta Plus™ ZB Series Filters. Meriden, USA, 2016
10. Baldwin WW, Kubitschek HE (1984) Buoyant density variation during the cell cycle in microorganisms. J Bacteriol 158(2):701–704
11. Alfa Laval (2014) Disc stack separator technology. Brochure

Chapter 17

Purification of Recombinant Glycoproteins from *Pichia pastoris* Culture Supernatants

David Johannes Wurm and Oliver Spadiut

Abstract

Pichia pastoris is a common host organism for the production of recombinant proteins. While unglycosylated recombinant proteins derived from this yeast can be purified efficiently by only a few conventional chromatography steps, the purification of glycosylated recombinant proteins is a very challenging process due to the intrinsic feature of the yeast of hypermannosylation. The resulting vast glycosylation pattern on the recombinant target protein masks its physicochemical properties hampering a conventional downstream process. Here, we describe a fast and efficient two-step chromatography strategy, where both steps are operated in flow-through mode, to purify recombinant glycoproteins from *P. pastoris* culture supernatants.

Key words *Pichia pastoris*, Recombinant protein production, Glycosylation, Downstream process, Flow-through chromatography

1 Introduction

Due to the ability of performing posttranslational modifications and secreting the target protein to the cultivation broth, the methylotrophic yeast *Pichia pastoris* is widely used for recombinant protein production in biotechnology and biopharmaceutical industry. The expression and subsequent purification of unglycosylated proteins recombinantly produced by this yeast are straightforward. Established state-of-the-art protein chromatography techniques, like ion exchange chromatography [1–3] and affinity chromatography [4], can be used to purify the target protein from the cell-free cultivation broth. However, if the target product describes a glycoprotein, the yeast has the tendency of hypermannosylation triggered by the native enzyme α-1,6-mannosyltransferase (Och1p) [5, 6]. The resulting vast and heterogeneous glycosylation pattern on the surface of the recombinant protein masks its physicochemical properties hampering an efficient downstream process [7, 8]. To date, this issue has been tackled by (1) engineering the *P. pastoris* strain to allow reduced and more homogeneous protein glycosylation and

Brigitte Gasser and Diethard Mattanovich (eds.), *Recombinant Protein Production in Yeast*, Methods in Molecular Biology, vol. 1923, https://doi.org/10.1007/978-1-4939-9024-5_17,

(2) chemical or enzymatic deglycosylation of the target protein before purification. Several studies have shown the possibility of engineering the native glycosylation pathway in *P. pastoris* and resulting changes in protein glycosylation [9–11]. Although this strategy could facilitate the subsequent purification of the recombinant target protein, *P. pastoris* strains with a modified native glycosylation pathway are physiologically impaired and thus not useful for industrial protein production processes [12]. Another strategy comprises a deglycosylation step of the hypermannosylated target protein before purification. This can either be done chemically, e.g., by trifluoromethanesulfonate (TMFS) [1, 13], or enzymatically, e.g., by PNGase F [1] or endoglycosidase H [14], or a combination thereof [15]. However, chemical deglycosylation describes a cumbersome procedure which is harmful to the target protein [16], and enzymatic deglycosylation also goes along with several drawbacks, as (1) the target protein must be denatured to eliminate steric hindrances at the glycoslyation sites and thus allow quantitative deglycosylation, (2) the required glycosidic enzymes describe an additional cost factor especially when it comes to large scales, and (3) the added glycosidic enzymes have to be removed again to obtain a pure target protein.

Due to abovementioned impediments, different downstream strategies to purify the hypermannosylated target protein itself have been tested. However, the addition of an affinity tag [17] as well as approaches comprising several consecutive steps of salt precipitation, hydrophobic interaction, ion exchange, and size-exclusion chromatography only gave very low recovery yields [3, 7, 18] resulting from hypermannosylation and the consequent masking of the protein's physicochemical properties. Recently, however, we even took advantage of this phenomenon by developing an efficient two-step chromatography approach for glycoproteins recombinantly produced by *P. pastoris* comprising a mixed-mode hydrophobic charge induction chromatography step (HCIC) and an anion exchange step using a monolith (AEC) [8, 19] both operated in flow-through mode.

2 Materials

All solutions have to be prepared with deionized water and analytical grade reagents (*see* **Note 1**) and have to be filtered through a 0.2 μm cutoff filter before use. Prepare and store all reagents at room temperature.

2.1 Diafiltration

The cell-free protein solution is diafiltrated in the appropriate binding buffer using a crossflow filtration device (Centramate 500S; Pall, Austria) and a membrane with an appropriate molecular weight cutoff (*see* **Note 2**).

2.2 Hydrophobic Charge Induction Chromatography (HCIC)

1. Resin: MEP HyperCel (Pall, Austria).
2. Binding buffer (HCIC-A): 20 mM NaOAc, 500 mM NaCl, pH 6.0. Weigh 1.18 g NaOAc and 29.22 g NaCl. Dissolve in water and adjust the pH. Make up to 1000 mL with water.
3. Elution buffer (HCIC-B): 50 mM Tris–HCl, 1 M NaCl, pH 8.0. Weigh 121.1 g Tris–HCl and 58.5 g NaCl and dissolve in water. Adjust the pH and make up to 1000 mL with water.
4. Cleaning agent: 0.8 M NaOH. Weigh 32 g NaOH and add water up to 1000 mL.
5. Storage agent: 20% EtOH, 1 M NaCl. Weigh 58.5 g NaCl and add water until the salt dissolves. Add 208 mL ethanol (96% v/v) and make up to 1000 mL with water.

2.3 Anion Exchange Chromatography (AEC)

1. Resin: 8 mL CIM-DEAE monolithic column (BIA Separations; Slovenia).
2. Binding buffer (AEC-A): 50 mM Tris–HCl, pH 8.0. Weigh 121.1 g Tris–HCl and dissolve in water. Adjust the pH and make up to 1000 mL with water.
3. Elution buffer (AEC-B): 50 mM Tris–HCl, 1 M NaCl, pH 8.0. Weigh 121.1 g Tris–HCl and 58.5 g NaCl and dissolve in water. Adjust the pH and make up to 1000 mL with water.
4. Cleaning agent: 1 M NaOH, 1 M NaCl. Weigh 40 g NaOH and 58.5 g NaCl, dissolve in water, and add water up to 1000 mL.
5. Storage agent: 20% EtOH. Measure 208 mL ethanol (96% v/v) and make up to 1000 mL with water.

3 Methods

All chromatography runs are performed at room temperature on an ÄKTApurifier system (GE Healthcare; Sweden; *see* **Note 3**) in flow-through mode.

3.1 Hydrophobic Charge Induction Chromatography (HCIC)

1. After fermentation, the cell-free cultivation broth is diafiltrated in binding buffer HCIC-A using a crossflow filtration device (*see* **Note 4**) and a membrane with an appropriate molecular weight cutoff (*see* **Note 2**).
2. The HCIC resin is equilibrated with five column volumes (CV) of HCIC-A. Then the crude extract is loaded—the binding capacity of this resin is >40 mg/mL. Since the hyperglycosylated product is located in the flow-through, it has to be collected for further processing. After the crude extract is loaded, a post-load wash with H_2O is performed.

3. Since the product is located in the flow-through, no elution is required. The bound host cell proteins can be removed from the column by a cleaning in place (CIP) step with 5 CV 0.8 M NaOH. The column is then stored in storage agent.

All steps are performed at a flow rate of 55 cm/h (*see* **Note 5**), and fractions of the flow-through of a volume of 0.5 CV are collected.

3.2 Anion Exchange Chromatography (AEC)

1. HCIC flow-through fractions containing the hypermannosylated target protein are diafiltrated in AEC-A (*see* **Note 4**).
2. The monolithic column is equilibrated with 5 CV of binding buffer AEC-A, before the crude extract is loaded (the binding capacity lies at around 100 mg/mL). Again, the flow-through, containing the hyperglycosylated product of interest, is collected.
3. Bound host cell proteins are removed from the column with 5 CV cleaning agent. Then the column is stored in 20% EtOH.

All steps are performed with a flow rate of 16.8 cm/h except the cleaning step which is done at 33.6 cm/h (*see* **Note 6**). Fractions of the flow-through of a volume of 0.5 CV are collected.

3.3 Data Analysis

To evaluate the efficiency of the applied two-step flow-through purification approach, the purification factor (PF; *see* **Note 7**) and the recovery yield of both the protein content and, if applicable, the enzyme activity in percentage (R%; *see* **Note 8**) are calculated in Eqs. 1 and 2. The suffixes "pre" and "post" indicate the respective values before and after a purification step [19]:

$$\mathrm{PF} = \frac{\text{specific activity}_{\text{post}}}{\text{specific activity}_{\text{pre}}} = \frac{\dfrac{\text{volumetric activity}_{\text{post}}}{\text{protein concentration}_{\text{post}}}}{\dfrac{\text{volumentric activity}_{\text{pre}}}{\text{protein concentration}_{\text{pre}}}} \quad (1)$$

$$\mathrm{R\%} = 100 \times \frac{\text{volumetric activity}_{\text{post}} \times \text{volume}_{\text{post}}}{\text{volumetric activity}_{\text{pre}} \times \text{volume}_{\text{pre}}} \quad (2)$$

Besides, the success of each purification step can be visualized on SDS gels (an example is shown in Fig. 1).

The success of the presented flow-through purification approach for hyperglycosylated proteins recombinantly produced by *P. pastoris* depends on the number of glycosylation sites and the degree of hypermannosylation (an example is shown in Fig. 2 for the HCIC step of different horseradish peroxidase isoenzymes varying in the amount of glycosylation sites) and has to be tested for each protein of interest.

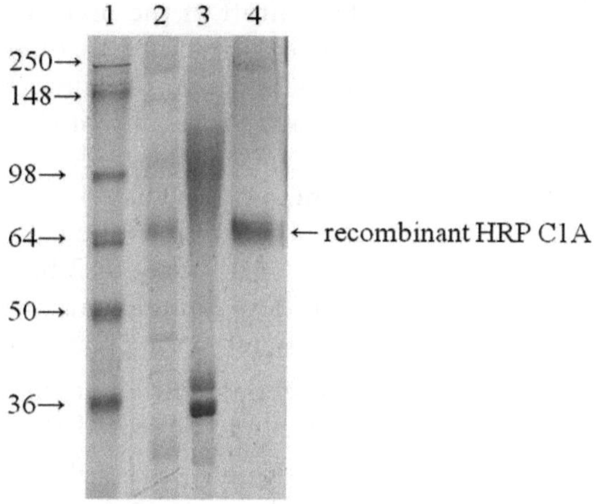

Fig. 1 SDS gel of the different HCIC fractions for the horseradish peroxidase isoenzyme C1A carrying eight glycosylation sites recombinantly produced in *P. pastoris*. Lane 1, protein marker (PageRuler Prestained Ladder); lane 2, cell-free cultivation broth after fermentation; lane 3, eluted HCIC fraction; lane 4, HCIC flow-through fraction. Approximately 5 μg protein were loaded per lane

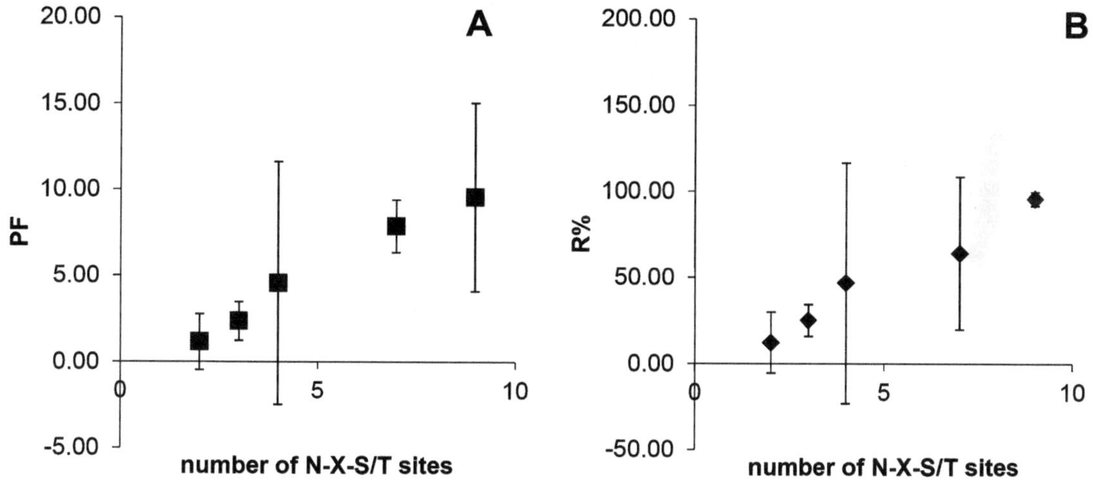

Fig. 2 Correlation of the number of N-glycosylation sites of different horseradish peroxidase isoenzymes and the success of the HCIC purification. (**a**) purification factor (PF); (**b**) recovery yield of HRP activity in percentage (R%) [19]

4 Notes

1. It is important to weigh the appropriate amount of reagent in a beaker and completely dissolve it in only an aliquot of deionized water. After adjusting the pH by either 2 M HCl or 2 M NaOH, fill up with water to the final volume.

2. The cutoff of the used ultrafiltration membrane should be at least threefold smaller than the molecular weight of the recombinant target protein. The most important variable for operating the ultrafiltration is the transmembrane pressure (TMP). It is the force for liquid transport through the membrane and calculated as the average applied pressure on the membrane minus the permeate pressure. We are recommending a TMP of about 2 bar during the tangential flow filtration to achieve high flow rates, and therefore save process time, but also preserve the protein from shear forces.
3. Between the different chromatography steps, the protein solution should be stored at +4 °C in the refrigerator.
4. Make sure to properly diafiltrate the protein solution before each chromatography step. As a rule of thumb, three to five volumes of buffer exchange are sufficient for that. It is possible to follow the diafiltration process by monitoring the conductivity and the pH value in the protein solution. Furthermore, when using a crossflow filtration device, make sure to drain the system by back-flushing pressurized air through the permeate valve to minimize product loss.
5. The HCIC column should be operated at the given flow rate, since the flow rate significantly influences the equilibrium between the sample and the resin. The maximum pressure must not exceed the typical working pressure of the MEP HyperCel mixed-mode resin of 1 bar. It is essential that the complete chromatography system is free of air bubbles to avoid damage of the packed resin. If possible, the flow direction during the cleaning step should be changed to bottom-to-top.
6. The AEC column should be operated at the given flow rates, since the flow rate significantly influences the equilibrium between the sample and the resin. The maximum pressure must not exceed the maximum pressure of the 8 mL CIM-DEAE monolithic column of 20 bar. It is essential that the complete chromatography system is free of air bubbles during all processing steps to avoid damage of the monolithic resin. The flow direction must be strictly adhered to prohibit damage of the column.
7. The achievable PF depends on (1) the amount of glycosylation sites on the recombinant target protein and (2) the recombinant protein production strategy. Cultivating *P. pastoris* in shake flasks, where conditions are not controlled and limitations in nutrients and oxygen can occur, can lead to cell lysis, and thus the cell-free cultivation broth might contain a substantial amount of contaminating proteins. Typical PFs for recombinant proteins carrying more than five glycosylation sites produced in shake flasks lie in the range from PF 6.0 to

PF 16.0 [19]. When *P. pastoris* is cultivated in the controlled environment of a bioreactor under optimal conditions avoiding major cell stress, less contaminating proteins are found in the cell-free cultivation broth, which is why typical PFs for recombinant proteins carrying more than five glycosylation sites lie in the range from PF 2.0 to PF 5.0 [8].

8. In order to consider a purification step as valid and exclude the possibility of undesired and unplanned effects, like problems with the resin or wrongly prepared buffers, the total recovery yields, i.e., the sum of protein and, if applicable, biological activity in the eluate and the flow-through, should lie between 90 and 100% of the initially determined values before the respective purification step. If this criterion is not met, the resin, the buffers, and the sample have to be checked, and the purification step has to be repeated.

References

1. Apte-Deshpnade A, Mandal G, Soorapaneni S et al (2009) High-level expression of non-glycosylated and active staphylokinase from *Pichia pastoris*. Biotechnol Lett 31:811–817
2. Hermanrud CE, Pathiraja V, Matar A et al (2012) Expression and purification of non-N-glycosylated porcine interleukin 3 in yeast *Pichia pastoris*. Protein Expr Purif 82:70–74
3. Peraino J, Zhang H, Hermanrud CE et al (2012) Expression and purification of soluble porcine CTLA-4 in yeast *Pichia pastoris*. Protein Expr Purif 82:270–278
4. Li Z, Moy A, Sohal K et al (2009) Expression and characterization of recombinant human secretory leukocyte protease inhibitor (SLPI) protein from *Pichia pastoris*. Protein Expr Purif 67:175–181
5. Nagasu T, Shimma Y, Nakanishi Y et al (1992) Isolation of new temperature-sensitive mutants of *Saccharomyces cerevisiae* deficient in mannose outer chain elongation. Yeast 8:535–547
6. Nakayama K, Nagasu T, Shimma Y et al (1992) OCH1 encodes a novel membrane bound mannosyltransferase: outer chain elongation of asparagine-linked oligosaccharides. EMBO J 11:2511–2519
7. Morawski B, Lin ZL, Cirino PC et al (2000) Functional expression of horseradish peroxidase in *Saccharomyces cerevisiae* and *Pichia pastoris*. Protein Eng 13:377–384
8. Spadiut O, Rossetti L, Dietzsch C et al (2012) Purification of a recombinant plant peroxidase produced in *Pichia pastoris* by a simple 2-step strategy. Protein Expr Purif 86:89–97
9. Vervecken W, Kaigorodov V, Callewaert N et al (2004) In vivo synthesis of mammalian-like, hybrid-type N-glycans in *Pichia pastoris*. Appl Environ Microbiol 70:2639–2646
10. Jacobs PP, Geysens S, Vervecken W et al (2009) Engineering complex-type N-glycosylation in *Pichia pastoris* using GlycoSwitch technology. Nat Protoc 4:58–70
11. Hopkins D, Gomathinayagam S, Rittenhour AM et al (2011) Elimination of beta-mannose glycan structures in *Pichia pastoris*. Glycobiology 21:1616–1626
12. Krainer FW, Gmeiner C, Neutsch L et al (2013) Knockout of an endogenous mannosyltransferase increases the homogeneity of glycoproteins produced in *Pichia pastoris*. Sci Rep 3:3279
13. Li H, Hui X, Yang S et al (2013) High level expression, efficient purification and bioactivity assay of recombinant human platelet-derived growth factor AA dimer (PDGF-AA) from methylotrophic yeast *Pichia pastoris*. Protein Expr Purif 91:221–227
14. Kang LX, Chen XM, Fu L, Ma LX (2012) Recombinant expression of chitosanase from Bacillus subtilis HD145 in *Pichia pastoris*. Carbohydr Res 352:37–43
15. Stergiou C, Zisimopoulou P, Tzartos SJ (2011) Expression of water-soluble, ligand-binding concatameric extracellular domains of the human neuronal nicotinic receptor alpha4 and beta2 subunits in the yeast *Pichia pastoris*: glycosylation is not required for ligand binding. J Biol Chem 286:8884–8892

16. Tams JW, Welinder KG (1995) Mild chemical deglycosylation of horseradish-peroxidase yields a fully active, homogeneous enzyme. Anal Biochem 228:48–55
17. Radha S, Gunasekaran P (2009) Purification and characterization of keratinase from recombinant *Pichia* and *Bacillus* strains. Protein Expr Purif 64:24–31
18. Ogawa S, Shimizu T, Ohki H et al (2009) Expression, purification, and analyses of glycosylation and disulfide bonds of Stereum purpureum endopolygalacturonase I in *Pichia pastoris*. Protein Expr Purif 65:15–22
19. Krainer FW, Pletzenauer R, Rossetti L et al (2013) Purification and basic biochemical characterization of 19 recombinant plant peroxidase isoenzymes produced in *Pichia pastoris*. Protein Expr Purif 95:104–112

Chapter 18

LC-MS Analysis of (Glyco-)Proteins of *Pichia pastoris*

Clemens Grünwald-Gruber and Friedrich Altmann

Abstract

In this chapter, we will present two methods for comprehensive glycoprotein characterization that are particularly but not exclusively useful for *Pichia pastoris* glycoproteins. One approach is intact protein mass measurement, where deglycosylation may be used to determine the mass of the unmodified protein. The other method is the classical bottom-up approach, where peptides and glycopeptides are analyzed by reversed-phase chromatography and detected by electrospray ionization mass spectrometry. The choice of chromatography solvents with a high ionic strength simplifies the identification of peaks of a particular peptide's glycopattern as it leads to co-elution of neutral and charged, i.e., phosphorylated, glycoforms.

Key words Ammonium formate buffer, Mass spectrometry, Electrospray ionization (ESI), Reversed-phase chromatography, Monolithic protein column, Phosphorylation

1 Introduction

Mass spectrometry is a very powerful tool to analyze different kinds of samples. In this chapter, we will focus on proteins or peptides purified from *Pichia pastoris*, with a special focus on the analysis of glycosylated variants. *P. pastoris* exhibits high mannose-type glycosylation with occasionally occurring phosphorylation. The mass increments between the different glycoforms are 162 Da (hexose) or 80 Da (phosphorylation). N-glycans are comprised of a pentasaccharide core (2GlcNAc-3Man) with varying numbers of mannose residues attached. Hyperglycosylation has also been described for *P. pastoris* [1]. When O-glycans are present, they are often linear, but also branched mannose chains are found [1].

In general, for the LC-ESI-MS analysis, two different approaches termed "bottom up" and "top down" (*see* Fig. 1) are described. In the bottom-up approach, the proteins are first digested with a protease, and the resulting peptides are analyzed. One of the advantages is that peptide mixtures are easier to handle, and separation techniques (HPLC) are more efficient. It allows determining site-specific glycosylation (one peptide—one

Brigitte Gasser and Diethard Mattanovich (eds.), *Recombinant Protein Production in Yeast*, Methods in Molecular Biology, vol. 1923, https://doi.org/10.1007/978-1-4939-9024-5_18,

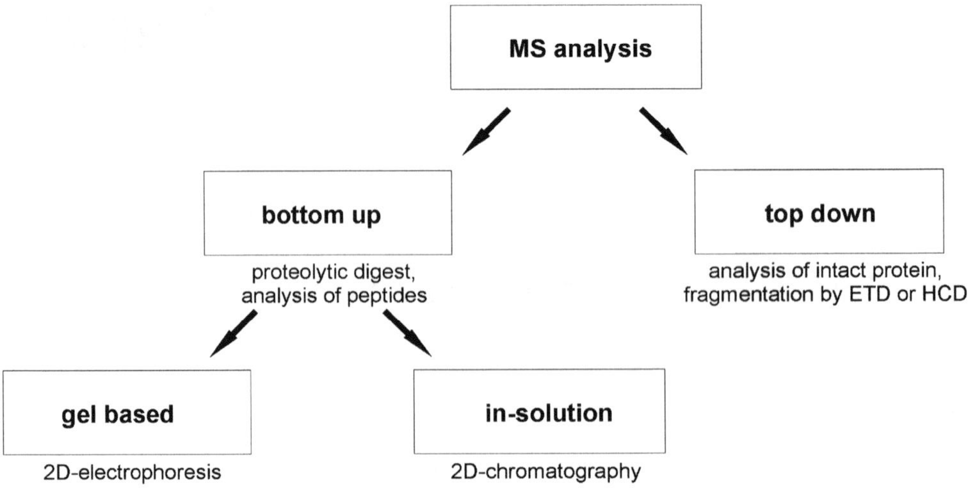

Fig. 1 Overview of techniques used in proteomic analysis

glycosylation site) which is not possible with an intact protein measurement if there are multiple glycosylation sites on a protein. The peptide mixture could be further deglycosylated to simplify (and automate) the glycopeptide analysis.

For peptide generation, gel-based and in-solution protocols are well established. The in-solution protocol is suitable to analyze all proteins present, while the in-gel approach provides the possibility to pre-fractionate a certain protein. In both cases an efficient and reproducible cell lysis protocol is needed. *P. pastoris* has a rigid cell wall, thus mechanical or enzymatic disruption is needed. Furthermore, detergents are used to ensure solubilization of hydrophobic proteins (like membrane proteins), but detergents must be removed completely before in-solution sample workups (no interferences with in gel-based approaches). Detergent removal could be done by protein precipitation, filter-aided sample preparation (FASP [2]), or the use of special detergents (e.g., RapiGest SF). Most frequently used is the precipitation with organic solvents (MeOH/chloroform removes SDS efficiently, but acetone can be used as well [3]) or acids (e.g., trichloroacetic acid).

To determine the site-specific glycosylation of a glycoprotein, a proteolytic digestion followed by analysis by mass spectrometry of the generated glycopeptides is done. In comparison with the analysis of released glycans, this approach offers many advantages: the glycoforms detected are in fact linked to the protein considered and not to a highly glycosylated impurity. Different sites are sometimes occupied by different glycoforms, and a very comprehensive view of the glycoprotein is possible, as all types of glycosylation and even other posttranslational modifications can be analyzed using one experiment. In addition, the peptide dominates the ionization efficiency, and thus it can be assumed that all the different glycoforms detected feature a very similar detection sensitivity, which is not the case if free oligosaccharides are analyzed [4]. The use of a buffered solvent system for the LC-ESI-MS measurement suppresses ion exchange processes which take place on many reversed-phase columns, which would lead to different elution positions for peptides of different charges (e.g., phosphorylated glycans). In the method presented, a 80 mM ammonium formate buffer is used. The narrow elution window of all glycoforms allows the identification by the exact mass, and thus glycoforms yielding poor MS/MS spectra are not ignored as is the case with strictly MS/MS-based identification routines. Deglycosylation (e.g., with PNGase A) and remeasurement serve to verify the peptide sequence inferred from the exact masses of glycoforms [5–8].

In the method described herein, no glycopeptide/glycoprotein enrichment step is used. At least for pure glycoproteins, we have never felt the need to improve the ratio of glycopeptides versus peptides prior to reversed-phase LC-ESI-MS, and thus we regard the extra handling step, e.g., by hydrophilic interaction chromatography or SPE cartridges [9], as usually dispensable. All the more, this step may introduce biases in the glycosylation pattern and certainly erases information on the occurrence of under-glycosylation. A major disadvantage of the glycopeptide approach is its inability to resolve structural isomers—only the glycan composition is determined [10].

The analysis of intact proteins provides another level of insight—all modifications and variants of a protein are revealed, and minor variants cannot be overlooked. On the other hand, it is next to impossible to determine the site of a certain modification. Thus, a combination of intact protein measurement and peptide analysis after proteolytic digestion is a good combination for a comprehensive protein analysis. A prerequisite for the intact measurement however is a rather pure or at least enriched protein sample.

2 Materials

2.1 Proteolytic Digestion (See Note 1)

1. Neat acetone.
2. Ammonium bicarbonate buffer (100 mM): Weigh 7.9 mg NH_4HCO_3 and dissolve in 1 mL of high-quality water (*see* **Note 2**).
3. Dithiothreitol (DTT) solution (10 mM): Weigh 1.54 mg DTT and dissolve in 1 mL 100 mM ammonium bicarbonate buffer (*see* **Note 3**).
4. Iodoacetamide solution (55 mM): Weigh 10.2 mg and dissolve in 1 mL 100 mM ammonium bicarbonate buffer.

5. Trypsin solution: Add the required amount of trypsin (e.g., sequencing grade modified trypsin, Promega—enzyme/substrate ratio 1:20 to 1:100).
6. Vertical shaker (e.g., Eppendorf ThermoMixer).
7. Vacuum concentrator.

2.2 Deglycosylation

1. PNGase F (peptide glycosidase F), recombinant (Roche Applied Science, Cat. Nr. 11365185001), reconstituted to have 1 U/μL (*see* **Note 4**).
2. PNGase F buffer: 50 mM ammonium acetate, pH 8.4 (*see* **Note 5**).
3. PNGase A (peptide glycosidase A, glycoamidase A) purified from almonds (Europa Bioproducts, Cambridge, GB) (*see* **Note 6**).
4. PNGase A buffer: 0.1 M sodium citrate of pH 5.0 (*see* **Note 7**).

2.3 LC-ESI-MS Measurement

2.3.1 Peptide Measurement

1. Ammonium formate buffer (80 mM, pH 3.0; LC solvent A): 3 mL of formic acid is added to roughly 800 mL HPLC-quality water, the pH is adjusted with 25% ammonia solution, and the buffer is then brought to 1 L with high-quality water. The final solution is degassed either by vacuum or by treatment in an ultrasonic bath (at least 10 min).
2. 80% Acetonitrile in ammonium formate buffer: Mix 800 mL acetonitrile (LC solvent B; use HPLC grade solvent, e.g., HiPerSolv LC-MS grade from VWR) with 200 mL 80 mM ammonia formate buffer pH 3—degas in an ultrasonic bath (at least 10 min).
3. Reversed-phase LC column (e.g., BioBasic C18, 5 μm particle size, 0.32 × 150 mm; Thermo Scientific).
4. LC-ESI-MS system: We use a Dionex UltiMate 3000 system combined with a Bruker maXis 4G MS instrument or Waters Ultima QTOF and the dedicated software.
5. ESI calibration mix (e.g., from Agilent).

2.3.2 Protein Measurement

1. 0.05% Trifluoroacetic acid (TFA): 0.5 mL of TFA is added to roughly 500 mL high-quality water, and the buffer is then brought to 1 L with high-quality water. The final solution is degassed either by vacuum or in an ultrasonic bath (at least 10 min).
2. 80% Acetonitrile in 0.05% TFA: Mix 800 mL acetonitrile (solvent B; use HPLC grade solvent, e.g., HiPerSolv LC-MS grade from VWR) with 200 mL of 0.05% TFA—degas in an ultrasonic bath (at least 10 min).
3. Thermo ProSwift™ RP-4H column (0.2 × 250 mm) LC column.

4. LC-ESI-MS system: We use a Dionex UltiMate 3000 system combined with a Bruker maXis 4G MS instrument or Waters Ultima QTOF instrument and the dedicated software.

3 Methods

3.1 In-Solution Digestion

All procedures are carried out at room temperature unless otherwise specified. The protocol is optimized for purified proteins. If a cell lysis has to be performed, a complete and reproducible extraction protocol should be used [11].

1. Redissolve your protein in 100 mM bicarbonate buffer (*see* **Notes 8** and **9**).
2. Add the same amount of DTT-containing buffer.
3. Disulfide bonds are reduced by incubation for 45 min at 56 °C.
4. Bring reaction tube to room temperature.
5. Add the same amount of iodoacetamide solution, and incubate for 30 min in the dark (to achieve carbamidomethylation of free thiol groups, *see* **Note 10**).
6. Continue with protein purification by acetone precipitation and digestion (*see* **Note 11**).
7. Add 4 volumes precooled acetone (−20 °C).
8. Incubate for 1 h at −20 °C.
9. Centrifuge the solution for 5 min at 10,000 rcf at 4 °C, and remove as much liquid as possible without disturbing the precipitate.
10. Speed-vac pellet and redissolve in appropriate buffer for protein digestion (e.g., for tryptic digestion in 100 mM ammonium bicarbonate buffer).
11. Add trypsin in an enzyme to substrate ratio 1:50, and digest overnight at 37 °C (*see* **Notes 12** and **13**).

3.2 Deglycosylation

To improve the MS/MS-based identification of (glyco-)peptides, deglycosylate a part of the sample, and apply the search engine to the deglycosylated peptides, which elute about 1 min after the respective glycopeptides. Use PNGase F as the first choice.

3.2.1 PNGase F

1. Peptides are dissolved in 10 μL PNGase F buffer, and 0.5 U of PNGase F (*see* **Note 14**) is added. Incubate overnight at 37 °C.

3.2.2 PNGase A

1. Peptides are dissolved in 10 μL PNGase A buffer, and 0.1 mU of PNGase A is added. Incubate overnight at 37 °C.

3.3 LC-ESI-MS Measurement

Peptides are redissolved in appropriate volume of H_2O (10–40 μL) to allow at least two measurements.

3.3.1 Peptide Measurement

1. Test the performance of your analytic system (e.g., with cytochrome C—peak shape, retention times, signal intensity, and background noise should be inspected).
2. Instrument calibration prior every queue ensures high mass accuracy.
3. Equilibrate chromatographic system with 5% solvent B.
4. Inject 5 μL to LC-ESI-MS system with an autosampler.
5. The recommended column has an inner diameter of 0.32; thus use a flow rate of 6 μL/min. To elute (glyco-)peptides, after a 5-min flush with starting conditions, a gradient is developed from 5% B to 42% B in 45 min, followed by a 15-min ramp to 90% B to facilitate elution of large peptides.
6. Perform the analysis in Auto MS/MS mode for ions of this mass range. If the sequence of the glycopeptide should be confirmed by MS/MS, choose a slightly higher collision energy as usual, to obtain also peptide-specific fragments [8, 12] (*see* **Note 15**).
7. Use Mascot (http://matrixscience.com/), GPM (http://www.thegpm.org/), or another data mining tool (e.g., http://ms.imp.ac.at/?goto=msamanda) to identify peptides in the PNGase-treated sample. Choose deamidation of Asn as deglycosylation converts Asn to Asp as variable modification.
8. The online tool NetNGlyc (website http://www.cbs.dtu.dk/services/NetNGlyc) helps to find all glycosylation sites that have to be analyzed. Then search for the peptides-containing glycosylation sites in the result list for the PNGase-treated sample, or proceed directly with **step 9**.
9. Now knowing the mass of the peptide moieties, prepare an Excel spreadsheet listing likely glycoforms with different charge states (Fig. 1 and Table 1).
10. Identify the glycopeptides by searching for the calculated masses in summed spectra of the elution region 1–4 min before the deglycosylated peptide or by extracting ion chromatograms for a few of these masses. MS/MS data could be used for searching for specific oxonium ions (204.08, 366.13) or the Y1 ion, i.e., the peptide with one GlcNAc attached (possibly multiply charged) [13] (*see* **Note 16**) (Fig. 2).
11. Sum up the spectrum of a narrow zone around the identified glycopeptide signal (make sure not to exclude any signal present).

Table 1
Example of a calculated mass list

	N-glycan composition		Peptide [M+H]$^+$: 1671.8085			
			Charge state			
	Hex	HexNAc	1+	2+	3+	4+
Y1 (for MS/MS spectra)	0	1	1874.888	937.948	625.634	469.477
Man8	8	2	3374.390	1687.698	1125.468	844.353
Man9	9	2	3536.443	1768.725	1179.486	884.866
Man10	10	2	3698.495	1849.751	1233.503	925.379
Man11	11	2	3860.548	1930.778	1287.521	965.892
Man12	12	2	4022.601	2011.804	1341.538	1006.406
Man13	13	2	4184.654	2092.830	1395.556	1046.919
Man14	14	2	4346.707	2173.857	1449.574	1087.432
Man15	15	2	4508.759	2254.883	1503.591	1127.945
Man16	16	2	4670.812	2335.910	1557.609	1168.458
Man17	17	2	4832.865	2416.936	1611.626	1208.972

The N-glycan structures or—precisely—compositions are named according to the number of mannose residues. Increment values for sugar residues were taken from http://web.expasy.org/glycomod/

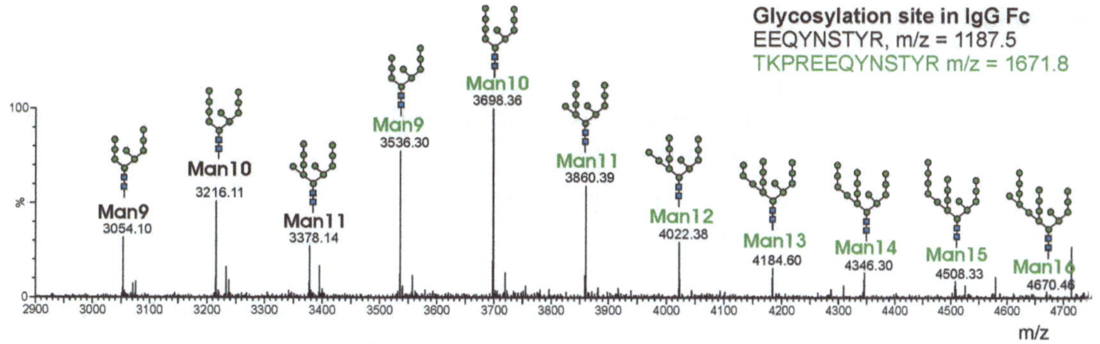

Fig. 2 Example of a glycopeptide spectrum. Mass spectrum for glycosylation EEQYNSTYR of an antibody produced in *Pichia pastoris*. A 1.4-min region around the peak for glycosylation site was summed. A missed-cleavage (TKPREEQYNSTYR) variant was detected as the main form. The peak heights of the deconvoluted spectrum can be used for the quantitation of the different glycan structures. It should be mentioned that the ammonium buffered solvent leads to ammonium adduct peaks with some glycopeptides. Note that only one of several possible isoforms is depicted

12. Perform a deconvolution to obtain a simplified glycopeptide profile, or use the obtained glycopeptide spectrum (mind the different charge states).
13. Assign glycan compositions to the multiple peaks in the deconvoluted profile (*see* **Notes 17–19**).
14. Perform relative quantitation by comparing peak heights of different glycoforms.

3.3.2 Protein Measurement

For optimal results, use purified samples only.

1. Perform **steps 1** and **2** as described in Subheading 3.3.2.
2. Equilibrate chromatographic system with 20% solvent B.
3. Inject to LC-ESI-MS system with an autosampler.
4. The recommended column has an inner diameter of 0.2 mm; thus use a flow rate of 8 μL/min. To elute (glyco-)proteins, a gradient from 20% B to 75% B in 25 min, followed by a 5-min ramp to 95% B, is developed, after an initial hold time of 5 min.
5. Perform an ESI-MS measurement in the protein tune (tuned for big molecules).
6. Sum up the spectrum of protein signal detected in the measurement (*see* **Note 20**).
7. Perform a MaxEnt deconvolution to obtain a simplified profile spectrum and assign peaks to the theoretical masses (*see* **Note 21**).

4 Notes

1. The procedure described is optimized for a tryptic digestion but works as well for other proteases (e.g., GluC or chymotrypsin).
2. Always prepare fresh to ensure pH value of around 8.1.
3. Disulfide bond reduction and carbamidomethylation need a slightly alkaline pH.
4. PNGase F is available from other suppliers, but different unit definitions of different suppliers should be considered. Mind that Roche defines the PNGase F activity units as 1 nmol/min. Glycerol-free formulations are preferable.
5. Other buffers (TRIS/HCl or phosphate) of pH values between 7.5 and 8.5 can also be used for PNGase F.
6. Ammonium acetate at a pH of about 5.0 is a suitable alternative. Use PNGase A if glycans may contain core α1,3-linked fucose, i.e., if non-vertebrate glycoproteins are analyzed. Other suppliers are, e.g., Roche Life Sciences, Proglycan, or Sigma-Aldrich.

7. Alternative: Ammonium acetate at a pH of about 5.0.
8. If your protein is dissolved in a different buffer, do a buffer exchange, or ensure a pH of around 8.
9. A protein concentration of ~1 mg/mL would be recommended.
10. Do not incubate longer to prevent reaction of methionines with IAM.
11. Quenching of the reaction by the addition of, e.g., 50 μL of 30 mM DTT, is rendered expendable by the precipitation step.
12. Alternatively, you could dry the pellet by flushing with nitrogen or air drying.
13. Often trypsin will be the protease of choice to get glycopeptides of appropriate size (~800–3500 Da) and chromatographic behavior. Protease action on cleavage sites adjacent to the glycosylation site is poorly predictable, and different degrees of missed cleavages may occur. Very large (and/or hydrophobic) and very small glycopeptides tend to become undiscoverable. A more complete coverage of glycosylation sites will usually be obtained with less specific proteases such as chymotrypsin (cleaves after F, Y, W, and also L, M, and sometimes H), but MS/MS-based (glyco-)peptide detection, possibly after deglycosylation, is compulsory. The highly specific proteases GluC, AspN, and LysC may also be used, and good results have been obtained with combinations, e.g., with trypsin plus GluC for erythropoietin analysis [14].
14. While some peptides occur in non-glycosylated form, it will usually require the use of peptide N-glycosidase to obtain deglycosylated peptides. These can be detected by highly efficient, usual peptide identification routines such as MASCOT or GPM. Thereby, unexpected peptides (missed cleavages, alternative cleavages, unexpected modifications, etc.) are easily identified [15].
15. As glycopeptides have a lower charge to mass ratio than unmodified peptides, the instrument shall be tuned for the mass region of 800–1600 amu.
16. Another option is the search for peptide-specific fragments obtained with higher-energy collisional dissociation or electron transfer dissociation MS [7, 10].
17. An Excel sheet with a large variety of glycopeptide masses and charge states may be helpful in this concern (Fig. 1).
18. Glycopeptide variants are eluting at essentially the same time (the hydrophilic glycan moiety does not contribute much to the binding to the C18 column). Even charged sugars, like

sialic acids, do not affect the retention time dramatically, when ammonium formate buffer is used [8].

19. Mass peaks not assigned can be examined in various ways: if this mass co-elutes with the identified glycopeptides or if sugar mass increments (for relevant charge state), connect the unknown peak with identified ones, or you could perform a GlycoMod (http://web.expasy.org/glycomod/) search using the assumed peptide moiety mass. Ideally, MS/MS data is used for the confirmation of the peak, but all too often, they are not of sufficient quality.
20. Proteins show a large number of different charge states and thus characteristic spectra.
21. Use the theoretical average mass of a protein to assign peaks.

References

1. Gasser B, Prielhofer R, Marx H et al (2013) *Pichia pastoris*: protein production host and model organism for biomedical research. Future Microbiol 8:191–208
2. Wiśniewski JR, Zougman A, Nagaraj N et al (2009) Universal sample preparation method for proteome analysis. Nat Methods 6:359–362
3. Botelho D, Wall MJ, Vieira DB et al (2010) Top-down and bottom-up proteomics of SDS-containing solutions following mass-based separation. J Proteome Res 9:2863–2870
4. Grünwald-Gruber C, Thader A, Maresch D et al (2017) Determination of true ratios of different N-glycan structures in electrospray ionization mass spectrometry. Anal Bioanal Chem 409(10):2519–2530
5. Huhn C, Selman MH, Ruhaak LR et al (2009) IgG glycosylation analysis. Proteomics 9:882–913
6. Zauner G, Selman MH, Bondt A et al (2013) Glycoproteomic analysis of antibodies. Mol Cell Proteomics 12:856–865
7. Pabst M, Altmann F (2011) Glycan analysis by modern instrumental methods. Proteomics 11:631–643
8. Pabst M, Chang M, Stadlmann J et al (2012) Glycan profiles of the 27 N-glycosylation sites of the HIV envelope protein CN54gp140. Biol Chem 393:719–730
9. Alagesan K, Khilji SK, Kolarich D et al (2017) It is all about the solvent: on the importance of the mobile phase for ZIC-HILIC glycopeptide enrichment. Anal Bioanal Chem 2:529–538
10. Hinneburg H, Hofmann J, Struwe WB et al (2016) Distinguishing N-acetylneuraminic acid linkage isomers on glycopeptides by ion mobility-mass spectrometry. Chem Commun (Camb) 52:4381–4384
11. Rußmayer H, Buchetics M, Gruber C et al (2015) Systems-level organization of yeast methylotrophic lifestyle. BMC Biol 13:80
12. Scott NE, Parker BL, Connolly AM et al (2011) Simultaneous glycan-peptide characterization using hydrophilic interaction chromatography and parallel fragmentation by CID, higher energy collisional dissociation, and electron transfer dissociation MS applied to the N-linked glycoproteome of *Campylobacter jejuni*. Mol Cell Proteomics 10:M000031-MCP000201
13. Irungu J, Go EP, Zhang Y et al (2008) Comparison of HPLC/ESI-FTICR MS versus MALDI-TOF/TOF MS for glycopeptide analysis of a highly glycosylated HIV envelope glycoprotein. J Am Soc Mass Spectrom 19:1209–1220
14. Jez J, Castilho A, Grass J et al (2013) Expression of functionally active sialylated human erythropoietin in plants. Biotechnol J 8:371–382
15. Wu SW, Pu TH, Viner R et al (2014) Novel LC-MS(2) product dependent parallel data acquisition function and data analysis workflow for sequencing and identification of intact glycopeptides. Anal Chem 86:5478–5486

Chapter 19

Assessing the Quality of Recombinant Products Made in Yeast

Karola Vorauer-Uhl and Gabriele Lhota

Abstract

The product quality of recombinant proteins is of major importance for their intended purpose. The initial characterization of both simple and complex products should be performed as soon as practical. However, to comply with this high standard, appropriate selection of complementary methods is required. Therefore, conventional and sophisticated techniques are available, providing diverse information about the product quality.

In this chapter methods are presented, which enable the determination of the overall protein quality, their aggregation and peptide composition. Methods applied for the determination of posttranslational modifications such as glycan analysis are not described. In this regard, chromatographic, high-resolution technologies for the integrity of proteins as well as Western blot with specific detection methods are introduced, and individual strengths and perceived limitations are highlighted.

Key words Protein quality, Size exclusion chromatography, Peptide map, Charge variations, Electrophoresis, Western blot

1 Introduction

Higher eukaryotes are of essential importance for the production of complex proteins. In this respect, yeasts with the advantage of rapid growth rates in simple mineral media and therefore lower costs are excellent candidates for this purpose. Furthermore, their potential to secrete sufficient amounts of simple and complex proteins associated with a variety of posttranslational modifications, including methylation, sulfation, phosphorylation, lipid addition, and glycosylation, makes them interesting hosts. Compared to other higher eukaryotes, similar posttranslational potency is described except for gamma-carboxylation as shown in Table 1 [1].

Brigitte Gasser and Diethard Mattanovich (eds.), *Recombinant Protein Production in Yeast*, Methods in Molecular Biology, vol. 1923, https://doi.org/10.1007/978-1-4939-9024-5_19, © Springer Science+Business Media, LLC, part of Springer Nature 2019

Table 1
Basic characteristics of different host strains

Characteristics	*E. coli*	Yeast	Mammalian cells
Doubling time	Rapid (30 min)	Rapid (90 min)	Slow (24 h)
Complexity of growth medium	Minimum	Minimum	Complex
Cost of growth medium	Low	Low	High
Expression level	High	Low to high	Low to moderate
Extracellular expression	Secretion to periplasm	Secretion to medium	Secretion to medium
Posttranslational modifications			
Protein folding	Refolding usually required	Refolding may be required	Proper folding
N-linked glycosylation	None	High mannose	Complex
O-linked glycosylation	No	Yes	Yes
Phosphorylation	No	Yes	Yes
Acetylation	No	Yes	Yes
Acylation	No	Yes	Yes
Gamma-carboxylation	No	No	Yes

Depending on the protein of interest, the choice of expression system, optimized clone screening, and the establishment of proper cultivation strategies are decisive to obtain the best quality. When and how the quality needs to be analyzed depends on several considerations: (1) intended application, (2) amounts of available protein, (3) project milestones, and last but not least (4) accessibility of methods and regulatory demands. In the light of these considerations, several applicable methods are introduced, while finally the project team has to decide when they need to be implemented [2, 3]. In this chapter methods will be presented which are established to analyze the overall quality of the expressed proteins, their aggregation and peptide composition. Methods for posttranslational modifications, such as full characterization of glycan structures, are not described here.

2 Chromatographic Methods

2.1 Size Exclusion Chromatography for the Analysis of Protein Aggregates and Degradation Products

2.1.1 Background

The complexity of proteins requires the monitoring of product quality during process development and production. The quality attributes can be measured by a set of broad complementary techniques addressing different quality features.

Aggregation, including dimers and multimers, but also degradation into the protein subunits are of essential interest. While numerous methods are applicable for this approach, size exclusion chromatography (SEC) performed with HPLC has been favored because of both, its speed and reproducibility [4]. For individual confirmation, comparative techniques are suggested. Based on the size of the protein, sedimentation velocity analytical ultracentrifugation (SV-AUG), asymmetric flow field fractionation (AFF), and multi-angle light scattering (MALS) are recommended.

While these techniques are frequently used, the dominant method is still SEC that was first introduced by Synge and Tiselius in 1950 and further developed in the following decades [5]. Lindqvist and Storgards (1955) reported the first separation of biomolecules with SEC, where they separated peptides from amino acids [6]. The application of soft polymeric resins for the early methods was less convenient due to the fact that these materials tend to compress under pressure and flow, which limits the chromatographic application. Meanwhile, numerous porous resins are on the market, where different types of sorbent configurations are available [7]. Besides these types, monoliths with low protein binding capacities and a broad separation range up to 670,000 Da, with most of the resolving power available for MW less than 66,000 Da [8].

Ideally, the separation of proteins and peptides is based on the Stokes radii of the analytes in solution. With a set of known analytes, a size-based calibration for the estimation of the molecular weight can be established by plotting log MW vs. the retention volume of the individual MW, which can be calculated more or less accurately. By normalizing the x-axis to the volume of the column when empty, the interstitial volume fraction, intra-particle fraction, and the stationary phase volume can be determined.

It has been widely demonstrated and discussed that the molecular weight range and slope are highly depending on the pore size and the packing resin. That means in practical work, the pore size and/or geometry restricts access of molecules based on their Stokes radius and because proteins vary in shape. The Stokes radius does not exactly correlate with the molecular weight. This circumstance might hinder the precise determination of the MW but enables estimation. Furthermore, it should be mentioned that non-ideal absorption might alter the retention volume causing an additional error. Whether the mobile-phase optimization minimizes this non-ideal adsorption needs to be individually tested (*see* **Note 1**).

The column resins and configuration mainly affect the thermodynamics, kinetics, and resolution. Furthermore, the instrumentation, either HPLC or FPLC, might also significantly alter the dispersion due to different design and configuration. Individual optimization is recommended (*see* **Note 2**). The predominant mode of detection is still UV. However, alternative detectors such as light scattering and fluorescence are of growing interest. In UV detection, wavelength of 280 nm gives greater responses for aromatic amino acids, while UV wavelength of 214 nm provides higher sensitivity for the amide bonds. A combination of both wavelengths is possible by using dual wavelength detection, which can be successfully applied for purity profiling. If more information about the protein, size, and shape is required, multi-scattering detectors are recommended, while SEC-MALS provides information about the size, shape, and concentration. Fluorescence is typically applied to improve the selectivity and/or sensitivity [9–13]. Mass spectroscopy (MS) is another method for obtaining molecular weight information. However, the mobile phases, used for SEC, are not in all cases compatible with MS. The most suitable mobile phases are ammonium formate and ammonium acetate, although they are not physiologic. To overcome these difficulties, denaturing mobile phases containing organic solvents and ion-pairing reagents have been developed.

Due to the fact that proteins are prone to interact with surface-charged sites of the stationary phase, method optimization is necessary. Thereby, salt concentration, modifiers in the mobile phase, pH value, flow rate, and sample load need to be adjusted with the column configuration, resin, and particle size. Advantages and limitations are shown in Table 2.

Table 2
Summary of applicability and limitations of size exclusion chromatography

Advantages	Perceived limitations
Amenable to diverse analyte or sample types	Lack of an ideal, universal detector
Reproducible qualitative and quantitative	Interference with other proteins—purity
Applicable for crude supernatant when low background noise is given	Interference with salts, detergents, organic solvents, sulfones, etc.
Automated operation	High viscosity (DNA, carbohydrates) alters the separation
Medium separation power—protein and related substances	Limited sample throughput Not combinable with all detectors

2.1.2 Examples

The following application is an example for the determination of IgG and its aggregates, respectively. Due to the fact that the selection of a SEC column needs to be carefully adjusted to the molecular weight of the analytes in the sample, for this example a column for separating proteins in the range from 10 to 500 kDa is used. Usually, these column types have a distinct pH range (here pH 2.5–7.5), a defined maximal salt concentration for the running buffer (here 0.5 M) and in contrast to other HPLC columns, relatively low pressure limits (2.5 MPa). This should be considered when defining running buffer and flow rate. Additionally, ionic and hydrophobic interaction between the biomolecule and the column should be minimized, because separation should be based on size or hydrodynamic radius of the analyte. This can be achieved by adding modifiers to the mobile phase, which alter the ionic strength (NaCl) or hydrophobicity of the buffer (e.g., acetonitrile). For stabilization of the protein structure and prevention of unwanted protein-column interaction, the basic amino acid arginine is widely used, as shown in this example. Sodium azide in the buffer is appropriate to avoid microbiological growth in the solvents because of its toxicity to bacteria. The individual retention time of different molecular weights is carried out with an appropriate molecular weight standard. In principle, IgG and IgG aggregation analysis is also feasible without any purification step (e.g., cell culture supernatants) prior to SEC-HPLC. In this case, method development with special focus on matrix effects has to be performed carefully [14, 15].

Product: mAb (IgG 1) [14].

Host: Not specified.

Aim: IgG and aggregates.

Method: SEC-HPLC.

Sample matrix: Purified.

Equipment: HPLC system consisting of a pump, auto-injector, and DAD detector (214, 230, and 280 nm). Analysis of IgG and aggregates is performed on a TSKgel G3000SW column (7.8 mm × 300 mm).

Method description:

1. Mobile phase: 0.2 M L-arginine, 0.2 M sodium phosphate, and 0.005% sodium azide, pH 7.0.
2. Flow rate: 0.6 mL/min for 30 min.
3. Data evaluation is performed relative to the analyzed MW marker; aggregates elute in front of the native form followed by smaller fragments.

2.2 Peptide Mapping

2.2.1 Background

Peptide mapping is considered as a comparative fingerprint technique for the estimation of the protein quality. Peptide mapping confirms the primary sequence and potential modifications but can be applied only for captured or purified fractions. After selective cleavage, either enzymatically or chemically, the released peptides are thoroughly separated [16, 17]. Additional information and sequence identification beyond that, obtained at the non-digested protein level, is achieved. Careful inspection of the complete characterization strategy is required to generate successful peptide maps due to the fact that a profile may consist of more than 100 peaks. The selection of a chromatographic technique to separate peptides and generate peptide maps depends on the protein, experimental objectives, and anticipated outcome. The resolving power and stability of reversed-phase chromatography (RPC) makes this technique the predominant HPLC technique for peptide mapping separations. In most applications, low pH and elevated temperature are used. Therefore, columns with appropriate pH stability and minimal silanol effects are chosen.

Sample preparation includes (1) selection of an appropriate cleavage agent, (2) alkylation/reduction, (3) digestion, and (4) enrichment/cleanup. In almost all cases, samples need to be properly prepared for digestion [18, 19]. In particular, depletion, enrichment, dialysis, or desalting strategies are applied to adjust the environmental conditions for digestion. The cleavage type used (enzymatic or chemical) is strongly related to the protein sequence. Although several techniques and related protocols are available, the most common cleaving agent is trypsin due to its well-defined specificity. Trypsin hydrolyzes only the peptide bonds in which the carbonyl group is followed either by an arginine (Arg) or lysine (Lys). For most of the proteins, reduction and alkylation are recommended prior the digestion. Reduction reduces disulfide bonds (e.g., DTT, 45 min, 60 °C), while alkylation caps SH groups (e.g., iodoacetamide, 1 h, in the dark).

Nevertheless, it needs to be noted that trypsin may digest itself in a process called autolysis. Ca^{++}, naturally present in most samples, binds at the Ca^{++}-binding loop in trypsin and prevents autolysis. Meanwhile, modified trypsin is commercially available with reduced autolytic property. The optimal cleavage pH is in the range of 7.5–8.5, and digestion is normally performed at 37 °C. To provide an optimal pH, usually either a 50 mM triethyl ammonium bicarbonate (tABC) or 12.5 mM ammonium bicarbonate (ABC) buffer is added. Alternatively, 2-amino-2-hydroxymethyl propane-1,3-diol (Tris) buffer can be used, but it should be taken into account that Tris buffer is incompatible with MS analysis, such as MALDI and ESI-MS [20–22].

Prior to peptide mapping, cleanup and/or enrichment are usually required for the successful analysis of peptide maps. There are many types of methods to accomplish cleanup and enrichment optimized to the sample type and targeted objective [23].

Successful method development includes several steps: (1) initial gradient conditions (column length, mobile-phase composition, flow rate, temperature, and detection), (2) initial separation optimized for retention (k′), and (3) adjustment of the gradient range (*see* **Note 3**). Each conventional detection system can be used where sensitivity and buffer compatibility are considered. However, mass spectroscopy detection is the favored method for determining the individual mass of the peptides and therefore confirms the specific amino acid sequences covered by the peptide map, as well as the identity of the protein. In addition, conformation of N-terminal and C-terminal peptides and screening of posttranslational modification are proposed.

2.2.2 Examples

Peptide mapping is an analytical, qualitative method, which should be applied as a comparative method, in the following way that the peptide pattern of the target protein is compared to a reference substance pattern. Therefore, the cleavage procedure has to be selected carefully. Hence, the reproducibility of the subsequent separation and the detectability of the fragments are of significant importance to get reliable data for structural characterization, determination of protein modifications, and protein identity [24, 25].

Product: Yeast enolase [25]

Host: *Saccharomyces cerevisiae*.

Aim: Peptide map.

Method: UPLC/MS.

Sample matrix: Purified.

Equipment: An appropriate HPLC/UPLC system equipped with a mass spectrometer for detection is used. Peptide separation is carried out on an ACQUITY UPLC 2.1 × 100 mm, BEH300Å, 1.7 μm Peptide Separation Technology Column.

Sample preparation:

1. Dissolve yeast enolase in 100 mM NH_4HCO_3 (5 μg/μL).
2. Add to 50 μL protein solution 5 μg of sequencing-grade trypsin.
3. Incubate at 37 °C overnight.
4. Dilute the digest to 1.5 pmol protein/μL with 5:95 acetonitrile (ACN)/water containing 0.1% formic acid (FA).

Method description:

1. Solvent A: 0.1% formic acid in water.
2. Solvent B: 0.1% formic acid in ACN.
3. Column temperature: 40 °C.
4. Flow rate of 0.2 mL/min.

5. Separation: 0–60 min.
6. Gradient: 0–50% B.
7. Mass accuracy reference: 100 femtomole (GLu1)-fibrinopeptide B (GFP) in 50:50 ACN/water containing 0.1% formic acid.
8. Data acquisition: positive ion V-mode – alternating the collision cell energy between low- (5 V, transfer cell energy 5 V) and elevated-energy setting (energy ramped from 20 to 40 V, transfer cell energy 10 V).
9. Scan time: 0.5 s in both acquisition modes (1 s total duty cycle).
10. Capillary voltage, 3.0 kV; source temperature, 100 °C; cone voltage, 37 V; cone gas flow, 10 L/h.
11. Sampling of the lock spray channel: every 1 min.
12. Minimum resolution: 10,000.
13. Calibrate to mass error of less than 3 ppm using a 100 femtomole GFP infusion before the experiments.

Product: Erythropoietin (rh-EPO) [24]

Host: *Pichia pastoris*.

Aim: Peptide map.

Method: HPLC-MS.

Sample matrix: Purified.

Equipment: HPLC system equipped with a mass spectrometer and Advion TriVersa NanoMate. Peptide separation is carried out on a Phenomenex Jupiter® Proteo column, 4.6 × 250 mm.

Sample preparation:

1. Denature and reduce 150 μg rhEPO in 6 M guanidine hydrochloride and 20 mM dithiothreitol (DTT) at 50 °C for 1 h.
2. Cool sample to room temperature.
3. Alkylate by incubating with 50 mM iodoacetamide (IAM) for 45 min.
4. Remove excess IAM by incubation with 5 mM DTT at room temperature for 10 min.
5. Bring sample to 25 mM ammonium bicarbonate pH 7.8 buffer by a buffer exchange step through a 5 kDa MW cutoff spin column.
6. Add 5 μg trypsin to the sample and incubate overnight (~16 h) at 37 °C.

Method description:

1. Solvent A: 0.1% formic acid in water.
2. Solvent B: 0.1% formic acid in acetonitrile.
3. Column temperature: 30 °C.

Table 3
Gradient for rh-EPO peptide map by rp-HPLC

Time (min)	Solvent (%B)	Function
0–70	2–35	Separation
70–80	35–98	Separation

4. Flow rate of 1000 μL/min and divert 300 nL/min to the mass spectrometer through the NanoMate (*see* **Note 4**).
5. Gradient is shown in Table 3.
6. MS spray voltage setting is 2.0 kV.
7. Capillary temp: 115 °C.
8. Capillary voltage: 5 kV.
9. Tube lens: 77 V.
10. Scan range: 200–2000 m/z.
11. Normalized collision energy: 25%.
12. Activation time: 30 ms.

2.3 Charge Variations of Proteins

2.3.1 Background

Charge variations of proteins differ within the type of modifications. While certain proteins are directly affected by posttranslational modifications, others are modified by conformational changes and/or variation of local charge distribution. Variations are either caused by the host and/or cultivation but may also occur during the downstream sequences. Potential mechanism are as follows: (1) clipping of C-terminal lysine or (2) nonenzymatic reaction of glycation in which reducing sugars covalently link amines of lysine residues, (3) cyclization of N-terminal glutamines to form pyroglutamic acids leading to positive charge loss by the conversion of the N-terminal amine to a neutral amide, and (4) deamidation, an additional degradation pathway that modifies asparagine residues to aspartic acid and/or isoaspartic residues in a nonenzymatic mode, resulting in the appearance of a negative charge. Furthermore, other posttranslational modifications affect the local charge distribution of proteins such as methionine oxidation or aspartic acid isomerization, leading to the insertion of an extra methyl group into the backbone of the protein to form isoaspartic acid [26]. The modifications of charge profiles can potentially affect the structure, but also the biological activity of proteins, leading to significant reduction in receptor- and/or antigen-binding affinity.

For the characterization of protein charge variants, the method of choice is currently the ion exchange chromatography (IEX). Within the last decades, numerous protocols have been developed to analyze recombinant proteins by this technique [27].

For analytical purposes, ion exchange chromatography on different types of HPLC is recommended. The separation of ionizable molecules enables the analysis of similar types of molecules that would be hampered by other techniques. This method can be applied for crude culture supernatants and purified samples. Depending on the pH of the environment, proteins may carry a net positive charge, a net negative charge, or no charge. Therefore, suitable buffers for this analysis should be ±0.5–1.5 pH values different to the pI of the protein of interest. After loading an impure protein sample onto an ion exchange chromatography column, the column is washed to remove undesired proteins and other impurities, followed by elution either by a salt gradient and/or a change in pH [28].

The charged salt ions compete with bound proteins for the charged resin functional groups. Proteins with few charged groups will elute at low salt concentrations, whereas proteins with many charged groups will have greater retention times and elute at high salt concentrations [29].

Furthermore, pH gradients are also be used for elution [30]. In this case, proteins will elute when the pH gradient reaches the pI, because at that value the individual net charge is neutral. The elution from an anion exchange resin is performed by a decreasing pH gradient, while an increasing pH gradient is chosen for elution from cation exchangers. Since it is very difficult to generate reproducible and accurate linear pH gradients, a stepwise gradient is recommended. Furthermore, the pH gradient can be combined with a salt gradient to refine separation. However, it might be considered that some proteins precipitate at a pH similar to the pI. For these proteins, elution with a pH gradient is not recommended.

Ion exchange chromatography resins are composed of positively or negatively charged functional groups that are covalently bound to a solid matrix such as cellulose, agarose, polymethacrylate, polystyrene, and polyacrylamide. The latter three matrices allow higher flow rates.

For many protein workflows, protein folding and stability is a concern. In these scenarios, the selection of an anion or a cation exchanger depends on the protein of interest's stability. Some proteins are stable in both scenarios; thus they can be analyzed above and below their pI applied either onto an anion or cation exchanger. Other proteins are stable only above or below the pI. Furthermore, weak or strong IEX can be used. The number of charges on a strong ion exchanger remains constant regardless of the buffer pH, and they retain their selectivity and capacity over a wide pH range. Examples for strong ion exchangers are quaternary ammonium (Q), sulfonate (S), and sulfopropyl (SP) resins. In contrast, weak ion exchangers display pH-dependent function, and they deliver optimal performance only within a small pH range. When the pH of the buffer no longer matches the acid dissociation constant (pKa) of the resin functional group, these resins suffer significant capacity loss. Weak anion exchangers function poorly above a pH of 9, and weak cation exchangers begin to lose their ionization below pH 6. Commonly used weak ion exchange resins are diethylaminoethyl (DEAE) or carboxymethyl (CM) resins. Although strong ion exchangers are often preferred due to their performance, weak ion exchangers can be powerful separation tools whenever strong ion exchangers fail because the selectivity of weak and strong ion exchangers often differs.

The ionic form of a support refers to the counterion that is adsorbed onto the resin's functional groups. Common counterions for anion and cation exchangers are Na^+ and Cl^-, modulating the strength of the interaction with a given resin. The lower the selectivity of a counterion for the support, the more readily it can be exchanged for another ion of similar charge. Similarly, elution buffer containing a counterion with a relatively lower selectivity for the support will displace proteins from the column resin less readily during elution. For special applications, this difference can be exploited by the addition of Li^+, Br^-, and SO_4^{2-} to improve resin selectivity.

Furthermore, particle size of the stationary phase affects the separation in terms of resolution. Smaller particles provide higher resolution but typically also require lower flow rates. However, high-resolution media are commonly suggested for analytical purposes. Larger particles permit higher flow rates but yield lower resolution. Unlike particle size, flow rate affects both resolution and capacity. Longer residence times increase both the capacity and the resolution of a resin.

Although IEX represents an excellent tool, some considerations should be mentioned: (1) the composition of loading, wash, and elution buffers is essential for reliable results, that is, if a buffer contains the wrong counterion, it can prevent binding of certain isoforms to the resin, and (2) the charged species in buffers should have the same sign as the charged species of the IEX resin. For example, if commonly used phosphate buffer sample is used, this is not appropriate for anion exchange chromatography because the phosphate ion interacts strongly with the positively charged anion exchange resins. Common buffers for anion and cation exchange chromatography are shown in Table 4.

The application of IEX has some other remarkable limitations: (1) binding is caused on electrostatic interactions after equilibration of the column with low-salt buffers (therefore, after the sample injection, the buffer is exchanged which may cause loss of individual isoforms); (2) IEX permits high flow rates, causing very long run times; and (3) working outside of the optimal pH range, resins rapidly lose capacity and, more importantly, resolution.

Table 4
Recommended buffers for anion and cation exchange chromatography used for the analysis of charged protein/peptide isoforms

IEX type	Buffer	pH range
Cation	Acetic acid	4.8–5.2
Cation	Citric acid	4.2–5.2
Cation	HEPES	6.8–8.2
Cation	Lactic acid	3.6–4.3
Cation	MES	5.5–6.7
Cation	MOPS	6.5–7.9
Cation	Phosphate	6.7–7.6
Cation	PIPES	6.1–7.5
Cation	TES	7.2–7.8
Cation	Tricine	7.8–8.9
Anion	Bicine	7.6–9.0
Anion	Bis-Tris	5.8–7.2
Anion	Diethanolamine	8.4–8.8
Anion	Diethylamine	9.5–11.5
Anion	L-histidine	5.5–6.0
Anion	Imidazole	6.6–7.1
Anion	Pyridine	4.9–5.6
Anion	Tricine	7.4–8.8
Anion	Triethanolamine	7.3–8.3
Anion	Tris	7.5–8.0

2.3.2 Example

In the following example, a weak cation exchange column from the Dionex ProPac series is used. The separation on the negatively charged surface is performed at pH values below the isoelectric point (pI) of the protein whereas of positively charged stationary phase above the pI of the protein. Here the stationary phase consists of 10 μm nonporous, solvent-compatible ethylvinylbenzene-divinylbenzene copolymer beads, which are surrounded by a highly hydrophilic, neutral polymer to reduce non-specific interactions between the surface and the biopolymer. The carboxylic acid groups, which are grafted on the surface, are responsible for the weak cation exchange mechanism [14].

Product: mAb (IgG 1) [14].

Host: Not specified.

Table 5
Gradient for mAb charge distribution by IEX-HPLC

Time (min)	Solvent (%B)	Function
0–15	20	Equilibration
15–30	20–60.5	Separation
30–33	100	Column cleaning

Aim: Charge distribution.

Method: IEX-HPLC.

Sample matrix: Purified.

Equipment: HPLC system equipped with a binary pump, autoinjector, and an UV detector (280 nm). For separation a Dionex ProPac WCX-10 column (4 × 250 mm) is used.

Method description:

1. Solvent A: 20 mM sodium citrate, pH 5.5.
2. Solvent B: 20 mM sodium phosphate and 95 mM sodium chloride, pH 8.0.
3. Column temperature: ambient.
4. Flow rate of 1 mL/min.
5. Gradient is shown in Table 5.
6. Assign peaks compared to a standard; in cation exchange chromatography, acidic variants of mAb elute in front of neutral variants followed by basic variants.

3 Qualitative Electrophoresis and Western Blot Analysis

3.1 Background

The determination of product quality, early in product development, is still performed with electrophoresis. Although many new advanced methods are available, gel electrophoresis in various forms and dimensions (1D or 2D) is still favored, due to the following reasons: (1) it is well established; (2) many individual protocols are available; (3) crude culture supernatants, clarified broth of disrupted cells, and pre-purified and final purified products can be analyzed; and (4) numerous reagents and equipment are available in most of the laboratories. Nevertheless, it should be argued that the quality of the results is strongly dependent on (1) the protein matrix, (2) the amount of the tested protein, and (3) the availability of appropriate standards. Due to these limitations, the obtained results may significantly vary in accuracy and precision [31].

Whether the purity of the sample is not sufficient or the accuracy needs to be evaluated, additional Western blot procedures are recommended. The technical term "blotting" describes a technique to transfer biological samples from a gel onto a membrane and their subsequent detection on the surface of the membrane. If the detection is performed with a specific antibody recognizing its antigen, the technique is called immunoblotting, first introduced by Towbin et al. in 1979 [32]. By this technique, qualitative and semiquantitative results can be achieved. Quantitative results can be obtained only under specific conditions (*see* **Note 5**). Western blot first requires separation of the macromolecules by gel electrophoresis, followed by the transfer onto a membrane. Mainly nitrocellulose or polyvinylidene difluoride (PVDF) is used. The protein transfer is either performed by capillary transfer or by transfer using wet, semidry, and dry electroblotting methods [33, 34]. After the transfer of the macromolecules, the membrane is blocked to prevent non-specific binding. Specific detection varies widely, at which direct and indirect detection can be distinguished. The rarely used direct detection method is performed with an enzyme- or fluorophore-conjugated primary antibody to detect the antigen. The more favored technique, the indirect detection, uses an unlabeled primary antibody that first binds to the antigen and subsequently is detected with an enzyme- or fluorophore-conjugated secondary antibody. In cases where the secondary antibody is coupled with an enzyme, the detection is performed with an appropriate substrate, which produces a detectable signal. Chromogenic substrates produce precipitates on the membrane resulting in colorimetric changes. The most sensitive detection is the chemiluminescent technique that produces light as a by-product of the reaction with the enzyme conjugated to the antibody. The light output is either measured by film techniques or digital imaging instruments based on charge-coupled device (CCD) cameras. Alternatively, fluorescence-tagged antibodies are used, where finally the fluorescent signal is measured. This technique is of growing popularity as it affords the multiplex detection of multiple proteins on a single blot [35]. Whatever system is used, the intensity of the signal should correlate with the abundance of the antigen on the membrane [36]. Advantages und limitations are summarized in Table 6.

Besides the determination of the integrity of proteins, also the evaluation of posttranslational modifications can be performed. In the following chapter, glycosylation and phosphorylation are exemplified.

3.1.1 First Example: Detection of Glycan Motives

The vast variety and the large extent of potential glycosylation motives complicate full and detailed analysis of protein glycosylation. The complexity is seen as even a single protein can undergo a number of N- and O-glycosylation, but additionally the same glycosylation site can be occupied by different glycans in different copies of a protein, a phenomenon termed microheterogeneity. Therefore, reliable, detailed characterization of glycans

Table 6
Advantages and limitations of Western blot analysis and the suggested detection systems for protein quality

Direct detection on membranes	
Advantages	*Perceived limitations*
Requires only one antibody	Label may interfere with target binding
Eliminates problems with secondary antibody cross-reactivity	Potential for high background if antibody specificity for target is weak
	Conjugated primary antibodies may be costly
Indirect detection on membranes	
Advantages	*Perceived limitations*
Signal amplification by secondary antibody Vast selection of conjugated secondary antibodies One secondary antibody may be used for a number of different primary antibodies Use of secondary antibody does not inhibit primary antibody target binding Use of labeled secondary antibodies provides options for multiple detection methods	Non-specific staining may increase background Additional steps are required when using indirect method—more time-consuming
Western blot	
Advantages	*Perceived limitations*
Powerful tool for protein detection	Imprecise: Loading of samples, differences in transfer of protein to the membrane, antibody lots, etc. Accuracy: Time for all incubations including equilibration of gel in transfer solution, blocking, antibody, and substrate incubations
Qualitative and semiquantitative Quantitative only under certain conditions Digital capture and image analysis for sensitive analysis	Standardization: Buffer quality can affect transfer of proteins Wet transfer provides a higher transfer quality than dry blot systems Reproducibility: Normalization factors out or in consistency with protein concentration resulting from loading errors or sample-to-sample variation

often requires the use of several methods. In yeast, both N- and O-glycan structures can be synthesized. However, yeast expression systems have mainly mannose-containing glycans, which may significantly differ to other cell types. Due to the fact that separation of proteins by SDS-PAGE is performed by the molecular weight of the protein, each band contains a high variety of glycan structures. Nevertheless, if only basic information, whether a protein is glycosylated or not, is aimed, a simple staining method such as the periodic acid-Schiff (PAS) reaction is applied. Periodic acid oxidizes two vicinal diol groups to form an aldehyde that reacts with the Schiff reagent to give a magenta color [37, 38]. This chromogenic gel staining is most commonly performed with acid fuchsin, which can be detected fluorescently at 535 nm and is two- up to fourfold more sensitive than visible staining methods [39]. Some alternatively available fluorescent stains utilize periodate oxidation to attach a fluorescent hydrazide; thus no further reduction is required. These fluorescent dyes are highly sensitive, detecting <18 ng of glycoproteins [40]. However, these periodate oxidation methods may be problematic, since non-specific reactions can occur with proteins that possess endogenous oxidizable moieties, for example, aldehydes or ketones. In addition, the staining intensity is lower for proteins with fewer glycosylation sites and "poorer" glycan trees than for heavily glycosylated proteins. Two other gel stains that should be mentioned are alcian blue and Stains-all [40, 41], which are used for detecting proteoglycans, glycosaminoglycans, and negatively charged glycoproteins. However, also these stains tend to non-specific reactions with phosphoproteins and other negatively charged proteins.

More specific detection is effectuated by affinity-based procedures either to detect saccharides or specific glycan moieties. A typical workflow includes (1) resolving the protein sample on SDS-PAGE; (2) transfer onto a nitrocellulose or polyvinylidene fluoride (PVDF) membrane; (3) blocking; (4) incubation with a specific lectin, labeled with a group such as digoxigenin (DIG) or biotin; (5) attachment of a secondary antibody or of avidin/streptavidin; and (6) the antibody or avidin conjugated to an enzyme catalyzes a color-producing reaction (alkaline phosphatase) or a more sensitive luminescence-producing reaction (horseradish peroxidase) [42, 43]. The main advantage is seen that various lectins found in animals, microorganisms, and plants are on the market which bind more or less specific mono- or oligosaccharides. Alternatively, specific antibodies, e.g., against O-β-GlcNAc, are applied to specifically detect this unique glycan moiety. Conventional secondary antibodies enforce the detection by color or luminescence reaction [44].

Click chemistry is the newer approach that utilizes specific enzymes to label and isolate glycosylated proteins. For example, a protein that is O-β-GlcNAcylated or carries O-α- and/or N-glycans with a terminal GlcNAc (either purified or in a protein mixture) is incubated with GALT (β-1,4-galactosyltransferase), which specifically adds an azidogalactose to GlcNAc. The modified O-GlcNAc is then allowed to react with a fluorescent alkyne. This sample is resolved on SDS-PAGE and the modified protein visualized by UV, using a gel imager. Similarly, OGlcNAcylated peptides can be labeled and further analyzed by mass spectrometry (MS).

3.1.2 Second Example: Phosphorylation

Approximately 30% of the proteins in eukaryotic cells are phosphorylated. It is therefore obvious that for certain proteins, where phosphorylation is assumed, this analysis should be established. Western blot is performed in the same procedure as for other applications using antibody detection mode. Most phospho-specific antibodies are quite sensitive and can readily detect the phosphorylated protein in a routine sample (e.g., 10–30 μg whole-cell extract or the pure supernatant is used). Phospho-specific antibodies (affinity-purified polyclonal or monoclonal antibodies) are monospecific for a target protein that is phosphorylated on a specific tyrosine, threonine, or serine residue. Meanwhile, more than 300 phosphorylation site-specific antibodies are commercially available [45]. Because the measured levels of a phosphoprotein may change with treatment or through gel loading errors, researchers often utilize an antibody that detects the total level of the cognate protein (regardless of phosphorylation state) to determine the phosphorylated fraction relative to the total fraction and to serve as an internal loading control. Both, chemiluminescent and colorimetric detection, are commonly used.

3.2 Examples

In the first example, a 10% tricine SDS-PAGE is performed, which is appropriate for the separation of peptides and low-molecular-weight proteins. As tricine gels do not contain SDS, they require SDS in the sample and running buffer. Buffers for electrophoresis are in principle commercially available, but to reduce the costs self-made buffers can also be applied. Therefore, the suppliers often provide suitable composition for particular needs. For molecular weight estimation, it is obligatory to apply a molecular weight marker, consisting of a protein mixture with well-defined molecular standards. Additionally, it is advantageous to use a pre-stained marker, preferentially examining the separation quality. In particular, staining methods such as Brilliant Blue R-250 and silver are recommended, which differ in sensitivity. The detection limit is about 100 ng protein for Brilliant Blue R-250 and 1 ng for silver stain. The staining method, described in the second example, is as sensitive as silver stain and therefore is named Coomassie colloidal blue silver [46], although silver is not included in any solutions [47–50].

3.2.1 Product: Insulin precursor (IP) [48]

Host: *Pichia pastoris*.

Aim: Qualitative analysis of IP production.

Method: SDS-PAGE—10%-polyacrylamide gel electrophoresis (tricine buffer system).

Sample matrix: Clarified culture broth.

Equipment: Electrophoresis chamber, suitable power supply.

Sample preparation:

1. Mix 40 μL sample with an equal volume of loading buffer (250 mmol/L tricine, pH 8.5, 10% β-mercaptoethanol, 2% SDS, 1% glycerol, and 0.1% bromophenol blue).
2. Vortex for 1 min and boil for 15 min.
3. Vortex again.
4. Clarify the sample by centrifugation (max. Speed, 15 min, RT).

Method description:

1. Load sample on the gel (20 μL per lane).
2. Prepare running buffer according to the manufacturer's instruction, or use a purchased buffer.
3. Use running conditions according to the applied electrophoresis system.
4. Run the gel until the colored tracking dye (here bromophenol blue) reaches the bottom of the gel.
5. Visualize the separated polypeptides with Brilliant Blue R-250 staining [50].
6. Fix the gel in fixing solution (50% methanol, 10% glacial acetic acid) for 1 h to overnight, and change the solution once.
7. Stain the gel in staining solution (0.1% Coomassie Brilliant Blue R-250, 50% methanol and 10% glacial acetic acid) for 20 min.
8. Destain the gel in destaining solution (40% methanol and 10% glacial acetic acid), and change the solution several times until background is fully destained.
9. Store destained gel in storage solution (5% glacial acetic acid).
10. Perform all steps of the staining procedure with gentle agitation.

Alternatively, polypeptides can be visualized with silver stain [49]:

1. Fix the gel in fixing solution (40% ethanol, 10% glacial acetic acid) at least for 20 min or overnight.
2. Incubate gel in incubation solution (30% ethanol, 6.8% sodium acetate, 0.25% glutaraldehyde, 0.2% $Na_2S_2O_3$ $5H_2O$) for 20 min.
3. Wash gel three times with deionized water approximately 5 min per wash step.
4. Incubate gel in silver solution (0.1% $AgNO_3$, 0.02% formaldehyde) for 30 min.

5. Incubate gel in developing solution (2.5% Na_2CO_3, 0.01% formaldehyde) until the desired band intensity of proteins is achieved.
6. Stop development with stopping solution (0.73% EDTA-Na_2) for 10 min.
7. Wash gel three times with deionized water approximately 5 min per wash step.
8. Put gel in drying solution (30% ethanol, 10% glycerine).

3.2.2 Product: Hepatitis B surface antigen (HBsAg) [47]

Host: *Pichia pastoris* (intracellular production of HBsAg).

Aim: Qualitative analysis of total cell lysates.

Method: SDS-PAGE: 15% SDS-polyacrylamide gel and Western blot.

Sample matrix: Cell lysate.

Equipment: Electrophoresis chamber, suitable power supply.

Sample preparation:

1. Mix 50 μL cell lysates with an equal volume of a modified SDS sample buffer (5% SDS (w/v), 50% β-mercaptoethanol (v/v), 0.5 mol/L DTT, 3% glycerol (v/v)).
2. Vortex for 1 min and boil for 15 min.

Method description:

1. Load 10 μL sample per lane.
2. Prepare running buffer according to the manufacturer's instruction, or use a purchased buffer.
3. Perform electrophoresis on a 15% SDS-polyacrylamide gel at 150 V for 80 min.
4. Visualize polypeptide bands with Coomassie colloidal blue silver stain.
5. Fix gels in fixing solution (30% methanol, 10% acetic acid) for 30 min up to 1 h.
6. Wash gels 4 times with deionized water, 15 min per wash step.
7. Stain gel in staining solution (0.12% Coomassie Blue G-250, 10% ammonium sulfate, 10% phosphoric acid, 20% methanol) for at least 3 h or overnight.
8. Destain gel with deionized water and change water 3–4 times.

Western blot analysis:

1. Pretreat PVDF membranes with methanol, and wash with transfer buffer (Towbin buffer: 25 mM Tris, 192 mM glycine, pH 8.3, 20% methanol (vol/vol), occasionally SDS 0.025–0.1%, depending on the protein).

2. Electro-transfer separated proteins from the gel to PVDF membrane at 100 V for 90 min.
3. Block PVDF membrane in blocking buffer (phosphate buffered saline containing 1% Tween 20 (v/v), 5% skim milk (w/v), and 2% polyvinylpyrrolidone (w/v)) for 1–3 h at room temperature.
4. Briefly pan membrane in wash buffer (PBS-Tween 20).
5. Incubate membrane with purchased rabbit polyclonal anti-HBsAg antiserum, diluted in blocking buffer according to the manufacturer's instruction.
6. Wash 3 times in wash buffer for 5 min.
7. Incubate membrane with anti-mouse IgG-alkaline phosphatase, and conjugate (e.g., diluted in 5% skim milk in PBS-T according to the manufacturer).
8. Wash 2 times in wash buffer for 5 min.
9. Wash for 5 min in staining buffer (0.1 M Tris, 0.1 M NaCl, 5 mM $MgCl_2$ $6H_2O$, pH 9.5).
10. Prepare staining solution: add 12 μL nitro-blue tetrazolium (NBT) stock solution.
11. (70 mg NBT in 1 mL 70% dimethylformamide) and 12 μL 5-bromo-4-chloro-3′-indolyphosphate (BCIP) stock solution (70 mg BCIP in 100% dimethylformamide) to 10 mL staining buffer (*see* **Note 6**).
12. Incubate membrane in staining solution to visualize polypeptide bands.
13. Rinse membrane 2 times with deionized water for 2 min.
14. Dry membrane.

4 Notes

1. The slope of the calibration curve measures the selectivity of the stationary phase. As the pore size distribution of the particle narrows, the slope becomes shallower, which results in a greater selectivity to discriminate between analytes that are similar in size.
2. "Non-well-behaved" SEC-HPLC systems can lead to increased broadening as compared to expected values. Typically, the HPLC systems are optimized to minimize the impact on dispersion utilizing tubing with low inner diameters and low lengths. Adjustment of the column inner diameter is recommended, which is typically larger than 7.5 mm. In several studies, it became obvious that negative impact on resolution and accuracy is in particular caused by extra system volume

from injectors, guard columns, detectors, and connectors. Since the commercialization of submicron columns, also SEC columns with sub-2 μm particles are available which take advantage of the improved low-dispersion instrumentation.

3. To minimize run time by eliminating variation of selectivity, consider gradient shape to further improve the separation and adjustment of column conditions. When band spacing and selectivity are optimized, varying run time and/or column length should be verified to improve resolution and/or analysis speed.
4. The device can be coupled to any conventional HPLC system. It can be operated in different modes. In this example, the effluent is split post column with a portion directed to a conventional LC/MS interface, while the remainder of the split effluent is directed to collect at time segments into a multi-well plate (homepage: https://advion.com/products/triversa-nanomate) [51].
5. For accurate quantification, the minimal prerequisite is a high quality standard. The best choice is a certified reference material. Furthermore, unbiased data should be produced, which are not affected by impurities or degradation products. Additionally, quantification implies internal loading controls which provide the verification that samples are uniformly loaded across the gel, confirm consistent transfer from gel to membrane, and enable comparison of relative protein levels among samples. In this process, called normalization, variations, such as sample-to-sample variability, can be mathematically compensated and correct this variation. However, in this process, important attributes need to be considered to obtain reliable data. Numerous literature examples are available, which address specific questions to compensate the limitations of this technology. Due to the circumstance that accurate normalization is becoming increasingly important, new standards such as multi-protein normalization but also application of bioinformatics strengthen the quality of Western blots as a sufficient method for product and process development.
6. NBT and BCIP substrates are also available as ready-made solutions which facilitate the workflow, since staining solution has always to be freshly prepared.

References

1. Lin-Cereghino GP, Lin-Cereghino J, Ilgen C, Cregg JM (2002) Production of recombinant proteins in fermenter cultures of the yeast *Pichia pastoris*. Curr Opin Biotechnol 13:329–332. https://doi.org/10.1016/S0958166902003300
2. Ferrer-Miralles N, Domingo-Espín J, Corchero J et al (2009) Microbial factories for recombinant pharmaceuticals. Microb Cell Factories 8:17. https://doi.org/10.1186/1475-2859-8-17

3. Martinez JLL, Liu L, Petranovic D, Nielsen J (2012) Pharmaceutical protein production by yeast: towards production of human blood proteins by microbial fermentation. Curr Opin Biotechnol 23(6):965–971. https://doi.org/10.1016/j.copbio.2012.03.011
4. Hong P, Koza S, Bouvier ESP (2012) A review size-exclusion chromatography for the analysis of protein biotherapeutics and their aggregates. J Liq Chromatogr Relat Technol 35:2923–2950. https://doi.org/10.1080/10826076.2012.743724
5. Synge RL (1950) Fractionation of hydrolysis products of amylose by electrokinetic ultrafiltration in an agaragar jelly. Biochem J 24:41–42
6. Lindqvist B, Storgards T (1975) Molecular-sieving properties of starch. Nature 175:511–512
7. Coutinho FMB, Rcia M, Lica A et al (1997) Porous structure and swelling properties of styrene– divinylbenzene copolymers for size exclusion chromatography. J Appl Polym Sci 65:1257–1262. https://doi.org/10.1002/(SICI)1097-4628(19970815)65
8. Li Y, Tolley HD, Lee ML (2010) Size-exclusion separation of proteins using a biocompatible polymeric monolithic capillary column with mesoporosity. J Chromatogr A 1217:8181–8185. https://doi.org/10.1016/j.chroma.2010.10.067
9. Striegel AM (2016) Viscometric detection in size-exclusion chromatography: principles and select applications. Chromatographia 79:945–960. https://doi.org/10.1007/s10337-016-3078-0
10. Manning MC, Manning RR, Holcomb RE et al (2014) Review of orthogonal methods to SEC for quantitation and characterization of protein aggregates. BioPharm Int 27:1–2
11. Folta-Stogniew E (2006) Oligomeric states of proteins determined by size-exclusion chromatography coupled with light scattering, absorbance, and refractive index detectors. In: New and Emerging Proteomic Techniques. Humana Press, Totowa, NJ, pp 97–112. https://doi.org/10.1385/1-59745-026-X:97
12. Folta-Stogniew E, Williams KR (1999) Determination of molecular masses of proteins in solution: implementation of an HPLC size exclusion chromatography and laser light scattering service in a core laboratory. J Biomol Tech 10:51–63
13. Tarazona MP, Saiz E (2003) Combination of SEC/MALS experimental procedures and theoretical analysis for studying the solution properties of macromolecules. J Biochem Biophys Methods 56:95–116. https://doi.org/10.1016/S0165-022X(03)00075-7
14. Ambrogelly A, Liu YH, Li H et al (2012) Characterization of antibody variants during process development: the tale of incomplete processing of N-terminal secretion peptide. MAbs 4:701–709. https://doi.org/10.4161/mabs.21614
15. Bouvier ESP, Koza SM (2014) Advances in size-exclusion separations of proteins and polymers by UHPLC. Trends Anal Chem 63:85–94. https://doi.org/10.1016/j.trac.2014.08.002
16. Gulati D, Bongers J, Burman S (1999) RP-HPLC tryptic mapping of IgG1 proteins with post-column fluorescence derivatization. J Pharm Biomed Anal 21:887–893. https://doi.org/10.1016/S0731-7085(99)00119-3
17. Lee GF, Anderson DC (1991) Reversed-phase high-pressure liquid chromatographic tryptic peptide mapping for the comparison and study of monoclonal antibodies. Bioconjug Chem 2:367–374. https://doi.org/10.1021/bc00011a012
18. Lundell N, Schreitmüller T (1999) Sample preparation for peptide mapping – a pharmaceutical quality-control perspective. Anal Biochem 266:31–47. https://doi.org/10.1006/abio.1998.2919
19. Wang Y, Li H, Shameem M, Xu W (2016) Development of a sample preparation method for monitoring correct disulfide linkages of monoclonal antibodies by liquid chromatography-mass spectrometry. Anal Biochem 495:21–28. https://doi.org/10.1016/j.ab.2015.11.010
20. Hennessy TP, Boysen RI, Huber MI et al (2003) Peptide mapping by reversed-phase high-performance liquid chromatography employing silica rod monoliths. J Chromatogr A 1009:15–28. https://doi.org/10.1016/S0021-9673(03)00445-X
21. Krokhin OV, Craig R, Spicer V et al (2004) An improved model for prediction of retention times of tryptic peptides in ion pair reversed-phase HPLC. Mol Cell Proteomics 3:908–919. https://doi.org/10.1074/mcp.M400031-MCP200
22. Chakraborty AB, Berger SJ (2005) Optimization of reversed-phase peptide liquid chromatography ultraviolet mass spectrometry analyses using an automated blending methodology. J Biomol Tech 16(4):327–335
23. Wagner K, Miliotis T, Marko-Varga G et al (2002) An automated on-line multidimensional HPLC system for protein and peptide mapping with integrated sample preparation.

Anal Chem 74:809–820. https://doi.org/10.1021/ac010627f

24. Gong B, Burnina I, Stadheim TA, Li H (2013) Glycosylation characterization of recombinant human erythropoietin produced in glycoengineered *Pichia pastoris* by mass spectrometry. J Mass Spectrom 48:1308–1317. https://doi.org/10.1002/jms.3291
25. Xie H, Gilar M, Gebler JC (2009) Characterization of Protein Impurities by Peptide Mapping with UPLC/MS^E Application Note 1–6, Waters Application Note, Waters Corporation, http://www.waters.com/webassets/cms/library/docs/720002809en.pdf
26. Leblanc Y, Ramon C, Bihoreau N, Chevreux G (2017) Charge variants characterization of a monoclonal antibody by ion exchange chromatography coupled on-line to native mass spectrometry: case study after a long-term storage at +5 °C. J Chromatogr B 1048:130–139. https://doi.org/10.1016/j.jchromb.2017.02.017
27. Fekete S, Veuthey JL, Guillarme D (2017) Achievable separation performance and analysis time in current liquid chromatographic practice for monoclonal antibody separations. J Pharm Biomed Anal 141:59–69. https://doi.org/10.1016/j.jpba.2017.04.004
28. Ahamed T, Nfor BK, Verhaert PDEM et al (2007) pH-gradient ion-exchange chromatography: an analytical tool for design and optimization of protein separations. J Chromatogr A 1164:181–188. https://doi.org/10.1016/j.chroma.2007.07.010
29. Fekete S, Veuthey JL, Guillarme D (2015) Comparison of the most recent chromatographic approaches applied for fast and high resolution separations: theory and practice. J Chromatogr A 1408:1–14. https://doi.org/10.1016/j.chroma.2015.07.014
30. Rea JC, Moreno GT, Lou Y, Farnan D (2011) Validation of a pH gradient-based ion-exchange chromatography method for high-resolution monoclonal antibody charge variant separations. J Pharm Biomed Anal 54:317–323. https://doi.org/10.1016/j.jpba.2010.08.030
31. Ghosh R, Gilda JE, Gomes AV (2014) The necessity of and strategies for improving confidence in the accuracy of western blots. Expert Rev Proteomics 11:549–560. https://doi.org/10.1586/14789450.2014.939635
32. Towbin H, Staehelin T, Gordon J (1979) Electrophoretic transfer of proteins from polyacrylamide gels to nitrocellulose sheets: procedure and some applications. Proc Natl Acad Sci 76:4350–4354. https://doi.org/10.1073/pnas.76.9.4350
33. Blancher C, Jones A (2001) SDS-PAGE and Western blotting techniques. In: Metastasis research protocols. Humana Press, Totowa, NJ, pp 145–162. https://doi.org/10.1385/1-59259-136-1:145
34. Kurien BT, Scofield RH (2009) Introduction to protein blotting. Methods Mol Biol 536:9–22. https://doi.org/10.1007/978-1-59745-542-8_3
35. Mishra M, Tiwari S, Gomes AV (2017) Protein purification and analysis: next generation Western blotting techniques. Expert Rev Proteomics 14:1037–1053. https://doi.org/10.1080/14789450.2017.1388167
36. Bakalova R, Zhelev Z, Ohba H, Baba Y (2005) Quantum dot-based western blot technology for ultrasensitive detection of tracer proteins. J Am Chem Soc 127:9328–9329. https://doi.org/10.1021/ja0510055
37. Devine PL, Warren JA (1990) Glycoprotein detection on immobilon PVDF transfer membrane using the periodic acid/Schiff reagent. BioTechniques 8:492–495
38. Thornton DJ, Carlstedt I, Sheehan JK (1996) Identification of glycoproteins on nitrocellulose membranes and gels. Mol Biotechnol 5:171–176. https://doi.org/10.1007/BF02789065
39. Packer NH, Ball MS, Devine PL, Patton WF (2002) Detection of glycoproteins in gels and blots. In: Protein protocols handbook. Humana Press, Totowa, NJ, pp 761–772
40. Roth Z, Yehezkel G, Khalaila I (2012) Identification and quantification of protein glycosylation. Int J Carbohydr Chem 2012:1–10. https://doi.org/10.1155/2012/640923
41. Gauci VJ, Wright EP, Coorssen JR (2011) Quantitative proteomics: assessing the spectrum of in-gel protein detection methods. J Chem Biol 4:3–29. https://doi.org/10.1007/s12154-010-0043-5
42. Sato T (2014) Lectin-probed Western blot analysis. Methods Mol Biol 1200:93–100. https://doi.org/10.1007/978-1-4939-1292-6_8
43. Badr HA, AlSadek DMM, Mathew MP et al (2015) Lectin staining and Western blot data showing differential sialylation of nutrient-deprived cancer cells to sialic acid supplementation. Data Brief 5:481–488. https://doi.org/10.1016/j.dib.2015.09.043
44. Zachara NE, Vosseller K, Hart GW (2011) Detection and analysis of proteins modified by O-linked *N* -acetylglucosamine. Curr Protoc Mol Biol Chapter 17:Unit 17.6. https://doi.org/10.1002/0471142727.mb1706s95

45. Mandell JW (2003) Phosphorylation state-specific antibodies: applications in investigative and diagnostic pathology. Am J Pathol 163:1687–1698
46. Candiano G, Bruschi M, Musante L et al (2004) Blue silver: a very sensitive colloidal Coomassie G-250 staining for proteome analysis. Electrophoresis 25:1327–1333. https://doi.org/10.1002/elps.200305844
47. Gurramkonda C, Adnan A, Gäbel T et al (2009) Simple high-cell density fed-batch technique for high-level recombinant protein production with *Pichia pastoris*: application to intracellular production of Hepatitis B surface antigen. Microb Cell Factories 8:13:1–13:8. https://doi.org/10.1186/1475-2859-8-13
48. Gurramkonda C, Polez S, Skoko N et al (2010) Application of simple fed-batch technique to high-level secretory production of insulin precursor using *Pichia pastoris* with subsequent purification and conversion to human insulin. Microb Cell Factories 9(3):1–11. https://doi.org/10.1186/1475-2859-9-31
49. Morrissey JH (1981) Silver stain for proteins in polyacrylamide gels: a modified procedure with enhanced uniform sensitivity. Anal Biochem 117:307–310
50. Neuhoff V, Stamm R, Eibl H (1985) Clear background and highly sensitive protein staining with Coomassie Blue dyes in polyacrylamide gels: a systematic analysis. Electrophoresis 6:427–448. https://doi.org/10.1002/elps.1150060905
51. Advion (2017) https://advion.com/. Accessed 14 Dec 2017

Chapter 20

Quantification of Recombinant Products in Yeast

Karola Vorauer-Uhl and Gabriele Lhota

Abstract

Quantification of various proteins expressed in yeast can be performed by different methods. In this respect, classical as well as advanced techniques can be applied, where the analysis of crude supernatants is of special interest in screening but also manufacturing.

The following chapter addresses the analytical background of the introduced methods followed by specific recommendations for the quantification of different products of industrial interest. The method portfolio includes electrophoresis, chromatography, and ELISA as classical techniques, but also biosensor-based, microfluidic and automated, miniaturized methods are introduced. Furthermore, individual strengths and perceived limitations are summarized.

Although prominent examples are described, it should be noted that individual modifications are required according to host and cultivation mode.

Key words Protein quantification, Electrophoresis, HPLC, ELISA, Bio-layer interferometry, Microfluidics CD technology, Surface plasmon resonance, Electrochemiluminescence assays, Microchip-based assays

1 Introduction

Biotechnology has traditionally employed pro- and eukaryotic organisms as production systems. In microbial production, predominantly yeasts and bacteria share beneficial features although some limitations for the production of human proteins are evident. The increasing attractiveness of yeasts for both basic research and industrial manufacture requires efficient production strategies and full characterization of the various products based on established, but even more in future, innovative new analytical perspectives [1–4].

Yeast platforms for recombinant protein production have several explicit advantages:

1. Rapid growth rate, coupled with ease of high cell density fermentation.
2. High levels of productivity in protein-free medium.

Brigitte Gasser and Diethard Mattanovich (eds.), *Recombinant Protein Production in Yeast*, Methods in Molecular Biology, vol. 1923, https://doi.org/10.1007/978-1-4939-9024-5_20, © Springer Science+Business Media, LLC, part of Springer Nature 2019

3. Absence of endotoxin and bacteriophage contamination.
4. Ease of genetic manipulation of well-characterized yeast expression vectors.
5. Absence of known human pathogenicity in the spectrum of lytic viruses.
6. Diverse posttranslational modifications that include polypeptide folding, glycosylation, methylation, acylation, proteolytic adjustment, and targeting to subcellular compartments.
7. The ability to produce secreted proteins that can be purified from growth medium without harvesting the yeast cells themselves [5].

In screening, cultivation and manufacturing various analytical methods for qualitative as well as quantitative analysis are needed. In respect to different scenarios such as (1) small and variable amounts of secreted protein of interest, (2) small volumes in early stages, (3) variable media composition, and (4) variable matrices caused by impurities, sufficient analytical methods need to be implemented for clone screening, media optimization, and process control purposes. Thereby, optimized strain selection and process-related product quantification but also verification of product quality are enabled. To fulfill these high standards, analytical methodologies need:

1. To be carefully assessed and developed according to the intended purpose.
2. To be reliable under different conditions.
3. To be selective and/or sensitive.
4. To be optimized for a broad dynamic working range.
5. To be reproducible and robust.
6. To be of high throughput and automatization.
7. To have data formats transferable into databases.
8. To have an efficient time and effort ratio to obtain information.

In the following chapter, classical, but also advanced, analytical methods, the theoretical background of each technique, applications, and selected examples is introduced. As it can be seen, even for one particular protein, different methods are possible, and the setup often depends on the equipment and established technologies already available in a certain laboratory or institution. Therefore, no overall recommendations can be given.

2 Classical Quantification Methods

2.1 Electrophoresis

2.1.1 Background

Protein gel electrophoresis is a well-established technique based on the movement of charged molecules in response to an electric field. The migration rate (units/cm^2/Vs) defines the rate of movement of the individual molecules. Meanwhile, numerous techniques are applicable for crude culture supernatants of different origins. The separation is based on the complex relationship between the physical characteristics of both, the electrophoresis system and the protein. The individual performance depends on the strength of the electric field, the temperature of the system, the pH, ion type, and concentration of the buffer as well as size, shape, and charge of the proteins [6, 7]. A typical workflow includes method selection, sample preparation, gel and buffer preparation, separation, and protein detection combined with data evaluation.

In research, but also manufacture, polyacrylamide gel electrophoresis (PAGE) is commonly applied and optimized for the intended application. Polyacrylamide gels are stable, chemically inert, electrically neutral, hydrophilic, and transparent for optical detection at wavelengths above 250 nm [6]. No particular gel type and/or buffer is suitable for all proteins. In principle, native and denatured PAGE can be distinguished, of which the latter one is favored [8]. In this case denaturing buffer systems [9] come into place. The anionic detergent sodium dodecyl sulfate (SDS) is non-covalently bound to proteins in a stoichiometric ratio of 1.4 g SDS per 1 g protein, leading to an overall negative charge of the protein. The caused rod shape of the protein enforces the migration primarily by the size enabling the molecular weight estimation. In case of the analysis of subunits, which are bridged by disulfide bonds, reductive conditions can be applied. Thiol-reducing agents which disrupt intra- and intermolecular disulfide bonds, such as 2-mercaptoethanol (ß-ME) or dithiothreitol (DTT), are predominately used. Disruption of hydrophobic interactions between and within proteins can be performed either with nonionic (NP-40, Triton X-100) or zwitterionic (CHAPS, sulfobetaines) detergents [10, 11]. In addition, chaotropic agents that disrupt hydrogen bonds as well as hydrophobic interactions, both between and within proteins, such as urea or thiourea are in use (*see* **Note 1**). The optimization of qualitative and quantitative protein electrophoresis that offers the highest resolution in the size range of interest may require some experimentation prior to use in routine analysis. Therefore, selected protein standards in the intended size range are suggested as well as protein content analysis and dilution series of unknown samples. In completion, Western blot analysis is recommended to test the protein integrity in a qualitative manner.

Depending on the aim and the available imaging equipment, different staining techniques are offered [12, 13]. Samples can be

Table 1
List of the most relevant protein staining dyes used in electrophoresis

Staining dye	Attributes/characteristics
Coomassie (Brilliant or R-250) blue stain	Most popular, anionic dye; for almost all proteins, good quantitative linearity, medium sensitive, favorable for mass spectroscopy for additional identification, adequate price
Fluorescent stain	High sensitivity, wide linear range (four orders of magnitude), simple and robust, but more expensive
Silver stain	Highest sensitivity, but low dynamic range, time-consuming, less reproducible, not sufficient reproducibility for quantification
Negative stain	Very rapid; proteins appear as clear areas; zinc and copper stains do not require fixation; less applicable for quantification
Specific protein stains	Specific for different protein classes, such as phosphoproteins
	Qualitative more than quantitative, preferable for identification

stained prior and/or after the separation, for qualitative and/or quantitative analysis (*see* **Note 2**). This enables more or less sensitive, specific, and unspecific labeling (Table 1). Data analysis is performed with different imaging devices producing electropherograms and may employ sophisticated software for data processing. State-of-the-art imaging systems are, among others, flatbed densitometers, charge-coupled device (CCD) cameras, or laser-based scanner systems.

Electrophoresis is commonly applied in basic research, but due to further developments and automatization combined with advanced methods, it is still of great importance in the manufacture of recombinant proteins. Due to the advantages but also the known limitations (Table 2) regarding high throughput, standardization, analysis, and data management, improved technical innovations have been established to fulfill the needs especially in R&D and manufacturing. Meanwhile, automated devices of several suppliers are commercially available. Most of them use microfluidic devices to overcome the electrophoretic limitation [14]. All of them are intended to analyze small volumes of samples, and analysis of several samples in parallel significantly reduces processing time. Furthermore, they apply more sensitive staining dyes for quantification and provide software solutions for analysis and data management. Detection is preferentially performed with fluorescence dyes enabling sensitive quantification of proteins with an appropriate dynamic working range. This approach implies efficient separation of the target protein and ideally requires the availability of a defined standard for the quantitative calibration.

Table 2
Summary of performance considerations in electrophoresis

Advantages	Perceived limitations
Crude supernatant	Non-specific
Various gel formats available	Nonselective
Different staining opportunities	Protein disintegration possible
Small sample volume	Interference with other proteins—purity
Qualitative/quantitative	Interference with salts, detergents, organic solvents, sulfones, etc.
	High viscosity (DNA, carbohydrates)
	Limited sample throughput Time-consuming
	Standardization Operator-dependent analysis Data management and documentation

2.1.2 Example

In this example a standardized workflow for SDS-PAGE and Western blot analysis, performed with the Amersham WB analyzer, is described [15]. The applied analyzer consists of two instrument units, one for electrophoresis and scanning to separate SDS-coupled proteins in a polyacrylamide gel and one for the full Western blot workflow including transfer, probing, and drying. The system is based on fluorescence detection and can be used to scan gels after electrophoresis as well as polyvinylidene (PVDF) cards after probing and drying in a Western blot experiment. For pre-labeling the fluorescent dye, Cy5 is used, which binds via its NHS ester to the lysine residues and the N-terminus of the proteins. There is no need of post-staining, and the obtained signal represents the total protein content. Depending on the protein, a limit of detection in the pg range is attainable, and calibration can be usually done in the ng range. Additionally, the target protein can be identified via a secondary antibody labeled with Cy3. Standardized protocols help to deliver consistent and quantifiable data. The use of Western blot analysis to detect a specific protein is highly recommended and confers an advantage over SDS-PAGE alone because the identity of the target protein can be determined with a higher degree of confidence. In this context it is advisable to run samples under reduced (DTT in the loading buffer) and non-reduced conditions. For example, by adding DTT, heavy and light chain of an antibody can be identified, after cleaving of the intramolecular disulfide bonds. However, prior to starting the analysis, it is recommended to perform a Bradford protein assay to estimate the total protein concentration in the supernatant. Several suppliers offer this simple colorimetric assay as a ready-to-use test kit, which has the advantage of getting fast results, which enables the selection of an adequate sample pre-dilution within the calibration range [15–17].

Product: Monoclonal antibody (mAb) [16, 17].

Host: Not specified.

Aim: Standardized quantification of antibody.

Method: SDS-Page, Western blot.

Sample matrix: Culture supernatant.

Equipment: Amersham Western blotting (WB) system.

Sample preparation:

1. Dilute each sample from culture media accordingly with 0.15 M NaCl.
2. Pre-labeling with Cy5 is performed according to an Amersham standard protocol: mix 17 μL labeling buffer (Tris-based buffer containing SDS, pH 8.7), 2 μL sample, and 1 μL Amersham WB Cy5 dye, and let the labeling reaction take place for 30 min at room temperature.
3. For reduced samples, add 20 μL Amersham WB loading buffer (Tris-Cl, SDS loading buffer) with 40 mM DTT to the sample, and heat 3–5 min at 95 °C.
4. For non-reduced samples, use Amersham WB loading buffer directly without DTT.
5. Dilute mAb standard (7.8–500 ng), and pre-label as described above.

Method Description:

Electrophoresis:

1. Load an Amersham WB molecular weight marker (10–225 kDa, labeled with both Cy3 and Cy5), samples, and a dilution series of the mAb standard (20 μL each) onto an Amersham WB gel card 8–18% (gel card for resolution of proteins ranging between 3.5 and 225 kDa).
2. Default electrophoresis settings of 600 V, and a maximum current of 50 mA is recommended.
3. Scan the gel after completed electrophoresis at a sensitivity setting of 5, and evaluate the data with Amersham WB software.

Western Blot Analysis:

1. Transfer samples (after electrophoresis) to a PVDF membrane using the Amersham WB tank transfer protocol: perform the transfer in a buffer with 192 mM glycine, 25 mM Tris, and 20% ethanol with default settings of 100 V for 30 min.
2. Use the Amersham WB default protocols to perform the blocking and probing procedures: first block the membranes in 3% BSA for 10 min.

3. Target the antibody by rabbit antihuman κ antibody (diluted 1:20,000 in PBS-T).
4. Use Amersham WB Cy3-labeled anti-rabbit IgG as secondary antibody for detection (diluted 1:2500 in PBS-T).
5. The probing volumes are 10 μL, and probing times are 60 and 30 min for primary and secondary antibodies, respectively.
6. Dry membranes after probing, for 10 min in an Amersham WB dryer.
7. Scan membranes in an Amersham WB scanner in Cy3 (Ex 550 nm/Em 570 nm) and Cy5 (Ex 650 nm/Em 670 nm) channels at a sensitivity setting of 2, and evaluate data with the Amersham WB software.

2.2 ELISA: Enzyme-Linked Immunosorbent Assay

2.2.1 Background

In 1971, the enzyme-linked immunosorbent assay (ELISA) was first described by Engvall and Perlmann [18, 19]. This technique is applicable to detect a specific target protein or antigen in crude culture supernatants or even cell extracts and has meanwhile replaced the formerly used radioimmunoassay, which has potential health risks. The development of antigen-specific monoclonal antibodies [20] accelerated the acceptance and applicability of this technology. Furthermore, the possibility to chemically link enzymes to antibodies, which enables a measurable signal in solutions containing appropriate substrates, strengthens the ELISA platform. With the development of fluorescence technology, signal generation using fluorophore-labeled antibodies has also become prevalent, especially in multiplex arrays.

2.2.2 Procedure

Although many variants of ELISA have been developed and are used in different configurations, they all include the following elements:

1. Coating/capture: direct or indirect immobilization of antigens to the surface of polystyrene microplate wells.
2. Plate blocking by addition of irrelevant protein (such as bovine serum albumin or casein) or other molecules (the most common nonionic detergent blocker is Tween 20; normal concentrations are about 0.01–0.1%) to cover all unsaturated surface-binding sites of the microplate wells.
3. Probing/detection by incubation with antigen-specific antibodies that selectively bind to the antigens.
4. Detection of the signal that is generated by a direct or secondary tag on the specific antibody.

Washing steps ensure that only specific (high-affinity) binding events are maintained to cause a signal at the final step (*see* **Note 3**) [21].

Coating of an antigen which represents the target protein in product quantification can be performed either traditionally (direct coating) or by capture (indirect coating) in recommended buffer (*see* **Note 4**). If the antigen is present at very low concentrations or does not adhere well to the plastic, alternative sandwich or capture ELISA is advantageous (*see* **Note 5**). Sandwich ELISAs are very popular for complex protein samples (*see* **Note 6**). For this approach, two different antibodies are required; the first antibody (bound to the plate) is called the capture antibody or coating antibody, whereas the second antibody detects the immobilized antigen (detection antibody). Combinations of monoclonal (mainly as the coating antibody) and polyclonal antibodies (detection antibody) can be used (*see* **Note 7**). Blocking buffers usually consist of formulations to prevent non-specific binding of proteins to the plate [22]. An optimal blocking buffer maximizes the signal-to-noise ratio but does not react with the antibodies or target protein (*see* **Note 8**). Some assays may benefit from the addition of a surfactant such as Tween 20 to the blocking solution. Surfactants might minimize hydrophobic interactions between the blocking protein(s) and the antigen or the antibodies. Typically, a final concentration of 0.05% (v/v) Tween 20 is used (*see* **Note 9**) [23, 24]. Both direct and indirect detection can be performed, with indirect detection often being preferred because in most of the assays, this technique is more sensitive. Nevertheless, direct detection is generally faster than indirect detection, and potential background signal from secondary antibody cross-reactivity with the coating antibody is also reduced (*see* **Note 10**).

The increasing number of stable fluorophores in the visible and infrared ranges has made fluorescent signal detection very popular. For this approach, either the detection antibody or the secondary antibody is labeled. When a multiplexing technique is applied, detection antibodies from different species are used in order to distinguish the separate signals [25]. The detection limit is typically approx. 100 pg/well.

2.2.3 Detection

For enzymatic detection two enzymes are preferably used. Alkaline phosphatase (AP) is a large enzyme used in a minority of assays. Its size (140 kDa) limits the amount of signal that can be generated. It is rather difficult to conjugate more than one or two molecules of the enzyme to each molecule of an antibody. AP is less sensitive and is also prone to stability issues unless stored and handled correctly. Alternatively, horseradish peroxidase (HRP) is a more commonly used enzyme. The small size of HRP (40 kDa) allows more molecules to be coupled to antibodies, and this can boost signal generation. Enzymatic signal generation requires the catalysis of a substrate to produce a colored or fluorescent compound or chemiluminescence (visible light) [26]. Colorimetric substrates form a soluble, colored product that accumulates over time relative to the amount of enzyme present in each well (*see* **Note 11**). When the desired color intensity is reached, the product absorbance is either measured directly or in some cases a stop solution is added. Colorimetric substrates for horseradish peroxidase are TMB (3,3′,5,5′-tetramethylbenzidine), OPD (o-phenylenediamine dihydrochloride), and ABTS (2,2′-azino-bis (3-ethylbenzothiazoline-6-sulphonic acid)), while for alkaline phosphatase pNPP (para-nitrophenylphosphate) is commonly used.

For small concentrations or sometimes for higher diluted samples, signal amplification is required to improve the signal-to-noise ratio. A commonly used technique is based on the biotin-streptavidin binding [27]. Biotinylation (biotin labeling) typically results in multiple biotin tags per detection antibody molecule, thus allowing more than one streptavidin molecule to bind to each moiety. Additionally, enzyme-conjugated streptavidin, appropriately diluted in wash buffer, is added. Finally, appropriate substrate solution is added to each well. After incubation at room temperature, color change is attained. The combined effect of this multiple labeling results in higher number of enzymes in the final, very stable immune complex. This increases the catalysis of the substrate and gives stronger signals. The used microplates are typically made of polystyrene, but also other materials, such as polypropylene, polycarbonate, and nylon, are available (*see* **Note 12**). Many plates are now gamma-irradiated to generate a positive charge, which facilitates enhanced coating.

Meanwhile, several pre-activated plates exist for specific, oriented binding of the coating antibody [28]. These plates are pre-coated with protein A or protein G. However, they are not recommended for sandwich ELISAs because of potential cross-reactivity with detection antibodies and/or secondary antibodies. For biotinylated samples or biotinylated coating antibodies, plates that are pre-coated with streptavidin are ideal. Some assays require direct immobilization of a histidine-tagged protein, in which nickel- or copper-coated plates are suitable (*see* **Note 13**). For group-specific attachment of molecules via amines or thiols to form a covalent bond, maleic anhydride- or maleimide-activated plates are provided, which enables direct attachment of peptide antigens that do not coat well by passive adsorption due to their small size.

2.2.4 Data Evaluation

If a quantitative result is needed, the simplest way to proceed is to average the triplicate, in minor applications the duplicate, of the standards readings and deduct the reading of the blank control sample. Next, plot the standard curve to find the line of best fit. A variation is to plot the data using semilog, log/log, log/logit, and its derivatives—the 4 or 5 parameter logistic models. Any dilutions made need to be adjusted for at this stage. Typically, the working

Table 3
Factors that impact ELISA signal generation

Factor	Variable characteristic
Assay plate	Material, well shape, pre-activation
Coupling buffer	Composition, pH
Capture antibody	Specificity, concentration, affinity, incubation time, and temperature
Blocking buffer	Composition, concentration, cross-reactivity
Target antigen	Conformation, stability, available epitope(s), matrix effects
Detection antibody	Specificity, concentration, affinity, incubation time and temperature, cross-reactivity
Enzyme conjugate	Type of enzyme, type of conjugate, activity, concentration, cross-reactivity
Wash step	Buffer composition, volume, duration, frequency
Substrate	Sensitivity, lot-lot consistency, age
Signal detection	Filters, imaging instrument, exposure time

range should be 0.1–1000 ng/mL. ELISAs belong to the most sensitive immunoassays available. The typical detection range is 0.1–1 fmole or 0.01–0.1 ng. However, the test sensitivity depends upon the particular characteristics of the antibody-antigen interaction.

By these examples the high variability and great number of applications become evident. Factors which significantly affect the assembly of the immune complexes are listed in Table 3, while the advantages and known limitations of the ELISA technique are summarized in Table 4.

2.2.5 Examples

In the first described sandwich ELISA, detection antibody is conjugated with horseradish peroxidase (HRP). The added OPD is a chromogenic substrate, which reacts with horseradish peroxidase conjugates, resulting in an orange soluble end product, and can be read spectrophotometrically. Development of the color is dependent on the amount of present HRP. Further dilution of the sample may be required, if color develops too quickly. This reaction is terminated by addition of sulfuric acid altering the pH of the system. Upon stopping, OPD changes in color from yellow to orange, and OD readings of OPD substrates increase by more than twice compared to the corresponding nonstop values. Alternatively, peroxidase substrates such as ABTS (2,2′-azinobis [3-ethylbenzothiazoline-6-sulfonic acid]-diammonium salt) or TMB (3,3′,5,5′-tetramethylbenzidine) are more frequently used. TMB is meanwhile recommended because it is not carcinogenic and no hazardous stopping solution is required. Furthermore, it is

Table 4
Summary of applicability and limitations of ELISA techniques

Advantages	Perceived limitations
Crude supernatant	Capture failure and unpredictable signal intensities may occur because large immune complexes are formed
Various plate formats available Various antibodies available	Selected antibodies may recognize many or all different isoforms but also only some variants
Different detection opportunities	High background signal and low or absent signal require additional optimization work Matrix effects/interferences
Small sample volume	Standard curves need to be appropriate for different samples Poor standard curve linearity Poor dynamic range
Qualitative/quantitative Automatization Sensitive applications	Nonautomated assays are time-consuming Reproducibility needs to be verified Cost-effective when assaying many samples
Fairly robust	

two times more sensitive than OPD. It produces a blue color measurable at a wavelength of 650 nm. For low-concentration assays, the reaction can be stopped with sulfuric acid [29, 30].

Product: 3D6 single-chain Fv-Fc anti-HIV-1 antibody [29]

Host: Pichia pastoris.

Aim: Quantification of secreted antibody.

Method: Sandwich ELISA.

Sample matrix: Culture supernatant.

Equipment: 96-well immunosorbent plates, microplate reader (492 nm/620 nm).

Method description:

1. Coat 96-well immunosorbent plates with 0.33 μg/mL goat antihuman IgG (γ-chain specific) antibody, diluted in coating buffer (0.1 M Na_2CO_3/$NaHCO_3$, pH 9.6) at 4 °C overnight.
2. After incubation step, wash the plates three times using washing buffer (phosphate-buffered saline, containing 0.1% Tween 20, pH 7.4).
3. Affinity purified 3D6scFv-Fc can be used as a standard at a starting concentration of 100 ng/mL.
4. Dilute standards and samples serially in washing buffer containing 1% bovine serum albumin, and apply onto the pre-coated plates.
5. Allow incubation for 1 h.

6. Incubate captured 3D6scFv-Fc with 0.5 μg/mL horseradish peroxidase-conjugated goat antihuman IgG (γ-chain specific) antibody for 1 h.
7. Use 100 mg/mL o-phenylenediamine diluted in 0.15 M citric acid buffer, pH 5.0 containing 0.02% H_2O_2 for staining.
8. Stop the reaction with 25% H_2SO_4.
9. Measure absorption at 492 nm (620 nm reference wavelength) using a microplate reader.

Product: Human albumin (HSA) [29]

Host: Pichia pastoris.

Aim: Quantification of HSA.

Method: Sandwich ELISA.

Sample matrix: Culture supernatant.

Equipment: Human Albumin ELISA Quantitation Set containing affinity purified human albumin coating antibody (1 mg mL^{-1}), human reference serum (22 mg mL^{-1}), HRP-conjugated human albumin detection antibody (1 mg mL^{-1}), 96-well plates, TMB One Component HRP Microwell Substrate, microplate reader.

Method description:

1. Dilute coating antibody 1:1000 in coating buffer (0.05 M carbonate-bicarbonate, pH 9.6) to a concentration of 1 μg/mL immediately prior to its addition to the wells (per plate: 10 μL antibody per 10 mL buffer).
2. Transfer 100 μL to each well, and incubate plate overnight at 4 °C or 1 h at room temperature on a rotary shaker (*see* **Note 14**).
3. Wash plate with wash buffer (50 mM Tris, 0.14 M NaCl, 0.05% Tween 20, pH 8.0) using a plate washer.
4. Add 200 μL blocking buffer (50 mM Tris, 0.14 M NaCl, 1% BSA, pH 8.0) to each well.
5. Incubate for 30 min at room temperature on a rotary shaker.
6. Wash each well five times with wash buffer.
7. Thaw 400 μL "ready-to-use" HSA standard aliquot (400 ng/mL) (*see* **Note 15**).
8. Dilute samples in dilution buffer (50 mM Tris, 0.14 M NaCl, 1% BSA, 0.05% Tween 20, pH 8.0) within the concentration range of the standards (*see* **Note 16**).
9. Pipet 100 μL dilution buffer from row B to G of the 96-well plate.
10. Transfer 200 μL blank (dilution buffer) standard or sample to assigned wells in row A (line 1 for blank, lines 2 and 3 for standards, lines 4–12 for samples).
11. For the dilution series, take 100 μL/well from row A, and dilute it 1:2 until the last row is reached.
12. Incubate for 1 h at room temperature on a rotary shaker.

13. Wash each well four times.
14. Dilute HRP conjugate 1:30,000 in dilution buffer.
15. Transfer 100 μL to each well.
16. Incubate 60 min at room temperature on a rotary shaker.
17. Wash each well five times.
18. Add 100 μL of TMB One Component HRP Microwell Substrate to each well.
19. Develop the enzymatic color reaction for 10–15 min in the dark at room temperature (the substrate reaction yields a blue solution).
20. Stop TMB reaction by applying 100 μL stop solution (2 M H_2SO_4) to each well, and tap plate gently to mix (color changes from blue to yellow) (*see* **Note 17**).
21. Read plate at 450 nm for TMB substrate using a microplate reader.
22. For data analysis, average the duplicate absorbance values for each standard, control, and sample and subtract the average blank from each value.
23. Create a standard curve using software that generates a four-parameter logistic (4-PL) curve fit, and calculate HSA concentration of samples (*see* **Note 18**).
24. Multiply obtained values by the dilution factor to determine the concentration of undiluted samples.

Product: Hepatitis B surface antigen (HBsAg) [30]

Host: Pichia pastoris.

Aim: Determination of soluble HBsAg.

Method: Sandwich ELISA.

Sample matrix: Cell extract.

Sample preparation:

1. Centrifuge aliquots of induced cells at 10,000 × *g* at 4 °C.
2. Wash with sterile water.
3. Resuspend cells in breaking buffer (25 mmol/L sodium phosphate, 150 mmol/L sodium chloride, 1 mmol/L EDTA, 1 mmol/L DTT) using 6.0 mL breaking buffer/g wet cells for 18 h at 4 °C.
4. Disrupt cells in a homogenizer (4 passes at chamber pressures of 13,000 psi).

Method description:

1. Clarify cell extracts.
2. Dilute extracts into a series of fivefold with 20 mmol/L PBS (pH 7.2).

3. Add 100 μL of the diluted extracts to each well of a 96-well plate coated with anti-HBsAg polyclonal antibody.
4. Incubate the 96-well plate for 1 h at 37 °C.
5. Empty and tape dry the plate.
6. Wash with PBS containing 0.02% Tween 20 5 times.
7. Add 100 μL horseradish peroxidase (HRP)-conjugated mouse anti-HBsAg to each well.
8. Incubate for 1 h at 37 °C.
9. Empty and tape dry the plate.
10. Wash five times as described above.
11. Add 100 μL of substrate (OPD), and allow the plate to develop for 30 min at 37 °C.
12. Stop the reaction by adding 50 μL of 2 mol/L H_2SO_4.
13. Read each well of the plate at a double wavelength of 450 nm/620 nm.
14. Make a reference curve with a series of standard dilution of HBsAg (0~4 ng).

2.3 HPLC: High-Performance Liquid Chromatography

2.3.1 Background

High-performance liquid chromatography (HPLC) is well established to separate, identify, and quantify individual components. It is feasible for purified analytes but also crude culture supernatants. HPLC operates under the basic principle that the separation of compounds in a sample into its constituent parts depends on the relative affinities of different molecules for the mobile phase and the stationary phase used for the separation. The column is packed with a stationary phase composed of irregularly or spherically shaped particles, a porous monolithic layer, or a porous membrane. The injected sample is forced through the column by an appropriate liquid with adjusted composition and flow rate. Based on the overall performance, conventional HPLC, u-HPLC (ultra-HPLC), and micro HPLC systems (μHPLC), which needs special equipment, can be distinguished [31, 32]. Depending on the available equipment and in particular the stationary phase and the related flow rate, two performances are applied: conventional systems create pressures between 6000 and 9000 psi, while u-HPLC systems reach pressures above 20,000 psi and are characterized by higher speed, resolution, and sensitivity making them ideally suited for use with MS (mass spectroscopy). U-HPLC systems were designed with (1) low dwell volume that is a term that refers to gradient separations, (2) extra column volume, (3) smaller particle size (sub-2 micron), and (4) dispersion that is minimized. More theoretical plates can reach better separation. For HPLC columns up to 2000 and more plates are ideal [33]. Small injection volumes for improved resolution and symmetric, sharp peaks are recommended (*see* **Note 19**). The tube is packed with tiny particles up to 5 micron followed by pumping the liquid phase through the column. Columns might be either made of glass, synthetic material, or polished stainless steel and are between 50 and 300 mm long.

The interaction of different molecules involves several chemical and/or physical interactions. For the analysis of supernatants, guard columns are recommended to remove small interfering particles and/or contaminants so that protection of the analytical column is ensured (*see* **Note 20**). Once separated, the molecules will be eluted and are detected by a flow-through device, called detector. Various detection principles are available such as UV spectroscopy, fluorescence, refractive index, mass spectrometry, light scattering, and electrochemical detectors [34]. In certain circumstances more than one detector can be operated in serial (*see* **Note 21**).

Instrumentation of HPLC

The HPLC instrumentation might be very diverse, consisting of different components:

1. Up to four mobile phase reservoirs as stock of mobile phase (s) for complex analysis.
2. A degasser, which degasses the mobile phase(s).
3. A pump in different designs, suitable as solvent delivery system, enabling the flow of the mobile phase.
4. The sample valve injector as sample delivery system, which introduces the sample to stationary phase.
5. The column for separation.
6. The column compartment, used to control the temperature of the column.
7. A detector, to detect each eluted component.
8. Data acquisition, to convert data from the detector to meaningful results.
9. A waste collector of liquid waste [35].

Mode of Separation

1. Normal Phase

   Polar stationary phases, typically silica, combined with relative nonpolar mobile phases (hexane, methylene chloride, chloroform, and diethyl ether). Suitable for somewhat polar to highly polar samples. Molecular weight range 10^0–10^4.

2. Reverse Phase

   The stationary phase is nonpolar to somewhat polar in nature, while the mobile phase is relatively polar and uses gradients of water and methanol or acetonitrile. It works on the principle of hydrophobic interactions. Samples are nonpolar to somewhat polar. Molecular weight range: 10^0–10^4.

3. Size Exclusion (Gel Permeation)

   The column is filled with small, porous, silica, or polymeric particles. Analytes are separated roughly to their molecular size, more precisely according to the hydrodynamic radius, resulting in broader peak widths compared to other techniques. The solvent is polar or nonpolar. Molecular weight range 10^0–10^6.

4. Ion Exchange

   Ion exchange resins are made of insoluble, high-molecular-weight functionalized material. This technique is applicable to highly polar to ionic molecules. The mobile phase is an aqueous buffer, where both pH and ionic strength can be adjusted. Sometimes organic solvents are added to moderate the solvent strength. Molecular weight range 10^0–10^4.

5. Affinity

   It is the most selective type, utilizing the specific interaction between one kind of solute molecule and a second molecule immobilized on the stationary phase.

Elution Techniques

1. Isocratic Elution

   In this case one composition of the mobile phase throughout the whole procedure is used. This elution is a good choice for simple separations but is limited to an increasing peak width by time that is linear with the number of theoretical plates.

2. Gradient Elution

   The mobile phase composition is changed during the separation process. A gradient elution decreases the retention of the later-eluting components so that they elute faster, giving narrower peaks. Stepwise and linear gradients are most popularly applied for complex samples where a better separation is required.

Effective Method Development

Analytical method development is a critical process when using HPLC for pharmaceutical analysis. Although this technique is commonly applied, both advantages and limitations are recognized (Table 5). A method needs to (1) satisfactorily separate the desired components, (2) generate accurate and constant results over time, and (3) be reproducible and robust. The most critical steps are (a) sample preparation, (b) the analysis, and (c) standardization. All individual parameters should be investigated before the final method optimization is performed (*see* **Note 22**). Method validation can finally be performed according to the ICH guidelines. Due to the high flexibility of this technique, individual working ranges and limits are adjustable. Nevertheless, it can be considered that for

Table 5
Summary of applicability and limitations of HPLC analysis

Advantages	Perceived limitations
Applicable for diverse analyte/sample types—crude supernatant	Lack of an ideal universal detector General costs
Excellent accuracy and precision	Relatively difficult for novices
Qualitative and quantitative	Time-consuming standardization
High flexibility	Operator-dependent analysis
Automated operation	Interference with other proteins—purity
High separation power for proteins and related substances	Interference with salts, detergents, organic solvents, sulfones, etc.
Sensitive detection	High viscosity (DNA, carbohydrates)
Coupling with mass spectroscopy	Limited sample throughput
Highly reproducible	Large quantities of expensive organics
Compared to other chromatographic methods—quick and efficient	Data management and documentation Complex troubleshooting
Short analysis time for single compounds	Long running time for multi-compound analysis

many applications this technique is less sensitive than other quantification methods and, however, more selective.

2.3.2 Examples

The following examples give an overview of the applicability of HPLC for the analysis of different products from yeasts. Although in the examples no guard columns are used, they are highly recommended to protect the analytical column against particulates. Additionally, gradients need to be adapted because the complex sample matrices can have an enormous impact on the separation. For the calibration of a quantitative HPLC method, either a purified material of the analyte or when available a USP reference standard should be used. Equal injection volumes and concentrations for standards and samples are recommended. For detection DAD detection system is favorable, for recording wavelengths in parallel, e.g., 280 and 214 nm (*see* **Note 23**) [36–41].

Product: Insulin precursor (IP) [36]

Host: Pichia pastoris.

Aim: Quantification of secreted IP in clone screening.

Method: Reversed-phase HPLC (RP-HPLC).

Sample matrix: Clarified culture broth.

Equipment: HPLC system equipped with a binary pump system, an auto-injector, a column oven, and an UV-VIS detector

(280 nm). As analytical column, a SUPELCOSIL™ LC-304, 5 μm, 300 Å, 4.6 × 250 mm, is applied.

Sample preparation:

1. Filtrate an aliquot of a *P. pastoris* culture broth through a syringe filter (0.2 μm).
2. Mix a filtered aliquot with an equal volume of solvent A.
3. Human insulin CRS from European Pharmacopoeia (EP) as standard is used.

Method description:

1. Solvent A (0.15% (v/v) TFA in Milli-Q water).
2. Solvent B (0.15% (v/v) TFA in acetonitrile).
3. Elution is performed with a flow rate of 1 mL/min and a gradient shown in Table 6 and illustrated in Fig. 1.

Product: Hepatitis B surface antigen (HBsAg) [38]

Host: Pichia pastoris (intracellular production of HBsAg).

Aim: Determination of total HBsAg.

Method: Reversed-phase HPLC (RP-HPLC).

Sample matrix: Cell lysate.

Equipment: HPLC system consisting of a binary pump, an auto-injector, column oven, and an UV-VIS detector (214 nm). A SUPELCOSIL™ LC-DB-18, 5 μm, 4.6 × 150 mm column, is used for separation.

Sample preparation:

Preparation of Cell Lysate

1. Determine the cell concentration via optical density (600 nm) with a spectrophotometer.
2. Centrifuge samples corresponding to 100 OD units (1 mL) at 15,000 × *g* for 15 min to get a cell pellet.

Table 6
Gradient for IP quantification by RP-HPLC

Time (min)	Solvent (%B)	Function
0–6	10	Equilibration
6–41	10–43	Separation
41–43	43–100	Column cleaning
43–53	100–10	Readjustment
53–60	10	Equilibration

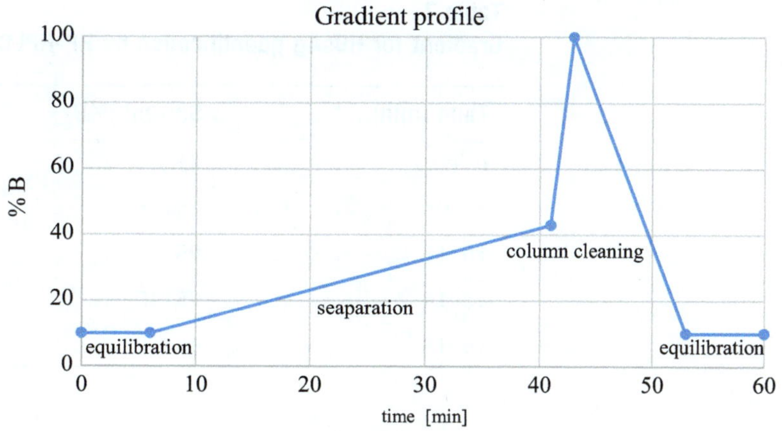

Fig. 1 Applied gradient for IP quantification by RP-HPLC illustrating the equilibration, separation, and cleaning of the column

3. Wash the cell pellet with 1 mL 10 mmol/L phosphate buffer, pH 7.2.
4. Mix it with 0.6 g glass beads of 0.45 mm diameter suspended in 500 μL of lysis buffer (10 mmol/L phosphate buffer (pH 8), 5 mmol/L EDTA, 500 mmol/L NaCl, 8% (v/v) glycerol).
5. The cells are lysed by 20 cycles of 1 min vortexing at maximum speed followed by 4–5 min resting on ice.
6. After the final cycle, transfer the lysate into a fresh tube.
7. Rinse the glass beads twice with lysis buffer, and pool with the initial lysate to obtain a final volume of 1 mL.
8. Suspend a 100 μL aliquot of the cell lysate in an equal volume of solubilization reagent (8% (w/v) SDS, 50% (v/v) β-mercaptoethanol, 1 mol/L DTT).
9. Boil for 15 min.
10. Clarify sample by centrifugation and filtrate through a syringe filter (0.2 μm).

Method description:

1. Solvent A (0.15% (v/v) TFA in Milli-Q water).
2. Solvent B (2-propanol: acetonitrile in 80:20 (v/v)).
3. Column temperature: 70 °C.
4. Gradient elution is performed at a flow rate of 1 mL/min as shown in Table 7.

Product: Human epidermal growth factor (hEGF) [39]

Host: Hansenula polymorpha.

Aim: Quantification of hEGF.

Method: Reversed-phase HPLC (RP-HPLC).

Sample matrix: Culture supernatants.

Table 7
Gradient for HBsAg quantification by RP-HPLC

Time (min)	Solvent (%B)	Function
0–4	45	Equilibration
4–10	45–95	Separation
10–13	95	Column cleaning
13–14	95–45	Readjustment
14–18	45	Equilibration

Table 8
Gradient for hEGF quantification by RP-HPLC

Time (min)	Solvent (%B)	Function
0–5	0–15	Separation
5–40	15–50	Separation

Equipment: HPLC system consisting of a binary pump, an autoinjector, and an UV-VIS detector (215 nm). A Hypersil C18 column, 5 μm, 4.6 × 150 mm column, is used for separation.

Sample preparation:

1. Centrifuge culture supernatants and filtrate through a syringe filter (0.2 μm).

Method description:

1. Solvent A (0.1% (v/v) TFA in Milli-Q water).
2. Solvent B (0.1% (v/v) TFA in acetonitrile).
3. After equilibration, gradient elution is performed at a flow rate of 1 mL/min shown in Table 8.

Product: mAb [41]

Host: Pichia pastoris.

Aim: Quantification of mAb.

Method: Protein A HPLC.

Sample matrix: Culture supernatant.

Equipment: HPLC system consisting of a binary pump, an autoinjector, and an UV-VIS detector (280 nm). For mAb quantitation, a POROS A 20 column, 2.1 × 30 mm, is used.

Table 9
Alternatives for buffer B for mAb quantification by protein A HPLC

Buffer B
100 mM citrate, pH 2.5
100 mM acetate, pH 2.5
100 mM phosphate, pH 2.5
100 mM glycine, pH 2.5
12 mM HCl, pH 1.9

Table 10
Gradient for mAb quantification by protein A HPLC

Time (min)	Solvent (%B)	Function
0–0.5	0	Equilibration/wash; 14 column volumes
0.51–1.5	100	Elution; 29 column volumes
1.51–3.0	0	Re-equilibration; 43 column volumes

Sample preparation:

1. Centrifuge and filtrate (syringe filter 0.2 μm) culture supernatant samples before loading onto the column.
2. Dilute samples, if required, in equilibration buffer.
3. For quantitation of the specific mAb, standard curves should be generated from purified material.

Method description:

1. Equilibration/wash buffer (buffer A), e.g., 50 mM phosphate, 150 mM NaCl, pH 7.0.

   In principle, the molar concentration of phosphate in buffer A can range from 25 to 100 mM with a pH ranging from 6.6 to 7.5 with or without sodium chloride up to 150 mM and has to be adapted on the specific requirements.
2. Elution buffer (buffer B): as buffer B either the same buffer as buffer A with a pH ranging from 2.0 to 3.5 is used or alternatives shown in Table 9,
3. Flow rate is 3 mL/min with gradient shown in Table 10 and illustrated in Fig. 2 (*see* **Note 24**).

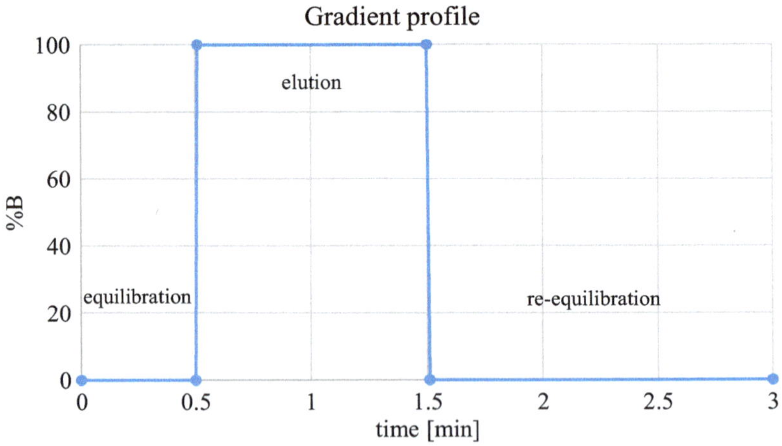

Fig. 2 Elution profile including equilibration, elution, and re-equilibration for mAb quantification by protein A HPLC

3 Advanced Methods for Quantification

3.1 Bio-layer Interferometry

3.1.1 Background

Over the last years, the implementation of biosensor technologies for qualitative and quantitative assays has markedly advanced. One promising biosensor platform, using bio-layer interferometry (BLI), is investigated by FortéBio; meanwhile PALL provides a real-time measuring performance based on a label-free technology. In particular, quantification and measurement of biomolecular interactions are intended [42]. This optical analytical technique utilizes the interference pattern of white light reflected from two surfaces: (1) a layer of immobilized protein on the biosensor tip and (2) an internal reference layer. The binding between a ligand immobilized on the biosensor tip surface and an analyte in solution produces an increase in optical thickness at the biosensor tip surface, which results in a wavelength shift ($\Delta\lambda$) as shown in Fig. 3. This change in thickness of the biological layer is a direct indicator for the specific interaction of the molecules, and by a certain calibration, the increase of signal intensities is used for quantification. Interactions are measured in real time, providing the ability to monitor binding specificity, rates of association and dissociation, or concentration profiles, with high precision and accuracy.

This non-fluidic technique is either performed in 96- or 384-well plate formats. Biosensors with different ligands (e.g., streptavidin; proteins A, G, and L; nickel-charged Tris-nitriloacetic acid (Tris-NTA) for his-tagged proteins; antihuman IgG Fc capture (AHC)) coated on a biocompatible matrix are commercially available. The dynamic range for a standard assay is assessed between 1.3 and 150 μg/mL. Signal enhancement, such as metal chelator (DAB), combined with a horseradish peroxidase-tagged antibody, is described.

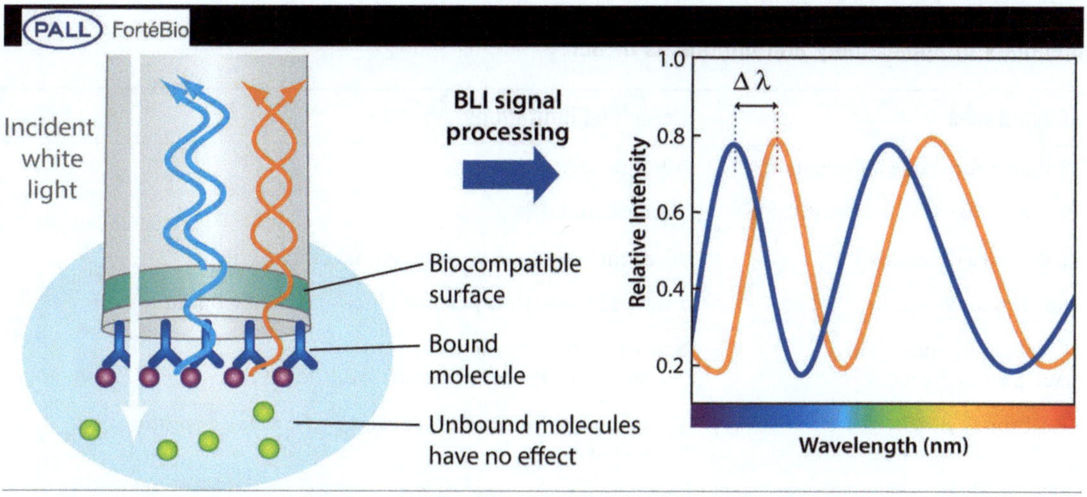

Fig. 3 Schematic drawing of the bio-layer interferometry principle of an individual tip surface. A layer of immobilized protein on the biosensor tip and an internal reference layer are shown on the left side. Any change in the number of molecules bound to the biosensor tip causes a shift in the interference pattern that can be measured in real time. The binding between a ligand immobilized on the biosensor tip surface and an analyte in solution produces an increase in optical thickness at the biosensor tip, which results in a wavelength shift, $\Delta\lambda$ as shown on the right side of the figure. Modified with permission from Pall FortéBio, USA

Biomolecules in crude lysates, culture supernatants, as well as purified samples can be analyzed [43]. Dilution of samples and standards needs to be performed externally by hand or automated, prior to transfer into the microwell plates. Due to microplate format, automatization for high-throughput purposes can be achieved by additional equipment. Test samples after cell removal, e.g., by centrifugation, but also with cell-containing suspensions can be applied if they are not interacting with the surface. Molecules and particles in the solution do not contribute to the signal. Only attached molecules are measured. During assay development unspecific binding needs to be carefully evaluated. Additionally, various buffers for individual washing steps and other required solutions, such as blocking solvents, are also placed in defined wells. The assay itself is subsequently performed automatically. Different sensors can be regenerated, according to established regeneration procedures, to minimize sensor costs. A summary of the applicability of bio-layer interferometry is given in Table 11.

3.1.2 *Example*

In principle, mAb concentration is calculated via the binding rate of antibodies to protein A sensor tips. A typical calibration range is between 0.25 and 300 μg/mL depending on the antibody as shown in Fig. 4. IgG standards for calibration should be as similar to the molecule to be analyzed, and also the matrix should match as closely as possible to the sample matrix, e.g., by diluting standards in freshly prepared media. Generally, the assay setting consists of an

Table 11
Summary of applicability and limitations of BLI

Advantages	Perceived limitations
Suitable for crude supernatants	Medium dynamic range
Various sensor formats available	Equipment costs
Label-free detection	Medium sensitivity without signal enhancement
Small sample volume	Interference with other proteins—unspecific binding
Qualitative/quantitative Accurate and precise	Sensor costs—regeneration Noise and drift optimization—additional time
Real time	Detectability for small molecules ($\leq$10 kDa)—optimization needed
Non-fluidic	
Significant throughput and flexibility	

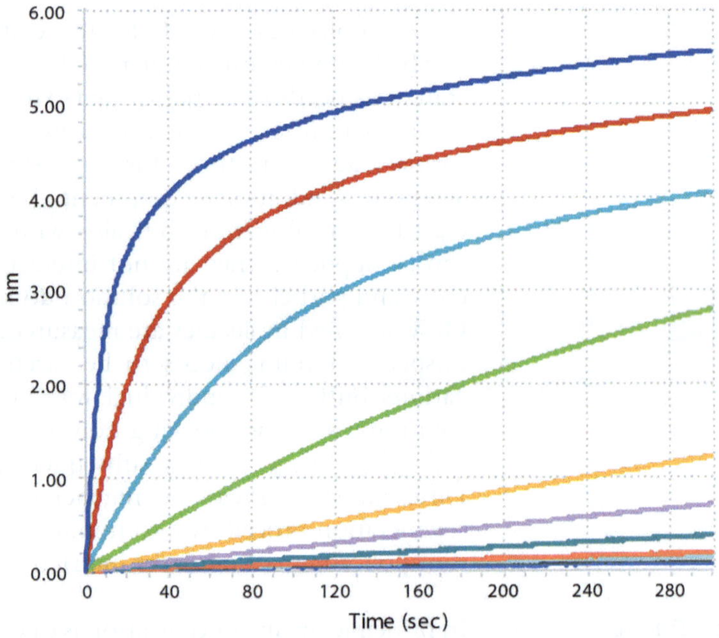

Fig. 4 Concentration-dependent binding curves from IgG standards pre-diluted in media to protein A biosensors. Standard concentration ranges between 0.25 (orange), 0.5 (dark green), 1 (lilac), 3 (yellow), 10 (light green), 30 (cyan), 100 (red), and 300 (blue) μg/mL. Reproduced with permission from Pall FortéBio, USA

equilibration, mAb loading, and regeneration step for the protein A sensors. Usually, biosensor tips can be reused up to ten times [44, 45].

Product: mAb [45].

Host: Not specified.

Aim: Quantification of mAb.

Method: Bio-layer interferometry (BLI) with protein A biosensors.

Sample matrix: Serum-free cell culture supernatants.

Equipment: Octet QK system.

Sample preparation: Centrifugation of culture supernatants.

Method description:

1. Use black 96-well plates for the measurements and sensor equilibration, respectively.
2. Equilibrate sensor tips by dipping in 200 μL equilibration buffer for 10 min (PBS, 0.1 Vol% Tween 20).
3. Prepare a second 96-well plate by pipetting in one row 200 μL regeneration buffer (10 mM glycine-HCL, pH 2.5) and in a second row 200 μL equilibration buffer (e.g., rows 11 and 12).
4. Transfer 50 or 25 μL culture supernatant starting from row 1.
5. Add 150 or 175 μL equilibration buffer, respectively, to each sample well in order to obtain a total volume of 200 μL.
6. Proceed with calibration standard in the same way, e.g., add to 50 μL standard (pre-diluted in media) 150 μL equilibration buffer.
7. Transfer the well plate in the well plate holder of the instrument.
8. Shake plates at 1000 rpm for 5 s to mix samples before each measurement.
9. Determine binding rates over a time interval of 300 s.
10. Calculate binding rates with the Octet software.
11. After each measurement, biosensor tips may be regenerated by washing in regeneration buffer (row 11) subsequently in equilibration buffer (row 12) for 5 s.

3.2 Microfluidics CD Technology

3.2.1 Background

Alternatively to the non-fluidic sensor-based technique, nanoliter microfluidics CD technology has been recently developed [46]. This technology is established as an open platform using miniaturized chambers located on a rotating CD (96 chambers per CD). Each chamber is filled with 15 nL affinity capture materials coated on modified beads (Fig. 5). Different equipment formats for one or more CDs are available for automated immunoassays at nanoliter-scale with parallel processing.

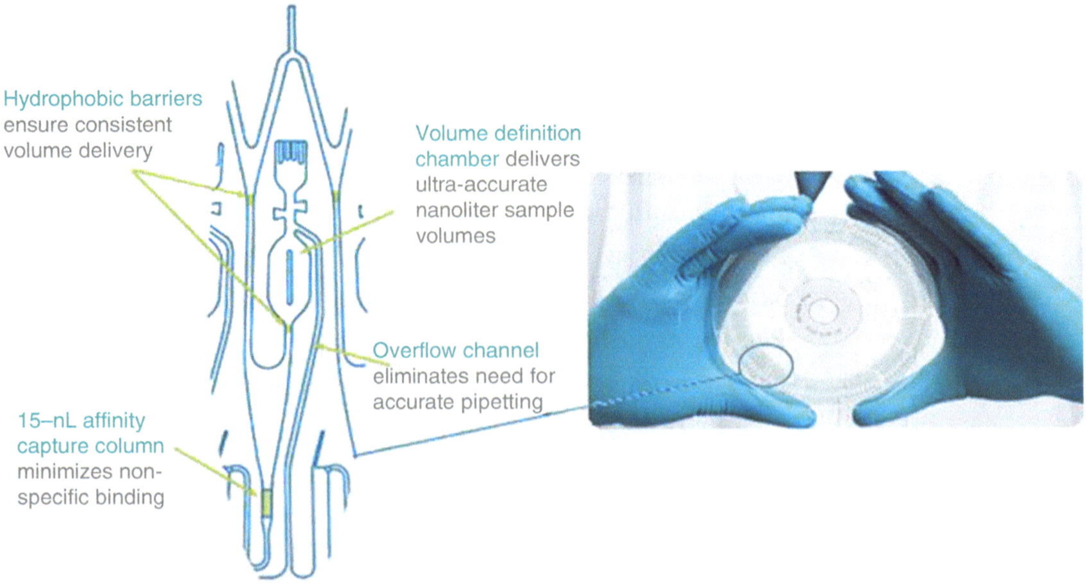

Fig. 5 Sketch of the nanoliter-scale device from Gyrolab Systems. Reproduced with permission from Gyros Protein Technologies, Sweden

The assay workflow includes precise, automated control of centrifugal and capillary forces, which steer liquid flow-through. Samples are diluted externally and transferred into special microwell plates with a protection slide to prevent evaporation of the small volumes (10–max. 100 μL). Transfer of the samples into the reaction chambers is performed via a volume definition chamber that delivers accurate sample volumes, with hydrophobic breaks to ensure consistency in volume delivery. An overflow channel eliminates the need for manual pipetting accuracy, and a 15 nL affinity capture column mitigates non-specific binding. Once CDs are loaded into the workstation, up to five discs in series, the platform performs parallel sample processing via high-sensitivity, laser-induced fluorescence detection. The CDs are spun at controlled speeds to ensure optimal binding and uniform conditions for all assays.

Although this technique has been established rather for biological samples than for biotechnological samples, this technique became of great interest due the PAT (process analytical technology) initiative. A very broad quantification range (3–4 logs) with sensitivity down to 1 ng/mL is offered. Therefore, this technology is feasible both for clone screening and also product manufacture. Advantages and perceived limitations are shown in Table 12.

Table 12
Summary of applicability and limitations of microfluidics CD technology

Advantages	Perceived limitations
Crude supernatant—with little need for sample pretreatment	Interference with other proteins—assay dependent Data export is complex
Various gel formats available	High costs
Different staining opportunities	Semiautomated combined with external sample preparation
Small sample volume	Evaporation risk
Reduced reagent consumption	
High sensitivity and precision Reduced matrix interference Broad dynamic range—reduced dilution Robust and rugged Fast total assay time Adaptability to automation	

3.2.2 Example

The subsequent example describes a protocol to quantify intact human IgG (IgG1, IgG2, IgG4) in cell supernatants. For this semi-automated assay standards, pre-diluted samples, capture reagent, detection reagent, and wash solution have to be provided in a 96-well plate. Subsequent pipetting steps as well as incubation and detection are operated automatically. The advantage of miniaturization is that extremely low sample volumes are required. The provided kit (Gyrolab-hulgG-k) includes ready-made reagents and can be used together with the CDs, Bioaffy 20 HC (300–1000,000 ng/mL, 20 nL sample volume, 112 microstructures) or Bioaffy 1000 HC (1–20,000 ng/mL, 1000 nL sample volume, 96 microstructures). Furthermore, a Bioaffy 200 HC (10–1000 ng/mL, 200 nL sample volume, 112 microstructures) is available. Solutions are evenly distributed by rotation of the CD and pervade the microstructures. Combination of different CD formats enables quantification of human IgG over six orders of magnitude.

In principle, a biotinylated derivative of protein A (capture reagent) is introduced into a microstructure in the CD to saturate a capture column packed with porous beads coupled with streptavidin, followed by sample application, where intact IgG is captured in the capture column. Subsequently an Alexa Fluor 647 labeled F$(ab')_2$ fragment of antihuman IgG (detection reagent) is added. The corresponding signal in the capture column represents the total response from the sample [47].

Product: Human IgG (IgG1, IgG2, IgG4) [47].

Host: Not specified.

Aim: Quantification of human IgG.

Method: Microfluidics CD technology.

Sample matrix: Cell supernatants.

Equipment: Gyrolab workstation.

Sample preparation:

1. Centrifuge samples at 12,000 × *g* for 4 min to sediment any aggregates or particulates.
2. Remove supernatant carefully.

Method description:

1. Transfer 50 μL of the ready-to-use capture reagent, detection reagent, and wash buffer (PBS—0.01% Tween 20) directly to the microtiter plate according to the instructions.
2. Dilute samples in sample dilution buffer at least 1 + 1 by volume, but further dilution steps may be necessary depending on the working range of the particular version of IgG titer assay.
3. Select an appropriate IgG molecule at known concentration as standard, or use standard from Gyrolab which contains a human monoclonal IgG1 (4 mg/mL) in 50 μL.
4. Dilute standards also in sample dilution buffer.

Automated test sequence (approx. 1.5 h):

1. Wash step: microstructures are rinsed with wash buffer.
2. Application of the capture reagent including a short incubation.
3. Wash step.
4. Sample application including a short incubation.
5. Wash step.
6. Recording of the background signal for following fluorescent detection.
7. Application of the detection reagent including a short incubation.
8. Wash step.
9. Detection of the fluorescence signal.
10. Data evaluation is performed by the Gyrolab Evaluator.

3.3 Surface Plasmon Resonance

3.3.1 Background

Surface plasmon resonance (SPR), a biosensor-based technology, is currently of great interest for the analysis of various compounds, including proteins and peptides. Liedberg et al. initially published the optical phenomenon of surface plasmon resonance over 30 years ago [48]. It was first shown that the interaction of molecules in a label-free state can be measured in which the binding

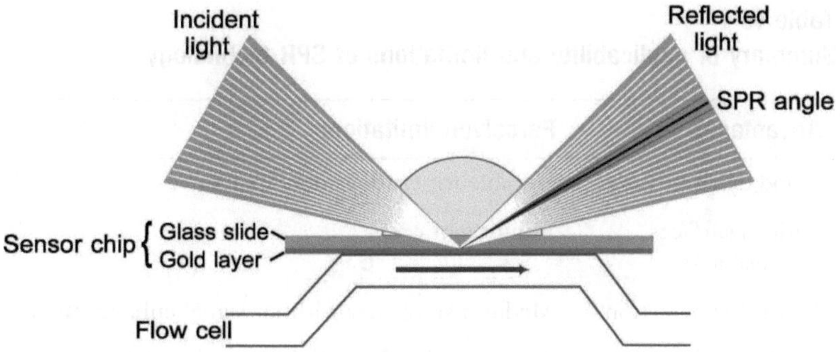

Fig. 6 Principle of surface plasmon resonance (SPR). Reproduced with permission from GE Healthcare Life Sciences

response is given in resonance units (RU) proportional to the molecular mass [49]. Biacore first commercialized this technique by offering a sensor chip platform in a microfluidic device, which controls the flow of buffer and samples across the sensor surface (Fig. 6).

In principle, SPR occurs in a thin contacting film placed at the interface between two media of different refractive indices. In Biacore systems (GE Healthcare Life Sciences), a 50 nm gold layer is sandwiched between a glass layer and the sample solution. Gold layers are preferred surfaces for SPR due to a well-defined reflectance minimum and due to the fact that it is amenable to covalently attach surface matrix layers. Furthermore, gold is mostly inert and suited for various buffer systems. For the efficient loading of various molecules, different surface layers are commercially available. The most popular surface is carboxymethylated dextran (CM). By certain modifications, capturing of the molecules from a solution can be optimized [50]. Covalent coupling can be performed either via free amino, thiol-disulfide, or aldehyde groups. Unbound and non-covalently bound material can be removed either with buffer or injection of suitable regeneration and cleaning solutions. The most critical step in assay development is the adjustment of proper immobilization chemistry. Direct and indirect techniques are available, and numerous techniques have been published [51].

For the application of SPR, one fundamental element is to distinguish the active protein form from the inactive form. This feature is of great importance because it allows the quantification of the protein of interest not only in a pre-purified form but also in crude, complex matrices [52, 53].

The detection is based on measuring the internal reflection of plane-polarized light from a near-infrared LED with a diode array detector. If a change in mass occurs, the angle of light at which SPR occurs shifts due to a change in refractive index near the sensor surface. One RU corresponds to 0.0001° shift in SPR angle, or

Table 13
Summary of applicability and limitations of SPR technology

Advantages	Perceived limitations
Good dynamic range	Suitable for crude supernatants
Various surfaces available	Equipment costs
Label-free detection	Medium sensitivity without signal enhancement
Small sample volume Mass detection	Unspecific binding might be a challenge—increased number of references demanded
Qualitative/quantitative	Sensor costs—regeneration recommended
Real time	Noise and drift optimization
Microfluidic	Interference of low-molecular-weight molecules
High sensitivity Automatization	Clogging due to the microfluidic system might occur
Accurate and precise	

1 pg/mm^2 of molecule on the surface. Since only the evanescent wave penetrates the sample, measurement of colored, turbid, and opaque solutions is possible.

SPR is a well-established technique for detection and monitoring biomolecular interactions in real time. Used orthogonally, it is powerful and complementary in basic research, drug discovery and development, and downstream bioprocessing. Advantages and perceived limitations are shown in Table 13.

3.3.2 Example

The example given below is applicable for proteins containing terminal his-tags, which can be attached to SPR-NTA sensor chips. Ligands are captured on the carboxymethylated dextran matrix which is pre-immobilized with nitrilotriacetic acid (NTA). NTA chelates present Ni^{2+}, creating coordination sites that bind to histidine residues in the ligand tag. Therefore, the NTA sensor chip has to be activated with Ni^{2+} ions (loading buffer), followed by binding of subsequent injected analyte (running buffer), which is finally detected (*see* **Note 25**). Thereafter, the chip surface is regenerated with EDTA solution (stripping buffer); after which a new analysis cycle with the Ni^{2+} solution pulse can be started [54–56].

Product: Single-chain Fv antibody fragment (ScFv4813; his-tagged) [54].

Host: Pichia pastoris.

Aim: Quantification.

Method: Surface plasmon resonance.

Sample matrix: Culture supernatant.

Equipment: Biacore 2000 system.

Sample preparation:

1. Clarify culture supernatants by centrifugation (2000 × *g*, 5 min at room temperature).

Method description:

1. Prepare the Biacore instrument with running buffer (10 mM HEPES, pH 7.4, 0.15 M NaCl, 0.005% (v/v) with a flow rate of 5 μL/min).
2. Each sample cycle consists of subsequent injections as follows:
3. 5 μL stripping buffer (running buffer containing 0.35 M EDTA at pH 8.3).
4. 10 μL of loading buffer (running buffer containing 500 μM $NiCl_2$).
5. 20 μL of the diluted sample (conc. Below 0.2 μM) in running buffer.

For quantification, the availability of a suitable standard is essential. An example for a concentration-dependent analyte binding is shown in Fig. 7 (*see* **Note 26**).

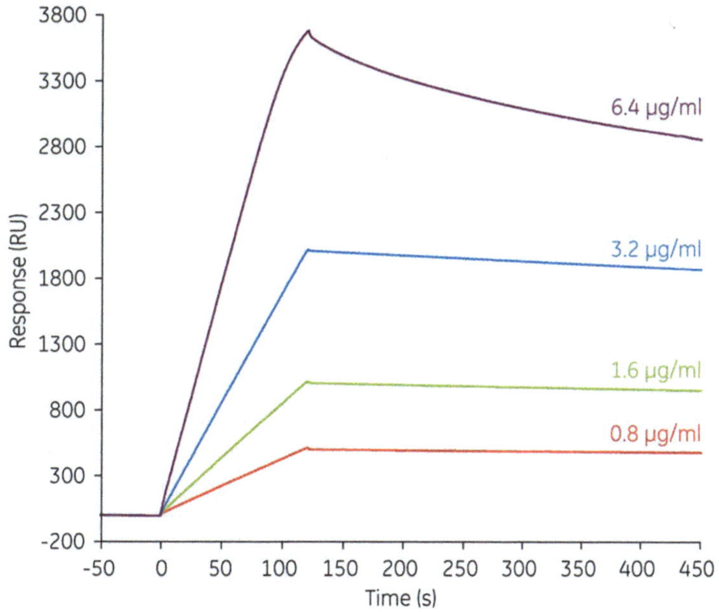

Fig. 7 Injection of different concentrations of histidine-tagged green fluorescent protein (0.8–6.4 μg/mL) over Sensor Chip NTA. Reproduced with permission from GE Healthcare Life Sciences

3.4 Electrochemiluminescence Assay

3.4.1 Background

The electrochemiluminescence technique, namely, the electrochemiluminescence immunoassay or ECLIA, has been established for drug development and clinical application but might be also used for biotechnological applications [57, 58].

Electrochemiluminescence labels generate light when stimulated by electricity. In the appropriate chemical environment, this light signal is used to measure proteins and other biomedical molecules. Analogous to chemiluminescence it does not require the use of external light sources. The combination of chemiluminescence and electrochemistry implies potential advantages. Meanwhile, various systems are launched, which can be easily adapted to biotechnological applications.

The first commercial instrument was produced by IGEN International, Inc. (technology later acquired by Roche Diagnostics Corp.) in 1994. This system uses magnetic microbeads modified with probe molecules, such as antibodies or nucleic acids, to capture analyte molecules in a sample labeled with $Ru(bpy)_3^{2+}$. Before the measurement is performed, the unbound $Ru(bpy)_3^{2+}$ labels are flushed from the cell. The bound $Ru(bpy)_3^{2+}$ labels are measured in tripropylamine (TPrA) solution. Both $Ru(bpy)_3^{2+}$ and TPrA are oxidized at the electrode surface, and $Ru(bpy)_3^{3+}$ is reduced by TPrA to produce the excited state. However, TPrA has several disadvantages. It is toxic and volatile and needs to be used in high concentrations (usually up to 100 mM) to reach good sensitivity.

Therefore, a more environmentally friendly co-reactant, the 2-(dibutylamino)ethanol (DBAE), was introduced. The intensities of the $Ru(bpy)_3^{2+}$/DBAE system at Au and Pt electrodes were approximately 10–100 times greater compared to the previously used system, when 25 mM of each co-reactant was used. The sensitivity is about an order of magnitude better. DBAE is much less toxic and less volatile and better soluble in aqueous solutions. Alternatively, "reductive-oxidative" co-reactants are used. The Ru $(bpy)_3^{2+}/S_2O_8^{2-}$ system has a very negative potential. Unfortunately, it is difficult to create stable signals in aqueous systems because of serious hydrogen evolution upon applying very negative potentials. To overcome this problem, bismuth electrode with a high hydrogen evolution potential is provided as a sufficient solution.

Another prominent example of a well-recognized microwell plate technique has been launched by Meso Scale Discovery (MSD), which is available in a classical 96-well but also 384-well formats, equipped with electrodes at the bottom of each well (Fig. 8). The excitation caused by a SULFO-TAG label is measured through a CCD camera. The label is either conjugated to streptavidin or to the detection antibody. This technology is meanwhile well recognized due to the advantages of the broad dynamic range (3–4 logs), low detection limits (1.0–10,000 pg/mL), low sample volumes (10–25 μL), and the suitability in complex matrices. Advantages and perceived limitations are shown in Table 14.

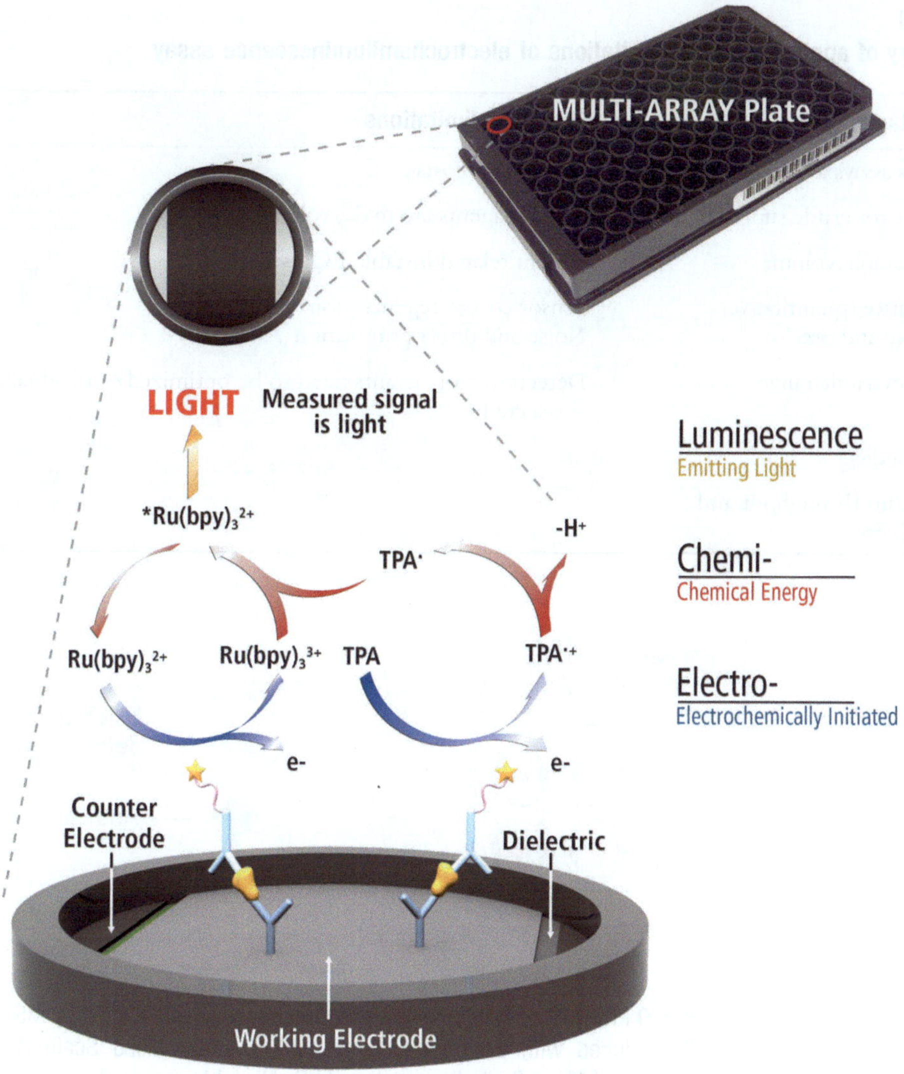

Fig. 8 Principle of electrochemiluminescence assay provided by MSD. Reproduced with permission from ©2006, 2009 Meso Scale Discovery, a division of Meso Scale Diagnostics, LLC. All rights reserved

3.4.2 Example

Meso Scale Discovery (MSD) provides a sandwich immunoassay for human insulin that detects insulin in cell supernatant, human serum, and plasma samples. In principle, in this single-plex assay, an antibody for a specific protein target is coated on one electrode (or "spot") per well. The assay kit consists of a multi-array 96-well plate that has been pre-coated with insulin antibody (capture antibody) and an anti-insulin antibody, labeled with an electrochemiluminescent compound, MSD SULFO-TAG label (detection antibody). Insulin in the sample binds to the immobilized capture antibody and the added detection antibody as schematically demonstrated in Fig. 9. Adding the MSD Read Buffer ensures the appropriate chemical environment for electrochemiluminescence measurement. This Tris-based buffer contains tripropylamine

Table 14
Summary of applicability and limitations of electrochemiluminescence assay

Advantages	Perceived limitations
Various assays available	Equipment costs
Suitable for crude supernatants	Some reagents are toxic, volatile
Small sample volume	System related instable ECL signals
Qualitative/quantitative Accurate and precise	Sensor costs—regeneration Noise and drift optimization
Broad dynamic range	Detection co-reactants need to be optimized/individually selected
Non-fluidic	
Significant throughput and flexibility	

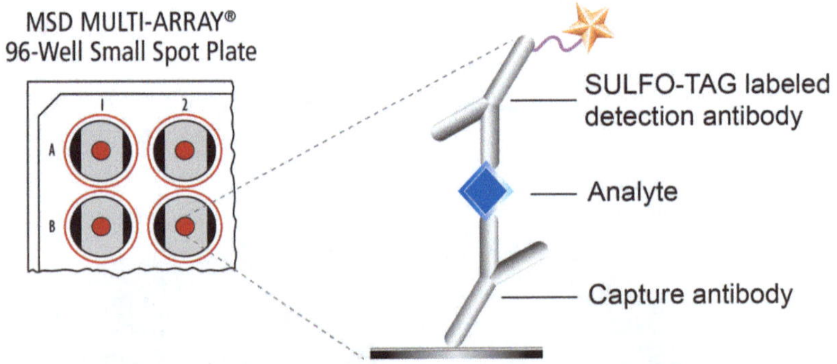

Fig. 9 Principle of insulin sandwich immunoassay established by MSD platform. Reproduced with permission from ©2006, 2009 Meso Scale Discovery, a division of Meso Scale Diagnostics, LLC. All rights reserved

(TPA) which should facilitate higher signals and sensitivity as well as lower background. After loading the plate in the instrument, a voltage is applied to the plate electrodes, which causes the labels bound to the electrode surface to emit light. The intensity of the emitted light is a quantitative measure of insulin present in the sample. Furthermore, an insulin calibrator and all buffer solutions, required for the assay, are provided in the kit [59].

Product: Insulin [59].

Host: Not specified.

Aim: Quantification.

Method: ECLIA.

Sample matrix: Cell supernatant, human serum, and plasma samples.

Equipment: MSD-SECTOR Imager.

Sample preparation: 25 μL sample.

Method description:

1. All buffer stock solutions delivered in the assay kit have to be diluted according to the manufacturer's recommendations.
2. The MSD plate is pre-coated with insulin antibody and is ready to use.
3. Dispense 150 μL of Blocker A solution into each well (mixture of proteins in a PBS buffer preventing non-specific binding, improving sensitivity, and reducing background signals).
4. Incubate at room temperature at 300–1000 rpm for 1 h.
5. Wash plate three times with PBS-T (0.05% Tween 20).
6. Dispense 25 μL/well detection antibody solution.
7. Immediately dispense 25 μL/well sample or calibrator.
8. Incubate the sealed plate at room temperature at 300–1000 rpm for 2 h.
9. Wash plate three times with PBS-T.
10. Dispense 150 μL/well Read Buffer.
11. Analyze the plate immediately after Read Buffer addition on SECTOR Imager.

3.5 Microchip-Based Assays

3.5.1 Background

Automatization of electrophoretic techniques is also an alternative issue to minimize manual manipulation. Basically, for product development, the analysis of proteins and related posttranslational modifications, such as glycan profiling, can be performed [60, 61]. A meanwhile well-established technology has been launched by PerkinElmer, Inc. (through its Caliper Life Sciences, Inc. division), but also comparable systems from other suppliers are on the market (Fig. 10).

To achieve high-throughput performance, chip-based microfluidic assays have been developed, which are proposed for the qualitative and quantitative analysis of proteins directly performed from supernatants, cell lysates, and column fractions, purified and partially purified samples. The commercially available systems enable the separation of proteins sizing from 14 to 200 kDa, applying samples directly from 96- and/or 384-well plates. Assays can be completed within a very short time, faster than 1 min, and repeated runs can be performed without changing the used chip. This can be achieved by sample preparation within tiny chambers and channels that are fabricated in quartz, glass, or plastic. Under the computer-controlled instrumentation for process control and data processing, chips have been optimized to achieve precise microfluidic transport of liquids and samples through the micro-channels. A typical workflow consists of the following steps: separation, staining, destaining, imaging, band detection, quantification, and basic data analysis.

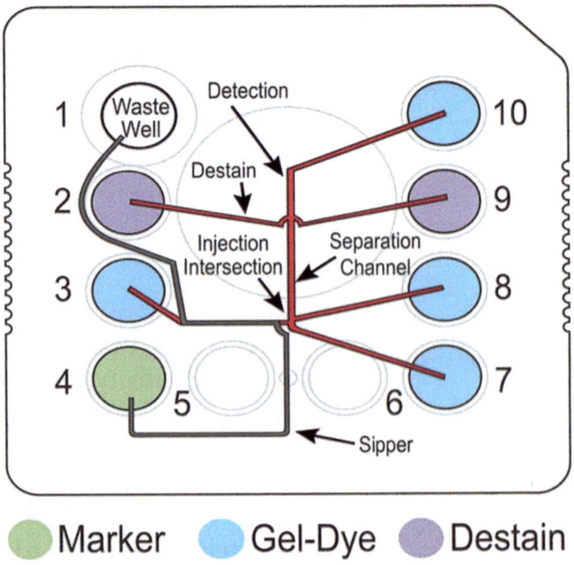

Fig. 10 Schematic diagram of the sipper LabChip® GXII Touch™ protein chip. ©2010–2017 PerkinElmer, Inc. All rights reserved. Printed with permission

Staining is preferentially performed with fluorescent dyes, which are directly incorporated into the sample components during separation. As the sample components pass through the microchannels, they are detected by laser-induced fluorescence as they pass by the position of the laser. Both the fluorescence signal and migration time are processed. Molecular weight sizing determination as well as quantification of individual compounds is aimed.

In advance to the protein quantification, glycan profiling assays have been developed (application notes are provided). The glycan profiling concept is based on a typical workflow: (1) glycan cleavage from purified proteins, (2) fluorophore labeling, (3) separation, and (4) identification. This technique can be used as a competitive assay for the determination of glycan binding sites and investigation of the core structures prior to the analysis of detailed glycan structures. Advantages and perceived limitations of microchip-based assays are shown in Table 15.

3.5.2 Example

The LabChip® GXII system is an automated platform for rapid analysis of proteins and nucleic acids. Each step (sample loading, electrophoresis, staining, destaining, and detection) is integrated in a microfluidic device. The heat-denatured proteins in high SDS concentration are loaded directly out of a 96- or 384-well plate through a capillary sipper. Protein chips are prepared by priming the channels with gel-dye and destaining solution, whereas the gel-dye solution serves as matrix for separation and as staining solution. Then a marker solution is pipetted onto the chip. Subsequent 170 nL of sample from the well plate are applied into the microfluidic channels. Simultaneously, the sample is diluted 2:1

Table 15
Summary of applicability and limitations of microchip-based assays

Advantages	Perceived limitations
Rapid, automated results	Changing pH gradient due to electrolysis
Detection of 0.8–1 μg/mL Small sample volumes (1–4 μL)	Streaming currents which counteract the external electric field or gas bubbles Quantification 5–10 μg/mL
Good reproducibility	Massively parallel setup is problematic due to the heat generated
Qualitative/quantitative	Handheld devices are challenging
Microfluidic	Separation distance—separation efficiency
Automatization	Instrumentation costs
Real time	

with a marker solution, which is used as a reference for migration time and determination of relative concentration of samples. A 40 pL sample plug of the marker-protein mixture is then automatically injected into the separation channel. Proteins are stained with a fluorescence dye, which is mixed with the gel and separated with resolution comparable to a 4–20% SDS-PAGE. Destaining is carried out by a dilution step, flushing SDS-free ions into the separation channel. Thus, free SDS micelles diffuse into the SDS-free fluid, which results in breakup of the micelles and a significant drop in the background fluorescence. SDS micelles bound to the protein remain intact, and protein bands are detected by using laser-induced fluorescence.

Depending on the analyte, good linearity can be achieved over the concentration range 8–2000 μg/mL as well as the ability to detect low-level impurities with a signal-to-noise ratio of 9:1 [62, 63].

Product: mAb [63].

Host: Not specified.

Aim: Purity and quantification.

Method: Microchip-based SDS electrophoresis.

Sample matrix: Crude samples.

Equipment: Caliper's LabChip GXII system.

Chip preparation:

1. Rinse wells of the chip three times by repeatedly pipetting and aspirating deionized water.
2. Prepare gel-dye: mix 520 μL gel matrix and 18 μL dye solution (mix for at least 20 s before use, and avoid light exposure) in a

sample tube; then centrifuge mixture in a spin filter for 5 min at 9300 rcf at room temperature.

3. Destain: centrifuge 250 μL gel matrix in a spin filter for 5 min at 9300 rcf at room temperature.
4. Marker is ready to use.
5. Prepare chip according to the schema shown in Fig. 10:

   1 × 120 μL + 3 × 75 μL gel-dye (positions 3, 7, 8, and 10)

   2 × 75 μL destain (positions 2 and 9).
6. 50 μL marker are adequate for 96 samples, and 120 μL marker are adequate for 384 samples (position 4).

Ladder and buffer preparation:

1. Spin 12 μL ladder in a sample tube.
2. Heat for 5 min at 100 °C or 10 min at 70 °C.
3. Add 108 μL deionized water directly after heating.
4. Transfer to ladder tube.
5. Transfer 750 μL wash buffer to buffer tube.
6. Put the tubes in the designated position.

Sample preparation:

1. Pipette 8 μL sample + 28 μL sample buffer in a 96-well plate.
2. Cover plate and spin down.
3. Heat plate 10 min at 70 °C or 5 min at 100 °C.
4. Centrifuge 5 min at 3000 rcf at room temperature.
5. After cooling, add 140 μL deionized water to denatured sample.
6. Centrifuge for 10 min at 3000 rcf at room temperature.
7. Ensure measurement within 1 h after water addition.
8. Data evaluation with the software of the specific instrumentation.

4 Notes

1. The individual application of these agents is strongly related with the characteristics of the protein, the accompanying proteins, and the aim of the analysis, which cannot be generalized.
2. For subsequent MS analysis, use fresh, previously unused containers for staining; otherwise you may detect false-positive proteins from previous uses.
3. In case of unacceptable high background it is expected that conventional wash steps are insufficient, additional washes should be performed, adding detergent or protein to the

wash buffer if appropriate or soaking for a few minutes between washes.

4. Carbonate/bicarbonate buffer at pH >9 is suggested for antigens with an isoelectric range narrow to this pH value, and alternative buffer systems should be evaluated.
5. For sample matrices, containing yeast extract and peptone, direct quantitative coating is only possible if a pre-dilution of 1:50 or more is applicable. Otherwise only a qualitative analysis can be performed.
6. The selectivity of commercially available antibodies needs to be carefully evaluated prior assay development. The selectivity is often not given in complex matrices.
7. Sandwich ELISAs sometimes require more optimization than traditional ELISAs, but usually the signal-to-noise (S/N) ratio is higher. If a monoclonal and a polyclonal antibody is available, the monoclonal should be coated, and the polyclonal should be used for the detection. The fine-tuning for the antibody selection can be performed with alternative methods, such as electrophoresis combined with Western blot to avoid false positive or negative results.
8. If cross-reactivity is observed, a different blocker should be tested, and if repeated cross-reactivity is observed, it may be advisable to switch to a nonmammalian protein blocker such as salmon serum or a protein-free blocking solution.
9. It needs to be evaluated whether the surfactants affect intramolecular hydrophobic interactions, which hamper correct quantification. For this approach different Tween variants with distinct properties are commercially available.
10. In principle, it is important to ensure that there is no reactivity or cross-reactivity between secondary antibody and coating antibody.
11. The individual time-dependent signal generation needs to be carefully adjusted to the applied standard for quantification and calibration purposes.
12. Fluorescent detection requires the use of an opaque black or white plate. Chemiluminescence detection requires the use of a black or white opaque plate and may also be measured from the top or the bottom.
13. Nickel or copper pre-coated plates can also be used to bind and orient capture antibodies as IgG molecules with histidine-rich sequence in their Fc domain.
14. Pre-coated plates may be stored at room temperature or at 4 °C overnight maximally. Longer periods will reduce OD values.
15. For each HSA ELISA plate, you need a 400 μL HSA standard aliquot, and therefore it is recommended to prepare ready-to-

use aliquots and store them at −20 °C. As described in the manufacturer's protocol, dilute 5 μL of HSA standard (22 mg/mL) in 11 mL dilution buffer, and perform another 1:25 dilution step (e.g., mix 2 mL of the first dilution with 48 mL dilution buffer) to get a HSA concentration of 400 ng/mL.

16. Best results are obtained when samples have a very similar concentration to the standard (400 ng/mL).
17. The incubation time of the TMB substrate depends on the intensity of color development. The highest standard should have an OD value of 2.0–2.5, and the lowest standard should be above background.
18. Standard curve can also be constructed by plotting the mean absorbance for each standard against concentration and drawing a best-fit curve through the obtained points. Linearization can be achieved by plotting the log of HSA concentrations versus the log of OD values, and best-fit line can be determined by regression analysis. This approach provides an adequate but less precise fit of the data.
19. For analytical application it is suggested to inject several μL up to 100 μL.
20. The use of guard columns is advantageous for crude samples but also samples with other complex matrices. They protect altering of the separation material and spend therefore costs and time; however it is essential that the guard columns have no separation capability.
21. The selection of a suitable detector depends on the detectability of the molecule, the choice of mobile phase, as well as the individual sensitivity. It needs to be mentioned that not all detection systems have a linear molecule response, which is important for quantification and validation aspects.
22. The following steps should be kept in mind for proper method development:
    (a) Define method objectives and understand the chemistry.
    (b) Determine the goals of the intended use of the method.
    (c) Understand the chemistry of the analytes and the drug product.
    (d) Develop preliminary HPLC conditions to achieve acceptable separations.
    (e) Develop a suitable sample preparation scheme for the drug product.
    (f) Appropriately standardize and use the relative response factors in calculations.
    (g) Optimize final method for robustness.

(h) Identify the "weaknesses" and optimize the method through experimental design.

(i) Perform method validation.

23. 280 nm is corresponding to absorbance of the aromatic amino acid residues and 214 nm for peptide bonds, respectively. Protein detection at 214 nm is more sensitive, but any organic compounds (any organic salts, amino acids, organic solvents, or detergents present in the running buffer) will also be detected at this absorbance wavelength. Those interference affects the baseline noise but also the quantification.

24. Stepwise performance is important to allow complete pH transition and stabilization of UV baseline. The elution is performed isocratic with 100% buffer B. Injection volumes for standards and samples should range between 20 and 100 μL, depending on sample concentration.

25. There are some aspects which should be considered when selecting the running buffer for SPR, such as avoidance of imidazole or other chelating agents and bivalent metal ions (Ca^{2+}, Zn^{2+}, Cu^{2+}) which can interfere with binding of Ni^{2+} to NTA.

26. When analyzing crude samples, it has to be evaluated, if non-analyte proteins bind to free Ni^{2+} on the chip surface, resulting in unwanted background responses.

References

1. Ahmad M, Hirz M, Pichler H, Schwab H (2014) Protein expression in *Pichia pastoris*: recent achievements and perspectives for heterologous protein production. Appl Microbiol Biotechnol 98:5301–5317. https://doi.org/10.1007/s00253-014-5732-5
2. Liu Z, Tyo KEJ, Martinez JL et al (2012) Different expression systems for production of recombinant proteins in *Saccharomyces cerevisiae*. Biotechnol Bioeng 109:1259–1268. https://doi.org/10.1002/bit.24409
3. Weinacker D, Rabert C, Zepeda AB et al (2013) Applications of recombinant *Pichia pastoris* in the healthcare industry. Brazilian J Microbiol 44:1043–1048. https://doi.org/10.1590/S1517-83822013000400004
4. Porro D, Sauer M, Branduardi P, Mattanovich D (2005) Recombinant protein production in yeasts. Mol Biotechnol 31:1–15
5. Nielsen KH (2014) Protein expression-yeast. Methods Enzymol 536:133–147. https://doi.org/10.1016/B978-0-12-420070-8.00012-X
6. Garfin DE (2009) Chapter 29: One-dimensional gel electrophoresis. Methods Enzymol 463:497–513. https://doi.org/10.1016/S0076-6879(09)63029-9
7. Garfin DE (1990) One-dimensional gel electrophoresis. Methods Enzymol 182:425–441
8. Nowakowski AB, Wobig WJ, Petering DH (2014) Native SDS-PAGE: high resolution electrophoretic separation of proteins with retention of native properties including bound metal ions. Metallomics 6:1068–1078. https://doi.org/10.1039/c4mt00033a
9. Laemmli UK (1970) Cleavage of structural proteins during the assembly of the head of bacteriophage T4. Nature 227:680–685
10. Zhou J, Dann GP, Shi T et al (2013) A simple sodium dodecyl sulfate-assisted sample preparation method for LC-MS-based proteomics. Anal Chem 84:2862–2867. https://doi.org/10.1021/ac203394r
11. Goldberg M, Expert-Bezancon N, Vuillard L et al (1996) Non-detergent sulphobetaines: a new class of molecules that facilitate in vitro

protein renaturation. Fold Des 1(1):21–27 Current Biology Ltd ISSN 1359-0278
12. Sundaram RK, Balasubramaniyan N, Sundaram P (2012) Protein stains and applications. Methods Mol Biol 869:451–464. https://doi.org/10.1007/978-1-61779-821-4_39
13. Kurien BT, Scofield RH (2012) A brief review of other notable protein detection methods on acrylamide gels. Methods Mol Biol 869:617–620. https://doi.org/10.1007/978-1-61779-821-4_56
14. Karlinsey JM (2012) Sample introduction techniques for microchip electrophoresis: a review. Anal Chim Acta 725:1–13. https://doi.org/10.1016/j.aca.2012.02.052
15. Amersham™ (2014) Automated Western blotting systems. Protein labeling and detection. Application note 29-1138-92 AB 09/2014
16. Amersham™ (2014) Analysis of therapeutic antibodies using Amersham™. WB system. Protein labeling and detection. Application note 29-1140-27
17. Amersham™ (2014) Quantitative fluorescence Western blot using Amersham™. WB system. Protein labeling and detection. Application note 29-1138-93
18. Engvall E, Perlmann P (1971) Enzyme-linked immunosorbent assay (ELISA). Quantitative assay of immunoglobulin G. Immunochemistry 8:871–874
19. Engvall E, Jonsson K, Perlmann P (1971) Enzyme-linked immunosorbent assay. II. Quantitative assay of protein antigen, immunoglobulin G, by means of enzyme-labelled antigen and antibody-coated tubes. Biochim Biophys Acta 251:427–434
20. Koehler G, Milstein C (2005) Continuous cultures of fused cells secreting antibody of predefined specificity. J Immunol 174:2453–2465
21. Lequin RM (2005) Enzyme immunoassay (EIA)/enzyme-linked immunosorbent assay (ELISA). Clin Chem 51:2415–2418. https://doi.org/10.1373/clinchem.2005.051532
22. Aydin S (2015) A short history, principles, and types of ELISA, and our laboratory experience with peptide/protein analyses using ELISA. Peptides 72:4–15. https://doi.org/10.1016/j.peptides.2015.04.012
23. Cheng HM (1996) Tween 20 selectively enhances naturally occurring anticardiolipin antibody binding in ELISA procedures. J Immunol Methods 191:87–88. https://doi.org/10.1016/0022-1759(96)00016-6
24. Cabral AR, Cabiedes J, Alarcón-Segovia D (1994) Tween 20 detaches cardiolipin from ELISA plates and makes anticardiolipin antibodies undetectable regardless of the presence of beta 2-glycoprotein-I. J Immunol Methods 175:107–114
25. Nielsen UB, Geierstanger BH (2004) Multiplexed sandwich assays in microarray format. J Immunol Methods 290:107–120. https://doi.org/10.1016/j.jim.2004.04.012
26. Zhang QY, Chen H, Lin Z, Lin JM (2012) Comparison of chemiluminescence enzyme immunoassay based on magnetic microparticles with traditional colorimetric ELISA for the detection of serum α-fetoprotein. J Pharm Anal 2:130–135. https://doi.org/10.1016/j.jpha.2011.10.001
27. Gao Z, Hou L, Xu M, Tang D (2015) Enhanced colorimetric immunoassay accompanying with enzyme cascade amplification strategy for ultrasensitive detection of low-abundance protein. Sci Rep 4:1–8. https://doi.org/10.1038/srep03966
28. Larsson A, Holmdahl R (1990) A microELISA useful for determination of protein A-binding monoclonal antibodies. Hybridoma 9:289–294. https://doi.org/10.1089/hyb.1990.9.289
29. Maccani A, Landes N, Stadlmayr G et al (2014) *Pichia pastoris* secretes recombinant proteins less efficiently than Chinese hamster ovary cells but allows higher space-time yields for less complex proteins. Biotechnol J 9:526–537. https://doi.org/10.1002/biot.201300305
30. Liu R, Lin Q, Sun Y et al (2009) Expression, purification, and characterization of hepatitis B virus surface antigens (HBsAg) in yeast *Pichia pastoris*. Appl Biochem Biotechnol 158:432–444. https://doi.org/10.1007/s12010-009-8527-x
31. Howard JW, Kay RG, Pleasance S, Creaser CS (2012) UHPLC for the separation of proteins and peptides. Bioanalysis 4:2971–2988. https://doi.org/10.4155/bio.12.283
32. Nováková L, Solichová D, Solich P (2006) Advantages of ultra performance liquid chromatography over high-performance liquid chromatography: comparison of different analytical approaches during analysis of diclofenac gel. J Sep Sci 29:2433–2443
33. Andrés A, Broeckhoven K, Desmet G (2015) Methods for the experimental characterization and analysis of the efficiency and speed of chromatographic columns: a step-by-step tutorial. Anal Chim Acta 894:20–34. https://doi.org/10.1016/j.aca.2015.08.030
34. Vehovec T, Obreza A (2010) Review of operating principle and applications of the

charged aerosol detector. J Chromatogr A 1217:1549–1556. https://doi.org/10.1016/j.chroma.2010.01.007

35. Fekete S, Veuthey JL, Guillarme D (2015) Comparison of the most recent chromatographic approaches applied for fast and high resolution separations: theory and practice. J Chromatogr A 1408:1–14. https://doi.org/10.1016/j.chroma.2015.07.014

36. Gurramkonda C, Polez S, Skoko N et al (2010) Application of simple fed-batch technique to high-level secretory production of insulin precursor using *Pichia pastoris* with subsequent purification and conversion to human insulin. Microb Cell Factories 9(31):1–11. https://doi.org/10.1186/1475-2859-9-31

37. Polez S, Origi D, Zahariev S et al (2016) A simplified and efficient process for insulin production in *Pichia pastoris*. PLoS One 11:1–15. https://doi.org/10.1371/journal.pone.0167207

38. Gurramkonda C, Adnan A, Gäbel T et al (2009) Simple high-cell density fed-batch technique for high-level recombinant protein production with *Pichia pastoris*: Application to intracellular production of Hepatitis B surface antigen. Microb Cell Factories 8(13):1–8. https://doi.org/10.1186/1475-2859-8-13

39. Heo J-H, Won HS, Kang HA et al (2002) Purification of recombinant human epidermal growth factor secreted from the methylotrophic yeast *Hansenula polymorpha*. Protein Expr Purif 24:117–122. https://doi.org/10.1006/prep.2001.1527

40. Ye J, Ly J, Watts K et al (2011) Optimization of a glycoengineered *Pichia pastoris* cultivation process for commercial antibody production. Biotechnol Prog 27:1744–1750. https://doi.org/10.1002/btpr.695

41. Applied Biosystems (2010) POROS A20 Analytical HPLC columns for the Quantitation of Monoclonal antibodies. Application Note 1-13

42. Moore JD, Perez-Pardo MA, Popplewell JF et al (2011) Chemical and biological characterisation of a sensor surface for bioprocess monitoring. Biosens Bioelectron 26:2940–2947. https://doi.org/10.1016/j.bios.2010.11.043

43. Petersen R (2017) Strategies using bio-layer interferometry biosensor technology for vaccine research and development. Biosensors 7 (49):1–15. https://doi.org/10.3390/bios7040049

44. Fortebio P (2009) MAb quantitation: protein A HPLC vs. protein A bio-layer interferometry. Application Note 15:1–21

45. Fortebio P (2009) High sensitivity detection of human IgG using protein A biosensors. Technical Note 15:1–9

46. Gilmore J, Islam M, Martinez-Duarte R (2016) Challenges in the use of compact disc-based centrifugal microfluidics for healthcare diagnostics at the extreme point of care. Micromachines 7:1–26. https://doi.org/10.3390/mi7040052

47. Gyros (2015) Instruction For Use Gyrolab™ huIgG Kit, D0024808/C:1-10

48. Liedberg B, Nylander C, Lundström I (1995) Biosensing with surface plasmon resonance—how it all started. Biosens Bioelectron 10:i–ix

49. Karlsson R, Larsson A (2004) Affinity measurement using surface plasmon resonance. Methods Mol Biol 248:389–415

50. Reinartz HW, Quinn JG, Zänker K, O'Kennedy R (1996) Bispecific multivalent antibody studied by real-time interaction analysis for the development of an antigen-inhibition enzyme-linked immunosorbent assay. Analyst 121:767–771

51. Mcdonnell JM (2001) Surface plasmon resonance: towards an understanding of the mechanisms of biological molecular recognition. Curr Opin Chem Biol 5:572–577

52. Nelson RW, Nedelkov D, Tubbs KA (2000) Biosensor chip mass spectrometry: a chip-based proteomics approach. Electrophoresis 21:1155–1163. https://doi.org/10.1002/(SICI)1522-2683(20000401)21:6<1155::AID-ELPS1155>3.0.CO;2-X

53. Hartmann-Petersen R, Gordon C (2005) Quantifying protein-protein interactions in the ubiquitin pathway by surface plasmon resonance. Methods Enzymol 399:164–177. https://doi.org/10.1016/S0076-6879(05)99011-3

54. Hellwig S, Emde F, Raven NPG et al (2001) Analysis of single-chain antibody production in *Pichia pastoris* using on-line methanol control in fed-batch and mixed-feed fermentations. Biotechnol Bioeng 74:344–352. https://doi.org/10.1002/bit.1125

55. GE Healthcare Life Sciences (2008) Sensor Chip NTA. Instruction 22-0519-97:1–11

56. GE Healthcare Life Sciences (2011) Sensor Chip NTA and NTA Reagent Kit Data file 29-0079-27 AA:14

57. Namba Y, Usami M, Suzuki O (1999) Highly sensitive electrochemiluminescence immunoassay using the ruthenium chelate-labeled antibody bound on the magnetic micro beads. Anal Sci 15:1087–1093. https://doi.org/10.2116/analsci.15.1087

58. Sanchez-Carbayo M, Espasa A, Chinchilla V et al (1999) New electrochemiluminescent immunoassay for the determination of CYFRA: analytical evaluation and clinical diagnostic performance in urine samples of patients with bladder cancer. Clin Chem 45:1944–1953
59. Meso Scale Discovery (2012) MULTI-ARRAY® Assay System Human Insulin Kit 17099-v5-2012May:1-15
60. Buhlmann C, Preckel T, Chan S et al (2003) A new tool for routine testing of cellular protein expression: integration of cell staining and analysis of protein expression on a microfluidic chip-based system. J Biomol Techniques 14:119–127
61. Pandey S, Lu CM, Herold DA (2008) Measurement of microalbuminuria using protein chip electrophoresis. Am J Clin Pathol 129:432–438. https://doi.org/10.1309/4JU0XQH62D3YLTGK
62. Caliper Life Sciences (2009) Antibody analysis using Caliper's LabChip GXII system. Application Note 400:1–4
63. Caliper Life Sciences (2009) Automated analysis of proteins using the LabChip 90 System. Application Note 100:1–4

INDEX

Brigitte Gasser and Diethard Mattanovich (eds.), *Recombinant Protein Production in Yeast*, Methods in Molecular Biology, vol. 1923, https://doi.org/10.1007/978-1-4939-9024-5, © Springer Science+Business Media, LLC, part of Springer Nature 2019

N

P

R

S

T

V

W

Y

MIX
Papier aus verantwortungsvollen Quellen
Paper from responsible sources
FSC® C105338

If you have any concerns about our products,
you can contact us on
ProductSafety@springernature.com

In case Publisher is established outside the EU,
the EU authorized representative is:
Springer Nature Customer Service Center GmbH
Europaplatz 3, 69115 Heidelberg, Germany

Printed by Libri Plureos GmbH
in Hamburg, Germany